Environmental Science: A Global Outlook

Environmental Science: A Global Outlook

Edited by Paige Tucker

SYRAWOOD
PUBLISHING HOUSE

New York

Published by Syrawood Publishing House,
750 Third Avenue, 9th Floor,
New York, NY 10017, USA
www.syrawoodpublishinghouse.com

Environmental Science: A Global Outlook
Edited by Paige Tucker

International Standard Book Number: 978-1-64740-139-9 (Hardback)

Cataloging-in-Publication Data

Environmental science : a global outlook / edited by Paige Tucker.
    p. cm.
Includes bibliographical references and index.
ISBN 978-1-64740-139-9
1. Environmental sciences. 2. Ecology. I. Tucker, Paige.
GE105 .E58 2022
333.7--dc23

TABLE OF CONTENTS

PREFACE

Environmental science is a multidisciplinary approach to the study of the environment through biological, physical and information sciences. Atmospheric sciences, ecology, environmental chemistry, and geosciences are the main components of environmental science. It is an essential field of scientific investigation due to the need for an interdisciplinary approach to analyze complex environmental problems. It includes the study of environmental system. It also studies the social sciences to understand human relationships with environment. Environmental science examines the effects of human activities on biophysical environment. It also focuses on protecting the environment. Environmental science also studies the interaction of chemical, physical and biological processes. This book is a compilation of chapters that discusses the most vital concepts and emerging trends in the field of environmental science. It aims to provide a detailed knowledge of the discipline. Scientists and students actively engaged in this field will find this book full of unexplored concepts.

After months of intensive research and writing, this book is the end result of all who devoted their time and efforts in the initiation and progress of this book. It will surely be a source of reference in enhancing the required knowledge of the new developments in the area. During the course of developing this book, certain measures such as accuracy, authenticity and research focused analytical studies were given preference in order to produce a comprehensive book in the area of study.

This book would not have been possible without the efforts of the authors and the publisher. I extend my sincere thanks to them. Secondly, I express my gratitude to my family and well-wishers. And most importantly, I thank my students for constantly expressing their willingness and curiosity in enhancing their knowledge in the field, which encourages me to take up further research projects for the advancement of the area.

Editor

Impacts of integrated soil and water conservation programs on vegetation regeneration and productivity as indicator of ecosystem health in Guna-Tana watershed: evidences from satellite imagery

Getachew Workineh Gella[1,2]*

Abstract

Background: Northern Ethiopian Highlands, including Guna-Tana watershed, have experienced profound natural resources degradation which are resulted from coupled natural and anthropogenic factors. To mitigate this problem, Ethiopian government has launched various soil and water conservation programs at different watersheds. Overall objective of this study was to analyze impacts of soil and water conservation programs on vegetation regeneration and ecosystem productivity at in Guna-Tana watershed. As prime data source, the study has utilized Moderate Imaging Spectrometer satellite bi-monthly Enhanced Vegetation Index, 8-day land surface temperature and annual Net Primary Productivity products of the past 17 years starting from 2000. Imagery was processed by using various image preprocessing and analytical techniques. Long-term trend was tested by using Sens slope estimator and Mann–Kendall's monotonic trend test. Analyzed trend was also segregated into slope and agroecology classes. More importantly, to supplement trend analysis, Vegetation Disturbance Index was developed.

Results: Results have showed that despite of long-term soil and water conservation programs, except small patches, vast expanses of the watershed have showed decrease in vegetation regeneration and primary productivity trend. This observed trend has also spatial variability across slope gradient and agroecological classes of the watershed.

Conclusion: Though there is tendency of increasing vegetation regeneration and productivity, its observed that significant positive change as a result of watershed conservation programs was very little. This indicates that for better regeneration of vegetation and maintenance of ecosystem health in a watershed, intervention programs should be revised and constraints should be assessed. Taking these into consideration, the study calls further implementation strategies which have accounted agroecology and livelihoods production system.

Keywords: Guna-Tana, Net primary productivity, ISWCP, Vegetation regeneration, Watershed

Background

Northern Ethiopian highlands are supporting large number of agricultural community whose livelihoods directly depends on farming. However, coupled with increasing population pressure and terrain characteristics of the area, there was reported massive land degradation manifested by soil erosion and resultant nutrient depletion, forest degradation and massive expansion of agricultural land use pattern with costs of other natural cover (Gete and Hurni 2001; Hurni et al. 2010; Woldeamlak and Solomon 2013). Similarly, studies undertaken specific to Guna-Tana watershed and Upper Blue Nile basin like (Hurni et al. 2005; Mellander et al. 2013; Seleshi et al.

*Correspondence: getawork71@gmail.com
[1] Department of Geography and Environmental Studies, Debre Tabor University, Debre Tabor, Ethiopia

2012, 2013, 2016; Adugnaw et al. 2016; Nigussie et al. 2017) have identified watershed level resource degradation problems emanated from anthropogenic and biophysical factors.

Since 1970s Ethiopian government have launched various soil and water conservation programs to curve resource degradation. Programs implemented at first phase during the Derg regime (1970s) and earlier phases of current government (1980s) were reported ineffective due to its top-down approach basically focused on physical soil erosion control measures (Osman and Sauerborn 2001; Essayas et al. 2014; Nigussie et al. 2015; Gebregziabher et al. 2016). Renovated watershed level soil and water conservation programs that have based community and its livelihoods as center piece of intervention like Community Based Natural Resources Management (CNRM), Integrated Watershed Soil and Water Conservation Program (ISWCP), Sustainable Land Management Program (SLM), Millennium Reforestation Program and Integrated Safety Net Programs (ISP) have been implemented nationally in many regions of Ethiopia including Guna-Tana watershed (Nigussie et al. 2015; Gebregziabher et al. 2016). In addition to these programs, every farm household is participating for 60 days annually in soil and water conservation works as community based campaign in Guna-Tana watershed.

The government repeatedly reported these soil and water conservation programs as effective in halting resource degradation scenarios and by reversing the trend while simultaneously improving rural livelihoods and ecosystem rehabilitation. Contrary to prospectus government reports, still there was natural vegetation degradation with opportunity cost of human induced land use pattern (Temesgen and Tesfahun 2014; Adugnaw et al. 2016). In these respect, some studies have reported as ISWCP have brought significant changes in agricultural productivity (Chisholm and Tassew 2012; Schmidt and Fanaye 2012), reduction of soil erosion and sedimentation (Kebede 2014; Nigussie et al. 2015; Molla 2016; Asnake 2017; Lemlem et al. 2017), vegetation change and positive hydrological responses (Fikir et al. 2009; Nyssen et al. 2010; Shimeles 2012), climate change adaptation (Meaza 2015), biomass recovery (Essayas et al. 2014; Lemlem et al. 2017). Though these studies have reported prospectus impacts of integrated soil and water conservation programs, still there is no any study undertaken to show long-term watershed level vegetation productivity and vegetation regeneration to see impacts of conservation programs on ecosystem health and productivity.

Remote sensing approaches were selected to harness its capability to provide spatially explicit account of ecosystem parameters (Xiaoming et al. 2010; Li et al. 2014) especially in places like Guna-Tana watershed where

there is acute shortage of in situ based hydro-meteorological observations and complete absence of flux tower observing stations for ecosystem health modeling. One of the basic attributes of vegetation to measure through remote sensing for ecosystem health assessment is vigor (Eve et al. 1999; Li et al. 2014). Hence, the study has used remotely sensed net primary productivity (NPP) and enhanced vegetation index (EVI), which are repeatedly used as ecosystem productivity, health and vegetation regeneration and degradation at different ecosystems and geographic regions (Potter et al. 1993; Li et al. 2012; Feng et al. 2013; Zhou et al. 2013; Pan et al. 2014; Binyam et al. 2015; Neumann et al. 2015; Chen et al. 2017). Vegetation vigor (EVI) measures health of vegetation while NPP measures amount of net carbon assimilated to ecosystems after supporting respiration and transpiration which is common ecosystem health parameter. Therefore, the prime objective of this study was to assess impacts of integrated soil and water conservation programs on vegetation productivity and regeneration by using long-term and frequently observed satellite imagery from 2000 to mid of 2017.

Methods
Study area
The study site is located within 37°30′–38°30′ East Longitude and 11°30′–12°30′ North Latitudes (Fig. 1). It encompasses 437,632 hectares of land with elevation gradient ranging from greater than 4000 m above mean sea level at the tips of Guna mountain towards 1700 m above mean sea level at the low-laying Fogera Plains with respective vegetation heterogeneity along elevation gradient. At some parts of the watershed, slope exceeds 70%, while at the outlets of watershed—Fogera Plain, there is almost level land. The study site has agro-ecological zones ranging from *Kola* (Tropical) towards *Wurch* (Afro-alpine) ecosystems, where human encroachment threatens mountain ecosystem to cause massive land use and land cover change (Adugnaw et al. 2016). At highland areas of the watershed alpine vegetation type is dominant while in lowland plains, broadleaved and riverine vegetation is common (Zerihun 1999). Because of degradation of natural forests, the community has adopted eucalyptus as alternative vegetation in private form plots and homesteads. Precipitation measurements obtained from climate hazards group infrared precipitation with stations (CHIRPS) products indicates that, long-term mean annual total rainfall ranges from 1250 mm/year at outlets of the watershed while it ranges to 15,440 mm/year at inlets of the watershed around peaks of Mount Guna. Due to this, Guna-Tana watershed is an area that contributes large amount of runoff and sediment for Blue Nile River system through Gumera and Ribb rivers

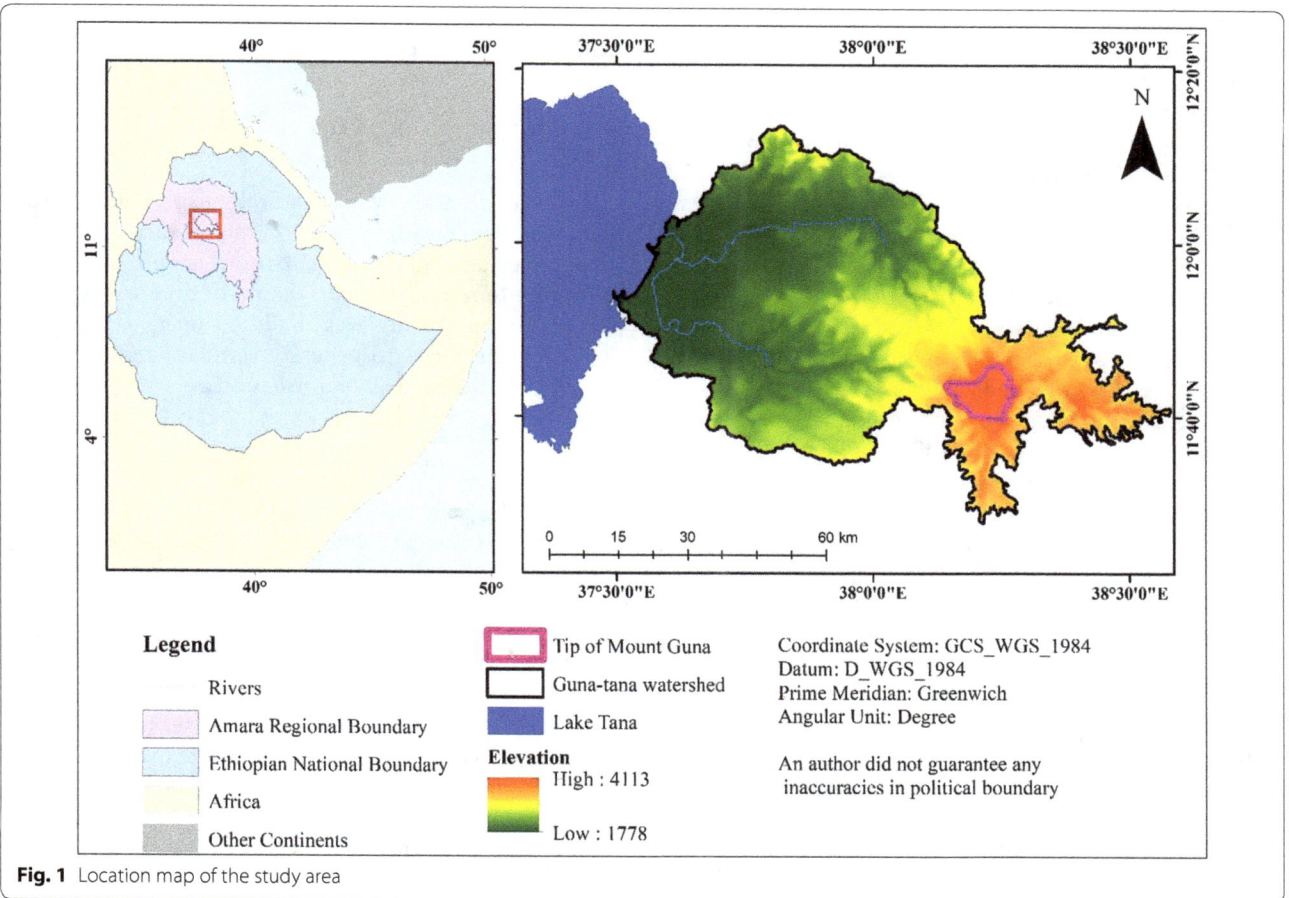

Fig. 1 Location map of the study area

(Setegn et al. 2008; Chebud and Melesse 2009). In its lower plains, there is encroaching irrigation activity and repeated flooding incidence caused by flood hazard from upper courses of the watershed (Woubet and Dagnachew 2011). Within the watershed there are five soil classes. These were Luvisols, Leptisols, Cambisols, Nitosols and vertisols with areal coverage of 67.9, 1.8, 9.1, 0.4, 20.7% of the area respectively. The watershed is experiencing advancing population increments in terms of number and density that, a single square kilometer is resided by 165 persons (CSA 2004). This made an area one of densely populated site in Ethiopia (CSA et al. 2006).

Data

The study has primarily utilized different remotely sensed imagery from different sources which are described as follows. Surface Radar Topographic Mission (SRTM) Digital Elevation Model (DEM) obtained from United States Geological Survey (USGS) website to delineate watershed boundary, identify slope gradient and agroecological classes. To see vegetation regeneration as result of watershed level integrated soil and water conservation

programs, 368 tiles of 16-day Moderate Imaging Spectrometer (MODIS) EVI (MOD13Q1) composite for the past 16 years starting from third decade of 2000 GC to mid of 2017 was accessed from Land and Atmospheric Archive System Distributed Active Archive Center (LAADS DAAC) website (ftp server). Enhanced Vegetation Index was used as robust remote sensing based vegetation vigor and regeneration index because of it does not affected by minor soil moisture differences. To see long-term changes in vegetation productivity as indicator of ecosystem health, long-term annual MODIS NPP data (MOD17A3) was downloaded from similar site with MOD13Q1. Various studies that have integrated in situ measurements from field at different ecosystems and climatic zones were validated MODIS NPP data and reported as reliable and freely available remotely sensed vegetation and ecosystem productivity data (Fensholt et al. 2006; Turner et al. 2006; Sjöström et al. 2013; Indiarto and Sulistyawati 2014; Sung et al. 2016). It incorporates principle of light use efficiency and ecological modeling approach where vegetation photosynthesis, carob fixation capacity and biome based conversion efficiency was accounted well (Running et al. 2004). Principally, the product is composited based on conceptualizations provided by (Monteith 1972). More importantly, vegetation disturbance accounting index was generated from integration of EVI and land surface temperature data. Therefore, in addition to EVI and NPP datasets, 736 tiles of MODIS 8 day land surface temperature composite (MOD11A2) was accessed from LAADS website.

Processing and analytical approach

Watershed boundary, which served as area of interest for this study was delineated by using ArcGIS hydrology toolbox. Downloaded MOD12A2, MOD13Q1 and MOD17A3 data was provided in Earth Observation Hierarchical Data Format (EOS-HDF) with sinusoidal projection system. These was made ready for Geographic Information Systems (GIS) softwares through format conversion and reprojection process. Format converted and re-projected data was rescaled from its quantized format by using rescaling factor provided with EOS-HDF metadata. From rescaled MODIS full scene imagery, study area of interest was extracted by using watershed boundary. More importantly these datasets have undergone a series of quality control like filtering of no data values, and removal of waterbodies from the imagery by using metadata attributes archived within the dataset.

As MODIS 16-day EVI was composited from daily EVI maximum values throughout the year, it has noise emanated from outliers and cloud contamination. This noise was removed by using Fast Fourier Transform (FFT) technique. Fourier transform converts spatial domain image into frequency domain images (real frequency, imaginary frequency, and power spectrum frequency) by using forward transformation method. Then, noise free image was constructed by using filtered frequency domain images through reverse Fourier transform (Rocchini et al. 2013). Mathematically, the representation of Fourier transforms for discrete functions in two-dimensional space can be expressed as weighted sum of sines and cosines and it is given as

$$F(u,v) = \frac{1}{NM} \sum_{x=0}^{N-1} \sum_{y=0}^{M-1} f(x,y) e^{-2\pi i \left(\frac{ux}{N} + \frac{vy}{M}\right)}$$

where u and v are spatial frequency, $F(u,v)$ is frequency domain function, $f(x,y)$ is spatial domain function, i is $\sqrt{-1}$, N is the number of pixels in the x-direction and M is the number of pixels in the y-direction. That is x indices go from 0 to N − 1 and y indices go from 0 to M − 1.

Fourier sequences F_u were multiplied by a low pass filter (w), giving a filtering signal wF_u in the frequency domain. After filtering is applied, usually to reduce noises associated with high frequency components, final

corrected image in spatial domain was reconstructed by using the inverse transform which is given as:

$$f(x,y) = \sum_{u=0}^{N-1} \sum_{v=0}^{M-1} wF(u,v) e^{2\pi i \left(\frac{ux}{N} + \frac{vy}{M}\right)}$$

As vegetation greenness naturally experience seasonal behavior synchronized with climatic variables, before time series analysis, seasonal pattern should be removed to reduce temporal autocorrelation of time series images. De-seasoning activity was made by using standardized Z score time series to generate standardized anomalies, which is mathematically represented as:

$$Z = \frac{EVI_{mean} - EVI_i}{EVI_\delta}$$

where EVI_{mean} is long-term mean EVI, EVI_i is EVI at time i, and EVI_δ is standard deviation of EVI time series within 17 years. Though there need removal of serial correlation by pre-whitening method for Mann Kendal trend test, based on recommendation provided by Yue and Wang (2002) on the nature of data the analysis was made on standardized anomalies without pre-whitening (Bayazit and Önöz 2008). As MODIS NPP dataset is annual composite and seasonality is accounted during compositing phase, there is no need of de-seasoning operation.

Presence of log-term trend in EVI and NPP within the watershed was examined by using Mann–Kendall's nonparametric significance test (Helsel and Hirsch 1992). Mathematically, with image time series x_t ($t = 1, 2, 3...n$), each value of the series (x_t) is compared with all subsequent values (x_{t+1}) and a new image series Z_i was created for trend test as provided in Machiwal and Jha (2012)

$$Z_i = 1 \; for \; x_t > x_i$$

$$Z_i = 0 \; for \; x_t = x_i$$

$$Z_i = -1 \; for \; x_t < x_i$$

$$S = \sum_{i=1}^{n-1} \sum_{t=i+1}^{n} Z_i$$

When S is a large positive number, later-measured values tend to be larger than earlier values and an upward trend is indicated. When S is a large negative number, later values tend to be smaller than earlier values and a downward trend is indicated in selected indices. When the absolute value of S is small, no trend is indicated. The test statistic Mann–Kendall τ (*Thau*) was computed as:

$$\tau = \frac{S}{n(n-1)/2}$$

which has a range of -1 to $+1$ and is analogous to the correlation coefficient in regression analysis. The null hypothesis of no trend is rejected when S and τ are significantly different from zero. The rate of change either increasing or decreasing in EVI and NPP was calculated using the Sen's slope estimator, which is provided Donald et al. (2011) as:

$$S_{slop} = \mathcal{M}\left(\frac{Y_j - Y_i}{X_j - X_i}\right)$$

$\mathcal{M}$ was the median for all $i < j$ and $i = 1, 2, ..., n - 1$. and $j = 2, 3,..., n$; in other words, computing the slope for all pairs of data that were used to compute S. The median of those slopes is the Sen slope estimator.

To supplement vegetation trend analysis yearly vegetation disturbance was computed by integrating MODIS land surface temperature and EVI. The study has adopted vegetation disturbance index designed by Mildrexler et al. (2007) and Mildrexler et al. (2009) to monitor instantaneous and non-instantaneous vegetation disturbance occurred either by anthropogenic or natural factors like wildfire and hurricane. The index is presented as:

$$DI_i = \frac{\left(\frac{LST_{imax}}{EVI_{imax}}\right)}{multiyear\ mean\left(\frac{LST_{max}}{EVI_{max}}\right)_{(i-1)}}$$

where DI_i is vegetation disturbance for year i, LST_{imax} annula maximum temperature composite for year i and EVI_{imax} is annual maximum EVI composite for year i. The denominator is long-term mean of ratios of yearly maximum temperature and EVI with exclusion of the year under consideration. Before vegetation disturbance generation, EVI values less than 0.25 were coded as no data values. This was recommended by Huete et al. (2002) to remove non-vegetated surfaces from computation. Then annual maximum LST and EVI compositing was made on time series imagery. During compositing process years with incomplete datasets (2000 and 2017) were excluded. For the purpose of detecting presence of either positive or negative vegetation change, yearly index variabilities greater than ± 1 standard deviations of long-term mean index were considered as disturbance. With this premise, yearly disturbance value lower than long-term disturbance is considered as increasing vegetation change and the reverse is true. for the purpose of monitoring vegetation and biomass change as a result of integrated watershed management, the index was implemented by Essayas et al. (2014) in Blue Nile Watershed.

To look into effectiveness of soil and water conservation programs across slope and agroecological classes of the watershed, slope in present rise was computed from SRTM DEM. Similarly, agroecological classes were generated from DEM of the watershed by using operational classifications provided in Hurni et al. (2016). Analytical outputs have yielded, timeseries trend, intercept and significance image layers. Overall processing and analysis was made by using HDF View, IDRISI macromodeler to automate workflows for batch processing, ArcGIS spatial analysis toolbox and R raster package for image analysis. Overall processing and analytical approach was presented in Fig. 2.

Results and discussion
Long-term changes in vegetation regeneration
Undertaken analysis showed that vegetation greenness in Guna-Tana watershed was flourishing in normal rainy seasons where there are sufficient crops covered precipitation and agricultural fields. With recession of precipitation, vast expanses of the study area experiences minimal EVI (Fig. 3) where precipitation deficit affects vegetation vigor.

Long term trend analysis showed that most parts of the study area (59.9%) have experienced decreasing vegetation greenness pattern while small areal extension of Guna-Tana watershed (9.6% of the total area) have experienced statistically significant increasing vegetation greenness (P < 0.05). The remaining parts of the area have shown both increasing and decreasing tendency which is not statistically significant within past 17 years (Table 1).

When we consider its spatial pattern of change, statistically significant increasing vegetation greenness pattern was observed at some patches of the watershed. These were at the outlets of watershed and outskirts of Guna Mountain (Fig. 4). This indicates that, soil and water conservation activities implemented at the watershed have spatial variability in terms of intensity and sustainability of the program. Vast expanses of lowland areas have experienced significant vegetation reduction. Lowland areas of the watershed were repeatedly reported as expansion of intensive agriculture which costs vegetation regeneration especially in dry season where crops were harvested and the field was left bare. As reported by Essayas et al. (2014), though there were massive watershed level plantation activities, large parts of Blue Nile basin have experienced decreasing biomass recovery except Lake Tana sub-basin (which is part of Guna-Tana watershed) that have positive vegetation regeneration. This was also attributed with conversion of large area to irrigated landscape with establishment of Kogga irrigation scheme and expansion of eucalyptus vegetation at privately owned farm plots.

In other way, contrary to upper parts of the watershed, soil and water conservation activities at the lower areas of the watershed was not basically focused on reforestation and area closure activities. More importantly, at

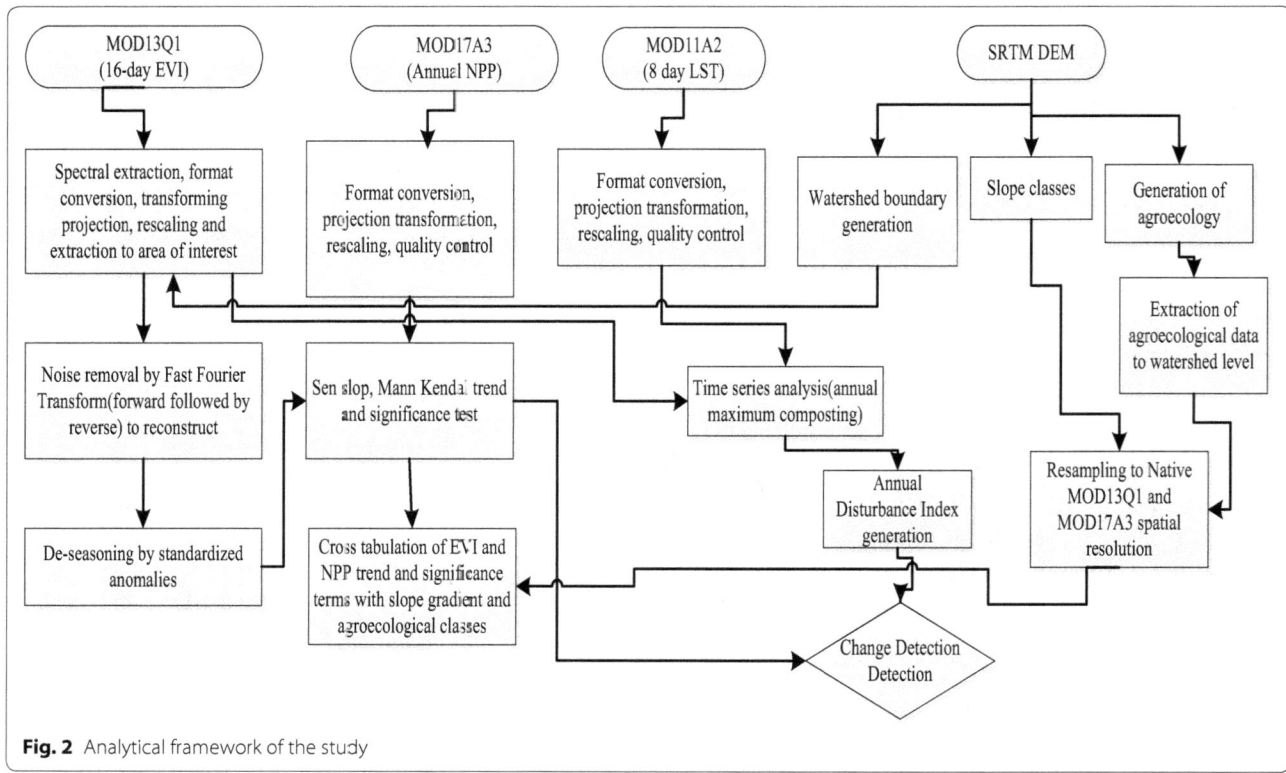

Fig. 2 Analytical framework of the study

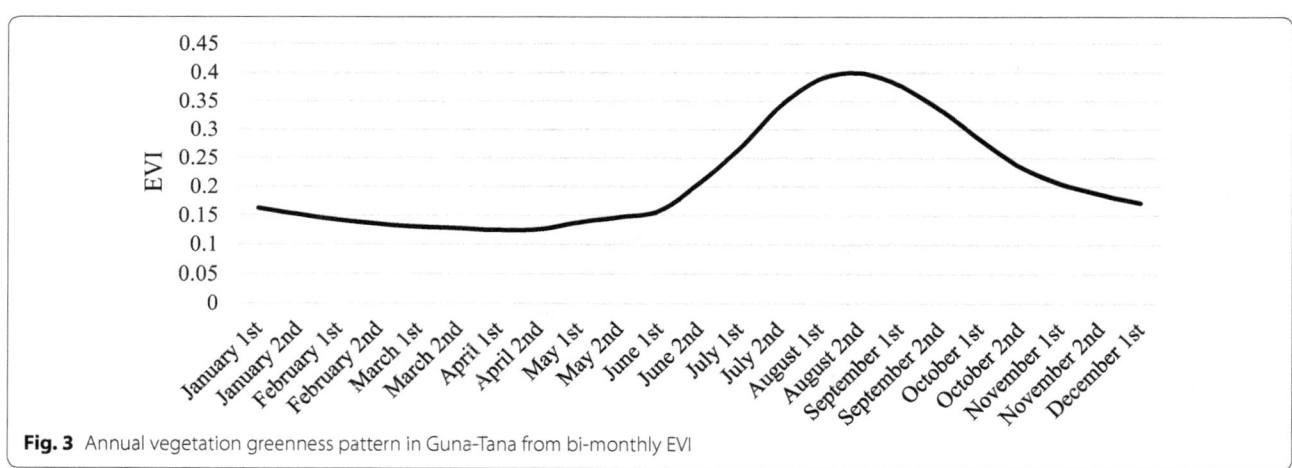

Fig. 3 Annual vegetation greenness pattern in Guna-Tana from bi-monthly EVI

Table 1 Watershed level vegetation change trend (2000–2017)

Trend	Area in hectares	% of total watershed area
Significant increase	42,047.0	9.6
Significant decrease	262,324.9	59.9
Non-significant increase	53,549.3	12.2
Non-significant decrease	79,721.6	18.2
Total	437,643.0	100.0

Mount Guna and its surrounding, there is strict conservation measures including closure activities and relocation of resident farm households for alpine ecosystem regeneration and biodiversity conservation. A study by Teferi et al. (2015) also reported that long-term spatiotemporal variability of either vegetation greenness or browning was caused by differences by watershed management practices, establishment of plantation agriculture and settlement programs that clear natural vegetation.

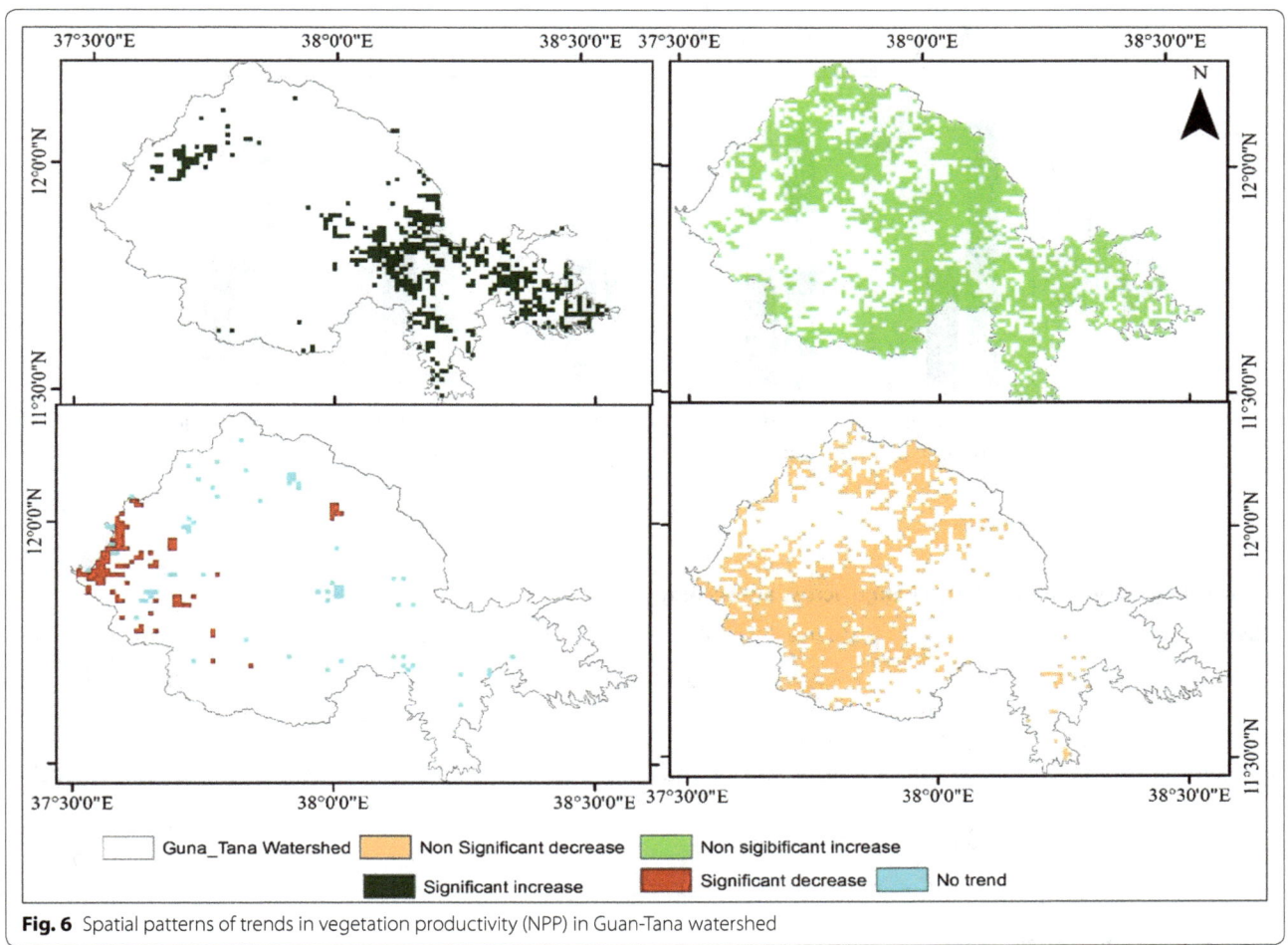

Fig. 6 Spatial patterns of trends in vegetation productivity (NPP) in Guan-Tana watershed

Table 3 Vegetation change across slope gradient

Observed trend	Low slope		Medium slope		Steep slope	
	Area (ha)	% age	Area (ha)	% age	Area (ha)	% age
Significant increase	25,296.5	60.7	11,483.7	27.5	4925.2	11.8
Significant decrease	152,977.9	58.6	70,908.5	27.2	36,991.4	14.2
Non-significant increase	28,948.5	54.4	15,707.1	29.5	8521.2	16.0
Non-significant decrease	40,978.8	51.9	24,010.9	30.4	13,949.4	17.7

variability (as its deviation is not greater than ± long-term disturbance values). Similarly, vegetation degradation is also under the natural range of variability except in 2005, 2006 and 2016 at very small number of pixels. The spatial nature of vegetation disturbance within analysis period is presented in Fig. 5.

As indicated in Fig. 5, red colors indicated positive disturbance values which connotes tendency of decreasing vegetation while tendency of increasing vegetation regeneration have negative values with green color. As per

this, northern, north eastern and eastern highlands of the watershed have showed persistent tendency of increasing vegetation but it was not beyond natural variability. During field observation it was evidenced that picks of Guna mountain were subjected for area closure conservation where vegetation is under regeneration. In the lowland parts of the watershed there is intensive agriculture where large part of biomass is crop residue and is not permanently recycled to ecosystem, since it is used for animal fodder. Similar to this finding, a study by Essayas et al.

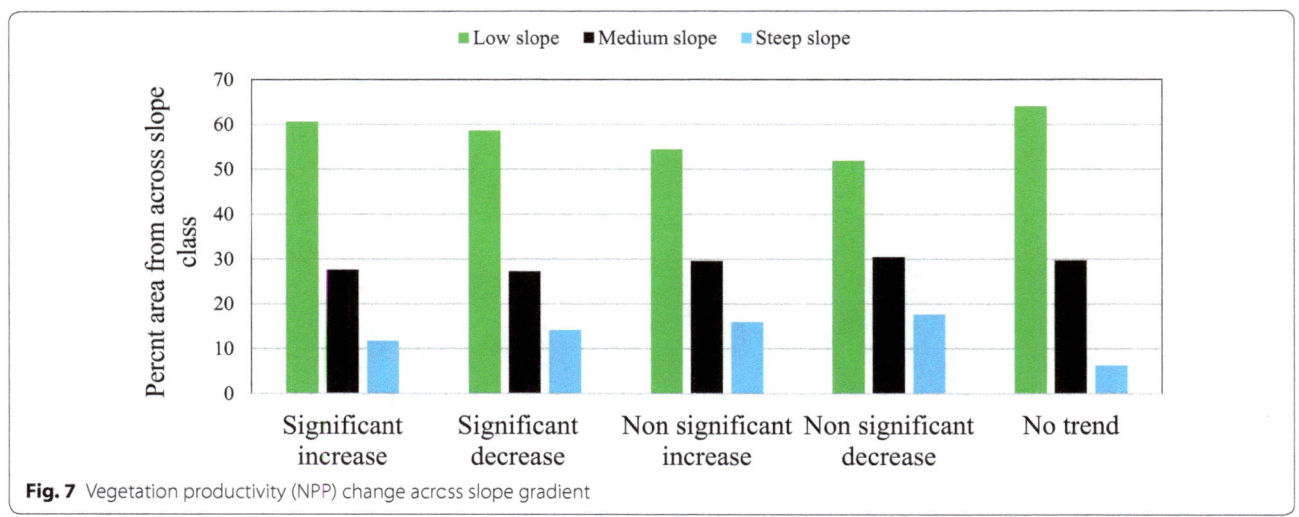

Fig. 7 Vegetation productivity (NPP) change across slope gradient

Table 4 Vegetation productivity change across agroecology

Observed trend	WD	% area	DE	% area	WU	% area	Hi WU	% area
Significant increase	5745.6	11.2	39,228.3	76.3	6339.9	12.3	99.1	0.2
Significant decrease	10,698.6	100.0	–	–	–	–	–	–
Non-significant increase	106,193.8	45.0	114,316.8	48.4	12,778.9	5.4	2674.7	1.1
Non-significant decrease	119,369.0	90.1	11,986.4	9.0	891.6	0.7	297.2	0.2
No change	4061.5	61.2	2377.5	35.8	198.1	3.0	–	–

WD is for Woinda dega, DE for Dega, WU for Wurch, Hi WU for High Wurch

Table 5 Vegetation disturbance

Year	Minimum deviation from long term mean	Maximum deviation from long term mean	Year	Minimum deviation from long term mean	Maximum deviation from long term mean
2003	− 0.69	0.88	2010	− 0.74	0.77
2004	− 0.67	0.51	2011	− 0.51	0.71
2005	− 0.59	1.42	2012	− 0.54	0.43
2006	− 0.64	1.54	2013	− 0.51	0.56
2007	− 0.49	0.98	2014	− 0.52	0.60
2008	− 0.49	0.98	2015	− 0.46	0.59
2009	− 0.54	0.58	2016	− 0.46	2.67

(2014) in broader Blue Nile Watershed has claimed that part of Guna-Tana watershed has not experienced vegetation regeneration greater than natural variability. As results presented in Fig. 8, except small patches of the watershed, the disturbance is within range of natural variability.

Discussions

The results of this study brought some clues on previous soil and water conservation programs implemented in Guna-Tana watershed. Firstly, increasing in vegetation greenness has spatial variability that it is realized in places where area closure was implemented. Studies undertaken in some parts of Ethiopia claimed that area closure is found effective for positive vegetation reclamation and biomass fixation in the soil (Mekuria and Aynekulu 2011; Wolde and Mastewal 2013; Abera and Fitih 2015). In other words, collective physical soil and water conservation measures and community based reforestation programs were encountering problems of implementation in terms of technical specifications and

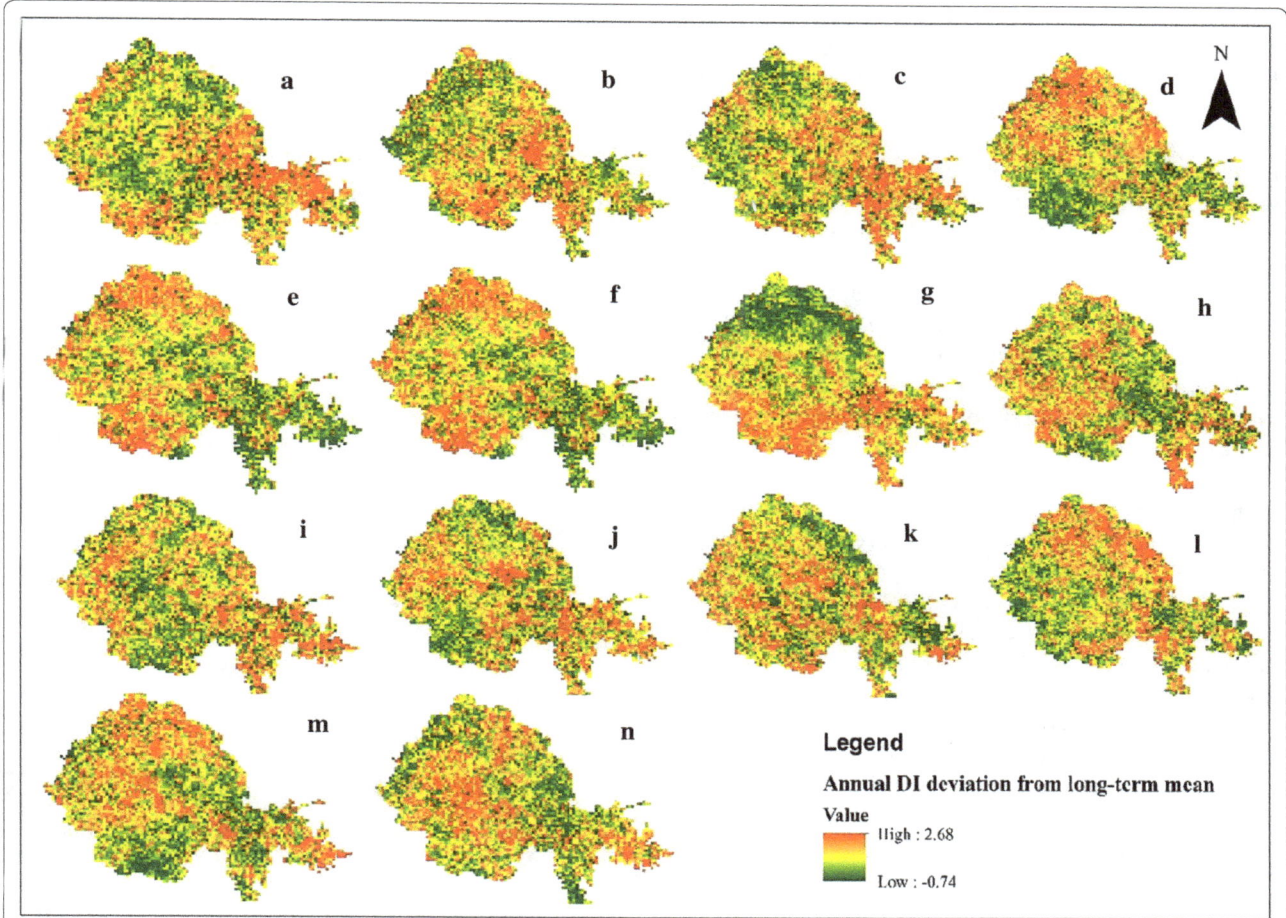

Fig. 8 Spatial patterns of vegetation disturbance in Guna-Tana watershed (**a–n** represent vegetation disturbance index anomaly from long-term mean disturbance from 2003 to 2016)

community acceptance (Waga et al. 2007; Abera and Fitih 2015). This hinders vast expanses of Guna tana watershed for not experiencing profoundly significant vegetation regeneration and changes in NPP which is a prime indication of ecosystem health. A study by Simane and Zaitchik (2014) supported the findings of this study. Accordingly, CNRM in northern Ethiopian highlands is reported unsustainable and even natural resources management institutions were at risk of unsustainability. Lack of institutional capacity and livelihood problems were challenging implementation process. In some watersheds where non-governmental organizations respond to support soil and water conservation programs and livelihood support system, there was intended watershed rehabilitation and positive biomass change (Essayas et al. 2014; Gebregziabher et al. 2016). This indicates that beyond inception of soil and water conservation programs in the watershed for rehabilitation and vegetation restoration,

institutional capacitation and community based awareness and mainstreaming of conservation activities with rural livelihoods.

Conclusion

Though there are various soil and water conservation programs implemented in Guna-Tana watershed, within past 17 years, large areas of the watershed have not experienced uniform changes in vegetation regeneration and watershed level vegetation productivity. Very small parts of the watershed experienced statistically significant increasing vegetation regeneration trend while vast expanses have showed decreasing trend. Similarly, ecosystem productivity (NPP) has also showed similar trends with indices used to show vegetation regeneration. Analysis segregated to slope classes has showed that watershed areas with medium and steep slope have experienced decreasing pattern with large share of area than increasing trend. This was also evident in Wurch (*Alpine*) and High Wurch (*Afro-alpine*) agroecologies which are fragile and sensitive ecosystems. Generally speaking, except very small patches of the watershed, vegetation regeneration and watershed level ecosystem productivity trend was decreasing in vast expanses of Guna-Tana watershed. These spatial differences might be resulted from variability of watershed level intervention strategies in terms of intensity, conservation type and sustainty of the program. Results from yearly vegetation disturbance index is also an indication of unsustainty of soil and water conservation programs. In some year conservation activities were implemented in a firm way while in other year integrated soil and water conservation programs implementation was loosened. As result, positive vegetation disturbance and ecosystem rehabilitation shows a dwindling pattern of change which is not significantly greater than natural range of variability. More importantly, sensitivity of vegetation regeneration for climate variability might have hindered expected vegetation changes in the watershed. With these concluding remarks, the following recommendations were forwarded for further work. To envision positive vegetation change, ecosystem productivity and reduction of soil erosion, integrated soil and water conservation programs should focus on integrated use of area closure, biological soil and conservation strategies and reforestation measures which are tailored with respective agroecology, slope gradient and production system of the watershed. More importantly, community based soil and water conservation activities that simultaneously integrated livelihoods approach should be fostered to sustain conservation programs for long-time. Further studies should also be undertaken to study plot level effectiveness of soil and water conservation intervention strategies by using high resolution satellite imagery and field measurements. More importantly, adopted soil and water conservation programs acceptability by the community in a sustained way should also investigated by house hold level survey studies.

Abbreviations

CNRM: Community Based Natural Resources Management; CSA: Central Statistical Agency; EVI: Enhanced Vegetation Index; EOS-HDF: Earth Observation Satellites-Hierarchical Data Format; FFT: Fast Fourier Transform; ISWC: Integrated Soil and Water Conservation Programs; ISP: Integrated Safety Net Programs; LAADS DAAC: Land and Atmospheric Archive System Distributed Active Archive Center; LST: Land Surface Temperature; MODIS: Moderate Imaging Spectrometer; MRP: Millennium Reforestation Program; NPP: Net Primary Productivity; SLM: Sustainable Land Management; SRTM DEM: Surface Radar Topography Mission Digital Elevation Model.

Author details

[1] Department of Geography and Environmental Studies, Debre Tabor University, Debre Tabor, Ethiopia. [2] Guna Tana Integrated Field Research and Development Center, Debre Tabor, Ethiopia.

Acknowledgements

Special thanks to United States Geological Survey for freely availing satellite imagery, Mr. Abiyu Tefera (computer expert at Department of Geography and Environmental Studies, Remote Sensing and GIS laboratory) for bulk downloading of time series satellite imagery. The author acknowledges R open source community for sharing image analysis packages and respective documentations. The author appreciates critical comments from anonymous reviewers that helped to improve the manuscript, Mr. Ashenafi Melese (Jimma University), Mr. Daniel Asfaw and Mr. Baymot for their proof read of the manuscript.

Competing interests

The author declares no competing interests.

Funding

The research has not received any fund from public or private entities.

References

Abera A, Fitih A (2015) Lessons from watershed based climate smart agricultural practices in Jego Gudedo watershed, Ethiopia. Int J Sci Technol Res 4(12):80–85

Adugnaw B, Frankl A, Jacob M, Lanckriet S, Hendrickx H, Nyssen J (2016) Afro-alpine forest cover change on Mt. Guna (Ethiopia). In: Geophysical research abstracts, vol 18 (EGU2016-1992)

Asnake M (2017) Assessing the effectiveness of land resource management practices on erosion and vegetative cover using GIS and remote sensing techniques in Melaka watershed, Ethiopia. Environ Syst Res 6:16. https://doi.org/10.1186/s40068-017-0093-6

Bayazit M, Önöz B (2008) To pre-whiten or not to pre-whiten in trend analysis? Hydrol Sci J 52(4):611–624. https://doi.org/10.1623/hysj.53.3.669

Binyam TH, Maeda ED, Heiskanen J, Pellikka P (2015) Reconstructing pre-agricultural expansion vegetation cover of Ethiopia. Appl Geogr 62:357–365

Chebud AY, Melesse MA (2009) Numerical modeling of the groundwater flow system of the Gumera sub-basin in Lake Tana basin, Ethiopia. Hydrol Process 23:3694–3704

Chen T, Huang Q, Liu M, Li M, Qu L, Deng S, Chen D (2017) Decreasing net primary productivity in response to urbanization in Liaoning Province, China. Sustainability 9:162. https://doi.org/10.3390/su9020162

Chisholm N, Tassew W (2012) Managing watersheds for resilient livelihoods in Ethiopia. OECD, Development Co-operation Report 2012. OECD Publishing, Paris. https://doi.org/10.1787/dcr-2012-15-en

Ciais P, Reichstein M, Viovy N, Granier A, Ogee J, Allard V, Aubinet M (2005) Europe-wide reduction in primary productivity caused by the heat and drought in 2003. Nature 437:529–533

CSA (2004) National population statistical abstract. CSA, Addis Ababa

CSA, EDRI, IFPRI (2006) Atlas of Ethiopian rural economy. CSA, EDRI, IFPRI, Addis Ababa

Donald WM, Jean S, Steven AD, Jon BH (2011) Statistical analysis for monotonic trends, Tech Notes 6, November 2011. Developed for U.S. Environmental Protection Agency by Tetra Tech, Inc., Fairfax, VA, 2011. https://www.epagov/polluted-runoff-nonpoint-source-pollution/nonpointsource-monitoringtechnical-notes

Eve M, Whitford GW, Havstadt MK (1999) Applying satellite imagery to triage assessment of ecosystem health. Environ Monit Assess 54:205–227

Essayas KA, Fasikaw, AZ, Amy SC, Seifu AT, Muhammed E, William DP, Tammo SS (2014) Monitoring state of biomass recovery in the Blue Nile Basin using image-based disturbance index. In: Melesse AM et al. (eds) Nile River Basin. Springer International Publishing, Cham. https://doi.org/10.1007/978-3-319-02720-3_13

Feng X, Fu B, Lu N, Zeng Y, Wu B (2013) How ecological restoration alters ecosystem services: an analysis of carbon sequestration in China's Loess Plateau. Sci Rep 3:2846. https://doi.org/10.1038/srep02846

Fensholt R, Sandholt I, Rasmussen SM, Stisen S, Diouf A (2006) Evaluation of satellite based primary production modelling in the semi-arid Sahel. Remote Sens Environ 105:173–188

Fikir A, Nurhussen T, Nyssen J, Atkilt G, Amanuel Z, Mintesinot B, Deckers J, Poesen J (2009) The impacts of watershed management on land use and land coverdynamics in Eastern Tigray (Ethiopia). Resour Conserv Recycl 53:192–198

Gebregziabher G, Abera DA, Gebresamuel G, Giordano M, Langan S (2016) An assessment of integrated watershed management in Ethiopia. International Water Management Institute (IWMI), Colombo. (IWMI Working Paper 170). https://doi.org/10.5337/2016.214

Genene TM, Abiy G (2014) Review on overall status of soil and water conservation system and its constraints in different agro ecology of southern Ethiopia. J Nat Sci Res 4(7):59–69

Getachew E, Getachew F, Mulatie M, Assefa MM (2016) Evaluation of technical standards of physical soil and water conservation practices and their role in soil loss reduction: the case of Debre Mewi watershed, north-west Ethiopia. In: Melesse MA, Abtew W (eds) Landscape dynamics, soils and hydrological processes in varied climates. Springer International Publishing, Cham. https://doi.org/10.1007/978-3-319-18787-7_35

Gete Z, Hurni H (2001) Implications of land use and land cover dynamics for mountain resource degradation in the northwestern Ethiopian highlands. Mt Res Dev 21(2):184–191

Hao W, Guohua L, Zongshan L, Xin Y, Meng W, Li G (2015) Impacts of climate change on net primary productivity in arid and semiarid regions of China. Chin Geogr Sci. https://doi.org/10.1007/s11769-015-0762-1

Helsel RD, Hirsch MR (1992) Statistical methods in water resources: studies in environmental science 49. Elsevier Science Publishing, New York

Huete A, Didan K, Miura T, Rodriguez EP, Gao X, Ferreira GL (2002) Overview of the radiometric and biophysical performance of the MODIS vegetation indices. Remote Sens Environ 83:195–213

Hurni H, Kebede T, Gete Z (2005) The implications of changes in population, land use, and land management for surface runoff in the upper Nile basin area of Ethiopia. Mt Res Dev 25(2):147–154

Hurni H, Solomon A, Amare B, Berhanu D, Ludi E, Portner B, Birru Y, Gete Z (2010) Land degradation and sustainable land management in the highlands of Ethiopia. In: Hurni H, Wiesmann U (eds) Global change and sustainable development: a synthesis of regional experiences from research partnerships, vol 5. Geographica Bernensia, Bern, pp 187–207

Hurni H, Berhe WA, Chadhokar P, Daniel D, Gete Z, Grunder M, Kassaye G (2016) Soil and water conservation in ethiopia: guidelines for development agents, 2nd edn. Bern: Centre for Development and Environment (CDE), University of Bern, with Bern Open Publishing (BOP). https://doi.org/10.7892/boris.80013

Indiarto D, Sulistyawati E (2014) Monitoring net primary productivity dynamics in Java island using MODIS satellite imagery. Asian J Geoinform 14(1):1–14

Kebede W (2014) Effects of soil and water conservation measures and challenges for its adoption: Ethiopia in focus. J Environ Sci Technol 7(4):185–199

Lemlem T, Suryabhagavan KV, Sridhar G, Legesse G (2017) Land use and land cover changes and soil erosion in Yezat watershed, North Western Ethiopia. Int Soil Water Conserv Res 5(2):85–94. https://doi.org/10.1016/j.iswcr.2017.05.004

Li A, Bian J, Lei G, Huang C (2012) Estimating the maximal light use efficiency for different vegetation through the CASA model combined with time-series remote sensing data and ground measurements. Remote Sens 4:3857–3876. https://doi.org/10.3390/rs4123857

Li Z, Xu D, Guo X (2014) Remote sensing of ecosystem health: opportunities, challenges, and future perspectives. Sensors 14:21117–21139. https://doi.org/10.3390/s141121117

Machiwal D, Jha KM (2012) Hydrologic time series analysis: theory and practice. Springer, Dordrecht

Meaza H (2015) the role of community based watershed management for climate change adaptation in Adwa, Central Tigray Zone. Int J Weather Clim Change Conserv Res 1(1):11–35

Mekuria W, Aynekulu E (2011) Enclosure land management for restoration of the soils in degraded communal lands in Ethiopia. Land Degrad Dev. https://doi.org/10.1002/ldr.1146

Mellander P-E, Gebrehiwot SG, Gardenas AI, Bewket W, Bishop K (2013) Summer rains and dry seasons in the upper blue Nile basin: the predictability of half a century of past and future spatiotemporal patterns. PLoS ONE 8(7):e68461. https://doi.org/10.1371/journal.pone.0068461

Mildrexler JD, Zhao M, Heinsch AF, Running WF (2007) A new satellite-based methodology for continental-scale disturbance detection. Ecol Appl 17:235–250

Mildrexler JD, Zhao M, Heinsch AF, Running WF (2009) Testing a MODIS global disturbance index across North America. Remote Sens Environ 113:2103–2117

Molla MA (2016) Integrated watershed management and sedimentation. J Environ Prot 7:490–494

Monteith J (1972) Solar radiation and productivity in tropical ecosystems. J Appl Ecol 9:747–766

Neumann M, Zhao M, Kindermann G, Hasenauer H (2015) Comparing MODIS net primary production estimates with terrestrial national forest inventory data in Austria. Remote Sens 7:3878–3906. https://doi.org/10.3390/rs70403878

Nigussie H, Tsunekawa A, Nyssen J, Poesen J, Tsubo M, Derege TM, Schu B, Enyew A, Firew T (2015) Soil erosion and conservation in Ethiopia: a review. Prog Phys Geogr 1–25. https://doi.org/10.1177/0309133315598725

Nigussie H, Atsushi T, Jean P, Mitsuru T, Derege TM, Ayele AF, Jan N, Enyew A (2017) Comprehensive assessment of soil erosion risk for better land use planning in river basins: case study of the Upper Blue Nile River. Sci Total Environ 574:95–108

Nyssen J, Clymans W, Descheemaeker K, Poesen J, Vandecasteele I, Vanmaercke M, Zenebe A, Camp M, Haile M, Haregeweyn N, Moeyersons J, Martens K, Gebreyohannes K, Deckers J, Walraevens K (2010) Impact of soil and water conservation measures on catchment hydrological response: a case in north Ethiopia. Hydrol Process 24:1880–1895. https://doi.org/10.1002/hyp.7628

Osman M, Sauerborn P (2001) Soil and water conservation in Ethiopia. J Soils Sediments 1(2):117–123

Pan S, Tian H, Dangal RS, Ouyang Z, Tao B, Ren W, Lu C, Running S (2014) Modeling and monitoring terrestrial primary production in a changing global environment: toward a multiscale synthesis of observation and simulation. Adv Meteorol. https://doi.org/10.1155/2014/965936

Potter SC, Randerson TJ, Field BC, Matson AP, Vitousek MP, Mooney AH, Klooster AS (1993) Terrestrial ecosystem production: a process model based on global satellite and surface data. Glob Biogeochem Cycles 7(4):811–841

Rocchini D, Metz M, Ricotta C, Landa M, Frigeri A, Neteler M (2013) Fourier transforms for detecting multitemporal landscape fragmentation by remote sensing. Int J Remote Sens 34(24):8907–8916. https://doi.org/10.1080/01431161.2013.853896

Running SW, Nemani RR, Heinsch FA, Zhao M, Reeves M, Hashimoto H (2004) A continuous satellite-derived measure of global terrestrial production. Bioscience 54:547–560

Schmidt E, Fanaye T (2012) Household and plot level impact of sustainable land and watershed management (SLWM) practices in the Blue Nile. Ethiopia Strategy Support Program II (ESSP II) ESSP II Working Paper 42

Seleshi Y, Ermias T, Griensven A, Uhlenbrook S, Mul M, Kwast J, Zaag P (2012) Land use change and suitability assessment in the Upper Blue Nile Basin under water resources and socioeconomic constraints: a drive towards a decision support system. In: International Environmental Modelling and Software Society (iEMSs) 2012 international congress on environmental modelling and software managing resources of a limited planet, Sixth Biennial Meeting, Leipzig, Germany

Seleshi Y, Zaag P, Mul M, Uhlenbrook S, Ermias T, Griensven A, Kwast J (2013) Coupled hydrologic and land use change models for decision making on land and water resources in the Upper Blue Nile Basin. In: Geophysical research abstracts, vol 15 (EGU2013-12838-1)

Seleshi GY, Marloes LM, Griensven A, Ermias T, Priess J, Schweitzer C, Zaag P (2016) Land-use change modelling in the Upper Blue Nile Basin. Environments 3:21. https://doi.org/10.3390/environments3030021

Setegn GS, Srinivasan R, Dargahi B (2008) Hydrological modelling in the Lake Tana Basin, Ethiopia using SWAT Model. Open Hydrol J 2:49–62

Shimeles D (2012) Effectiveness of soil and water conservation measures for land restoration in the Wello area, northern Ethiopian highlands. PhD dissertation Submitted to University of Bonn, Germany

Simane B, Zaitchik FB (2014) The sustainability of community-based adaptation projects in the Blue Nile highlands of Ethiopia. Sustainability 6:4308–4325. https://doi.org/10.3390/su6074308

Sjöström M, Zhao M, Archibald S, Arneth A, Cappelaere B, Falk U, De Grandcourt A, Hanan N, Kergoat L, Kutsch W, Merbold L, Ardö J (2013) Evaluation of MODIS gross primary productivity for Africa using eddy covariance data. Remote Sens Environ 131:275–286. https://do..org/10.1016/j. rse.2012.12.023

Sung S, Nicklas F, Georg K, Lee KD (2016) Estimating net primary productivity under climate change by application of global forest model (G4M). J Korean Soc People Plants Environ 19(6):549–558. https://doi. org/10.11628/ksppe.2016.19.6.549

Tabari H, Taye TM, Willems P (2015) Statistical assessment of precipitation trends in the upper Blue Nile River basin. Stoch Environ Resour Risk Assess. https://doi.org/10.1007/s00477-015-1046-0

Teferi E, Uhlenbrook S, Bewket W (2015) Inter-annual and seasonal trends of vegetation condition in the Upper Blue Nile (Abay) Basin: dual-scale time series analysis. Earth Syst Dyn 6:617–636. https://doi.org/10.5194/ esd-6-617-2015

Temesgen G, Tesfahun F (2014) Evaluation of land use/land cover changes in east of lake Tana, Ethiopia. J Environ Earth Sci 4(11):49–53

Turner PD, Rifts DW, Cohen BW, Gower TS, Running WS, Zhao M, Costa HM, Kirschbaum AA, Ham MJ, Saleska RS, Ahl ED (2006) Evaluation of MODIS NPP and GPP products across multiple biomes. Remote Sens Environ 102:282–292

Waga M, Deribe G, Tilahun A, Matta D, Jermias M (2007) Challenges of collective action in soil and water conservation, the case of Gununo watershed, Southern Ethiopia. In: African crop science conference proceeding, vol 8, pp 1541–1545

Wolde M, Mastewal Y (2013) Changes in woody species composition following establishing enclosures on grazing lands on the lowlands of northern Ethiopia. Afr J Environ Sci Technol 7(1):30–40. https://doi.org/10.5897/ AJEST11.378

Woldeamlak B, Solomon A (2013) Land-use and land-cover change and its environmental implications in a tropical highland watershed, Ethiopia. Int J Environ Stud 70(1):126–139. https://doi.org/10.1080/00207233.201 2.755765

Woubet G, Dagnachew L (2011) Flood hazard and risk assessment using GIS and remote sensing in Fogera Woreda, Northwest Ethiopia. In: Melesse MA (ed) Nile River Basin. Berlin: Springer Science + Business Media. https://doi.org/10.1007/978-94-007-0689-7_9

Xiaoming F, Bojie F, Xiaojun Y, Yihe L (2010) Remote sensing of ecosystem services: an opportunity for spatially explicit assessment. Chin Geogr Sci 20(6):522–535

Yue S, Wang CY (2002) Applicability of pre-whitering to eliminate the influence of serial correlation on the MannKendall test. Water Resour Res 38(6):WR00861

Zerihun W (1999) Forests in the vegetation types of Ethiopia and their status in the geographical context. Paper presented at forest genetic resources conservation strategy development workshop, June 1999. Addis Ababa, Ethiopia

Zhengchao R, Huazhong Z, Hua S, Xiaoni L (2011) Spatio–temporal distribution pattern of vegetation net primary productivity and its response to climate change in Buryatiya Republic, Russia. J Resour Ecol 2(3):257–265. https://doi.org/10.3969/j.issn.1674-764x.2011.03.009

Zhou W, Li LJ, Mu JS, Gang CC, Sun GZ (2013) Effects of ecological restoration-induced land-use change and improved management on grassland net primary productivity in the Shiyanghe River Basin, north-west China. Grass Forage Sci 69:596–610. https://doi.org/10.1111/gfs.12073

2

Phosphorous status and adsorption characteristics of acid soils from Cheha and Dinsho districts, southern highlands of Ethiopia

Bereket Ayenew[1*], Abi M. Tadesse[2], Kibebew Kibret[3] and Asmare Melese[4]

Abstract

Background: Though soils in the study areas are characterized by higher iron and aluminum oxides, and low available P contents, study on P adsorption characteristics is limited. The purpose of this experiment was to evaluate adsorption properties of selected soils and determine the standard phosphorous requirements of the soils. In this experiment, separately weighed 2 g soil samples were equilibrated in 50 ml of 0.01 M $CaCl_2$ solution containing KH_2PO_4 at rates of 0, 1, 2.0, 4.0, 8.0, 16.0, 24.0, and 32.0 mg P L^{-1}.

Results: The Freundlich model was found to be the best model for the description of the P adsorption characteristics of the soils. The Freundlich coefficient K_f ranged from 123.32 to 315.31 mg P kg^{-1}. The Goha-1 soil had the highest K_f (315.31 mg P kg^{-1}) as Ketasire had lowest K_f (123.32 mg P kg^{-1}) values. The value of SPR_f was ranged from 50.50 to 154.02 mg P kg^{-1} for soils of the study area. Highly significant ($P \leq 0.01$) correlation was observed between the Freundlich adsorption parameters and soil physicochemical properties.

Conclusion: The standard P requirement of the studied soils was higher than the blanket P fertilizer rate recommendations in Ethiopia. Lack of inadequate knowledge about internal and external P requirement of each crop might have decreased yield in the study areas as it could have resulted in under-application of fertilizer P. The presence of high correlation between the adsorption parameters and the soil properties suggested the indices' prominent role in explaining P adsorption characteristics of the soils. Since higher dose of P is required by soils in the study area because of fixation, alternative P management strategies is needed to reduce P adsorption and enhance P availability.

Keywords: Fertilizer, Freundlich model, Isotherm, Oxyhydroxides, Phosphorous requirement

Background

Phosphorus (P) in soils is important because adequate availability of this nutrient among other nutrients is required for plant growth and crop production. However, in acidic soils, leaching of elements such as Ca, Mg, K and Na, silicates and carbonates, and transformations of Fe and Al into oxides or oxyhydroxides creates new functional groups for P adsorption (Jesse et al. 2016). Phosphorus (P) deficiency is particularly widespread in rain-fed upland farming systems throughout the tropics and remains a major plant nutrient constraint (Asmare et al. 2015). Similar to the other agricultural soils of the tropics Ethiopian soils are generally low in P (Mamo and Haque 1987) and hence P is one of the limiting elements in crop production in the highlands of the country. Therefore, P application has become an essential part of crop production systems in order to provide adequate food and fiber for human consumption (Jalali 2007).

Phosphorous reacts readily with metallic species and in highly weathered acidic soils, Fe and Al oxyhydroxides are the dominant species with which adsorption of P occurs principally via the formation of an inner-sphere

*Correspondence: bersofsam12@gmail.com
[1] School of Natural Sciences, Department of Chemistry, Madda Walabu University, Bale-Robe, Ethiopia

complex between orthophosphate anions and amorphous oxides [Fe_{ox} and Al_{ox} (Tan 2000; Sims and Pierzynski 2005; Ahmed et al. 2008; Rayment and Lyons 2011)]. These short range crystalline oxyhydroxides (amorphous) are well known to affect P sorption in soils (Janardhanan and Daroup 2010). Phosphorus adsorption depends on the nature and quality of sites available on the mineral surfaces and is therefore affected by high clay contents (Bahia et al. 1983). Studies also revealed that (Bedin et al. 2003; Moreira et al. 2006) the presence of large proportions of sesquioxides on clays, pH, exchangeable aluminum and organic matter influence phosphate adsorption and precipitation with iron and aluminum.

The extent of P adsorption and availability greatly varies from soil to soil owing to their differences in physicochemical properties (Muindi et al. 2015) and management (Moazed et al. 2010). Thus, understanding the P-sorption characteristics of soils are important for designing appropriate management strategies and predicting fertilizer requirements that are needed to be applied (Zhang et al. 2005). Most of the current management strategies rely on measuring P availability by leaching the soil with an extractant (Hansen and Strawn 2003). However, sorption reactions and phosphorous buffer capacity (PBC) of soils play an important role in both agronomic and environmental P management aspects. Sorption isotherms are much used to characterize the retention of P and the PBC of soil (Yli-halla et al. 2002).

Phosphorus availability to plants is mainly ruled by three factors: (i) the concentration of P in the soil solution ('intensity' factor, I), (ii) the amount of P in the solid phase that can be easily made available to plants ('quantity' factor, Q), and (iii) the capacity of soil to keep the P concentration in soil solution sufficiently high (P buffer capacity, PBC) (Sánchez-Alcalá et al. 2014). The capacity of the soil to adsorb P greatly influences plant responses to applied P fertilizers and the calibration of soil tests for P (Amrani et al. 1999). Therefore, knowledge of the ability of the soil to adsorb P fertilizer is required to provide an accurate estimate on the P fertilizer requirements of soils (Jalali 2007). This could be obtained by determining the amount of P adsorbed by soils at a concentration of phosphorus in solution known to be non-limiting to plant growth (Henry and Smith 2003).

Although most highland soils are characterized by soil acidity and high P fixation capacity due to intensive weathering and leaching attributed to high rainfall conditions (Desta 2015), little is known about soil P adsorption capacity and P fertilizer requirements of soils in Ethiopian highlands. Moreover, factors affecting the adsorption capacity of the soils and appropriate amount of P fertilizer required by the soils of the study area have not yet been investigated. Thus, the objective of the present work was to evaluate the P adsorption characteristics of soils and estimate the adequate amount of fertilizer P required by the soils and relate the coefficients of the adsorption models to the soil chemical properties.

Methods

Description of the study area

The study was conducted in Ethiopian highlands of Dinsho district in Bale zone and Cheha district in Gurage zone, Ethiopia (Fig. 1). Dinsho district lies between 6°58′40″ and 7°20′0″ N, and 39°44′0″ and 40°26′40″ E. Physiographically, most of the land area of the district is situated above 2000 m above sea level (masl). The district is classified into three agro-climatic zones: highlands (2300–2600 masl), midlands (1500–2300), and lowlands (1200–1500). The district has a bimodal rainfall distribution with mean annual rainfall of about 1150 mm. The maximum and minimum mean annual temperatures of the District are 17.5 and 6 °C, respectively. Wheat and barley are some of the major cereal crops grown in the area. The major reference soil groups in the district are Pellic Vertisols, Eutric Cambisols, Eutric Nitosols (now Nitosols), and Chromic Luvisols (FAO 1984).

Cheha district is situated between 8°32′0″ and 8°20′0″ N, and 37°41′20″ and 38°2′40″ E, at an elevation that ranges from 900 to 2812 masl. EIAR (2011) classified the area into three agro-ecological zones i.e. highlands (2300–3200 masl), midlands (1500–2300 masl), and lowlands (500–1500 masl) based on the bimodal rain fall system. The 10 years mean annual rainfall of the district is about 1268 mm. The mean annual maximum and minimum temperatures are 24.97 and 10.69 °C, respectively. The dominant soil types are Eutric Nitosols, Leptosols, and Pellic Vertisols (FAO 1984).

Site selection, soil sampling, and soil analysis

A preliminary soil survey and field observation was made using topographic map (1:50,000) of the study area. Soil pH (potentiometer), altitude (GPS), and slope (clinometer) were used as criteria for selection of soil sampling sites. Fifteen (15) sub-samples were collected from each sampling site to make one composite sample. Accordingly, twelve composite soil samples (0–15 cm), seven from Cheha district (Goha 1, Goha 2, Goha 3, Aftir, Abret, Kechot, and Moche), and five from Dinsho district (Doyomarufa 1, Doyomarufa 2, Tulu, Weni, and Ketasire) having pH values of less than 5.5 were obtained within an altitudinal range of 2000–3000 masl and slope less than 8% (Table 1). Soil pH was measured at field condition using portable pH meter to select soils having a pH of less than 5.5. The soil samples were put in plastic bag, tagged, and transported to laboratory for analysis during the 2014/15. Consequently, adequate

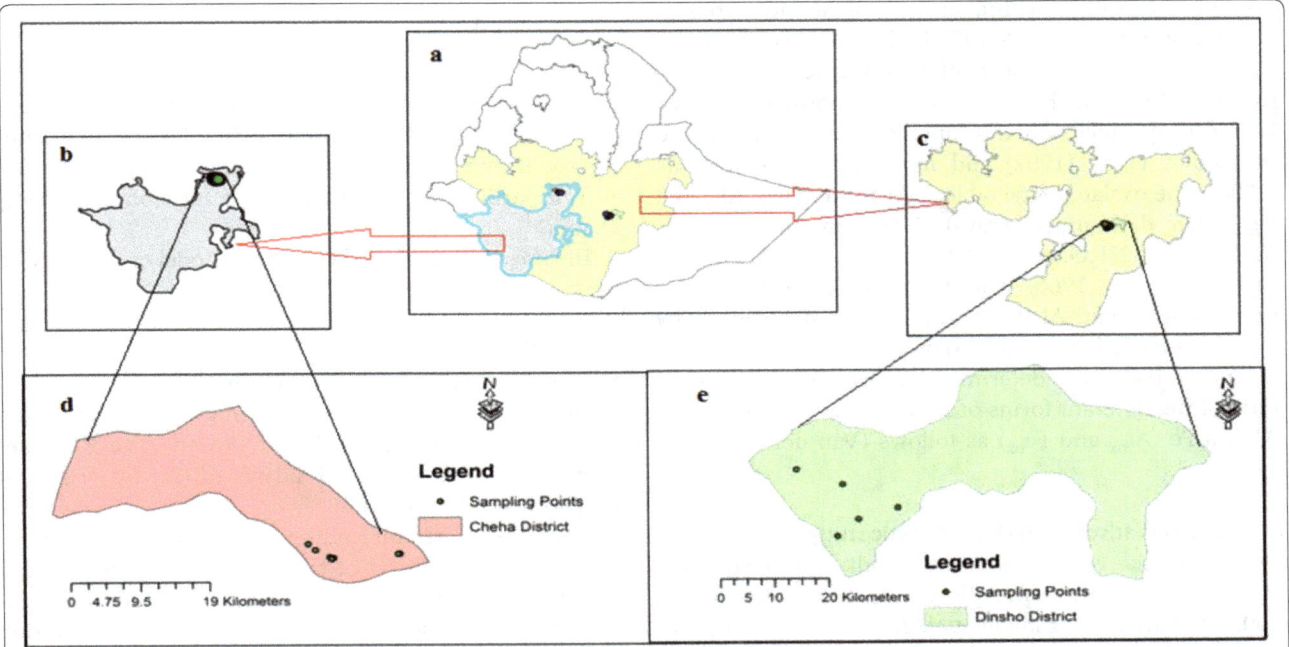

Fig. 1 Location Map of the Study Areas, **a** Oromia and SNNP regions in Ethiopia, **b** Cheha district in SNNP region, **c** Dinsho district in Oromia region, **d** Cheha district, **e** Dinsho district

Table 1 Sampling locations and site description

Sampling Sites	Longitude (E)	Latitude (N)	Altitude (masl)	Slope (%)
SC1	037°93′891″	08°033′45″	2426	4
SC2	037°94′389″	08°032′35″	2508	3
SC3	037°94′134″	08°030′78″	2498	2
SC4	037°91′180″	08°052′36″	2302	5
SC5	037°92′114″	08°043′58″	2401	3
SC6	038°02′572″	08°038′63″	2603	5
SC7	038°02′377″	08°038′23″	2563	5
SD1	039°51′934″	07°075′65″	2819	4
SD2	039°51′900″	07°072′28″	2832	5
SD3	039°52′387″	07°066′39″	2790	3
SD4	039°52′450″	07°071′01″	2768	2
SD5	039°52′714″	07°080′40″	2719	5

SC1 Goha 1, *SC2* Goha 2, *SC3* Goha 3, *SC4* Aftir, *SC5* Abret, *SC6* Kechot, *SC7* Moche, *SD1* Doyomarufa 1, *SD2* Doyomarufa 2, *SD3* Tulu, *SD4* Weni, *SD5* Ketasire

(about 1 kg) amount of composited soil samples were air dried and ground to pass through a 2 mm sieve for analysis of selected soil chemical and physical properties except organic carbon and total nitrogen in which case the samples were passed through a 0.5 mm sieve.

Analysis of soil physical and chemical properties

Soil particle size distribution was analyzed by the Bouyoucus hydrometer method (Day 1965). Soil bulk density was measured from three undisturbed soil samples collected using a core sampler following the procedure described by Jamison et al. (1950). Soil pH measured potentiometrically in H_2O and 1 M KCl solution at the ratio of 1:2.5 for soil: water and soil:KCl solutions using a combined glass electrode pH meter (Van Reeuwijk 1992). Cation exchange capacity (CEC) of the soils was determined by making use of the method suggested by Lavkulich (1981). Exchangeable acidity was determined by saturating the soil samples with 1 M KCl solution and titrating with 0.02 M NaOH as described by Rowell (1994). Soil organic carbon was determined by the dichromate oxidation method as described by Walkley and Black (1934). Available P was determined as Mehlich-3 P by shaking the soil samples with an extracting solution of 0.2 M $CH_3COOH + 0.25$ M $NH_4NO_3 + 0.015$ M $NH_4F + 0.013$ M $HNO_3 + 0.001$ M EDTA for 5 min (Mehlich 1984) and total soil P was determined using the method stated by Olsen and Sommers (1982).

The P fractions were successively extracted with 1 M NH_4Cl (available Pi), 0.5 NaHCO$_3$ (labile P_i and Po adsorbed on the soil surface), 0.1 M NaOH (moderately labile Pi and Po held more strongly by chemisorption to surfaces of Al and Fe oxides), Sonicate + 0.1 M NaOH

(Pi and Po adsorbed within surfaces of Al and Fe oxides of soil aggregates-occluded P), 1 M HCl (P associated to Ca, derived from primary mineral-apatite) and a mixture of HNO_3 and $HClO_4$ (residual P-non-labile, stable Po forms and relatively insoluble Pi forms) as described by Hedley et al. (1982) and modified by Chen et al. (2000). The oxalate extractable P, Al and Fe (P_{ox}, Al_{ox} and Fe_{ox}) were determined with 0.05 M ammonium oxalate $((NH_4)_2C_2O_4.2H_2O$, pH 3.3) for 2 h in the dark (Mckeague and Day 1966). Citrate bicarbonate dithionite-extractable Fe and Al (Fe_d and Al_d) were determined by the method of Mehra and Jackson (1960). The degree of P saturation (DPS) was determined as the percentage of the ratio of the different forms of P to the oxalate extractable Al and Fe (Al_{ox} and Fe_{ox}) as follows (Van der Zee et al. 1988).

Phosphorous adsorption characteristic study

In view of the results of several studies on adsorption characteristics of acidic soils of Ethiopia (Birru et al. 2003; Asmare 2014; Zinabu et al. 2015) and P adsorption kinetics of the studied soils (Bereket 2017) (unpublished) it was decided to use low initial P concentrations for adsorption experiments. Data for plotting P adsorption isotherms were obtained by equilibrating triplicate samples of 2 g of each soil. Following the procedure of Fox and Kamprath (1970), separate air-dried ground soil was equilibrated in 50 ml of 0.01 M $CaCl_2$ supporting electrolytes containing KH_2PO_4 at rates of 0.0, 1.0, 2.0, 4.0, 8.0, 16.0, 24.0, and 32.0 mg P L^{-1}. Three drops of toluene were added to each container to inhibit microbial activity. Three similar sets of containers for each soil were prepared and equilibration was carried out in 50 ml tube for 24 h at 25 °C. Each tube was covered with stopper to check the water losses through evaporation. Immediately after equilibration, the tubes were centrifuged and the supernatant solution was filtered through Wattman 42 filter paper and P content of the supernatant was determined using the ascorbic acid molybdenum blue color method (Watanabe and Olsen 1965). The quantities of P adsorbed by the soil were calculated from the difference between the initially applied P and the equilibrium soil solution P concentration (Yli-Halla et al. 2002). The P adsorption properties of the soil samples were studied with the Q/Ce (Q = adsorbed P and Ce = equilibrium P concentration) plot technique. Meanwhile, soil external P requirements were determined by substituting the desired P concentration into the fitted equations (Dodor and Oya 2000). The adsorption data were fitted to the Langmuir and Freundlich models and coefficients of the equations which best describe the sorption data has been used for a correlation study. The linear form of the Langmuir model could be written as:

$$1/Q = 1/Q_{max} + 1/(KL\,Q_{max}\,Ce)$$

where Q = amount of P adsorbed (mg P Kg^{-1}), Q_{max} = maximum amount of P adsorbed (mg P Kg^{-1} soil), Ce = equilibrium P concentration (mg P L^{-1}), K_L = affinity coefficient between phosphate ions and the surface of soil particles, which is related to the bonding energy (L mg $^{-1}$).

The linear form of the Freundlich model can be written as:

$$logQ = logKf + N\,logCe$$

where Q = amount of P adsorbed (mg P kg^{-1}), Ce = equilibrium solution P concentration (mg P L^{-1}) and K_f and N are fitting parameters (buffer power). The K_f values represents the amount of P adsorbed (mg kg^{-1}) at unit equilibrium concentration.

From the adsorption isotherms, the standard P requirement (SPR) (mg kg^{-1}) of the soil was determined. The SPR is the amount of P required to maintain an equilibrium concentration of 0.2 mg P L^{-1} in soil solution which has been shown to be a threshold for many crops. The PBC was derived by the first derivative of Q to Ce, from the non-linear Q-Ce curve (Ehlert et al. 2003) i.e. the slope of the curve at 0.2 mg P L^{-1} concentration of P (Morel et al. 2000).

The dQ/dCe of the Freundlich model is computed as:

$$dq/dCe = K_f \times N \times Ce^{N-1}$$

Similarly, the dQ/dCe of the Langmuir model is as follows:

$$dq/dC = Q_{max} \times K_L/(1 + C_e \times K_L)^2$$

Statistical analysis

The experimental data were statistically analyzed using SAS version 9.00 (SAS 2004). The fit of each adsorption model was evaluated by the determination coefficient (R^2) and root mean square error (RMSE) at a confidence level of 95%. Furthermore, the coefficients of the Freundlich model were correlated with the chemical properties of the studied soils.

Results and discussion

Physicochemical properties of the studied soils

The textural class of studied soils was predominantly clayey. In this regard, the clay content of the studied soils ranged from 36.6 to 49.8% and the highest clay content was recorded from SC3 soil, while the lowest was obtained from SD3 soil. The variation in percent clay among the soils studied could be attributed to land use history, the intensity of cultivation. On this subject Onweremadu et al. (2007) noted slight temporal textural differences in arable soils resulting from continuous

Table 2 Selected physicochemical properties of the soils

Soil	pH	Clay %	CEC (cmol$_c$ kg^{-1})	PAls %	OC %	DPS %	M-3P (mg kg^{-1})	Al$_{ox}$ (mmol kg^{-1})	Fe$_{ox}$ (mmol kg^{-1})	NH$_4$Cl + NaHCO$_3$ (Pi + Po) %	NaOH (Pi + Po) %	HCl-Pi %	Res-P %
SC1	4.65	48.2	19.15	8.04	1.58	4.13	8.9	152.37	97.7	1.17	48.77	11.76	27.07
SC2	4.79	44.6	23.88	3.43	1.86	5.31	10.14	139.07	78.54	1.40	43.34	11.19	27.84
SC3	4.73	49.8	20.88	6.18	1.93	4.65	9.73	145.96	88.18	1.09	46.76	12.08	29.06
SC4	4.81	46.6	26.98	1.3	1.97	5.41	9.15	130.07	79.96	1.18	44.46	11.15	26.82
SC5	5.13	48.0	22.81	1.84	2.11	6.49	10.79	126.37	77.46	1.19	44.17	11.40	26.78
SC6	4.76	43.4	29.54	1.46	2.00	6.58	13.94	131.85	76.39	1.23	43.41	11.00	28.09
SC7	4.91	45.0	25.03	1.96	1.91	5.59	10.38	134.37	81.75	1.13	44.46	11.03	27.01
SD1	5.30	37.8	33.7	0.83	2.39	11.68	23.26	90.0	66.21	2.15	22.94	35.65	29.81
SD2	5.01	43.6	29.8	1.07	2.37	8.88	18.33	91.85	82.29	1.48	22.88	35.75	30.34
SD3	5.12	36.6	34.03	0.71	2.58	18.62	24.35	71.89	51.75	2.10	23.17	35.55	29.74
SD4	5.36	44.2	29.14	0.75	2.56	10.49	22.29	90.37	59.43	1.87	23.06	35.71	29.90
SD5	5.45	38.2	30.66	0.68	2.52	13.69	21.34	78.15	63.36	2.00	23.43	35.60	29.53

M-3P Mehlich-3 P, *CEC* cation exchange capacity, *PAlcs* percent aluminum saturation, *OC* organic carbon, *Al$_{ox}$* oxalate extractable Al, *Fe$_{ox}$* oxalte extractable Fe, *NaOH-Pi* NaOH extractable inorganic P, *HCl-P* HCl extractable P

cultivation. The pH (H₂O) value of the soils varied between 4.65 and 5.45. Based on the rating suggested by Jones (2003), very strongly acidic soils accounted for 58%, while 42% of the studied soils could be categorized as strongly acidic suggesting the presence of substantial quantities of exchangeable hydrogen and aluminum ions which are associated with acidity (Table 2).

The cation exchange capacity of the studied soils ranged from 21.15 to 36.03 $cmol_c$ kg^{-1}. According to Hazelton and Murphy (2007) the CEC of the soils were in the range of moderate to high values. The variation in CEC among the soils might be due to the differences in OM content, clay mineral content and pH range of the soils (Table 2). In agreement with the current study Peinemann et al. (2002) reported that clay and organic matter are the main sources of CEC and the more clay and organic matter (humus) a soil contains, the higher its CEC. The exchangeable acidity and exchangeable Al of the studied soils were found to be in the range of 0.56–2.61 $cmolc$ kg^{-1} and 0.21–1.54 $cmolc$ kg^{-1}, respectively. The variability in exchangeable Al in soils was probably due to the different soil types and other soil chemical properties, such as pH, the organic matter content and CEC of the soils (Al Baquy et al. 2017).

The percentage of organic carbon in soils in the study area was ranged from 1.58 to 2.58% (Table 2). In general, as per the rating developed by Hazelton and Murphy (2007) OC content of soils of the study area were ranged from moderate with average structural condition and stability (1.00–1.80%) to high (1.80–3.00%) with a good structural condition and stability. The variation in percent organic carbon in soils of the study area could be attributed to the management practices. Many studies confirm that carbon retention in soil is influenced by crop management systems, such as crop rotation (Saljnikov-Karbozova et al. 2004), tillage (Saljnikov et al. 2009), residue management (Rasmussen et al. 1980) and fertilization and fertility (Saljnikov et al. 2005).

The amount of P extracted with Mehlich-3 ranged from 8.90 to 25.75 mg kg^{-1}. However, there was a considerable variation among soils of the districts in terms of the available P. As per the rating established by Ethio-SIS (2014), Mehlich-3 extractable P contents of studied soils could be qualified as very low (0–15 mg kg^{-1}) and low (15–30 mg kg^{-1}) for soils. Soils of the study area were appeared to have different concentrations of dithionite-citrate-bicarbonate (DCB) and oxalate extractable Fe, Al, and P. Consequently, Al_{ox} contents of the soils varied between 78.148 and 147.37 mmol kg^{-1}; Fe_{ox}, 73.75 and 174.69 mmol kg^{-1}; Pox, 5.51 and 12.29 mmol kg^{-1}; DPS, 4.13 and 18.62%. In view of the cut off point for loss of P due to runoff (Ige et al. 2005) all the studied soils, showed DPS values less than 20% indicating no risk of P loss from soil. These low values might be attributed to the higher adsorption capacity of these soils for applied P fertilizer.

The relative abundances of P forms in the studied soils were in the following order: Res-P > NaOH-Po > NaOH-Pi > HCl-P > (NaOH-Pi)sn > (NaOH-Po)sn > NaHCO₃-Po > NaHCO₃-Pi > NH₄Cl-P in soils from Cheha District and HCl-P > Res-P > NaOH-Po > NaOH-Pi > (NaOH-Po)sn > (NaOH-Pi)sn > NaHCO₃-Po > NaHCO₃-Pi > NH₄Cl-P for soils of Dinsho district. The labile P (Pi in NH₄Cl + Pi and Po in NaHCO₃) varied between 1.09 and 2.15% suggesting plant growth could be affected due to the presence of low biologically available P forms (Kiflu et al. 2017). The moderately labile P (Pi + Po in NaOH) called the Fe–Al associated P was varied between 22.88 and 48.77% and appeared to be very high compared to other fractions. The relatively higher abundance of Al and Fe bound P could be as a result of the presence of variable Al and Fe contents in soils at various stages of relative development and their reaction with soil P (Kiflu et al. 2017). On the other hand the P associated to Ca (HCl-Pi) varied between 11.00 and 35.75% which could be attributed to the difference in Ca contents of the studied soils and the residual-P varied between 26.78 and 30.34% (Table 2).

Phosphorous adsorption characteristics
Phosphorous adsorption isotherms

The preliminary investigation we carried out on phosphorous adsorption kinetics exhibited a period of 24 h for maximum P adsorption. Furthermore, the kinetic models also pointed out that the best linear regression was observed at lower initial P concentrations. Therefore, Phosphate adsorption isotherms of the studied soils were determined by plotting the equilibrium concentration of phosphate (Ce) against the amount of phosphate adsorbed (Q) which was resulted from 24 h equilibration period.

A graphical illustration of the adsorption isotherms of the studied soils with different curves followed a smooth plateau pattern depicting variation in data set, where shape of isotherms appeared to vary with soil characteristics was presented in Figs. 2, 3 and 4. It can be seen that the phosphorous adsorption and equilibrium P concentrations were found to increase with increasing levels of added P (0.5–32 mg P L^{-1}) in all the soils under investigation. In this regard Barrow (2015) reported that the amount of phosphate adsorbed by soil increases with increases in the solution concentration of phosphate. However, a sharp increase in phosphorous sorption was showed at lower concentrations and eventually approached a stable state at higher concentrations. This might be attributed to the unavailability of more adsorption sites for P to be adsorbed as P is

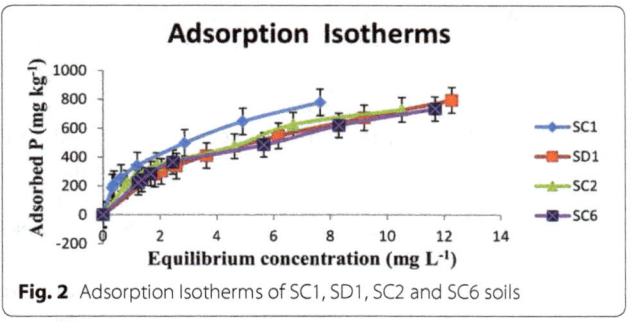

Fig. 2 Adsorption Isotherms of SC1, SD1, SC2 and SC6 soils

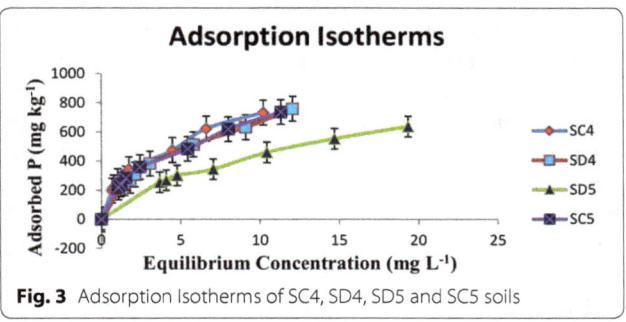

Fig. 3 Adsorption Isotherms of SC4, SD4, SD5 and SC5 soils

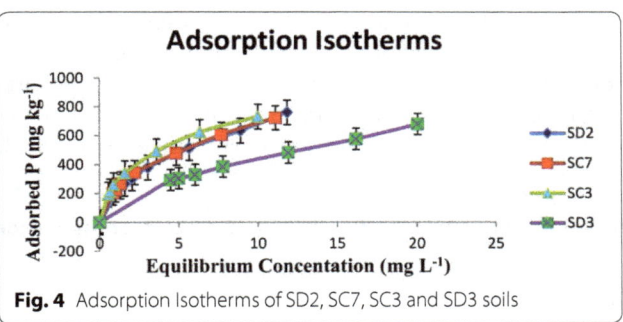

Fig. 4 Adsorption Isotherms of SD2, SC7, SC3 and SD3 soils

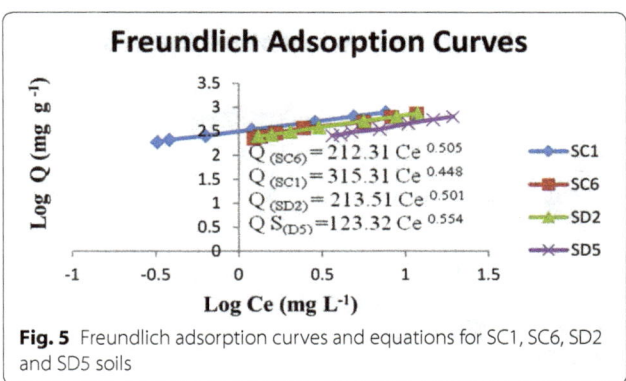

Fig. 5 Freundlich adsorption curves and equations for SC1, SC6, SD2 and SD5 soils

and oxides of Al or Fe (Bedin et al. 2003; Moreira et al. 2006).

Phosphorous adsorption indices

Fitting adsorption data to the adsorption equations was carried out by plotting the data in the linearized forms of the adsorption equations. The goodness of fit was evaluated by simple linear regression coefficients calculated for the linear transforms (Figs. 5, 6, 7). Plotting the adsorption data of the studied soils in the linear form of the Langmuir and Freundlich equations resulted in comparable and highly significant relationships between P in equilibrium solution and P adsorbed by soil.

The adsorption indices (coefficients) of linearized form of both models are presented in Table 3. Accordingly, the adsorption maxima (Q_{max}) of the soils were ranged from 680.22 to 1112.11 mg kg^{-1} at 24 h equilibration period (Table 3), indicating different concentration of the strong sites (Fe_{ox} and Al_{ox}) might have influenced the adsorption capacity of the soils (Chunye et al. 2009). The lowest (680.22 mg P kg^{-1}) and the highest (1112.11 mg P kg^{-1}) Q_{max} values were observed in SD1 and SC1 soils from the adsorption study respectively. The variations in Q_{max} might be associated with the amount of Fe_{ox} and Al_{ox} present in the soils. In agreement with the present study, Campos et al. (2016) examined tropical soils with Q_{max} ranging from 60 to 5500 mg kg^{-1}, and found that Al and Fe were critical ions in controlling P sorption in these soils. Moreover, very inconsistent Q_{max} values have been reported by several studies from different parts of the country, of which Birru et al. (2003) selected 16 soil samples from Northwestern Highlands of Ethiopia and reported very low value (5–108 mg P kg^{-1}) compared to the current study. On the other hand, soils under the present study appeared to have lower Q_{max} as compared to acidic soils in Wonago district, southern Ethiopia (Zinabu 2015). Likewise, the PBC_L determined at the standard P concentration in the 24 h of equilibration

further added to the soil. In concurrent with the present study Bai et al. (2012) also found that the P adsorption on the three soils consistently showed a sharp increase at lower initial P levels, and then approached a steady state at higher phosphorus levels. The slopes of the sorption curves showed that the amount of P adsorbed by the soils differed among the studied soils. Therefore, it is evident from these isotherms (Figs. 2, 3 and 4) that soils in the study area exhibited different adsorption characteristics. The probable reason for this variation in adsorption capacity may be the disparity among the soils in terms of soil properties which control the availability of P such as pH, clay content, organic matter,

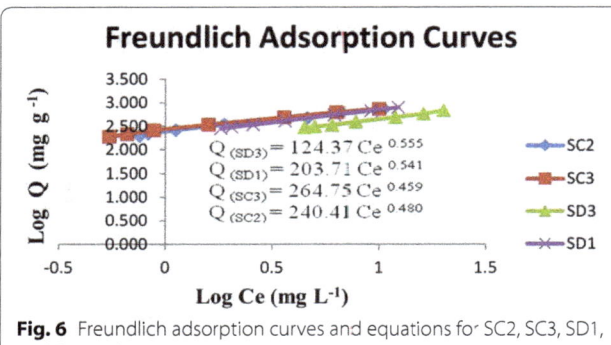

Fig. 6 Freundlich adsorption curves and equations for SC2, SC3, SD1, and SD3 soils

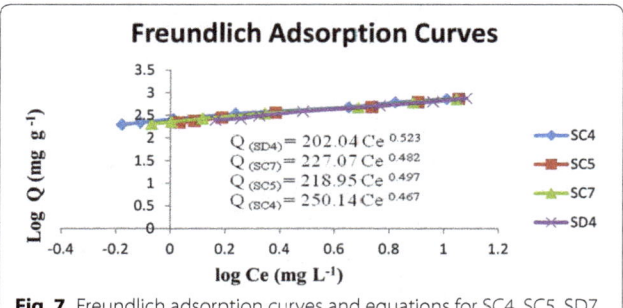

Fig. 7 Freundlich adsorption curves and equations for SC4, SC5, SD7, and SD4 soils

period ranged from 85 to 398 L kg^{-1} and was within the range found by Asmare (2014) and Zinabu et al. (2015) in different acidic soils of the country.

The result of the regression analysis of the data from the Langmuir model showed that the adsorption energy coefficient or binding energy constant of P on the adsorption sites (K$_L$) values ranged from 0.09 to 1.13 L mg^{-1} at 24 h equilibration period indicating the studied soils had variable adsorption energy coefficient. Lin and Banin (2005) reported 0.25–0.26 and 0.02–0.05 L mg^{-1} of the binding energy constants on the strong sites and weak sites, respectively, in the clayey sand with 0.02–0.32% of carbonate content. Therefore, the values measured in the present study roughly fall within the same magnitude as those for clayey soils. In agreement with, this study Asmare (2014) also noted that the binding energy of P on the adsorption sites of clayey acidic soils of Farta district in northwestern highlands of Ethiopia was ranged from 0.21 to 0.96 L mg^{-1}.

It is evident from the data that the energy of adsorption (K$_L$) value was found to be greater for soils that had higher contents of oxyhydroxides of Fe and Al, higher sorption capacity (PSC), higher exchangeable Al (exAl) and lower oxalate extractable Phosphorous (P$_{ox}$) (Tables 3).

The maximum buffering capacity (MBC) of the soil, which is the product of Q$_{max}$ and K$_L$, is a capacity factor that measures the ability of the soil to replenish phosphate ions to the soil solution (Asmare 2014). The MBC is potentially useful as a tool to characterize soil types especially for low concentration range because it describes both the nature and the capacity of P adsorption. According to the data presented in Table 3, the values of the MBC of the soils were ranged from 86.82

Table 3 The Freundlich and Langmuir model parameters, SPR, PBC, MBC, RMSE and r^2 values after 24 h equilibration period

Sampling Sites	Freundlich parameters						Langmuir parameters						
	K$_f$	N	RMSE	r^2	SPR$_{f(0.2)}$	PBC$_{f(0.2)}$	Q$_{max}$	K$_L$	RMSE	r^2	SPR$_{L(0.2)}$	PBC$_{L(0.2)}$	MBC$_{(0.2)}$
SC1	315.31	0.44	1.52	0.99	154.02	342.82	680.27	1.13	56.91	0.96	125.47	769.23	627.35
SC2	240.41	0.48	2.38	0.99	110.95	266.54	781.25	0.45	32.66	0.99	65.02	354.61	298.04
SC3	264.75	0.46	2.11	0.99	126.39	290.33	746.27	0.61	29.88	0.99	81.69	458.71	363.77
SC4	250.14	0.47	2.25	0.99	117.91	275.49	763.36	0.52	35.53	0.99	71.63	395.26	324.56
SC5	218.95	0.49	2.78	0.99	98.37	244.51	833.33	0.33	28.07	0.99	51.28	273.22	240.63
SC6	212.14	0.50	2.93	0.99	94.04	237.67	884.96	0.28	21.27	0.99	46.34	244.49	219.56
SC7	227.07	0.48	2.59	0.99	104.44	251.99	769.23	0.41	30.74	0.99	58.31	315.46	269.44
SD1	203.71	0.54	3.44	0.99	85.23	230.71	1112.11	0.19	18.22	0.98	39.92	207.04	192.43
SD2	213.51	0.50	3.01	0.99	95.33	238.81	847.46	0.29	33.44	0.98	46.99	248.76	221.93
SD3	124.37	0.55	6.43	0.99	50.85	141.29	943.39	0.09	14.88	0.97	17.69	90.17	86.82
SD4	202.04	0.53	3.23	0.99	86.13	228.14	925.93	0.24	19.61	0.98	42.05	220.26	200.71
SD5	123.32	0.55	5.88	0.99	50.50	140.06	869.56	0.11	12.39	0.98	18.62	95.15	91.11
Mean	216.31	0.502	3.21	0.99	97.85	240.69	846.34	0.39	27.80	0.98	55.42	306.03	261.36

K$_f$ amount of P adsorbed, N slope, RMSE root mean square error, SPR$_f$ standard P requirement from Freundlich model, PBS$_f$ phosphorous buffering capacity from Freundlich model, Q$_{max}$ maximum adsorption, K$_L$ affinity coefficient, SPR standared P requiremnts, PBC phosphorous buffering capacity, MBC maximum buffering capacity

Phosphorous status and adsorption characteristics of acid soils from Cheha and Dinsho...

23

to 627.35 L kg^{-1}. It can be seen that soils with higher adsorption capacity, Fe$_{ox}$ and Al$_{ox}$ concentration, clay percent and exchangeable Al were appeared to have higher maximum buffering capacity. The soils with the highest MBC (SC1 and SC3) have higher adsorption capacity, reflecting low P supply into soil solution for a longer period as compared to the less buffered soils which can supply ample P to soil solution. On the other hand, soils with low buffering capacity will have more P in soil solution and will enhance P mobility in such soils.

The Freundlich adsorption capacity (K$_f$), which is a measure of the reactive surface area (relative adsorption capacity) of the soil, is presented in Table 3. It has been shown that the K$_f$ values ranged from 123.32 (SD5) to 315.31 (SC1) mg kg^{-1} respectively. It may be seen that the K$_f$ value was higher for the high P adsorbing soils that adsorb more P per kg soil than the low P adsorbing soils indicating strong affinity of adsorbate (P) towards the adsorbent surfaces (soil). The variation observed in P sorption capacities (K$_f$) among soils from the assorted sites might be ascribed to differences in oxides of Al and Fe and clay content in soils (Obura 2008). Moreover, the K$_f$ values of the current study were in the range found by Birru et al. (2003) who reported 90–4915 mg P kg^{-1} for acidic soils of north western highland of Ethiopia.

Phosphorous buffer capacity (PBC) of soils plays an important role in both agronomic and environmental aspects of P management (Ehlert et al. 2003). Sorption isotherms are much used to characterize the retention of P and the PBC$_f$ of soil (Yli-halla et al. 2002). Accordingly, the phosphorous buffer capacity was obtained for soils of each district from the slope of the linear equation by plotting P adsorbed against P equilibrium concentration at 0.2 mg L^{-1}. The result of this study indicated that PBC$_f$ values determined at the standard P concentration of 0.2 mg P L^{-1} increased with increase in the K$_f$ values.

Model validation and interpretation

The P adsorption data fitted to the linearized Langmuir and Freundlich models are shown in Table 3. The coefficients of determination (R^2) for the Freundlich model (0.99–1.00) were almost slightly greater than the Langmuir model (0.98–0.99) in the studied soils. Furthermore, the average distance at which the observed values deviated from the regression line (RMSE) of the Freundlich model (5.4–9.8) was smaller than that of the Langmuir model (19.8–28.0) (Table 3). Hence, the Freundlich model can be considered as a superior model for the description of the P adsorption characteristics of the soils in this particular study because of its lower (RMSE) and slightly higher R^2 values. Moreover, the Freundlich model, although originally empirical, implies that the affinity for adsorption decreases exponentially with increasing saturation of the surface which is closer to reality than the assumption of constant bonding energy inherent in the Langmuir equation. Therefore, phosphorous sorption by the studied soil was well described by the Freundlich equation.

The Freundlich constants K$_f$ and N, which represent the intercept and slope of the log-transformed sorption isotherm, may be taken as a measure of the extent of adsorption and the energy of adsorption, for the thought-out soils respectively. The former could also be considered as capacity factor implying that a soil having larger K$_f$ value has superior adsorbing capacity than otherwise. The Freundlich coefficient K$_f$ ranged from 123.32 to 315.31 mg P kg^{-1} (Table 3). In this study, SC1 exhibited the highest K$_f$ (315.31 mg P kg^{-1}) value where as SD5 did show the lowest K$_f$ (123.32 mg P kg^{-1}). The order of soils according to their adsorption capacities is found to be: SC1 > SC3 > SC4 > SC5 > SC6 > SC2 > SC7 > SD2 > SD4 > SD1 > SD3 > SD5. The observed difference in K$_f$ value of the soils in the study area might be due to the disparity in the distribution of oxyhydroxides of Fe and Al. It has been noted that adsorption of P in soils is controlled to a large extent by the presence of amorphous Fe and Al (Agbenin 2003; Yan et al. 2013). These values are relatively high and comparable with those found by Asmare (2014) for the acidic soils of Farta district, northwestern highlands of Ethiopia which varied from 80 to 259 mg P kg^{-1}. On the other hand the soils were found to have lower K$_f$ value compared to the one reported and by Zinabu et al. (2015) in acidic soils from Bule district, southern Ethiopia, which varied from 479 to 487 mg P kg^{-1}. The exponent term (N) in the Freundlich relationship describing solid phase P and solution P at equilibrium is less than one for all of the soils (Table 3). This indicates that the relationship is not curvilinear and suggests that adsorption of P is controlling solution P concentrations in most of the soils (Bertrand et al. 2003).

Phosphorus buffer capacity was obtained for each soil from the slope of the linear equation by plotting P adsorbed against P concentration at equilibrium for the adsorption dominated part of the isotherm (0.5–32 mg P L^{-1}). Phosphorous buffer capacity (PBC$_f$) of the soils varied from 140.06 to 342.82 L kg^{-1} (Table 3). In the current study, all the studied soils appeared to have a very high PBC$_f$ (Moody and Bolland 1999), the soils from Cheha being higher. It can be inferred that more added fertilizer P could be adsorbed, as I or Q is increased (Jensen and Jakobsen 1970) in soil samples from Cheha district compared to soils from Dinsho district. Moreover, the soils with the highest PBC (SC1) have higher adsorption capacity and would maintain low P supply in soil solution for a longer period as compared to least (SD5) buffered soils which can supply ample P to soil solution.

On the other hand, soils with low buffering capacity will have more P in soil solution and will enhance P mobility in such soils. About 75% of the studied soils were within the range found by Asmare (2014) for eight acidic soils of northwestern highlands of Ethiopia.

Standard phosphorus fertilizer requirements (SPR)

The SPR_f of the soils calculated from the Freundlich equation (Table 3) at 0.2 mg P L^{-1} is an estimate of P sorption potential (Jackman et al. 1997; Wang et al. 2000). The amount of P adsorbed at an equilibrium concentration of 0.2 mg P L^{-1} was between 50.50 and 154.02 mg P kg^{-1} of soil. Fox (1981) and Afif et al. (1993) indicated that this concentration (0.2 mg P L^{-1}) is an adequate external P requirement for most crops. The value of SPR_f ranged from 94.04 to 154.02 mg P kg^{-1} for the soils of Cheha district. Similarly, the SPR_f of soils in Dinsho district varied from 50.50 to 95.33 mg kg^{-1}. The distinction in adsorption capacities of these soils might be ascribed to the difference in abundance and distribution of strongly reactive P adsorption sites on amorphous Fe and Al oxides (Chunye et al. 2009). In this regard different studies realized that Fe_{ox} and Al_{ox} were considered to be the key factors in regulating P adsorption due to its large surface area, variable-charge surface and high reactivity, especially in redox-changing environments (Makris et al. 2005; Wang et al. 2009; Yan et al. 2016). The higher values of SRP_f in case of Cheha soils indicate the need for application of more P fertilizers to maintain optimum crop production compared to soils collected from Dinsho district.

According to Sanchez and Uehara (1980), soils (SC2, SC3, SC4, SC5, SC6, SC7, SD1, SD2, SD3, SD4 and SD5) that adsorb less than 150 mg P kg^{-1} soil to meet the SPR_f value of 0.2 mg L^{-1} in soil solution were considered to be low adsorbing soil and the other adsorbing an amount exceeding this value was high P adsorbing ones (SC1). Furthermore, the results of the current study on the subject of SPR of these soils presented a comparable result with those found by Asmare (2014); Zinabu (2015) and Duffera and Robarge (1999) in different highly weathered acidic soils of Ethiopia (42–175 mg P kg^{-1}) in West Africa (Abekoe and Sahrawat 2001). As a case in point Duffera and Robarge (1999) reported that the amount of added P required maintaining a concentration of 0.2 mg P L^{-1} in solution (SPR) ranged from 50 to 201 mg P kg^{-1} for surface soil samples from non-cultivated and non-fertilized areas in Ethiopia. Likewise, Zinabu et al. (2015) has found that the SPR values of acidic soils of Bule and Wonago districts in southern Ethiopia ranging from 71.8 to 211 mg kg^{-1}.

The consumption of fertilizers in sub Saharan Africa was estimated to be 8 kg P ha^{-1} (Morris et al. 2007) and the blanket recommendation for cereal crops in Ethiopia was 20 kg P ha^{-1}. However, in the present study at the 24 h of equilibration time, the SPR_f of the soils ranged from 50.50 to 154.02 mg P kg^{-1} which was equivalent to the application of 115.64 to 352.71 kg P ha^{-1} and superior than the blanket recommendation in Ethiopia by about a factor of 6–18. The differences in the SPR_f of the soils in the study sites indicate that application of blanket P fertilizer rates for all study sites is not feasible. The blanket recommendation of P fertilizer for Ethiopian soils is therefore, inadequate for optimal crop production at all the sites. This is because it can supply at most only 20 kg P ha^{-1}, which is much below the soils SPR_f of 115.64–352.71 kg P ha^{-1}. Soils of the study area were found to have analogous SPR_f values with soils appraised by Asmare (2014), where the application of P fertilizers based on the blanket recommendation in the northwestern highlands of Ethiopia could result in a substantial yield deficit. Therefore, a mechanism has to be devised to increase availability of P by attenuating the P high adsorption capacity of the soils in order to increase productivity especially for soils with relatively high P fixing capacity (Cheha district).

The SPR_f (0.2 mg L^{-1}) levels for the studied soils that were obtained after 24 h equilibration periods were not remain similar and the soil solution P (intensity factor) depends on the adsorbed P (capacity factor) or the buffering capacity of the soils (Fox 1981). Consequently, the use of soil test P fertilizer recommendation based on the adsorption curves together with plant response for the applied P for different crop varieties should be done for the soils of the study area rather than using the usual conventional practice which could underestimate or overestimate the amount P fertilizer to be applied. Since the SPR_f of the soils calculated from the Freundlich equation (Table 3) is an estimate of P sorption potential (Wang et al. 2000) and an adequate external P requirement for most crops (Afif et al. 1993), diverse crop varieties in the study area could require different amount of P fertilizer. Thus it could be inferred that the external and internal P requirement of a crop as well as variety has to be investigated to find a reasonable amount of fertilizer to be applied to a particular soil with respect to crops and sites.

Correlation between Freundlich adsorption indices and soil properties

The correlation analysis was done between the Freundlich adsorption indices (K_f, PBC_f, and SPR_f) and the various soil properties (Clay, exAl, CEC, OC, P_{ox}, Al_{ox}, Fe_{ox}, Mhlich 3 P, NH_4Cl-P, NaOH-Pi and HCl-P) and presented in Table 4. Adsorption parameters were strongly and positively correlated with soil properties i.e., [r = 0.89, P ≤ 0.01(K_f and Fe_{ox}); r = 0.87, P ≤ 0.01

Table 4 Correlation table for Freundlich model coefficients and soil properties

	CEC	M-3 P	Al_{ox}	Fe_{ox}	OC	Clay	HCl-P	NaOHPi	NH_4Cl-P	P_{ox}	exAl
K_f	−0.81**	−0.79**	0.87**	0.89**	−0.86**	0.84**	−0.75**	0.62*	−0.84**	−0.91**	0.78**
PBC_f	−0.80**	−0.78**	0.86**	0.88**	−0.85**	0.84**	−0.74**	0.61*	−0.84**	−0.91**	0.77**
SPR	−0.84**	−0.83**	0.89**	0.91**	−0.89**	0.85**	−0.77**	0.65*	−0.86**	−0.91**	0.81**
N	0.90**	0.94**	−0.94**	−0.94**	0.93**	−0.90**	0.88**	−0.77**	0.93**	0.93**	−0.79**

CEC cation exchange capacity, *M-3 P* Mehlich-3 P, *Al_ox* oxalte extractable Al, *Fe_ox* oxalate extractable Fe, *OC* organic carbon, *HCl-P* HCl extractable P, *NaOH-Pi* NaOH extractable inorganic P, *NH4Cl-P* NH4Cl extractable P, *Pox* oxalte extractable P, *exAl* exchangeable Al

$(K_f$ and $Al_{ox})$; $r = 0.89$, $P \le 0.01$ $(Al_{ox}$ and $SPR_f)$; $r = 0.84$, $P \le 0.01$ $(K_f$ and Clay); $r = 0.62$, $P \le 0.05$ (NaOH-pi and $K_f)$; $r = 0.81$, $P \le 0.01$ (exAl and $SPR_f)$; $r = 0.91$, $P \le 0.01$ $(Fe_{ox}$ and $SPR_f)$; $r = 0.89$, $P \le 0.01(Al_{ox}$ and $SPR_f)$]. The adsorption parameters were positively and strongly correlated with Fe_{ox} and Al_{ox} indicating the presence of reactive and strong P adsorption sites on Fe and Al oxides (Chunye et al. 2009). The presence of high correlation between the adsorption parameters and the oxalate extractable Al and Fe suggest that those adsorption coefficients did play a prominent role in explaining P adsorption characteristics of the soils in terms of oxides of Fe and Al in the study area. It can also be seen from the correlation study that in soils where there was higher clay content, exAl, Al_{ox} and Fe_{ox}, concurrently higher SPR_f, PBC and K_f values were observed. It verified the expectation that higher phosphorous buffering capacity, fertilizer requirements and adsorption capacity is expected in soils where clay content, exAl and oxides of Fe and Al are dominating. This confirmed the significant influence of soil texture on P sorption capacity as reported by several authors (Yuan and Lucas 1982; Mozaffari and Sims 1994).

A significant and negative correlation was found between the soil properties (CEC, OC, P_{ox}, NH_4Cl-P, and HCl-P) and the Freundlich adsorption parameters. However, NaOH-Pi was strongly and positively $(p \le 0.01)$ correlated with adsorption coefficients. The compelling reason for strong positive association between NaOH-Pi and the adsorption parameters could be the greater contents of oxyhydroxides of Fe and Al which in turn vigorously stabilize NaOH-Pi in the soil (Yan et al. 2017). The significant and negative correlation between the organic carbon content and adsorption indices (K_f, SPR_f and PBC_f) $(r = -0.86$, $P \le 0.01$; $r = -0.89$, $P \le 0.01$, $r = -0.85$, $P \le 0.01$) respectively, for soils was concurrent with several reports (Juo and Fox 1977 and Daly et al. 2001). However, there are results indicating positive correlation as well (Wild 1950; Börling et al. 2001). Moreover, as case in point other research results showed that OC can have a significant and positive relationship (Ayaz et al.

2010) and significant negative relationship (Burt et al. 2002; Moazed et al. 2010) with the adsorption capacity of the soils. From this strong association between these parameters and soil properties it could be inferred that in soils where we found high CEC, OC, P_{ox}, NH_4Cl-P and HCl-P there would be low adsorption capacity, phosphorus buffering capacity and phosphorous fertilizer requirements. Thus, the coefficients k_f (amount adsorbed), PBC_f (Phosphorous buffering capacity), SPR_f (standard phosphorous requirement) and N (buffer power) of Freundlich adsorption model were best predicted from CEC, Mehlich-3 P HCl-P, NH_4Cl-P, clay, Fe_{ox}, Al_{ox}, NaOH-pi and exAl and P_{ox} and therefore put forth a profound influence on the rate of P adsorbed onto the soils.

Conclusions

The adsorption study was conducted to found the relationship between adsorbed P and solution P to appraise the P adsorption properties of the soils by making use of appropriate empirical model and evaluate the correlation between adsorption indices and soil properties. The Freundlich model could be considered as the best model for the description of the P adsorption characteristics of the soils in this particular study area. The Freundlich coefficient K_f (adsorption capacity) value ranged from 123.32 to 315.31 mg P kg^{-1} and depended on amorphous form of Fe and Al (i.e. Fe_{ox} and Al_{ox}). Consequently, the standard P requirements of the soils varied from 50.50 to 154.02 mg P kg^{-1} of soil, implying the blanket recommendation of P fertilizer for Ethiopian soils is therefore, inadequate for optimal crop production in the study area and was much below the soils' actual P requirements. Phosphorous buffer capacity (PBC_f) of the soils varied from 140.06 (SD5) to 342.82 L kg^{-1} (SC1) and all of the premeditated acidic soils (100%) was classified under very high (> 90) phosphorous buffering capacity soils and would maintain low P supply in soil solution for a longer period. Strong and positive significant relationship was observed between the Freundlich adsorption parameters (K_f, PBC_f and SPR_f) and oxalate extractable Al and Fe $(P \le 0.01)$. The Freundlich adsorption coefficients play a prominent role in explaining P adsorption characteristics of the soils in terms of oxides of Fe and Al in the study area. The amount of P adsorbed, Phosphorous buffering capacity, standard phosphorous requirement of the soils were best predicted from CEC, Mehlich-3 P HCl-P, NH_4Cl-P, clay, Fe_{ox}, Al_{ox}, NaOH-pi and exAl and P_{ox} and therefore put forth a considerate influence on the rate of P adsorbed onto the soils. Therefore, a mechanism has to be devised to increase availability of P by attenuating the P adsorption capacity of the soils in order to increase productivity especially for soils with relatively high P fixing capacity. The blanket P fertilizer rate recommendations in Ethiopia without the knowledge of external P requirement of each crop might have decreased yield in the study areas as it could have resulted in under-application of fertilizer P. Thus it could be inferred that the external and internal P requirement of a crop as well as variety has to be investigated to find a reasonable amount of fertilizer to be applied to a particular soil with respect to crops and sites.

Abbreviations

Q_{max}: adsorption maximum; Q: amount P adsorbed; CEC: cation exchange capacity; DPS: degree of phosphorous saturation; DCB: dithionite-citrate-bicarbonate; Ce: equilibrium P concentrations; EDTA: ethylene diaminetetraacetic acid; FAO: Food and Agriculture organization; K_f: Freundlich adsorption capacity; GPS: geographic position system; K_L: Langmuir energy of adsorption; MBC: maximum buffering capacity; Al_{ox}: oxalte extractable aluminium; Fe_{ox}: oxalte extractable iron; PBC: phosphorous buffering capacity; PSC: phosphorous sorption capacity; RMSE: root mean square error; SNNP: South Nation Nationalities and people; SPR: standard phosphorous requirement; SAS: statistical analysis system.

Authors' contributions

BA: collected, analyzed, interpreted the data and made the final write up which was part of his Doctoral thesis in Soil Science at Haramaya University, Ethiopia. AMT, KK and AM, as co-authors edited the final manuscript. All authors read and approved the final manuscript.

Author details

[1] School of Natural Sciences, Department of Chemistry, Madda Walabu University, Bale-Robe, Ethiopia. [2] Department of Chemistry, College of Natural and Computational Sciences, Haramaya University, Dire Dawa, Ethiopia. [3] School of Natural Resources Management and Environmental Sciences, Haramaya University, Dire Dawa, Ethiopia. [4] Department of Plant Science, College of Agriculture and Natural Resource Sciences, Debre Berhan University, Debre Berhan, Ethiopia.

Acknowledgements

The project was funded by SIDA (Swedish International Development Cooperation Agency) and Madawalabu University. Thus, we would like to thank these institutions. We would also like to thank the anonymous reviewers who contributed significantly to the improvement of the article.

Competing interests

The authors declare that they have no competing interests.

Funding

Swedish International Cooperation Development Agency (SIDA) and Ministry of Education provided fund only for sample collection and laboratory analysis.

References

Abekoe MK, Sahrawat KL (2001) Phosphate retention and extractability in soils of the humid zone in West Africa. Geoderma 102:175–187

Afif E, Matar A, Torrent J (1993) Availability of phosphate applied to calcareous soils of West Asia and North Africa. Soil Sci Soc Am J 57:756–760

Agbenin JO (2003) Extractable iron and aluminum effects on phosphate sorption in a savanna alfisol. Soil Sci Soc Am J 67:589–595

Ahmed MF, Kennedy IR, Choudhury ATMA, Kecskes ML, Deaker R (2008) Phosphorous adsorption in some Australine soils and influence of bacteria on the desorption of phosphorous. Commun soil sci Plant Anal 39:1269–1294

Al Baquy MA, Li JY, Chen YX, Khalid M, Xu RK (2017) Determination of critical pH and Al concentration of acidic Ultisols for wheat and canola crops. Solid Earth 8: 149–159. www.solid-earth.net/8/149/2017/. https://doi.org/10.5194/se-8-149-2017

Amrani M, Westfall DG, Moughli L (1999) Phosphate sorption in calcareous Moroccan soils as affected by soil properties. Commun Soil Sci Plant Anal 30(9–10):1299–1314

Asmare (2014) Phosphorus adsorption characteristics and kinetics of acid soils of Farta District, Northwestern highlands of Ethiopia. A PhD dissertation submitted to the School of Natural Resources Management and Environmental Sciences, School of Graduate Studies, Haramaya University

Asmare M, Heluf G, Markku YH, Birru Y (2015) Phosphorus status, inorganic phosphorus forms, and other physicochemical properties of acid soils of Farta district, Northwestern highlands of Ethiopia. Appl Environ Soil Sci. https://doi.org/10.1155/2015/748390 (Article ID 748390)

Ayaz M, Saleem A, Memon M (2010) Phosphorus adsorption parameters in relation to soil characteristics. J Chem Soc Pak 32:129–139

Bahia FAFC, Braga JM, Resende M, Ribeiro AC (1983) Relationship between phosphorus adsorption and mineralogical components of latosols clay fraction of the central plateau. Braz J Soil Sci 7:221–226

Bai JH, Wang JJ, Yan DH, Gao HF, Xiao R, Shao HB, Ding QY (2012) Spatial and temporal distributions of soil organic carbon and total nitrogen in two marsh wetlands with different flooding frequencies of the Yellow River Delta, China. Clean Soil Air Water 40(10):1137–1144

Barrow NJ (2015) A mechanistic model for describing the sorption and desorption of phosphate by soil. Eur J Soil Sci 66:9–18. https://doi.org/10.1111/ejss.12198-2

Bedin I, Furtini NAE, Resende AV, Faquin V, Tokura AM, Santos JZL (2003) Phosphate fertilizers and soybean production in soils with different phosphate buffer capacities. Braz J Soil Sci 27:639–646

Bereket (2017) Chemical forms of phosphorous and physicochemical properties of acidic soils of Cheha and Dinsho districts, southern highlands of Ethiopia. A PhD dissertation submitted to the School of Natural Resources Management and Environmental Sciences, School of Graduate Studies, Haramaya University

Bertrand I, Holloway RE, Armstrong RD, McLaughlin MJ (2003) Chemical characteristics of phosphorus in alkaline soils from southern Australia. Aust J Soil Res 41:61–76

Birru Y, Heluf G, Gupta VP (2003) Sorption characteristics of soils of the northwestern highlands of Ethiopia. Ethiop J Nat Resour 5:1–16

Börling K, Otabbong E, Barberis E (2001) Phosphorus sorption in relation to soil properties in some cultivated Swedish soils. Nutr Cycl Agroecosyst 59:39–46

Burt R, Mays MD, Benham EC, Wilson MA (2002) Phosphorus characterization and correlation with properties of selected bench mark soil of the United States. Commun Soil Sci Plant Anal 33:117–142

Campos MD, Antonangelo JA, Alleoni LRF (2016) Phosphorus sorption index in humid tropical soils. Soil Tillage Res 156:110–118

Chen GC, He ZL, Huang CY (2000) Microbial biomass phosphorus and its significance in predicting phosphorus availability in red soils. Commun Soil Sci Plant Anal 31:655–667

Chunye L, Zhigang W, Mengchang H, Yanxia L, Ruimin L, Zhifeng Y (2009) Phosphorus sorption and fraction characteristics in the upper, middle and low reach sediments of the Daliao river systems, China. J Hazard Mater 170:278–285

Daly K, Jeffrey D, Tunney H (2001) The effect of soil type on phosphorus sorption capacity and desorption dynamics in Irish grassland soils. Soil Use Manage 17:12–20

Day PR (1965) "Hydrometer method of particle size analysis," in methods of soil analysis. In: Black CA (ed) Agronomy Part I No. 9. American Society of Agronomy, Madison, pp 562–563

Desta HA (2015) Reclamation of phosphorus fixation by organic matter in acidic soils. Glob J Agric Sci 3:271–278

Dodor DE, Oya K (2000) Phosphate sorption characteristics of major soils in Okinawa, Japan. Commun Soil Sci Plant Anal 31:277–288

Duffera M, Robarge WP (1999) Soil characteristics and management effects on phosphorus sorption by highland plateau soils of Ethiopia. Soil Sci Soc Am J 63:1455–1462

Ehlert P, Morel C, Fotyma M, Destain JP (2003) Potential role of phosphate buffering capacity of soils in fertilizer management strategies fitted to environmental goals. J Plant Nutr Soil Sci 166(4):409–415. https://doi.org/10.1002/jpln.200321182

EIAR (Ethiopian Institute of Agricultural Research) (2011) Coordination of national agricultural research system, Ethiopia, English and Amharic Version. EIAR, Addis Ababa

EthioSIS (Ethiopian Soil Information System) (2014) Soil fertility status and fertility recommendation atlas for Tigray regional state, Ethiopia, July 2014. Addis Ababa, Ethiopia

FAO (Food and Agriculture Organization) (1984) Provisional soil map of Ethiopia. Land Use Planning Project. FAO, Addis Ababa

Fox RL (1981) External P requirements of crops. In: Stelly M (ed) Chemistry of the soil environment. American Society of Agronomy, Madison (Special Publication No. 40)

Fox RL, Kamprath EJ (1970) Phosphate sorption isotherms for evaluating the phosphate requirements of soils. Soil Sci Soc Am Proc 34:902–907

Hansen JC, Strawn DG (2003) Kinetics of phosphorus release from manure-amended alkaline soil. Soil Sci 168(12):869–879

Hazelton P, Murphy B (2007) Interpreting soil test results: what do all the numbers mean, 2nd edn. CSIRO Publishing, Clayton

Hedley MJ, Stewart JWB, Chauhan BS (1982) Changes in inorganic and organic soil phosphorus fractions induced by cultivation practices and by laboratory incubations. Soil Sci Soc Am J 46:970–976

Henry PC, Smith MF (2003) A single point sorption test for the routine determination of the phosphorus requirement of low to moderate P-fixing soils. S Afr J Plant Soil 20(3):132–140. https://doi.org/10.1080/02571862.2003.10634922

Ige DV, Akinremi OO, Flaten DN (2005) Environmental index for estimating the risk of phosphorus loss in calcareous soils of Manitoba. J Envt Qual 34:1944–1951

Jackman JM, Jones RC, Yost RS, Babock CJ (1997) Rietveld estimates of mineral percentages to predict phosphorous sorption by selected Hawaiian soils. Soil Sci Soc Am J 61:618–629

Jalali M (2007) Phosphorus status and sorption characteristics of some calcareous soils of Hamadan, western Iran. Environ Geol 53:365–374. https://doi.org/10.1007/s00254-007-0652-7

Jamison VC, Weaver HH, Reed IF (1950) A hammer-driven soil core sampler. Soil Sci 69:487–496

Janardhanan L, Daroup SH (2010) Phosphorous sorption in organic acids in south florida. Soil Sci Soc Am J 74:1597–1606

Jensen A, Jakobsen I (1970) The occurrence of vesicular-arbuscular mycorrhiza in barley and wheat grown in some Danish soils with different fertilizer treatments. Plant Soil 55:403–414

Jesse R, Fink A, Vasconcellos I, Tales T, Vidal B (2016) Iron oxides and organic matter on soil phosphorous availability. Cienc Agrotecnol 40(4):369–379

Jones JB (2003) Agronomic handbook: management of crops, soils, and their fertility. CRCPress LLC, Boca Raton

Juo ARS, Fox RL (1977) Phosphate sorption characteristics of some benchmark soils of West Africa. J Soil Sci 124:370–376

Kiflu A, Sheleme B, Schoenau J (2017) Fractionation and availability of phosphorus in acid soils of Hagereselam, Southern Ethiopia under different rates of lime. Chem Biol Technol Agric 4:21. https://doi.org/10.1186/s40538-017-0105-9

Lavkulich LM (1981) Methods manual, pedology laboratory. Department of soil science. University of British Columbia, Vancouver, British Columbia

Lin C, Banin A (2005) Effect of long-term effluent recharge on phosphate sorption by soils in a wastewater reclamation plant. Water Air Soil Pollut 164:257–273

Makris KC, Harris WG, O'Connor GA, El-Shall H (2005) Long-term phosphorus effects on evolving physicochemical properties of iron and aluminum hydroxides. J. Colloid Interf Sci 287:552–560

Mamo T, Haque I (1987) Phosphorus status of some ethiopian soils-I. Sorption characteristics. Plant Soil 102(2):261–266

McKeague JA, Day JH (1966) Dithionite and oxalate extractable Fe and Al as acids in differentiating various classes of soils. Can J Soil Sci 46:13–22

Mehlich A (1984) Mehlich-III soil test extractant: a modification of Mehlich 2. Commun Soil Sci Plant Anal 15:1409–1416

Mehra OP, Jackson ML (1960) Iron oxide removal from soils and clays by a dthionite–citrate system buffered with sodium bicarbonate. Clays Clay Miner 7:317–327

Moazed H, Hoseini Y, Naseri AA, Abbasi F (2010) Determining phosphorus adsorption isotherm in soil and its relation to soil characteristics. J Food Agric Environ 8:1153–1157

Moody PW, Bolland MDA (1999) Cadmium in soils and plants. Dev Plant Soil Sci 85:1–9

Moreira FLM, Mota FOB, Clemente CA, Azevedo BM, Bomfim GV (2006) Phosphorus adsorption in Ceará State soils, Brazil. Rev Ciên Agron 37:7–12

Morel C, Tunney H, Plénet D, Pellerin S (2000) Transfer of phosphate ions between soil and solution: perspectives in soil testing. J Environ Qual 29:50–59. https://doi.org/10.2134/jeq2000.00472425002900010007x

Morris M, Kelly V, Kopicki R, Byerlee D (2007) Fertilizer use in African agriculture: lessons learned and good practice guidelines. World Bank, Washington, DC

Mozaffari M, Sims JT (1994) Phosphorus availability and sorption in an Atlantic coastal plain watershed dominated by animal based agriculture. Soil Sci 157:97–107

Muindi EM, Mrema JP, Semu E, Mtakwa PW, Gachene CK, Njogu MK (2015) Phosphorus adsorption and its relation with soil properties in acid soils of Western Kenya. Int J Plant Soil Sci 4(3):203–211 (Article no. IJPSS.2015.021 ISSN: 2320–7035)

Obura PA (2008) Effects of soil properties on bioavailability of aluminium and phosphorus in selected Kenyan and Brazilian soils. Ph.D. thesis, Purdue University, Indiana

Olsen SR, Sommers LE (1982) Phosphorus. In: Page AL et al (eds) Methods of soil analysis. Part 2, 2nd edn. Agron. Monogr. 9. ASA and SSSA, Madison, pp 403–430

Onweremadu EU, Eshett ET, Osuji GE (2007) Temporal variability of selected heavy metals in automobile soil. Int J Environ Sci Technol 4:35–41

Peinemann N, Nilda MA, Pablo Z, Maria BV (2002) Effect of clay minerals and organic matter on the cation exchange capacity of silt fractions. J Plant Nutr Soil Sci 163(1):47–52. doi:https://doi.org/10.1002/(sici)1522-2624(200002)163:1<47:AID-JPLN47>3.0.CO;2-A

Rasmussen PE, Allmaras RR, Rohde RR, Roager NC (1980) Crop residue influences on soil carbon and nitrogen in a wheat-fallow system. Soil Sci Soc Am J 44:596–600

Rayment GE, Lyons DJ (2011) 'Method 4b1 pH of 1:5 soil 0.01 M calcium chloride extract-direct (without stirring during measurement)', 'method13A1 oxalate extractable iron, aluminium and silicon'. Method 9B1 bicarbonate extractable P (Colwell P)–manual colour' "method 9J1 phosphate sorption curve–manual colour' in soil chemical methods-Australasia. CSIRO, Publishing, Australia

Rowell DL (1994) Soil science: method and applications. Addison Wesley Longman, London

Saljnikov KE, Funakawa S, Akhmetov K, Kosaki T (2004) Soil organic matter status of Mollisols soil in North Kazakhstan: effects of summer fallow. Soil Biol Biochem 36:1373–1381

Saljnikov E, Hospodarenko H, Funakawa S, Kosaki T (2005) Effect of fertilization and manure application on nitrogen mineralization potentials in Ukraine. Zemljiste I biljka 54(3):221–230

Saljnikov E, Cakmak D, Kostic L, Maksimovic S (2009) Labile fractions of soil organic carbon in Mollisols from different climatic regions, vol LIII. Agrochimica, Bolzano, p 6

Sánchez-Alcalá I, Campillo MC, Torrent J (2014) Extraction with 0.01 M CaCl2 underestimates the concentration of phosphorus in the soil solution. Soil Use Manag 30(2):297–302. https://doi.org/10.1111/sum.12116

Sanchez PA, Uehara G (1980) Management considerations for acid soils with high phosphorus fixation capacity. In: Khasawneh FE, Sample EC, Kamprath EJ (eds) The role of phosphorus in agriculture. American Society of

Agronomy, Madison, pp 263–310

SAS (Statistical Analysis System) Institute) (2004) SAS/STAT user's guide. Proprietary software version 9.00. SAS Institute, Inc., Cary

Sims JT, Pierzynski GM (2005) Chemistry of phosphorus in soils. In: Tabatabai MA, Sparks DL (eds) Chemical processes in soils, Soil science society of America book series 8. Soil Science Society of America, Madison, pp 151–192

Tan KH (2000) Environmental Soil Science, 2nd edn. Marcel Dekker Inc., New York

Van der Zee SEATM, Nederlof MM, Van Riemsdijk WH, de Haan FAM (1988) Spatial variability of phosphate adsorption parameters. J Environ Qual 17:682–688

Van Reeuwijk LP (1992) Procedures for soil analysis, 3rd edn. International Soil Reference and Information Center (ISRIC), Wageningen. p 34

Walkley A, Black IA (1934) An examination of the Degtjareff method for determining soil organic matter and a proposed modification of the chromic acid titration method. Soil Sci 37:29–38

Wang XJM, Jackman Yost RS, Linquist BA (2000) Predicting soil phosphorous buffer coefficients using potential sorption site density and soil aggregation. Soil Sci Soc Am J 64:240–246

Wang Y, Shen ZY, Niu JF, Liu RM (2009) Adsorption of phosphorus on sediments from the Three-Gorges Reservoir (China) and the relation with sediment compositions. J Hazard Mater 162:92–98

Watanabe and Olsen (1965) Test of an ascorbic acid method for determining phosphorus in water and $NaHCO_3$ extracts from soil. Soil Sci Soc Am Proc 29:677–678

Wild A (1950) The retention of phosphate by soil. A review. J Soil Sci 1:221–238

Yan X, Wang D, Zhang H, Zhang G, Wei Z (2013) Organic amendments affect phosphorus sorption characteristics in a paddy soil. Agric Ecosyst Environ 175:47–53

Yan JL, Jiang T, Yao Y, Lu S, Wang QL, Wei SQ (2016) Preliminary investigation of phosphorus adsorption onto two types of iron oxide-organic matter complexes. J Environ Sci-China 42:152–162

Yan X, Zongqiang W, Qianqian H, Zhihong L, Jianfu W (2017) Phosphorus fractions and sorption characteristics in a subtropical paddy soil as influenced by fertilizer sources. Geoderma 295:80–85

Yli-Halla M, Hartikainen H, Vaatainen P (2002) Depletion of soil phosphorus as assessed by several indices of phosphorus supplying power. Eur J Soil Sci 53:431–438

Yuan TL, Lucas DE (1982) Retention of phosphorus by sandy soils as evaluated by adsorption isotherms. Soil Crop Sci. Soc. Fla Proc 11:197–201

Zhang H, Schroder JL, Fuhrman JK, Basta NT, Storm DE, Payton ME (2005) Path and multiple regression analyses of phosphorus sorption capacity. Soil Sci Soc Am J 69:96–106

Zinabu W (2015) Phosphorus sorption characteristics and external phosphorus requirement of Bulle and Wonago Woreda, Southern Ethiopia. J Biol Agric Healthc. www.iiste.org ISSN 2224–3208 (Paper) ISSN 2225-093X (Online). 5(5)

Zinabu W, Wassie H, Dhyna S (2015) Phosphorus sorption characteristics and external phosphorus requirement of Bulle and Wonago Woreda. Southern Ethiopia. Adv Crop Sci Tech. 3:2

3

Spatiotemporal trends of urban land use/land cover and green infrastructure change in two Ethiopian cities: Bahir Dar and Hawassa

Kassahun Gashu[1*] and Tegegne Gebre-Egziabher[2]

Abstract

Background: The spatiotemporal analysis of urban land use/land cover change (LULCC) helps to understand the dynamics of the changing environment of green infrastructure (GI) on the basis of sustainable city development. There are important links between spatiotemporal land use/land cover and GI change in urban areas. Therefore, the main objective of this study was to examine the spatiotemporal trends of urban land use/land cover and GI changes in Bahir Dar and Hawassa cities for the last four decades (1973–2015). Three different sets of Landsat satellite data were procured from EMA for Bahir Dar and Hawassa from 1973, 2000 and 2015 using Landsat 4 MSS, 7 TM and 8 OLI respectively. Based on this, using ERDAS Imagine (ver. 9.2) and Arc GIS (Ver.10.3) five LULCC classes were identified for analysis purpose.

Result: The results show that vegetation decreased by 30 and 14% in Bahir Dar and Hawassa respectively for the period 1973–2015, while built-up areas expanded by 10 and 24% respectively in the two cities. These land use changes have significant impacts on spatiotemporal trends of GI in urban areas. GI has increased in Bahir Dar and Hawassa in association with built-up area expansion and deliberate activity of city administrations with effective implementation of spatial plans of corresponding cities.

Conclusions: There is a growing concern about GI in cities. Policy makers and stakeholders should also decide on how to use the land at present and in the future. LULCC policymaking processes should aim to balance GI and other types of land use/land cover for sustainable urban development. Urban LULCC has important effects on the urban GI system.

Keywords: City planning, Green infrastructure, GIS, Landsat image, Land use land cover change, Remote sensing, Spatiotemporal

Background

With a little more than 50% of the human population living in urban areas, urbanization is now recognized as a major phenomenon (UN 2014; Zhang et al. 2013). Social scientists, urban planners, and geographers have investigated the unprecedented urban concentration from many perspectives, including the geography, demographics, economies and spatial evolution of cities (McIntyre et al. 2000) as well as urban green infrastructure (Mell

2014). The process of urbanization involves the growth of urban population and built-up areas. The share of world urban population is expected to increase to 66% by 2050, and of this about 90% will be concentrated in Africa and Asia (UN 2014). This population increase will lead to fast growth of built-up areas that consumes the surrounding productive land and encroaches on the necessary ecosystems. At the same time, the horizontal rapid expansion of built-up areas will lead to discontinuous suburbs with low density and uneven pattern (Tewolde and Cabral 2011; Varshney 2013).

As urbanization increases and urban areas continue to grow fast, there is a concern on urban environment

---
*Correspondence: Kg19me@gmail.com
[1] Department of Geography and Environmental Studies, University of Gondar, P.O. Box 196, Gondar, Ethiopia

and its quality because the quality of urban environment directly influences the social and economic development of the city (Masser 2007; He et al. 2010). Many scholars hold the view that the urban environment represents a highly complex area depicting a continuum of different spatial, temporal and spectral variability in land use and land cover (Haregeweyn et al. 2012). Spatial variability arises due to the varied landscape: temporal variations are attributed to periodic seasonal changes over the year while spectral variability is due to the great variety of materials and structures associated with the urban area (Zoran 2007). It is thus necessary to analyze the spatiotemporal patterns of LULCC in order to understand the urban ecology (McIntyre et al. 2000; Abebe and Megento 2016).

Land use/land cover change (LULCC) refers to the earth's territorial surface modification by human activities (Anderson et al. 1976; Meyer and Turner 1992; Lu et al. 2004; Muriuki et al. 2011; Ayalew et al. 2012). The process of LULCC affects biodiversity, climate, soil, and air in particular, and the ecosystem, in general, and it has become the greatest environmental concern for human beings to date (Long et al. 2007; Tsegaye et al. 2010; Hailemariam et al. 2016). LULCC is useful to understand environmental changes because it can provide a tool to assess ecosystem changes and their environmental implication at various temporal and spatial scales (Anderson et al. 1976; Haregeweyn et al. 2012).

Urban space consists of built-up areas that include variety of land uses in commercial, institutional and residential areas. It also consists of non-built area that is mostly dominated by greenery and open spaces (Moroney and Jones, 2006; Tzoulas et al. 2007; Mansor et al. 2010). Previous researches (Kong and Nakagoshi 2006; Phan and Nakagoshi 2007; Byomkesh et al. 2012) indicated that urban green spaces are those lands that are covered with natural or man-made vegetation but are present in built-up areas. However, the universally agreed definition is still arguable. Most developed countries have their own definition (Byomkesh et al. 2012). Therefore this research used as its working definition stated by the Ethiopian Ministry of Urban Development and Housing (MoUDH): green infrastructure typologies to include parks, sports fields, roadside and squares, plazas and festive areas, river and riverside areas, lakes and lakeside areas, watershed areas, urban agriculture development, woodlots and green belts (inside and surrounding forests), private compounds and surroundings, institutional compounds and surroundings (both governmental and non-governmental), communal housing compounds and surroundings (condominiums, real estate, etc.), religious institutions compounds and surroundings, neighborhood

open spaces, cemeteries, nursery sites, and green roofs and walls (MoUDH 2015).

Green space is sometime synonymous with green infrastructure, though the latter is more inclusive than the former. Green space helps reduce heat effects of buildings, provides shadow to pedestrians and ground and has the ability to improve air quality and the environment (Noor et al. 2013). The development of urban green infrastructure planning and management practices requires important information from LULCC studies (Yang et al. 2014). Previous studies implied that traditional investigation of urban environment was not considered GI (Miller and Hobbs 2002). The urban green infrastructure, however, enables urban residents to experience outdoors visually and kinetically. Green infrastructure network in any urban area is significant because it attempts to provide optimal experiential qualities to urban residents and to overcome the negative effects of living in the urban built environment (Mansor et al. 2010). Moreover, based on a deeper understanding of the relationships between the LULCC and GI change require that the underlying mechanisms, patterns, and processes of land conversion as well as the response of urban environment should be addressed throughout official decision-making processes (Zhang et al. 2013). The planners and decision-makers could fully evaluate the consequences of different land development scenarios and therefore have scientific basis with which to improve future planning and regulations of GI. In terms of GI, the spatiotemporal analysis of LULCC can help to understand the dynamics of the changing environment of GI and form the basis for sustainable development and provide a fundamental piece of information required for policy making and planning (Hu et al. 2008; Teferi et al. 2016).

Though LULCC is not a recent phenomenon in Ethiopia (Hailemariam et al. 2016), it is exacerbated by the scale, speed and long-term nature of urbanization and modernization (Msoffe et al. 2011). Existing studies on LULCC in Ethiopia have focused on land degradation and associated consequences due to expansion of cultivation and deforestation (Taddese 2001; Feoli and Vuerich 2002; Amsalu et al. 2007; Meshesha et al. 2010; Tsegaye et al. 2010). There is little effort to understand LULCC in relation to green infrastructure changes in urban areas.

This study highlights the important links between spatiotemporal land use/land cover and green infrastructure change in urban areas. In this research, green infrastructure is taken as one category of land use/land cover that is an interconnected network of multifunctional, predominantly un-built, spaces that support ecological and social activities (Kambites and Owen 2006; Tzoulas et al. 2007; MoUDH 2015). The transformation of land use/land cover types leads to a change in the structure and

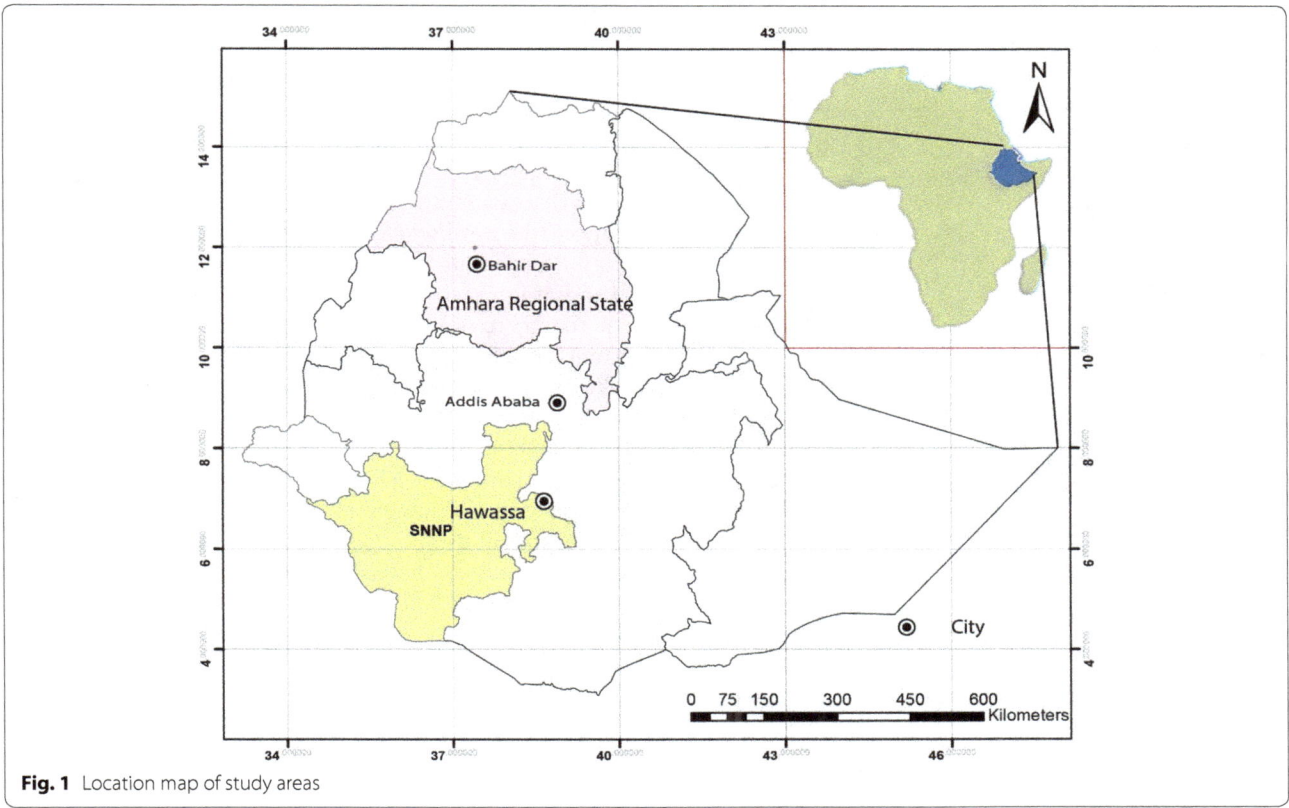

Fig. 1 Location map of study areas

function of green infrastructure services (Lei and Zhu 2017). The need to balance economic, social and ecological ecosystems is becoming increasingly urgent because LULCC is in the direction of rapid urbanization (Song et al. 2016). This study aims to investigate the rapid urban expansion on LULCC and GI, and its development and planning. Our research focused on (1) The rates of LULCC in Bahir Dar and Hawasa between 1973 and 2015, (2) LULCC trends during the 1973–2000 and 2000–2015 periods for both Bahir Dar and Hawassa, (3) Which land-cover types were most affected by the change process, and (4) The rates of changes and conversion from other land cover types to green spaces and urban areas over the period 1973–2015.

Methods

Study area

Bahir Dar and Hawassa are the capital cities of Amhara National Regional State and Southern Nations, Nationalities and Peoples' Regional State respectively. The former is located at 565 km to Northwest of Addis Ababa, on the southern shore of Lake Tana, the source of the Blue Nile (Abay) river, while the latter is located at 275 km to the south of Addis Ababa near the eastern shore of Lake Hawassa, one of the rift valley lakes in Ethiopia. Bahir

Dar is located at 11° 36′ North and 37° 23′ East and has an average elevation of 1801 m above sea level. It has a city administration with special zone status and nine subcities which have district status (Fig. 1). It is also the seat for the Bahir Dar Zuria district. Hawassa is located at 07° 03′ North and 30° 28′ East with an average elevation of 1708 m above sea level. Similar to Bahir Dar, Hawassa has a city administration status and has eight sub-cities with district status (Fig. 1). Hawassa is also a seat for Sidama administrative zone.

According to the National Meteorological Agency (NMA), Bahir Dar has an average annual temperature and precipitation of 19.6 °C and 1419 mm respectively (NMA 2013). It is situated in the *woina-dega*[1] agro-ecological zone and experiences uni-modal rainfall over a 3-month period from mid-June to mid-September. Hawassa has an average annual temperature and precipitation of 20.8 °C and 993.4 mm respectively (NMA 2013). It is one of the major urban areas of Ethiopia located inside the greater Ethiopian rift valley. It experiences uni-modal rainfall over a 3-month period from mid-June to mid-September and has *woina-dega* agro ecological zone.

---

[1] *Woina-dega* is a local term that defines mid altitude climate.

Table 1 Land use land cover change (LULCC) classification schemes used in this study

LULC class	Description
Urban built-up area	Includes areas with all types of artificial surfaces, including residential, commercial, and industrial land uses as well as transportation infrastructure
Vegetation	Includes areas with dense vegetation cover, such as those covered with shrubs forming closed canopies, trees and other vegetation that is relatively tall and dense, as well as areas covered with both indigenous and exotic trees
Water body	Includes lakes, rivers, ponds
Green spaces in built-up areas	An area of grass, trees, or other vegetation set apart for recreational or aesthetic purposes inside urban built environment. It includes urban parks, greenery, roundabouts, public squares and plaza, open Spaces, medians and sport fields
Crop land	Includes grazing areas, cultivated lands, community open lands and areas along the lake shore that are used for agricultural purposes when the lake level retreats following the long dry-season. Information obtained from local residents indicates that the units categorized in this category can generally be used in one way or another for agricultural purposes

These two cities are among the largest and the fastest growing urban centres in Ethiopia. The population of Bahir Dar city grew from 96,140 in 1994, the second census period, to 155,428 in 2007, the third census period (CSA 2007). The rate of growth between the two censuses periods was 3.7%. According to the CSA (2017), the population of Bahir Dar is estimated to be 348,429. The population of Hawassa was 69,169 in 1994 and it grew to 157,139 in 2007 showing a growth rate of 6% (CSA 1994, 2007). The CSA estimated the population of Hawassa in 2017 to be 315,267 (CSA 2017).

Bahir Dar and Hawassa cities were selected as research sites for this study in addition to rapid population increase is that both are lakeside cities, regional capitals, fast growing cities and have relatively better availability of green infrastructure as compared to other cities and towns in the country. According to Municipality of Hawassa (2015) and Municipality of Bahir Dar (2015) Hawassa and Bahir Dar has 21.96 and 17.44% GI coverage respectively.

Data
This study uses three different sets of Landsat satellite data for Bahir Dar and Hawassa over four decades (1973–2015). These satellite images were procured from the

Ethiopian mapping agency (EMA) in GeoTIFF file format projected in UTM projection and WGS 84 datum universal transverse Mercator (UTM), Zone 37° North coordinate system. The three Landsat satellite images with 30 m resolution were acquired for January 1973, January 2000 and January 2015. A study by Sadidy et al. (2009) pointed out that Landsat images with varying resolution are among the most widely used data sources in order to gain important input for mapping and planning projects. The Landsat images were geo-referenced to the digitized map of the corresponding area using first-order polynomial transformation and nearest neighborhood resembling (Yuan et al. 2005; Murat et al. 2006).

Data analysis
There are many change detection approaches for remotely sensed images (Yuan et al. 2005). Among these, the post-classification comparison method is particularly attractive due to its nature of clearly identified change (Hung and Wu 2005; Muttitanon and Tripathi 2005; Yuan et al. 2005). This study employs the post-classification method to detect changes.

LULC maps for both Bahir Dar and Hawassa for 1973, 2000 and 2015 were prepared for the study areas by independently supervised classifications using a maximum likelihood algorithm classifier. Hence, the five land-cover classes are as follows: urban built-up, vegetation, water body, green spaces, and crop land were mapped.

These five land use/land cover classification (Table 1) schemes were chosen considering the standard classes defined by the National Aeronautics and Space Administration (NASA) and the US Geological Survey (USGS) as well as the study detail and objectives (Mohan et al. 2011). Some studies (Thompson 1996) also outlined the need to contextualize LULCC classification systems for the local situation. The reason is that no universally accepted classification system exists as it is influenced by specific users' objectives and also often by geographic location. ERDAS Imagine (ver. 9.2) and Arc GIS (Ver.10.3) were used to perform LULCC classification in a multi-spectral approach. Satellite images with remote sensing techniques are used to show spatiotemporal trends of LULCC of the study areas. In order to determine the quality of the information derived from the data, accuracy assessment of classification was made for 1973, 2000 and 2015 images. We used the most obvious method of change detection (Lu et al. 2004; Lu and Weng 2007; Butt et al. 2015) which involves a comparative analysis of spectral classifications for times t_1 and t_2 produced independently (Mas 1999). The percentages of change detection of LULCC were calculated using the following equation:

$$\Delta = \frac{A_2 - A_1}{A_1} * 100 \qquad (1)$$

where, Δ is the land use/land cover change percentage, and A_1 and A_2 are initial and final.

LULCC areas respectively. In this equation, the positive values suggest a gain, whereas negative values imply a loss in extent. The LULCC rate was also computed using the formula suggested by Puyravaud (2003):

$$r = \frac{1}{\Delta t \ln(A_2 - A_1)} * 100 \qquad (2)$$

where, r is the annual rate of change in %, Δt is the time interval in years during the LULCC being assessed; ln is the base of the natural logarithm function; and A_1 and A_2 are initial and final LULCC areas respectively.

In the present study, each image of Landsat 4 MSS, Landsat 7 TM and Landsat 8 OLI for both cities were independently classified for the three-time periods (1973, 2000 and 2015). The ground referenced data were gathered by combining Google Earth data and GPS points during the field survey and the resulting samples were imported to the ERDAS Imagine software and the intersection files were generated.

LULCC can be summarized in a unique change statistic that quantifies the proportion of pixels that have changed in the overall study area independent of their classes. Field-based information supports the interpretation of the process of LULCC (Fig. 2a, b). In this study, supervised classification was carried out using the multi-date images to classify the images into clusters and to identify the type of potential changes. Post-classification comparison is used to detect LULCC among three images in Bahir Dar (Fig. 2a) and Hawassa (Fig. 2b). Object based supervised classification (Zhou and Troy 2008; Radoux et al. 2011; Robertson and King 2011; Chen et al. 2012; Hussain et al. 2013) was carried out using maximum likelihood algorithm method (MLC) for each image separately to test the accuracy assessment of the classification. Stratified random method is used for land use/land cover representation extracted from satellite images over classes of the area. Past and recent studies have identified image differences as being the most accurate change detection technique.

The accuracy assessment was done based on ground truth data and visual interpretation using 100 points (Butt et al. 2015). Statistical error matrices together with nonparametric Kappa index were used for comparison of reference data and classification result (Robertson and King 2011; Dabboor et al. 2014). The producer accuracy, user accuracy, overall accuracy, and Kappa coefficient were calculated for the classified map of 1973, 2000 and 2015 based on the formula given by Congalton and

Green (2009). The formula for computing producer accuracy, user accuracy, overall accuracy, and Kappa index coefficient is given as follows:

$$\text{producer's accuracy i} = \frac{nii}{Gii} \qquad (3)$$

$$\text{User's accuracy i} = \frac{nii}{Cii} \qquad (4)$$

$$\text{Over all accuracy} = \frac{\sum_{i=1}^{k} nii}{n} \qquad (5)$$

$$k = \frac{\sum_{i=1}^{k} nii - \sum_{i=1}^{k}(GiCi)}{n2 - \sum_{i=1}^{k}(GiCi)} \qquad (6)$$

where, i is the class number, n is the total number of classified pixels that are being compared to ground truth, nii is the number of pixels belonging to the ground truth class i, that have also been classified with a class i, Ci is the total number of classified pixels belonging to class i and Gi is the total number of ground truth pixels belonging to class i.

A nonparametric Kappa index is a measure of agreement between predefined producer ratings and user assigned ratings (Foody 2002). Using formulas, 3, 4, 5 and 6, the kappa index results indicated that all of the images met the minimum of 85% accuracy in LULCC analysis to each classified object that matches (intersects) a given reference object (Table 2).

Urban expansion analysis

The extent and direction of the cities' expansion for the years 1973, 2000, and 2015 were analyzed by superimposing the different time-series images and by calculating the corresponding areas in GIS software. The annual rates of urban area expansion (UAE) for the periods: 1973–2000, 2000–2015, and 1973–2015 are calculated for Bahir Dar and Hawassa using the following relationship (Valdkamp et al. 1992; Mohan et al. 2011) in a modified form:

$$\frac{UAx + n - UAx}{n * UAx} * 100 \qquad (7)$$

where: $UAx + n$ and UAx are the urban area in Ha at time $x + n$ and x, respectively, and n is the interval of the calculating period (in years).

In this study we also used, land consumption rate (LCR) as an index to evaluate the progressive spatial expansion of urban areas. The land consumption rate (LCR) is computed using the following formula (Fanan et al. 2011; Sharma et al. 2012):

$$\frac{UA}{P} \qquad (8)$$

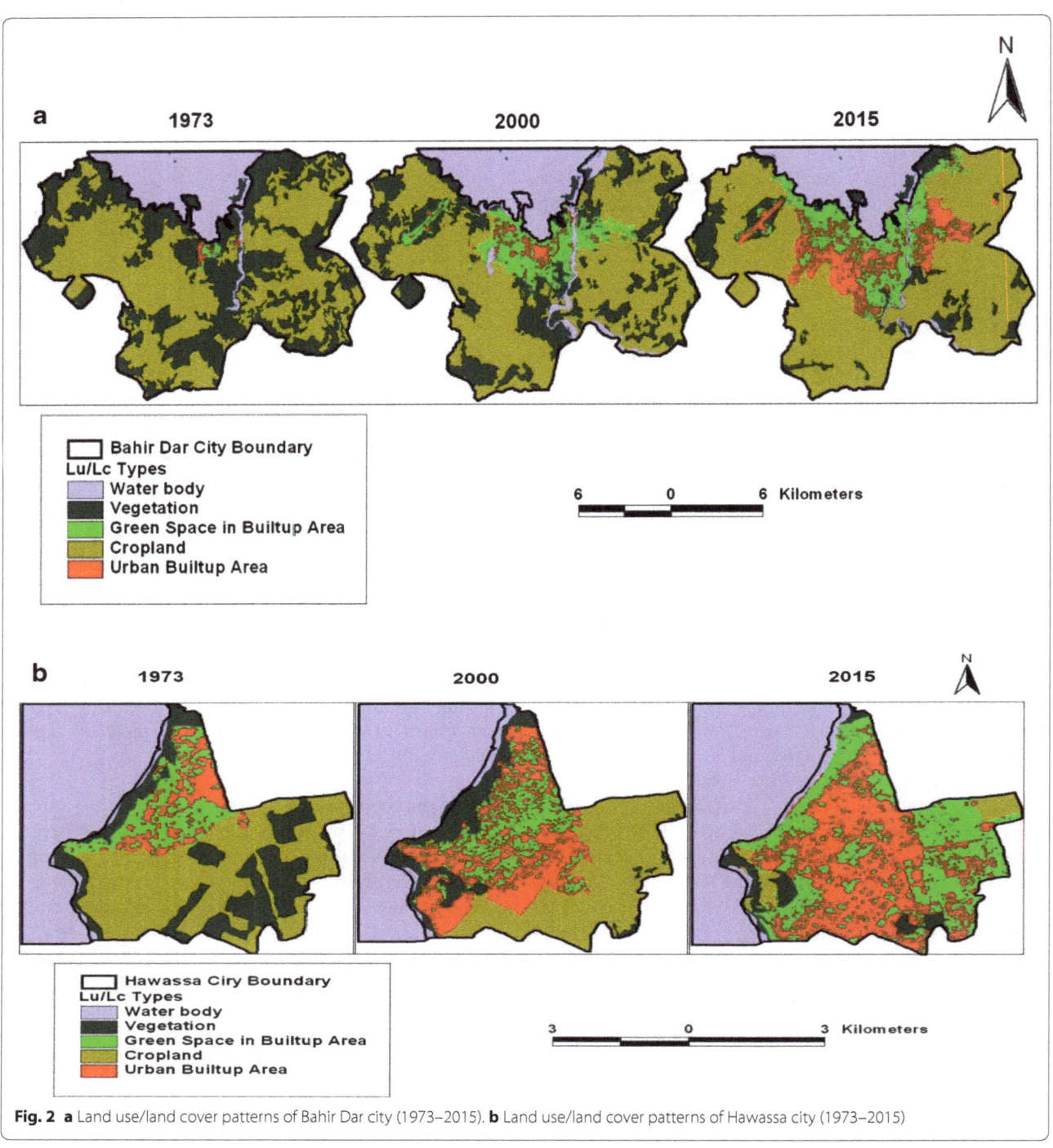

Fig. 2 a Land use/land cover patterns of Bahir Dar city (1973–2015). **b** Land use/land cover patterns of Hawassa city (1973–2015)

where: UA is area of the city (ha) and P is the population.

Results

Land use/land cover change and urban expansion

The major land cover areas presented in the images for Bahir Dar and Hawassa cities are given in Table 3a, b.

The classification scheme was created on the basis of the cover types in the study areas that were present in large quantities. These were the classes that were extracted as thematic classes from the images and for which area statistics were generated at local situation. The matrix indicates the amount of land in hectares and percentage of

Table 2 Accuracy assessment for classified images of Bahir Dar and Hawassa

Bahir Dar				Hawassa			
Reference year	Classified image	Overall classification accuracy (%)	Overall kappa coefficient	Reference year	Classified image	Overall classification accuracy (%)	Overall kappa coefficient
1973	Landsat 4 MSS[a]	86.75	0.75	1973	Landsat 4 MSS	87.37	0.85
2000	Landsat 7 TM[b]	95.5	0.93	2000	Landsat 7 TM	96.0	0.91
2015	Landsat 8 OLI[c]	98.0	0.96	2015	Landsat 8 OLI	95.45	0.92

[a] Multi-spectral Scanner

[b] Thematic Mapper

[c] Operational Land Imager

Table 3 LULCC pattern and change in Bahir Dar (1973–2015) and Hawassa (1973–2015)

LULC type	1973		2000		2015		Change					
							1973–2000		2000–2015		1973–2015	
	Ha	%	Ha	%	Ha	%	Ha	%	Ha	%	Ha	%
Bahir Dar												
Water body	3742.3	14.73	3710.0	14.60	3700.2	14.56	−31.7	−0.12	−10.2	−0.04	−42.1	−0.16
Vegetation	10,111.8	39.79	6447.1	25.37	2583.2	10.166	−3664.7	−14.42	−3863.9	−15.21	−7528.6	−29.63
Green space in built-up area	98.7	0.4	1762.9	6.9	2841.5	11.2	1664.2	6.55	1078.6	4.24	2742.8	10.79
Cropland	11,415.3	44.9	12,096.9	47.6	13,504.6	53.1	681.6	2.68	1407.7	5.54	2089.3	8.22
Urban built-up area	85.3	0.3	634.6	2.5	2739.6	10.8	549.4	2.16	2105.0	8.28	2654.3	10.44
Total	25,411.4	100.0	25,411.4	100.0	25,411.4	100.0	0.0	0.0	0.0	0.0	0.0	100.0
Hawassa												
Water body	1512.6	36.37	1514.9	36.43	1561.4	37.5	2.3	0.06	46.5	1.16	48.8	1.17
Vegetation	744.6	17.91	389.9	9.38	169.5	4.1	−354.7	−8.53	−220.5	−5.30	−575.1	−13.83
Green space in urban area	343.0	8.2	457.5	11.0	1049.6	25.2	114.5	2.75	592.2	14.24	706.6	16.99
Cropland	1281.9	30.8	1020.8	24.5	117.1	2.8	−261.1	−6.28	−903.7	−21.73	−1164.8	−28.01
Urban built-up area	276.5	6.6	775.5	18.6	1261.0	30.3	499.0	11.99	485.5	11.67	985.0	23.67
Total	4158.6	100.0	4158.6	100.0	4158.6	100.0	0.0	0.0	0.0	0.0	0	0.0

land use/land cover class changed to other type. The values were presented in terms of hectares and percentages as stated in formula 1 and 2.

The data presented in Table 3a shows that in 1973, the vegetation cover in Bahir Dar was 40% and this was reduced to 25% in 2000 and 10% in 2015. On the other hand, crop land increased from 45% in 1973 to 48% in 2000 and further to 53% in 2015. This change in land cover could indicate a shift from vegetation to cropland use. Table 3b depicts that in Hawassa both vegetation and crop land showed a decline. The vegetation cover declined from 18% in 1973 to 9% in 2000 and 4% in 2015. Similarly crop land declined from 31% in 1973 to 25% in 2000 and 3% in 2015. It should be noted that the water body located in the study areas namely Lake Tana in Bahir Dar and Lake Hawassa in Hawassa showed

no significant change (Table 3a, b). The changes in green spaces and urban area expansion are presented in detail.

The persistence values are the values which mean unchanged amount. The gain values computed by subtracting the persistence value from the total area of final year and the loss value also computed by negative of subtracting the persistence value from the total area of the initial year. Table 4 presents the persistence, gains, losses and net changes of different land use and land cover. Accordingly, in Hawassa, water body has shown a higher persistence accounting for 75.7% while cropland has shown a higher loss (55%). In addition, the land cover type which persisted least is vegetation (4.8%) and the land cover with least loss is water body (0.1%). In Bahir Dar, the land cover type with the highest persistence is cropland (60%) and that with the highest loss is vegetation (70%). Built up area has shown low persistence and

Table 4 Percentage of LULCC in (ha) in Hawassa and Bahir Dar during (1973–2015)

LULC types	Hawassa								Bahir Dar							
	Persistence		Gains		Losses		Net change		Persistence		Gains		Losses		Net change	
	Ha	%	Ha	%	Ha	%	Ha	%	Ha	%	Ha	%	Ha	%	Ha	%
Urban built up area	175.5	8.8	1085.6	50.16	−101.0	−4.7	985.5	45.5	42.5	0.3	2697.2	22.51	−42.9	−0.4	2654.3	22.2
Vegetation	96.6	4.8	72.9	3.368	−648.0	−29.9	−575.1	−26.6	1730.1	12.9	853.1	7.119	−8381.7	−69.9	−7528.6	−62.8
Water body	1510.2	75.7	51.2	2.364	−2.4	−0.1	48.7	2.3	3512.8	26.2	229.3	1.914	−187.5	−1.6	41.8	0.3
Green space in urban area	116.5	5.8	933.1	43.12	−226.5	−10.5	706.6	32.7	55.6	0.4	2785.9	23.25	−43.1	−0.4	2742.9	22.9
Cropland	95.6	4.8	21.4	0.99	−1186.2	−54.8	−1164.8	−53.8	8086.9	60.2	5417.6	45.21	−3328.0	−27.8	2089.6	17.4
Total	1994.4	47.96	2164.2	52.04	−2164.2	−52.04	0.0	0.0	13,428.0	52.84	11,983.2	47.16	−11,983.2	−47.16	0.0	0.0

Table 5 Horizontal urban expansion of Bahir Dar and Hawassa (1973–2015)

Year	Bahir Dar				Hawassa			
	Urban area (ha)	Change (Ha)			Urban area (ha)	Change (Ha)		
		1973–2000	2000–2015	1973–2015		1973–2000	2000–2015	1973–2015
1973	184	2214			619	613		
2000	2398		3181		1232		1079	
2015	5579			5395	2311			1692
% Change (ha year^{-1})		44.56	8.84	69.81		3.67	5.84	6.51

Horizontal urban expansion includes urban built up area and green spaces with in built-up area (see Fig. 3a, b)

losses but higher gain. In Bahir Dar and Hawassa cities the land cover types which gained more are built-up area (50%) and crop land (45%) respectively (Table 4 and Fig. 2a, b).

In general, the results show that 53% of Bahir Dar and 48% of Hawassa land use/land cover remained unchanged over the 1973–2015 periods. On the other hand, 47% of Bahir Dar and 52% of Hawassa land use/land cover changed during 1973–2015. This indicates that there is a higher change of LULCC in Hawassa than in Bahir Dar in the last four decades (Table 4 and Fig. 2a, b).

The driving factors for this rapid LULCC are the rapid growth of urban population and the horizontal expansion of urban areas (see below). In line with this, the population in Bahir Dar has more than tripled between 1994 and 2017 (96,140 in 1994 and 348,429 in 2017) and quadrupled in Hawassa between 1994 and 2017 (69,169 in 1994 and 315,267 in 2017). The Landsat images analysis reveals, however, that land cover change is faster since 2000 than it was during the 1990s. The following discussion focuses on two types of land use changes namely the urban expansion and the green space in both cities.

Urban expansion

Following the technique in formula 7, the annual rates of urban expansion are analyzed from two perspectives. The first is the expansion in LULCC as a result of the sprawl of each city, which is the horizontal expansion while the second is the changes in LULCC that occurred within the 1973 boundaries of the cities during the period 1973–2015. This type of change is referred to as intensification increases in the density of dwellings and other infrastructure within existing built-up areas.

The data presented in Table 5 show that the annual rates of urban expansion for Bahir Dar in the period 1973–2000, 2000–2015, and 1973–2015 were 45, 9, and 70% per year per hectares respectively. It is thus clear that urban expansion was much higher during the period 1973–2000 than 2000–2015. This could be due to the time gap within 1973–2000 (27 years) is longer

than time gap within 2000–2015 (15 years). Similarly, the data in Table 5 show that the rates of urban expansion in Hawassa for the periods 1973–2000, 2000–2015, and 1973–2015 were 4, 6, and 7% per year per hectares respectively, implying that annual rates of urban expansion is much higher for the period 2000–2015 than for the period 1973–2000. The intensification of built-up area in Bahir Dar and Hawassa for the past four decades (1973–2015) was 70 and 7% per year per hectares respectively. This shows that 5395 and 1692 ha of land were converted to urban uses from other land cover types in Bahir Dar and Hawassa respectively. Bahir Dar has a much higher average annual rate of urban expansion than Hawassa. This could be because Bahir Dar has a larger boundary than Hawassa, and this might have prompted the rapid conversion of other land uses to urban land use (Fig. 3a and b). On the other hand, although the political boundary is small, Hawassa is relatively close to Addis Ababa, the capital city, and has higher potential to attract businessmen and investments which are very important contributors to the fast growth of urban areas.

Using the technique presented in formula 8, the LCR result for Bahir Dar is 0.002, 0.003 and 0.015 for the years 1973, 2000 and 2015 respectively. Likewise, the LCR for Hawassa is 0.023, 0.005 and 0.009 for the years 1973, 2000 and 2015 respectively. It can be seen that the LCR result is in accordance with LULCC result and is higher for Bahir Dar except for the year 1973 and 2000.

Changes in green space in built-up area

Table 1 defined green space in urban area as an area of grass, trees, or other vegetation set apart for recreational or aesthetic purposes inside urban built environment. It includes urban parks, greenery, roundabouts, public squares and plaza, open spaces, medians and sport fields. It is clear that this type of land use is created by the city government as part of its land use planning schemes.

The data presented in Table 3a, b clearly show that green spaces in Bahir Dar and Hawassa have increased significantly between 1973 and 2015. In Bahir Dar,

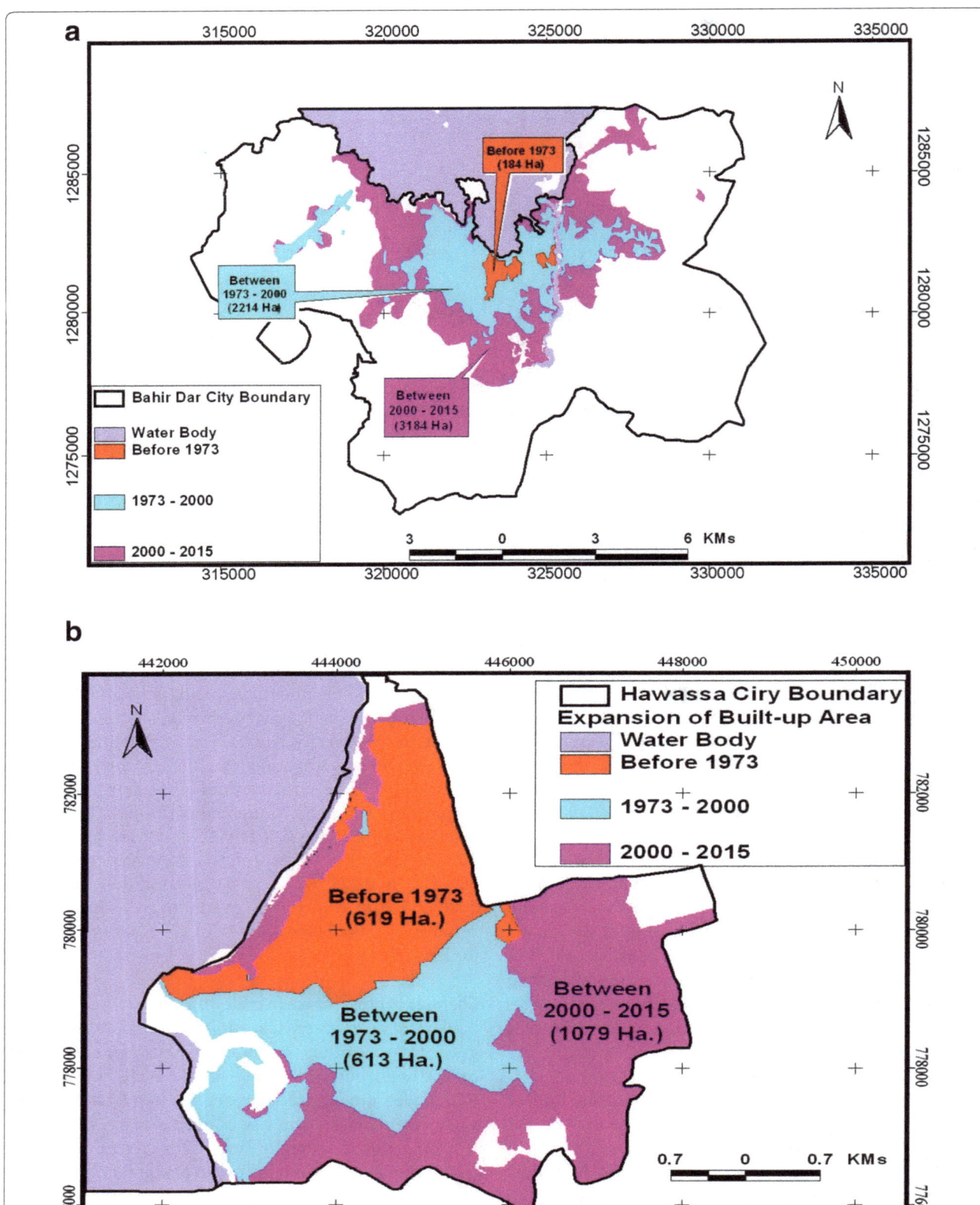

Fig. 3 **a** Horizontal expansion of Bahir Dar city (1973–2015). **b** Horizontal expansion of Hawassa city (1973–2015)

green space increased by 2742.8 ha while it increased by 706.6 ha in Hawassa between the years 1973 and 2015. The percentage increase is much higher for Bahir Dar than for Hawassa. This is because of both the small base and the higher additions of green spaces in Bahir Dar than in Hawassa. Hawassa, however, has a higher percent increase for the period 2000–2015 than Bahir Dar. The land cover types mostly changed to green infrastructure are vegetation and cropland.

In comparing green space and built-up area expansion, it can be seen that built-up area in Bahir Dar increased by 10% in 1973–2015 and this is proportional to green space increment of 11% of green space for the same period. The built-up area in Hawassa however increased (24%) more than green space (17%) though both have a rising tendency.

A comparative perspective between Hawassa and Bahir Dar shows that green space increment in Hawassa (33%) is by far more than Bahir Dar (23%) (Table 4). The reason could be that Hawassa has witnessed a decline in both vegetation and crop land which must have contributed to built-up and green spaces in the city. On the other hand, in Bahir Dar though vegetation has decreased, crop land has increased there by competing with the increase in built up and green space. Therefore care should be taken to conserve these lands.

Land use land cover types most affected

The land cover proportions obtained from the successive (enhanced) classifications revealed that in 1973, Bahir Dar was dominated by crop land (45% of total area), followed by vegetation (40%) (Table 3a). However, after a quarter of a century, in 2000, vegetation occupied only 25% of the total area, and crop land increased and occupied 48%. The change was further intensified after 2000 as vegetation was reduced to 10% and crop land was increased to 53%. Moreover, as presented in Table 4 urban built-up area increased by 45% ha per year during the period 1973–2015.

The data presented in Table 3b indicates that in 1973, Hawassa was dominated by crop land (31%) and vegetation (17%). However, in 2000 or after a quarter of a century, vegetation occupied only 9% and crop land decreased and covered 25%. The urban built-up area on the other hand increased alarmingly by 30%, with an expansion of 7% ha per year during the periods of 1973–2015. This implies that in both cities, urban-built up area is the land use type that showed marked increase while crop land and vegetation land have different trends. The Landsat images analysis confirmed that the major land cover conversions were from the vegetation cover classes to crop land and built-up classes (Fig. 2a, b). This is further confirmed by examination of net changes of the five

land use/land cover classes 1973–2015 due to intensification of the built-up area. The result shows that built-up increased by 22% in Bahir Dar, whereas the cropland land increased by net-change of 17% at the expense of vegetation which showed a decrease by net-change of 63%. A similar analysis for the periods 1973–2015 in Hawassa revealed that net-change in built-up area is 45%. During the entire study period (1973–2015), crop land and vegetation decreased by net-change of 54 and 27% respectively (Table 4). Green space is also one of the land use categories that showed a rapid increase with a higher-level gain than loss in both cities.

Discussion

This study showed that land use/land cover is imperative for understanding the GI conditions of urban areas. Land use/land cover can be used for planning and monitoring the status of GI. On the other hand, GI study requires land use/land cover change detection in order to understand GI within the setting of other land use/land covers. Some researches for instance, Li et al. (2015) indicated that LULCC can be an important indicator to link GI and human activities in urban ecosystems. Liu et al. (2014) also examine LULCC and urbanization effect on urban environment. Although, there are studies on farm land effects of LULCC or urbanization (Pauleit et al. 2005), this study has made the first attempt to explore the combined effect of LUCC and GI under the rapid urbanization on farm land in fast growing cities of Bahir Dar and Hawassa 1973–2015.

The green infrastructure concept has come into the table of discussion in the last few decades and is used for urban green environment improvement (Tzoulas et al. 2007). In this study, the foregoing data showed that green spaces in both study cities have increased. The increasing trend of green infrastructure in Bahir Dar and Hawassa during the period 1973–2015 was due to continued and drastic increment of built-up areas at the cost of other land cover types (vegetation and cropland). Some studies (Noor et al. 2013) indicated that the issue of green infrastructure has become major concern throughout the world particularly among developing countries due to the obvious negative impacts which occurred as the result of loss of green infrastructure in terms of visual quality, environmental quality and health quality with in fast growing cities and towns.

A study by Luck and Wu (2002) recognized that urbanization is one of the most important driving forces behind LULCC in Jinan city (China). Kong and Nakagoshi (2006) also reported that the driving forces are the policies that affect the development and management of urban GI. However, Byomkesh et al. (2012) noted, the causes of changes in GI, among other things, are rapid population growth driven by rural–urban migration, economic development and a lack of awareness among city managers and city dwellers. In addition, the use of political power to influence the illegal conversion and leasing of GI, a lack of appropriate rules and regulations to protect urban GI, limited budget for the management and maintenance of urban GI, are also factors that contribute to GI change. In their study about analysis of LULCC and urban expansion of Nairobi city, Mundia and Aniya (2005) noted the analysis of LULCC shows that the biggest challenge to city planners' is perhaps to maintain an internal balance between economic activity, population growth, infrastructure and services not limiting impacts on the natural environment. In order to maintain ecological balance and proper functioning of ecosystems, comprehensive GI planning and management strategy needs to be formulated. Educating the people to increase their awareness about the role and importance of GI for healthy environment is necessary. It is worth mentioning that comprehensive land use planning would contribute to enhance GI and sustainability and livability of urban development.

This study has presented the LULCC and dynamics of urban expansion which is demonstrated by the interplay between biophysical, location site and socio-economic characteristics in shaping the growth of both cities. The spatial expansion of both Bahir Dar and Hawassa cities is very rapid during the last 10–15 years. The driving forces to this urban expansion resulted from population growth, economic reform and industrialization (Meyer and Turner 1992; Morrisette 1992; Rockwell 1992; Sanderson 1992). Increase in investment brings fast economic development which leads accelerated urban area expansion because the development of the industrial parks (Grubler 1992) in both cities. Industrial development is a major driving force for urbanization (Xu et al. 2000). Expansion direction is also necessary in city management and study of LULCC. The direction of urban expansion is importantly controlled by the topographical and physical factors. Urban expansion directions and land use conversions analysis indicates that deliberate planning is largely important in Bahir Dar and Hawassa urbanization process. Bahir Dar city expands towards South, West and North-East but no more expansion towards the North because of Lake Tana and towards the South-East due to bezawit ridge. Hawassa city expands towards East, North and South-East but not towards the West and South because of Lake Hawassa and amoragedel ridge respectively. These areas of both cities are considered as green belts of corresponding cities. The horizontal expansion of urban and suburban areas requires more land and drives the conversion of surrounding rural areas to urban land use/land cover (Farooq and Ahmad 2008; Mohan et al.

2011). Horizontal expansion has strong effects on other urban land uses, such as crop/agricultural land, green space, and forest lands (Mohan et al. 2011).

The results from the analysis of the Landsat images show that for the two cities the land cover types that have significantly contributed to gains for built-up area and the corresponding green space are crop land and vegetation. These two-land use/land cover types have fed built-up areas and green spaces while they show a drastic decline in their coverage. This is consistent with previous research (Xu et al. 2000) crop lands are under great pressure from rapid urban expansion. As it is explained in tables above in both cities the expansion rate is fast and many agricultural lands are changed into urban areas. Severe arable land loss will have a significant impact on the county's further agricultural development. Obviously, dynamical monitoring of the expansion of urban areas is valuable for the sustainable development of the country. Rana (2011) noted rapid urbanization is always characterized by spatial extension in the periphery, which leads to exploitation of forest and crop land. This could be because of limited capacity of planning. The city authorities are facing huge lack in skilled manpower and sufficient resources to reach the detail plan stages (Islam 2002; Shafi 2003).

It is important to note that urban expansion and the loss of crop land have impacts on the surrounding farmers and the nearby water bodies in the study area. With regard to famers, loss of crop land and the associated urban expansion give rise to changes in the livelihood of farmers as they derive reduced income from farming (Haregeweyn et al. 2012). In relation to water bodies, the small decrease noted in both cities between 1973 and 2015 is associated with a retreat of the lakes caused by siltation and the subsequent use of this land for built-up areas. Field observation and satellite images analysis verify this because there are clear indicators of the retreat of Lake Tana and Lake Hawassa. Other related studies (Gashaw and Fentahun 2014; Wondrade and Tveite 2014; Teshale and Bantider 2015; Minale and Belete 2017) conducted in different parts of Ethiopia reported that the life of both artificial and natural lakes is threatened by a high sedimentation rate, with the sediment primarily being delivered from agricultural watersheds. At the city level, the results revealed that different anthropogenic activities had significantly affected the urban green infrastructure composition and configuration within the inner cities of both Bahir Dar and Hawassa.

Our findings are helpful for policy makers to better understand and address these complex relationships between urbanization, LUCC, and GI. It is important to develop improved land-use policies that balance LUCC, GI proportion and urbanization. The findings of this research have not only important policy implications for urban GI design and management, but also provide important information for other research areas such as urban environment and ecology.

Conclusions

The LULCC dynamics largely depend on dynamic relationships not only natural factors but also among population and policy/institutional factors. In this study we noted the spatiotemporal trends of urban land use/land cover and an aspect of green infrastructure change. Change detection is important to understand the magnitude and direction of change in any land use/land cover category in general and in green spaces in particular. Our result revealed that green infrastructure defined as urban parks, open spaces, greenery, roundabouts, public squares and plaza, medians and sport fields have increased in both cities during the period 1973–2015. Such increase is believed to be associated with urban expansion since the latter have increased in both cities. The mechanism is the implementation of land use planning at city level in order to cope up with the increasing urban expansion. The two cities could thus be taken as exemplary to other cities and towns in Ethiopia since the increase in green spaces is closely related to sustainable urban development.

In recent years there has been growing concern among planners about the green infrastructure in cities. In addition, policy makers and stakeholders should also decide on how to use the land at present and in the future. Therefore, LULCC policymaking processes should aim to balance green infrastructure and other types of land use/land cover for sustainable urban development. Generally, we can imply that urban land use/land cover have important effects on the urban green infrastructure system.

In conclusion, LULCC dynamics and GI analyses are imperative for understanding the landscape ecological conditions of urban environment. This study revealed that the vegetation and crop land are decreasing over the course of time due to the increasing pace of built-up area. This activity is causing the destruction of landscape ecological processes and the biodiversity in urban areas. Therefore, a comprehensive urban land use planning and GI management strategy should be implemented for proper functioning of the urban environment.

Authors' contributions
KG has conceived of the study and made contributions in the design, data collection and analysis, interpretation of results and revisions of the manuscript. TG-E has participated in the sequence alignment and critical commenting of the draft manuscript. He also participated in its design and coordination, and helped to draft and edits the manuscript. Both authors read and approved the final manuscript.

Author details
[1] Department of Geography and Environmental Studies, University of Gondar, P.O. Box 196, Gondar, Ethiopia. [2] Department of Geography and Environmental Studies, Addis Ababa University, P.O. Box 1176, Addis Ababa, Ethiopia.

Acknowledgements
We would like to thank the anonymous reviewers and the editor for their genuine comments and corrections which helps the paper to be in its present form. Special thanks to Ethiopian Mapping Agency (EMA) for accessing the satellite imageries.

Competing interests
The authors declare that they have no competing interests.

Funding
The authors would like to thank Addis Ababa University for financial support for this research for both researchers and University of Gondar for financial support to the first researcher.

References
Abebe MT, Megento TL (2016) The city of Addis Ababa from 'Forest City' to 'Urban Heat Island': assessment of urban green space dynamics. J Urban Environ Eng 10(2):254–262

Amsalu A, Stroosnijder L, de Graaf J (2007) Long-term dynamics in land resource use and the driving forces in the Beressa watershed, highlands of Ethiopia. J Environ Manag 83:448–459

Anderson JR, Hardy EE, Roach JT, Witmer RE (1976) A land-use and land-cover classification system for use with remote sensor data. US Geological Survey Professional Paper 964, Washington, DC

Ayalew D, Kassahun D, Woldetsadik M (2012) Detection and analysis of land-use and land-cover changes in the Midwest escarpment of the Ethiopian Rift Valley. J Land Use Sci 7(3):239–260

Butt A, Shabbir R, Ahmad SS, Aziz N (2015) Land use change mapping and analysis using Remote Sensing and GIS: a case study of Simly watershed, Islamabad, Pakistan. Egypt J Remote Sens Space Sci 2015(18):251–259

Byomkesh T, Nakagoshi N, Dewan AM (2012) Urbanization and green space dynamics in Greater Dhaka, Bangladesh. Landsc Ecol Eng 8:45–58

Chen G, Hay GJ, Carvalho LMT, Wulder MA (2012) Object-based change detection. Int J Remote Sens 33(14):4434–4457

Congalton RG, Green K (2009) Assessing the accuracy of remotely sensed data: principles and practices, 2nd edn. Taylor & Francis, Baco Raton

CSA (1994) Statistical report. Addis Ababa, Ethiopia

CSA (2007) Statistical report. Addis Ababa, Ethiopia

CSA (2017) Statistical abstract. Addis Ababa, Ethiopia

Dabboor M, Howell S, Shokr M, Yackel J (2014) The Jeffries–Matusita distance for the case of complex Wishart distribution as a separability criterion for fully polarimetric SAR data. Int J Remote Sens 35(19):6859–6873

Fanan U, Dlama KI, Olusey IO (2011) Urban expansion and vegetal cover loss in and around Nigeria's Federal Capital City. J Ecol Nat Environ 3(1):1–10

Farooq S, Ahmad S (2008) Urban sprawl development around Aligarh city: a study aided by satellite remote sensing and GIS. J Indian Soc Remote Sens 36:77–88

Feoli E, Vuerich L (2002) Processes of environmental degradation and opportunities for rehabilitation in Adwa, Northern Ethiopia. Landsc Ecol 17(4):315–325

Foody GM (2002) Status of land covers classification accuracy assessment. Remote Sens Environ 80:185–201

Gashaw T, Fentahun T (2014) Evaluation of land use/land cover changes in east of lake Tana, Ethiopia. J Environ Earth Sci 4(11):49–53

Grubler A (1992) Technology and global change: land-use, past and present. In: Meyer WB, Turner BL II (eds) Global land-use/land-cover change. Boulder, AIES

Hailemariam SN, Teshome S, Teketay D (2016) Land use and land cover change in the Bale Mountain Eco-Region of Ethiopia during 1985 to 2015. Land 5:41

Haregeweyn N, Fikadu G, Tsunekawa A, Tsu:o M, Tsegaye DM (2012) The dynamics of urban expansion and its impacts on land use/land cover change and small-scale farmers living near the urban fringe: a case study of Bahir Dar, Ethiopia. Landsc Urban Plann 106(2012):149–157

He C, Shi P, Xie D, Zhao Y (2010) Improving the normalized difference built-up index to map urban built-up areas using a semiautomatic segmentation approach. Remote Sens Lett 1(4):213–221

Hu H, Liu W, Cao M (2008) Impact of land use and land cover changes on ecosystem services in Menglun, Xishuangbanna, South west China. Environ Monit Assess 146:147–156

Hung M, Wu Y (2005) Mapping and visualizing the Great Salt Lake landscape dynamics using multi-temporal satellite images, 1972–1996. Int J Remote Sens 26:1815–1834

Hussain M, Chen D, Cheng C, Wei H, Stanley D (2013) Change detection from remotely sensed images: from pixel-based to object-based approaches, ISPRS. J Photogramm Remote Sens 80(2013):91–106

Islam N (2002) The Bangladesh urban environment (Editorial Notes). CUS bulletin on urbanization and development vol 43. Center for urban studies, Dhaka

Kambites C, Owen S (2006) Renewed prospects for green infrastructure planning in the UK. Plan Pract Res 21(4):483–496

Kong F, Nakagoshi N (2006) Spatial-temporal gradient analysis of urban green spaces in Jinan, China. Landsc Urban Plan 78:147–164

Lei C, Zhu L (2017) Spatio-temporal variability of land use/land cover change (LULCC) within the Huron river: effects on stream flows. Climate Risk Manag 19(2018):35–47

Li W, Bai Y, Zhou W, Han C, Han L (2015) Land use significantly affects the distribution of urban green space: case study of Shanghai, China. J Urban Plan Dev 141(3):A4014001

Liu Y, Huang X, Yang H, Zhong T (2014) Environmental effects of land-use/cover change caused by urbanization and policies in Southwest China Karst area—a case study of Guiyang. Habitat Int 44(2014):339–348

Long H, Tang G, Li X, Heilig GK (2007) Socio-economic deriving forces of land-use change in Kunshan, the Yangtze River Delta economic area of China. J Environ Manag 83(3):351–364

Lu D, Weng Q (2007) A survey of image classification methods and techniques for improving classification performance. Int J Remote Sens 28(5):823–870

Lu D, Mausel P, Brondízio E, Moran E (2004) Change detection techniques. Int J Remote Sens 25(12):2365–2401

Luck M, Wu J (2002) A gradient analysis of urban landscape pattern: a case study from the Phoenix metropolitan region, Arizona, USA. Landsc Ecol 17:327–339

Mansor M, Said I, Mohamad I (2012) Experiential contacts with green infrastructure's diversity and well-being of urban community. Asian J Environ Behav Stud 49:33–48

Mas JF (1999) Monitoring land-cover changes: a comparison of change detection techniques. Int J Remote Sens 20(1):139–152

Masser I (2007) Managing our urban future: the role of remote sensing and geographic information systems. Habitat Int 25:503–512

McIntyre N, Knowles-Yánez K, Hope D (2000) Urban ecology as an interdisciplinary field: differences in the use of urban between the social and natural sciences. Urban Ecosyst 4:5–24

Mell C (2014) Aligning fragmented planning structures through a green infrastructure approach to urban development in the UK and USA. Urban For Urban Green 13(2014):612–620

Meshesha D, Tsunekawa A, Tsubo M (2010) Continuing land degradation and its cause-effect in Ethiopia's Central Rift Valley. Land Degrad Dev 23(2):130–143

Meyer WB, Turner BL (1992) Human population growth and global land-use/cover change. Annu Rev Ecol Syst 23(1):39–51

Miller J, Hobbs R (2002) Conservation where people live and work. Conserv Biol 16:330–337

Minale AS, Belete W (2017) Land use distribution and change in lake Tana sub basin. In: Stave K, Goshu G, Aynalem S (eds) Social and ecological system dynamics. AESS interdisciplinary environmental studies and sciences series. Springer, Cham

Mohan M, Pathan S, Narendrareddy K, Kandya A, Pandey S (2011) Dynamics of urbanization and its impact on land use/land cover: a case study of megacity Delhi. J Environ Prot 2:1274–1283

Moroney J, Jones D (2006) Biodiversity space in urban environments: implications of changing lot size. Aust Plan 43(4):22–27

Morrisette PM (1992) Developing a political typology of global patterns of land and resource use. In: Meyer WB, Turner BL II (eds) Global land-use/land-cover change. Boulder, AIES

MoUDH (2015) Ethiopia national urban green infrastructure standard, Addis Ababa, Ethiopia

Msoffe FU, Kifugo SC, Said MY, Neselle MO, Gardingen PV, Reid RS, Ogutu JO, Herero M, de Leeuw J (2011) Drivers and impacts of land-use change in the Maasai Steppe of northern Tanzania: an ecological, social and political analysis. J Land Use Sci 6(4):261–281

Mundia CN, Aniya M (2005) Analysis of land use/cover changes and urban expansion of Nairobi city using remote sensing and GIS. Int J Remote Sens 26(13):2831–2849

Murat H, Selçuk R, Mustafa A (2006) Detection of spatial-temporal changes of development potential of Aksaray city using remote sensing and GIS. In: Shaping the change, XXIII FIG congress, Munich, Germany, October 8–13

Muriuki G, Seabrook L, McAlpine C, Jacobson C, Price B, Baxter G (2011) Land cover change under unplanned human settlements: a study of the Chyulu hills squatters, Kenya. Landsc Urban Plan 99(2011):154–165

Muttitanon W, Tripathi N (2005) Land use/cover changes in the coastal zone of Ban Don bay, Thailand using Landsat 5 TM data. Int J Remote Sens 26:2311–2323

NMA (2013) National meteorological station report. Federal democratic republic of Ethiopia, Addis Ababa

Noor NM, Abdullah A, Manzahari H (2013) Land cover change detection analysis on urban green area loss using GIS and remote sensing techniques. J Malays Inst Plan 2013(11):125–138

Pauleit S, Ennos R, Golding Y (2005) Modeling the environmental impacts of urban land use and land cover change—a study in Merseyside, UK. Landsc Urban Plan 71(2–4):295–310

Phan DU, Nakagoshi N (2007) Analyzing urban green space pattern and eco-network in Hanoi, Vietnam. Landsc Ecol Eng 3:143–157

Puyravaud JP (2003) Standardizing the classification of annual rate of deforestation. For Ecol Manag 177(1–3):593–596

Radoux J, Bogaert P, Fasbender D, Defourny P (2011) Thematic accuracy assessment of geographic object-based image classification. Int J Geographical Inf Sci 25(6):895–911

Rana MP (2011) Urbanization and sustainability: challenges and strategies for sustainable urban development in Bangladesh. Environ Dev Sustain 13:237–256

Robertson LD, King DJ (2011) Comparison of pixel- and object-based classification in land cover change mapping. Int J Remote Sens 32(6):1505–1529

Rockwell R (1992) Culture and cultural change as driving forces in global land-use/cover changes. In: Meyer WB, Turner BLI II (eds) Global land-use/land-cover change. Boulder, AIES

Sadidy J, Firouzabadi P, Entezari A (2009) The use of radar sat and land sat image fusion algorithms and different supervised classification methods to use map accuracy—case study. Sari lain-Iran. http://www.isprs.org/procedding/XXXVI/5-C55/papers-Sadidy_javad.pdf. Accessed 19 Oct 2015

Sanderson S (1992) Institutional dynamics behind land use change. In: Meyer WB, Turner BL II (eds) Global land-use/land-cover change. Boulder, AIES

Shafi SA (2003) Use of planning tools as a guide to balanced urban development. CUS Bulletin on urbanization and development, vol 44. Center for urban studies, Dhaka

Sharma L, Pandey PC, Nathawat MS (2012) Assessment of land consumption rate with urban dynamics change using geospatial techniques. J Land Use Sci 7(2):135–148

Song X, Chang K, Yang L, Scheffran J (2016) Change in environmental benefits of urban land use and its drivers in Chinese cities, 2000–2010. Int J Environ Res Public Health 13:535

Taddese G (2001) Land degradation: a challenge to Ethiopia. Environ Manag 27(6):815–824

Teferi E, Bewket W, Belay S (2016) Effect of land use land cover on selected soil quality indicators in the head water area of the Blue Nile basin of Ethiopia. Environ Monit Assess 2016(188):1–12

Teshale RR, Bantider A (2015) Land use land cover dynamics in Hawassa tabor and alemura ridge and its surroundings in the case of SNNPR, Ethiopia, Doctoral dissertation, Haramaya university

Tewolde MG, Cabral P (2011) Urban sprawl analysis and modeling in Asmara, Eritrea. Remote Sens 2011(3):2148–2165

Thompson M (1996) Standard land cover classification scheme for remote sensing application in South Africa. S Afr J Sci 92:34–42

Tsegaye D, Moe S, Vedeld P, Aynekulu E (2010) Land-use/cover dynamics in Northern afar rangelands, Ethiopia, agriculture. Ecosyst Environ 139(2010):174–180

Tzoulas K, Korpela K, Venn S, Yli-Pelkonen V, Ka´zmierczak A, Niemela J, James P (2007) Promoting ecosystem and human health in urban areas using green infrastructure: a literature review. Landsc Urban Plan 81(2007):167–178

UN (2014) World urbanization prospect, department of economic and social affairs, New York

Valdkamp E, Weitz A, Staritsky I, Huising E (1992) Deforestation trends in the Atlantic zone of Costa rica: a case study. Land Degrad Rehabil 3:71–84

Varshney A (2013) Improved NDBI differencing algorithm for built-up regions change detection from remote-sensing data: an automated approach. Remote Sens Lett 4(5):504–512

Wondrade N, Tveite DH (2014) GIS based mapping of land cover changes utilizing multi-temporal remotely sensed image data in Lake Hawassa Watershed, Ethiopia. Environ Monit Assess 186(3):1765–1780

Xu H, Wang X, Xiao G (2000) A remote sensing and GIS integrated study on urbanization with its impact on arable lands: Fuqing city, Fujian province, China. Land Degrad Dev 11:301–314

Yang J, Huang C, Zhang Z, Wang L (2014) The temporal trend of urban green coverage in major Chinese cities between 1990 and 2010. Urban For Urban Green 13(2014):19–27

Yuan F, Sawaya KE, Loeffelholz BC, Bauer ME (2005) Land cover classification and change analysis of the twin cities (Minnesota) metropolitan area by multitemporal Landsat remote sensing. Remote Sens Environ 98(2–3):317–328

Zhang H, Qi ZF, Ye XY, Cai YB, Ma WC, Chen MN (2013) Analysis of land use/land cover change, population shift, and their effects on spatiotemporal patterns of urban heat islands in metropolitan Shanghai, China. Appl Geogr 44:121–133

Zhou W, Troy A (2008) An object-oriented approach for analyzing and characterizing urban landscape at the parcel level. Int J Remote Sens 29(11):3119–3135

Zoran M (2007) Urban environmental quality assessment by satellite and in situ monitoring data. In: AIP conference proceedings 899, 407

4

Analysis of the long-term agricultural drought onset, cessation, duration, frequency, severity and spatial extent using Vegetation Health Index (VHI) in Raya and its environs, Northern Ethiopia

Eskinder Gidey[1,2,3*], Oagile Dikinya[1], Reuben Sebego[1], Eagilwe Segosebe[1] and Amanuel Zenebe[2,3]

Abstract

Background: Droughts cause serious effects on the agricultural and agro-pastoral sector due to its heavy dependence on rainfall. Several studies on agricultural drought monitoring have been conducted in Africa in general and Ethiopia in particular. However, these studies were carried out using the limited capacity of drought indices such as Normalized Difference Vegetation Index (NDVI), Vegetation Condition Index (VCI), and Deviation of Normalized Difference Vegetation Index (DevNDVI) only. To overcome this challenge, the present study aims to analyze the long-term agricultural drought onset, cessation, duration, frequency, severity and its spatial extents based on remote sensing data using the Vegetation Health Index (VHI) 3-month time-scale in Raya and its surrounding area, Northern Ethiopia. Both the MOD11A2 Terra Land Surface Temperature (LST) and eMODIS NDVI at 250 by 250 m spatial resolution and hybrid TAMSAT monthly rainfall data were used. A simple linear regression model was also applied to examine how the agricultural drought responds to the rainfall variability.

Results: Extremely low mean NDVI value ranged from 0.23 to 0.27 was observed in the lowland area than mid and highlands. NDVI coverage during the main rainy season decreased by 3–4% in all districts of the study area, while LST shows a significant increase by 0.52–1.08 °C. VHI and rainfall value was significantly decreased during the main rainy season. Agricultural drought responded positively to seasonal rainfall ($R^2 = 0.357$ to $R^2 = 0.651$) at $p < 0.01$ and $p < 0.05$ significance level. This relationship revealed that when rainfall increases, VHI also tends to increase. As a result, the event of agricultural drought diminished.

Conclusions: Remote sensing and GIS-based agricultural drought can be better monitored by VHI composed of LST, NDVI, VCI, and TCI drought indices. Agricultural drought occurs once in every 1.36–7.5 years during the main rainy season, but the frequency, duration and severity are higher (10–11 times) in the lowland area than the mid and highlands area (2–6 times) during the last 15 years. This study suggests that the effect of drought could be reduced through involving the smallholder farmers in a wide range of on and off-farm practices. This study may help to improve the existing agricultural drought monitoring systems carried out in Africa in general and Ethiopia in particular. It also supports the formulation and implementation of drought coping and mitigation measures in the study area.

Keywords: Agricultural drought, LST, VCI, NDVI, TCI, Rainfall, VHI, Remote sensing, GIS, Raya, Ethiopia

*Correspondence: eskinder14@yahoo.com
[1] Department of Environmental Science, University of Botswana, Private Bag UB 0704, Gaborone, Botswana
Full list of author information is available at the end of the article

Background

In arid and semi-arid regions, rain-fed agricultural production is mostly a risky practice because of its high sensitivity to climate extremes, including drought (Lei et al. 2016; Choi et al. 2013). Several studies have indicated that drought causes a significant decline in agricultural production and productivity all over the world. This can occur with no warning, without recognizing borders or economic and political differences (Kogan 2000). For instance, during the periods of 2001–2012, moderate-to-exceptional (ME), severe-to-exceptional (SE) and extreme-to-exceptional (EE) droughts covered about 17–35%, 7–15% and 2–6% of the total land mass of the world, respectively (Kogan et al. 2013). For example, the droughts of 2010 in Russia and 2011/12 in the USA produced considerable local and global economic impacts (Kogan et al. 2016). As a result, the balance of food supply and demand was significantly affected due to severe droughts at local, regional, and global scales (Van Hoolst et al. 2016; Song et al. 2004). In dry areas, where the rainfall pattern is highly variable, the most susceptible shock is realized (Maybank et al. 1995). Several regions of the world, particularly the main grain-growing countries (e.g., USA, China, Russia, India, and European Union) are thus experiencing an increase in the frequency and intensity of droughts incidence (Kogan et al. 2016; Owrangi et al. 2011).

In developed countries, drought monitoring and early warning systems are based on earth observation products and it is highly effective, while in the majority of African countries (including Ethiopia) the situation largely depends on the *in-situ* climate data only, which significantly affects the smallholder farmers. It also lacks the continuous spatial coverage needed to characterize and monitor the detailed spatial pattern of drought conditions (Gu et al. 2007). For instance, drought is a persistent problem in Botswana (Segosebe 1990) and other African countries. Efforts have been made to set up regional drought monitoring in the Southern African Development Community (SADC), the Great Horn of Africa (GHA), and the West African Permanent Interstate Committee on Drought Control in the Sahel (CILSS). All these monitoring systems are confined only to the selected regions and hence, they do not include the entire African countries and their initiatives are ineffective in majority part of the continent in providing provide real-time information on the past and future drought events (Vicente-Serrano et al. 2012). Vicente-Serrano et al. (2012) reported that many droughts affected developing countries, including Ethiopia, facing difficulties in monitoring droughts due to weak institutional structures, lack of technical capacity, limited progress in mobilizing stakeholder involvement and investment, and lack of in-depth understanding of the benefits of effective drought management for poverty alleviation and economic development and the lack of a preparedness culture. The drought has thus remained a bottleneck problem in the area. For instance, during the 1981–1984, several countries in the continent were under the spell of catastrophic drought events.

Ethiopia is one of the countries with frequent drought events due to poor and erratic rainfall availability where the problem is severe in the northern parts. Sholihah et al. (2016) reported that the incidence of El Nino phenomenon droughts has also been frequently occurring over the decades triggering several threats to the agriculture sector. Particularly, the arid and semi-arid area has been severely affected by the recurrent droughts. The cessation, duration, frequency, severity and spatial extent of agricultural drought in the area is high. Although, substantial growth in the major crop types (e.g., teff, barley, maize, wheat, and sorghum and others) were observed in terms of productivity and area coverage, yields are low when evaluated by international standards. This is because production is highly susceptible to weather shocks, particularly droughts (Se et al. 2011). Agricultural production, mainly in the poor area has remained highly dependent on the weather (Zhang et al. 2016). The challenges may also arise in the future as the natural resources largely over exploited due to rapid population growth. Vicente-Serrano et al. (2012) stated that the current population projections in the area also significantly increased in the regions where the area intermittently affected by the persistent water shortage leading to catastrophic drought. Umran Komuscu (1999) revealed that drought impacts are usually first apparent in agriculture, but gradually move to other water-dependent sectors. The agricultural drought was, therefore, occurring due to unfavorable precipitation. Agriculture is the first sector affected by the hydro-meteorological droughts because it adversely affects the growth of vegetation as well as crop production (Bhuiyan et al. 2006), but later moves to other water dependent sectors (Umran Komuscu 1999).

Agricultural drought is primarily expressed by the reduction of crop production and/or productivity due to erratic rainfall as well as insufficient soil moisture in the crop root zones (Sruthi and Aslam 2015; Alemaw and Simalenga 2015). However, the reliance on weather data alone is not adequate to monitor an area of drought, particularly when these data are untimely, sparse, and incomplete (Peters et al. 2002). The conventional ways of drought monitoring which depend only on weather stations (e.g., Ethiopia) lack continual spatial coverage to characterize and monitor the spatial pattern of drought incidences in depth (Gu et al. 2007).

Kogan (2000) reported that the recent advances in satellite technology improved the ability to monitor droughts. Remote sensing and GIS-based agricultural drought monitoring has thus attracted interest of various scientists such as agriculturalists, hydrologists, meteorologists, and environmentalists because it provides more accurate, flexible and reliable findings (e.g., spatio-temporal trends of drought) in drought studies. Seiler et al. (1998) noted that reliable, satellite-based drought indices are credible in detecting the spatial and temporal drought occurrence, which is highly important for conducting effective drought monitoring, and for alleviating the risk arises from drought. Likewise, those satellite observation products can complement the information gathered by traditional and ground-based drought assessment techniques that rely only on meteorological observations. However, it requires timely information about vegetation condition related to drought, flooding, or fire danger (Brown et al. 2015). This method of drought monitoring is feasible, highly accurate and cost–effective to assess large areas with different time-scale. It also provides real-time and dynamic information for terrestrial ecosystems, facilitating effective drought monitoring (Zhang et al. 2016). Bhuiyan (2004) stated that agricultural droughts reflect vegetation stress. Assessing the vegetation health status of a given area is paramount significant to characterize the incidence of agricultural drought, but it requires at least 10 years of satellite observation data and suitable drought index. Furthermore, the understanding, monitoring, and mitigating drought are becoming a very difficult task because of the intrinsic nature of the phenomenon (Vicente-Serrano et al. 2012). However, satellite observations overcome some limitations of station-based meteorological observations, providing the potential for cost–effective, spatially explicit and dynamic large-scale drought monitoring (Zhang et al. 2016). Likewise, satellite observation products (e.g., eMODIS NDVI, MOD11A2 LST) supported with advanced remote sensing drought indices such as Vegetation Health Index (VHI) can help to assess the incidence of agricultural droughts. Liu and Kogan (1996) stated that the seasonal and/or inter-annual droughts can be delineated by using the Vegetation Condition Index (VCI) and Temperature Condition Index (TCI) because both indices can help to generate VHI. Rhee et al. (2010) reported additional drought indices such as the Normalized Difference Drought Index (NDDI), the Normalized Difference Water Index (NDWI), and the Normalized Multiband Drought Index (NMDI) were introduced based on hyperspectral remote sensing data. These drought indices might be significant, but VHI has been the popular agricultural drought index. However, it requires both NDVI and LST data (Zhang et al. 2017; Choi et al. 2013).

Agricultural drought monitoring using VHI is therefore essential to provide reliable information. Studies showed that (e.g., Kogan and Guo 2016) the incident of droughts has continued with a significant agricultural production reduction or loss and other associated impacts such as malnutrition, human health deterioration, depletion of water resources, rising of food prices, population migration, and mortality. Therefore, there is a need to obtain synoptic information on a recurring and timely basis, drought-affected agricultural zones to identify area requiring immediate attention (Van Hoolst et al. 2016) and to mitigate the implication. In this study, the agricultural drought monitoring was conducted in 3-month time-scale (i.e., July–September). The specified time-scale is vegetation as well as the crop gestation period during the main rainy season. Similarly, Zhang et al. (2017) studied the drought phenomenon during vegetation growing seasons in the United States. Therefore, studying agricultural drought during the vegetative phase can provide better drought characteristics information. The novelty of this study is that it conducted VHI based long term agricultural drought monitoring in Africa in general and Ethiopia in particular. The objective of this study was to analyze the long-term agricultural drought onset, cessation, duration, frequency, severity, and its spatial extent using the VHI that integrates NDVI, VCI, LST and TCI in Raya and its surrounding area, Northern Ethiopia. The study is decisive for monitoring, understanding and managing the incidence of agricultural droughts through satellite earth observation data.

Methods
Study area
This study was undertaken in Raya and its environs (Northern Ethiopia) which is an intermountain plain area located at 39°24′40″ and 40°25′20″ longitude Easting and 12°7′20″ and 13°8′0″ latitude Northing (Fig. 1) (Gidey et al. 2017). It consists of 11 districts, namely Megale, Yalo, Gulina, Gidan, Kobo, Alaje, Alamata, Hintalo Wejirat, Ofla, Endamehoni, and Raya Azebo. The total area coverage of the study area is estimated at 14,532 km² of which (48%) falls in the southern Tigray region, 22% in Amhara and (30%) in the Afar region (Gidey et al. 2017). The area receives up to 558 mm of rainfall annually (Gidey et al. 2017). Rainfall is erratic and bimodal (Ayenew et al. 2013). In 2015, the highest temperature was observed since 1984. During this time, the maximum temperature (Tmax) and minimum temperature (Tmin) were 30.5 and 15.9 °C, respectively. The study area consists of four river basins such as Denakil basin, which covers about 10,265.8 km² (70.64%), Lake Ashinge 16.0 km² (0.11%), Abay (Blue Nile) 13.2 km² (0.09%), and Tekeze 4237.0 km² (29.16%). The mean elevation value of the

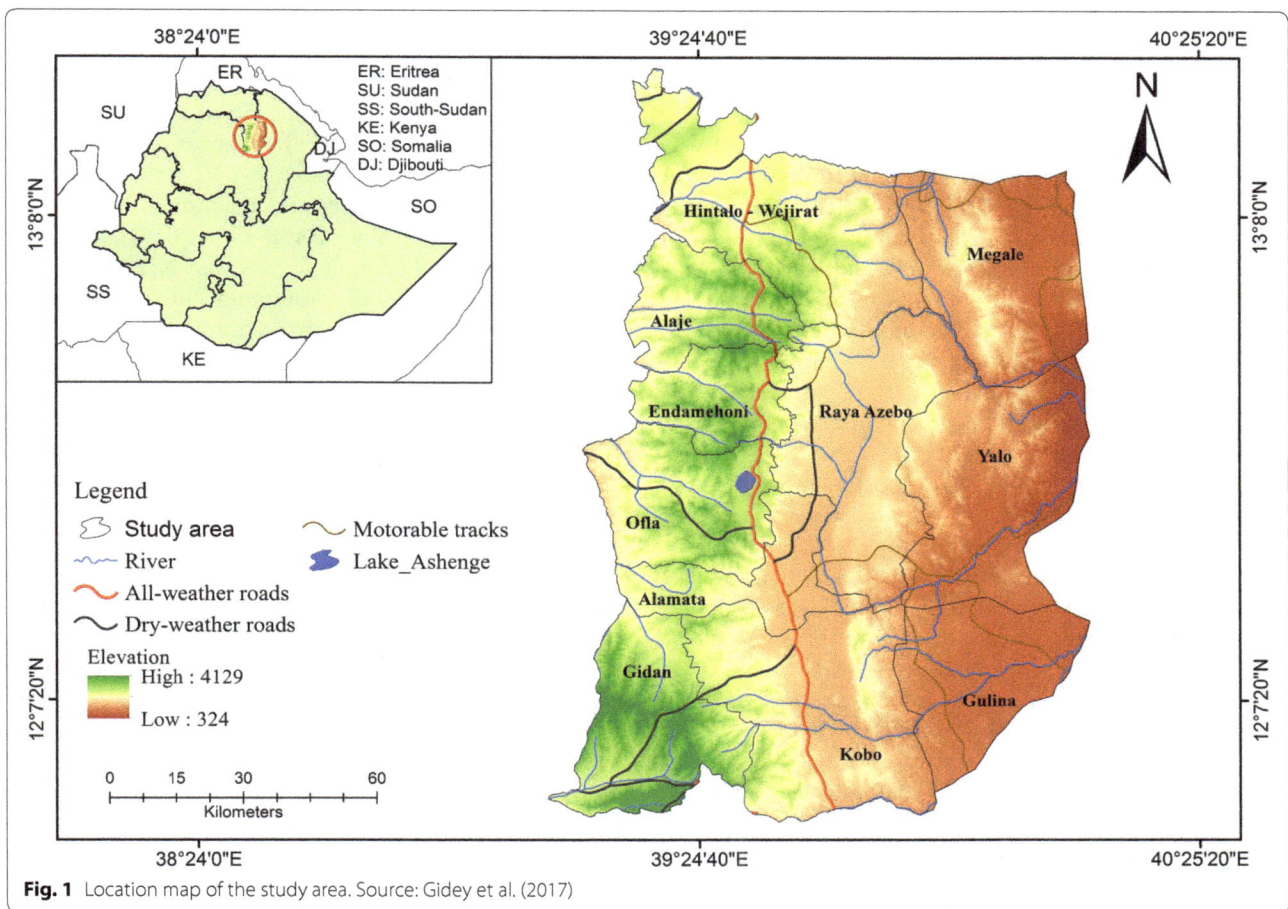

Fig. 1 Location map of the study area. Source: Gidey et al. (2017)

area is 1762 meters above sea level (m.a.s.l) (Gidey et al. 2017). Similarly, the slope of the study ranged from 0% (flat) to 395.3% (very steep slope). The soils of the study area i.e., eutric cambisols are the predominant soil type in the area covering about 4667.1 km^2 or 32.1%, while dystric gleysols cover only small portions of the site, i.e., nearly 1.1 km^2 or 0.001%, respectively (Gidey et al. 2017). The prominent land cover type is deciduous woodland which covers nearly 6097.6 km^2 (42.0%), while others e.g., Croplands cover 3362.2 km^2 (23.1%), open grassland with sparse shrubs 1517.4 km^2 (10.4%), deciduous shrub land with sparse tree 1298.1 km^2 (8.9%), sparse grassland 789.9 km^2 (5.4%), croplands with open woody vegetation 503.5 km^2 (3.5%), Bare soil 409.6 km^2 (2.8%), open grassland 202.3 km^2 (1.4%), closed grassland 197.0 km^2 (1.4%), mosaic forest/savanna 129.2 km^2 (0.9%), montane evergreen forest 14.5 km^2 (0.1%) and water bodies 11.1 km^2 (0.1%), respectively. According to the Raya Valley Livelihood Zone report (2007), the dominant crop types in the study area are sorghum, teff, and maize. Of all crops, sorghum and maize are widely used as a staple food by the community, while teff (Eragrostis tef) is largely produced

for both cash and food crops to improve their livelihoods. In the study area, the smallholder farmers prepare their lands during the months of May and June because July–September are main rainy season.

Data acquisition

Expedited MODIS (eMODIS)-TERRA NDVI

Tsiros et al. (2004) reported that the earth observation data could effectively be used to monitor drought onset, cessation and the vegetation's response to drought. In this study, the agricultural drought condition of the study area was investigated using the real-time and historical EROS Moderate Resolution Imaging Spectroradiometer Earth observation products. A multi-temporal smoothed monthly Terra expedited Moderate Resolution Imaging Spectoradiometer Normalized Difference Vegetation Index (eMODIS-NDVI) data from the period of 2001 to 2015 at 250 m spatial resolution were acquired from the Famine Early Warning Systems Network (FEWS NET) East-Africa region. The Terra eMODIS-NDVI data are better for agricultural drought monitoring than Aqua. The main reason is that the Aqua eMODIS data are more prone to noise than the Terra data, likely due to differences in the internal cloud mask used in MOD/MYD09Q1 or compositing rules (Brown et al. 2015).

Land Surface Temperature (LST)

In this study, the MOD11A2 LST and Emissivity Terra 8-day temporal resolution (later aggregated into monthly bases) data were obtained from the National Aeronautics and Space Administration (NASA)—United States Geological Survey (USGS) Land Process Distributed Active Archive Center (LP DAAC). The ultimate reasons to use the daytime (Terra) LST data were its temporal evolution. Frey et al. (2012) reported that the temporal evolution of LST acquired during the daytime is better to get in-depth information than the Aqua (night-time) because a significant change in LST change can be observed during the nighttime. However, in the nighttime (Aqua), LST largely remains stable; as a result, the restriction on time differences could be relaxed. The MODIS LST introduces a higher quality of LST data than AVHRR sensor due to its temporal and spatial variations and up-to-date algorithm such as time of acquisition, satellite view zenith and azimuth angle, quality flags for easy interpretation of the products (Frey et al. 2012). A total of 169 MOD11A2 LST (morning overpass or Terra) data product collection of 005 used to assess the LST condition of the study area from the period of 2001–2015. The daytime or Terra temperature of vegetation canopy is an essential characteristic (Kogan and Guo 2016). This data was used as an input to compute the TCI and VHI, which is an advanced and integrated agricultural drought-monitoring model.

Precipitation

Precipitation data are an extremely useful meteorological parameter in drought studies. In this study, the long-term monthly precipitation data were collected from the National Meteorological Agency of Ethiopia for the period 2001–2015. The data were mainly used to investigate the response of agricultural drought to rainfall.

Data processing and analysis
Expedited MODIS (eMODIS)-TERRA NDVI

eMODIS is a process for creating a community-specific suite of vegetation monitoring products based on the National Aeronautics and Space Administration's (NASA) Earth Observing System (EOS) Moderate Resolution Imaging Spectroradiometer (MODIS) and produced in the U.S. Geological Survey's (USGS) Earth Resources Observation and Science (EROS) Center (Jenkerson and Schmidt 2008). Jenkerson et al. (2010) reported that the eMODIS NDVI data are well suited for vegetation studies because the data were acquired with a frequent and repeated cycle. Besides, the same author

stated that the spatial resolutions of the data were better than the Advanced Very High Resolution Spectroradiometer (AVHRR) and SPOT-Vegetation products. Rhee et al. (2010) reported that the Normalized Difference Vegetation Index (NDVI) has been most widely used for drought monitoring. However, NDVI data alone cannot fully show the severity and magnitude of droughts (Kogan et al. 2013; Kogan and Guo 2016). Therefore, the multi-temporal analysis of eMODIS NDVI data supported by VCI and TCI can significantly improve the drought monitoring and early warning systems. Barbosa et al. (2006) reported that the satellite derived NDVI can be computed based on the red, which has low reflectance value and NIR high reflectance, portions of the wavelength. Predominantly, in non-drought periods, green and vigorous vegetation reflects little light in the visible (VIS) spectrum due to high light absorption by chlorophyll and much reflection in the near-infrared (NIR) part due to the specificity of light scattering by leaf internal tissues and water content (Kogan and Guo 2016). In this case, the healthy vegetation is strongly absorbed the visible incident solar (red) and it reflects less solar radiation in the visible spectrum. However, the unhealthy vegetation strongly reflects the near-infrared light. Hence, healthy and dense vegetation has the highest NDVI value typically > 0.5 than the unhealthy. Furthermore, the main reason to use eMODIS NDVI data in this study was that the eMODIS Terra data are corrected from molecular scattering, ozone absorption, and aerosols. Likewise, the eMODIS NDVI is good to measure the density of chlorophyll contained in vegetative cover (Swets 1999). Kogan (1995) revealed that NDVI data helps to assess the VCI development reflects both temperatures and precipitation conditions. The NDVI was mathematically computed as follows (Eq. 1):

$$NDVI = (NIR - RED)/(NIR + RED) \qquad (1)$$

where NIR = near-infrared reflectance and RED = visible-red reflectance.

In this study, the row eMODIS data were processed, rescaled and analyzed in ArcGIS 10.4.1 package to find out the real NDVI value of the study area as follows (Eq. 2):

$$eMODIS\ NDVI = Float\ (Smoothed\ eMODIS\ NDVI - 100)\ /\ 100 \qquad (2)$$

The value of eMODIS NDVI ranges from − 1.0 to + 1.0. The standard unit of eMODIS NDVI is NDVI ratio. The negative NDVI ratio shows less vigorous or unhealthy vegetation cover mainly occurred in a barren rock (rock outcrop), and sand, while the positive NDVI value depicts the healthy vegetation cover. NDVI values are much higher in healthy and dense vegetation than rocks, water, and bare soil (Kogan 1995). Similarly, sparse vegetation cover such as grasslands, bushes/shrubs may result in moderate NDVI values range from 0.2 to 0.5. High NDVI values (0.6–0.9) correspond with dense vegetation in the temperate and tropical forests or crops at their peak growth stage. The NDVI is thus a very good parameter for studying vegetation greenness, and mapping vegetation health or cover dynamics status in each satellite image pixel. In this study, the eMODIS NDVI data were used as input to compute the VCI only.

Vegetation Condition Index (VCI)

Several drought indices have been developed for assessing the drought characteristics such as intensity, duration, severity and spatial extent (Mishra, and Singh 2011) (e.g., VCI). The VCI which is derived from remote-sensing data has been used naturally allied with vegetation state and cover (Karnieli et al. 2010). The index is highly applicable for assessing the vegetation stress and/or to examine the response of vegetation. VCI quantifies the weather component (Singh et al. 2003) and portray precipitation dynamics as compared to the NDVI (Kogan 1990). This index helps to show the cumulative environmental impact on vegetation (Singh et al. 2003). The VCI permits not only the description of vegetation but also an estimation of spatial and temporal vegetation changes and weather impacts on vegetation (Kogan 1990). In this study, the smoothed monthly eMODIS NDVI data were used as input to compute the VCI model. Kogan (1995) pointed out that VCI has an excellent capability to identify drought and measure its time of onset, intensity, duration, and impact on vegetation. In this study, the VCI model was applied to examine the agricultural drought status of the study area as follows (Eq. 3):

$$VCI = 100 \times (NDVI_i - NDVI_{min})/(NDVI_{max} - NDVI_{min}) \quad (3)$$

where $NDVI_i$ = the current smoothed NDVI value of ith month, $NDVI_{min}$, and $NDVI_{max}$, is a multi-year (2001–2015) absolute minimum and maximum NDVI value for every pixel at a particular period.

Vegetation Condition Index values show how much the vegetation has advanced or deteriorated in response to weather. According to Kogan (1995), the value of VCI is measured in percentile ranged from 0 to 100. A high value of VCI signifies healthy and/or unstressed vegetation condition. It is thus the area is free of the agricultural drought incidence. The VCI value of 50–100% shows above normal or wet condition. This means that there is no drought, while values between 35 and 50 percent show area under the incidence of moderate drought (MD) and VCI value between 20 and 35 percent shows severe drought (SD) prevalence. Furthermore, the seasonal and/or annual VCI value 0–20% is showing very

severe agricultural drought event (VSD). However, in some cases, VCI model based on NDVI alone is not sufficient for agricultural drought monitoring (Kogan 1995; Sholihah et al. 2016). Hence, the combination of both VCI and TCI derived from MOD11A2 LST Terra data are significant to assess agricultural droughts. This study, therefore, applied VHI to analyze the long-term agricultural drought onset, cessation, duration, frequency, severity and its spatial extents.

Temperature Condition Index (TCI)

Land Surface Temperature (LST)   Land Surface Temperature (LST) described as the radiative skin temperature of the land derived from solar radiation.[1] This data used as an indicator of the energy balance at the Earth's surface and the so-called greenhouse effect in climate change studies (Frey et al. 2012). The MOD11A2 Terra v.005 LST and emissivity measures the ground temperature of the earth's surface. This helps to assess the overall vegetation health, soil moisture status and impact of thermal (Parviz 2016; Karnieli et al. 2010). In this study, the MOD11A2 Terra 8 days LST data initially acquired at a 1 km spatial resolution archived in Hierarchical Data Format–Earth Observing System (HDF–EOS). However, the MODIS Re-projection Tool (MRT) v 4.1 developed on March 2011 was applied to resample the 1 km MOD11A2 LST data in 250-m spatial resolution together with the eMODIS NDVI data. The MRT also used to convert the Hierarchical Data Format (HDF) into a GeoTIFF image format to carry out better analysis and interpretations on the MOD11A2 LST and eMODIS NDVI. In addition, the MRT tool was used to reproject the data from its Sinusoidal Projection type into Universal Transverse Mercator (UTM) projection zone 37 as the dominant part of Ethiopia relies on this projection type. The values of the MOD11A2 Terra LST data were computed by averaging all the valid pixels under clear-sky. The valid LST value ranges from 7500 to 65,535 (Wan 2006) and it was rescaled by 0.02 to get the correct LST value in Kelvin unit. Hence, the values of LST will be from 150 to 1310.7 Kelvin. In this study, the LST data were rescaled and converted into °C (degree Celsius) unit as follows (Eq. 4):

$$LST = (\varpi \times 0.02) - 273.15 \quad (4)$$

where LST = Land Surface Temperature in Degree Celsius (°C), ϖ = Row Scientific data (SDS).

The TCI is a thermal stress indicator used to determine temperature related drought situations. This satellite-derived index assumes that during the drought event soil

---

[1] http://lst.nilu.no/language/en–US/Home.aspx.

moisture diminished significantly and cause high vegetation stress. Kogan (1995) noted that computation of the TCI model is more likely similar to the VCI. However, the model has considerably improved to assess the response of vegetation to temperature. The TCI assumed that higher temperature has a tendency to cause deterioration or drought during the vegetative growth period, while low temperatures are largely favorable to vegetation during its development. Hence, low TCI values correspond with vegetation stress due to dryness or harsh weather by high-temperature condition (Karnieli et al. 2006; Bhuiyan 2004). The TCI was estimated using the following mathematical expression (Eq. 5):

$$TCI = 100 \times (LST_{max} - LST_i)/(LST_{max} - LST_{min}) \quad (5)$$

where LST_i = LST value of ith-month, LST_{max} and LST_{min} are the smoothed multi-year maximum and minimum LST.

Vegetation Health Index (VHI)

Rhee et al. (2010) reported that the recently developed drought indices (e.g., NMDI, NDWI, and NDDI) did not perform significantly better than NDVI with 1 km resolution in the arid region. Studies showed NDVI only is not capable to depict drought or non drought conditions. The VHI model has been found to be a robust agricultural drought-monitoring index and it has good efficiency to explore the spatial extent of agricultural severity drought. In the arid region, VHI was quite highly correlated with *in-situ* variables (Rhee et al. 2010). Karnieli et al. (2006) stated that the VHI was applied only in arid, semi-arid and sub humid climatic regions where water is the main limiting factor for vegetation growth. VHI is dependent on the weather and ecological conditions of the region (Singh et al. 2003). Seiler et al. (1998) reveal that the VHI combination of TCI and VCI is essential to characterize the spatial extent, the magnitude, and severity of agricultural droughts in a good agreement with precipitation patterns. Likewise, they are paramount significant to examine the effect of weather on vegetation and to exemplify the condition of crop development. Furthermore, both the VCI and TCI indices have used for estimation of vegetation health and drought monitoring (Singh et al. 2003; Jain et al. 2009). Hence, the vegetation stress due to dry and wetness condition was assessed to investigate the severity of agricultural droughts in the study area. Tsiros et al. (2004) and Parviz (2016) reported that the combination of both VCI and TCI the so-called VHI has shown satisfactory results in several parts of the globe when it is used for drought detection, assessment of weather impact and/or evaluation of vegetation condition. The VHI show the availability of moisture and temperature

or thermal condition in vegetation (Kogan 2001). This drought index has better performance for agricultural drought monitoring (Parviz 2016). Marufah et al. (2017) reported that VHI used to understand the duration, spatial distribution, and severity or category of agricultural drought. Studies showed that low VCI and TCI values or warm weather largely signifies stressed vegetation conditions and the prevalence of agricultural droughts. In this study, both the VCI and TCI components given an equal weight due to the reason that moisture and temperature contribution during the vegetative growth period not yet known (Kogan 2001). Similarly, Karnieli et al. (2006) reported that due to a lack of more accurate information on the influence of VCI and TCI on the VHI in Mongolia, the coefficient of the VHI equation was fixed at 0.5.

The VHI was mathematically computed as follows (Fig. 2) (Eq. 6):

$$VHI = a \times VCI + (1 - a) \times TCI \quad (6)$$

where VHI = Vegetation Health Index, a = 0.5 (contribution of VCI and TCI), VCI = Vegetation Condition Index, TCI = Temperature Condition Index.

Drought warning issued if the VHI values decrease below 40 (Kogan et al. 2013). The lower VHI indicated that the high incidence of drought whereas a higher VHI value show that wet or non-drought conditions (Table 1).

This study analyzed the onset, cessation, duration, and recurrence interval of agricultural drought. Studies showed that agricultural drought is striking when the VHI value is below 40 and ends if the values exceed 40 (Table 1). The agricultural drought duration of this study was also analyzed by the number of consecutive drought periods, i.e., the time-period between the onset and the end of the drought.

Coefficient of variation (CV) analysis

The coefficient of variation (CV) analyses was conducted to examine the seasonal VHI variability relative to the mean percent from the periods of 2001–2015. The coefficient of variation statistically computed as follows (Eq. 7):

$$CV(\%) = 100 \times \frac{\sigma}{\bar{x}} \quad (7)$$

where $CV(\%)$ = Coefficient of variation of VHI in percentage, σ = Standard deviation of VHI, $\bar{x}$ = long-term mean of VHI.

Regression analysis between VHI and rainfall

In this study, a regression analysis was carried out between agricultural drought as derived from VHI and rainfall only because there is no long-term record of crop yield data in the study area. Wilhite and Glantz (1985)

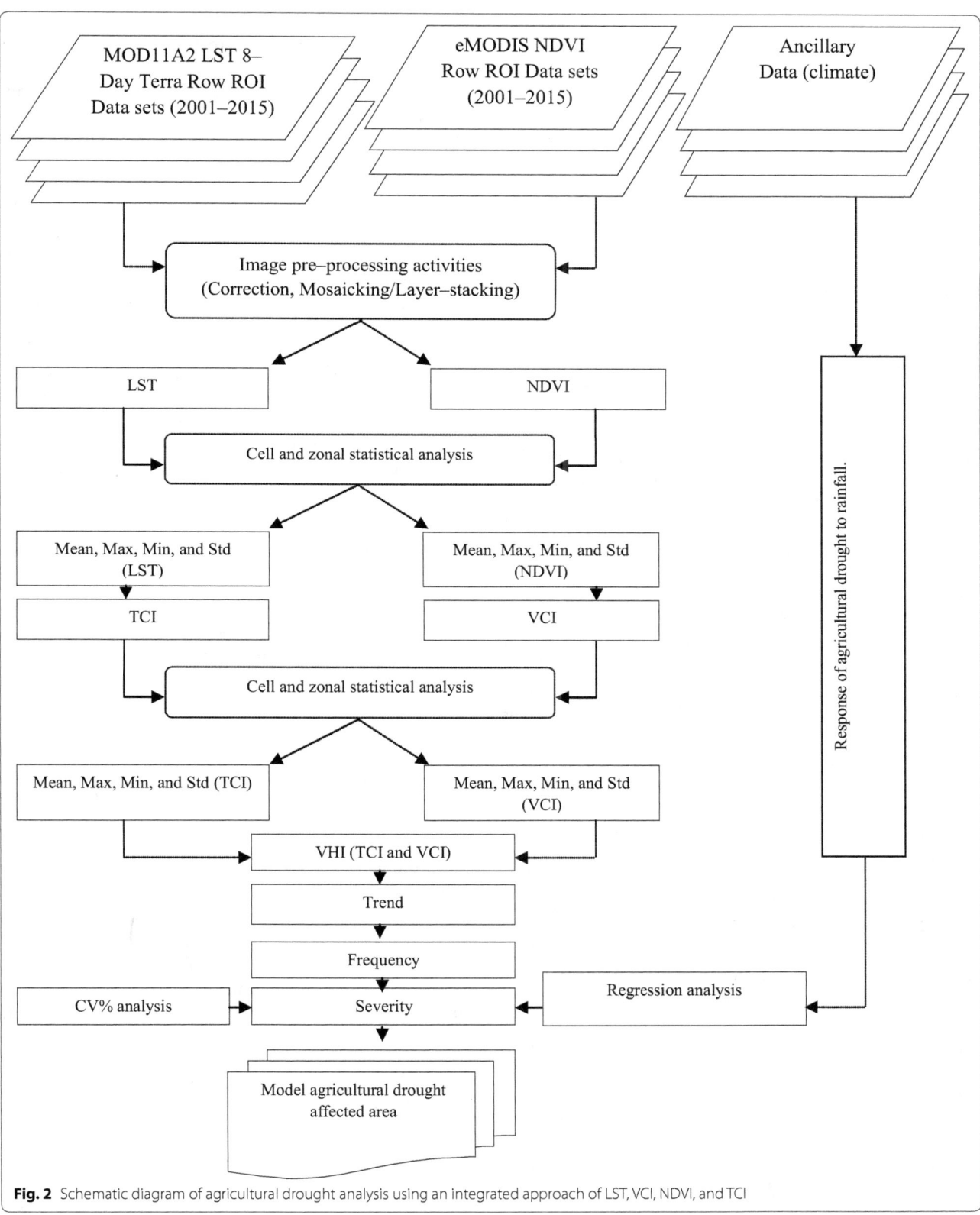

Fig. 2 Schematic diagram of agricultural drought analysis using an integrated approach of LST, VCI, NDVI, and TCI

Table 1 Agricultural drought severity by VHI (Source: Kogan 2001)

Level of severity	VHI values
Extreme drought	< 10
Severe drought	10–20
Moderate drought	200
Mild drought	30–40
No drought	> 40

reported that drought can occur in both high as well as low rainfall area. Therefore, it is useful to evaluate how the agricultural drought responded to rainfall because there is high rainfall variability in Raya and its environs. The regression analysis was conducted as follows (Eq. 8):

$$Y_i = \beta_0 + \beta_1 X_i + \varepsilon_i \qquad (8)$$

where Y_i = VHI for the ith period, X_i = seasonal rainfall, $\beta_0 + \beta_1 X_i$ = linear relationships between the independent and dependent variables, β_0 = Mean of Y_i when X_i = 0 (intercept), β_1 = Change in the mean of Y_i when X_i increases by 1 (slope), ε_i = Random error term.

Results and discussion
Long-term agricultural drought analysis
Figure 3 shows the multi-temporal trend of LST-NDVI, VCI-TCI, and VHI—rainfall for the period 2001 to 2015. The lowland area presented in Fig. 3a1–c1 reveals that the mean NDVI value was between 0.23 and 0.27 and this sparse NDVI value is extremely low when it is evaluated by scientifically accepted thresholds, while the LST was high and it ranges between 39.6 and 41.29 °C. Therefore, low NDVI values are mostly reached at high LST values because the vegetation is under high water stress. In the midland area shown in Figs. 3d1–f1 and 4a1 relatively better NDVI value ranged between 0.44 and 0.57 was observed, while the LST was between 30.3 and 34.97 °C. In this area, the LST value was relatively lower than the lowland area stated in Fig. 3a1–c1, but it is still an unfavorable condition for the vegetation high moisture stress. In the highlands area, good NDVI coverage ranges between 0.53 and 0.57 was observed. Besides, low LST value ranges between 22.85 and 24.6 °C was observed in the same area. High LST during the vegetation growing period may cause vegetation stress. Hence, the increase in surface temperature may significantly influence vegetation development (Karnieli et al. 2006). Singh et al. (2003) reported that NDVI becomes an important tool for vegetation cover and/or growth analysis. Generally, this study observed that NDVI coverage during the main rainy season decreased by 3–4% in all districts of the

study area. However, the LST shows a significant increase by 0.52–1.08 °C across all agro-ecologies as well as districts in the last 15 years (Fig. 5). The increase in LST and the decrease in NDVI contribute considerable moisture stress that can trigger the incidences of agricultural drought. Furthermore, Figs. 3a2–f2 and 4a2–e2 show the trend of VCI and TCI. The results showed that the stress of vegetation was due to rising surface temperature. In the lowland area, the values of VCI were between 37.18 and 44.48, while TCI was largely between 38.54 and 39.58. In the midland area, the values of VCI were between 53.77 and 62.65, while TCI was 52.57–64.4. In the highland area, the VCI value ranged between 63.94 and 67.87, while TCI was 66.63–68.88.

Furthermore, Figs. 3a3–f3 and 4a3–e3 indicated that VHI and rainfall value was significantly diminished during the main rainy season. This revealed that the incidence of agricultural drought became more frequent and severe because it is more sensitive to soil moisture, particularly the lowland and some parts of mid and highland area was seriously affected. For instance, the VHI value of the lowland area was between 38.38 and 40.55, while rainfall was about 274.42–379.87. In the midland area, better VHI values were observed ranged from 53.17–62.82. Moreover, in the highland area, the VHI value ranged between 66.47 and 70.65 was observed. Bhuiyan (2008) reported that during 1985 and 1986 monsoon season, VHI showed severe to extreme droughts in the western and some northern parts of Thar Desert, India. In the same region, mild to moderate droughts severity were also observed in the rest of the country. Moreover, the validity of the VHI as a drought detection index relies on the assumption that NDVI and LST at a given pixel will vary inversely over time, with variations in VCI and TCI driven by local moisture conditions (Karnieli et al. 2010).

Agricultural drought onset, cessation, duration, and recurrence interval analysis
Table 2 shows the seasonal agricultural drought onset, cessation, duration, and recurrence interval. The results reveal that agricultural drought occurred in a different time-period, duration and recurrence interval. It strikes all districts once in every 1.36–7.5 years during the main rainy season. Serious drought conditions during the crop growing season eventually affect crop yield (Rhee et al. 2010). For example, the districts of Yalo and Gulina were hit by the agricultural drought that started in 2004 and ends in 2009. This incidence was affecting the livelihood of the community for about 6 years and it was recorded as the highest drought period during the last 15 years (Table 2). Similarly, another drought event which covers the larger portion of the area was started in 2011 and ends in 2015. The duration of this drought event was

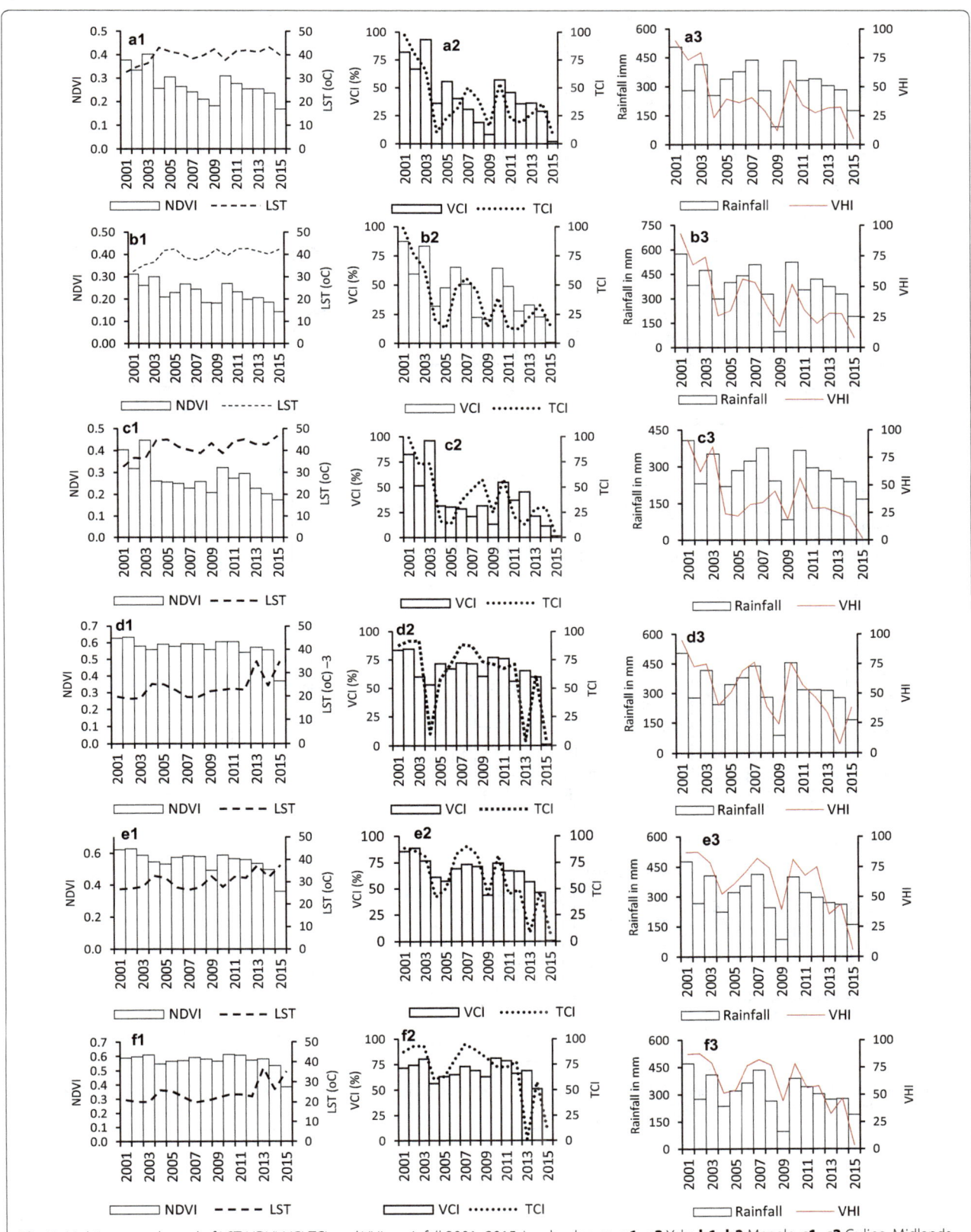

Fig. 3 Multi-temporal trend of LST-NDVI, VCI-TCI, and VHI—rainfall 2001–2015. Lowlands area: **a1**–**a3** Yalo, **b1**–**b3** Megale, **c1**–**c3** Gulina, Midlands area: **d1**–**d3** Raya Azebo, **e1**–**e3** Alamata, **f1**–**f3** Kobo

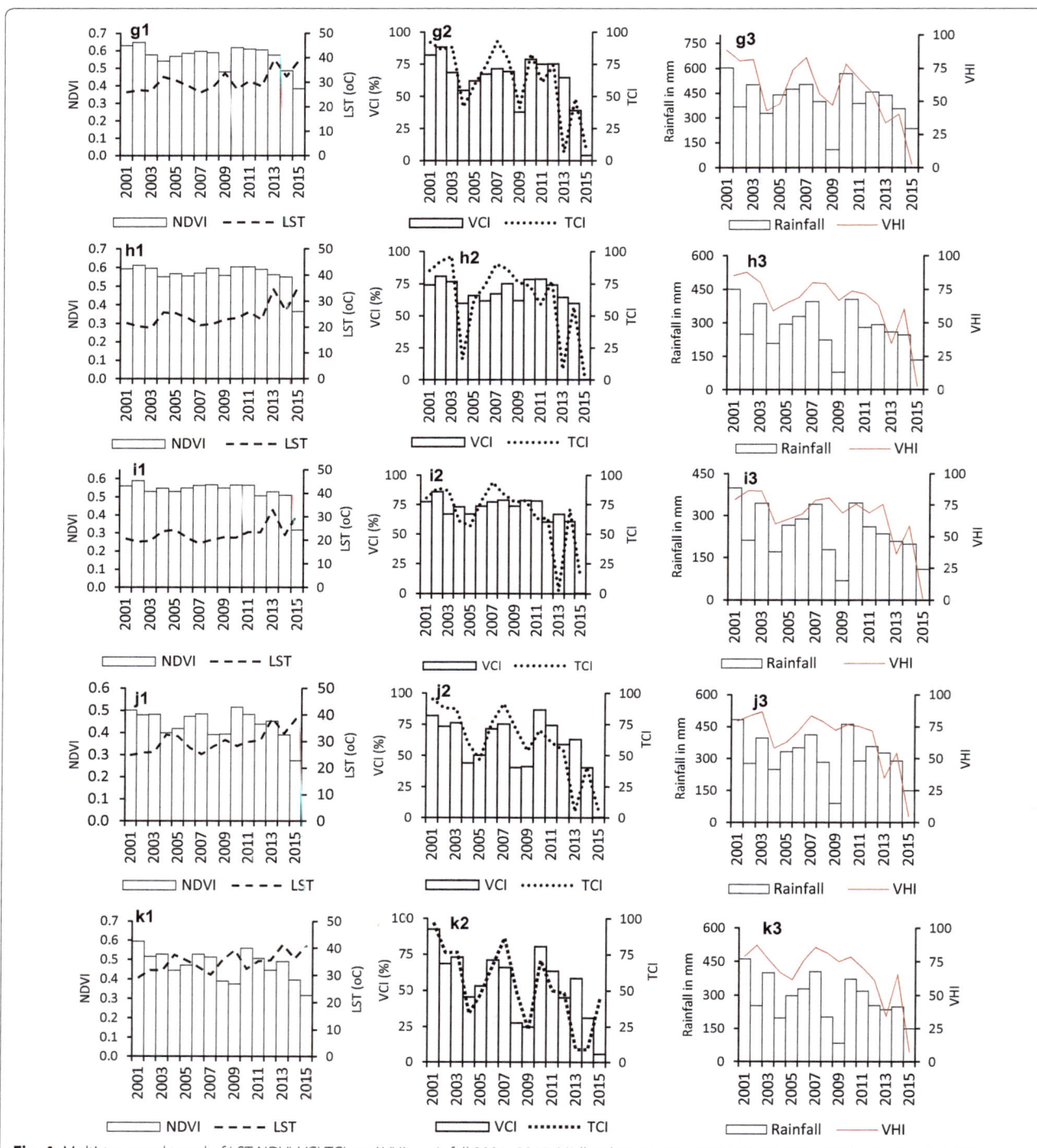

Fig. 4 Multi-temporal trend of LST-NDVI, VCI-TCI, and VHI—rainfall 2001–2015. Midlands area: **g1**–**g3** HintaloWejirat, Highlands area: **h1**–**h3** Endamehoni, **i1**–**i3** Ofla, **j1**–**j3** Alaje, **k1**–**k3** Gidan

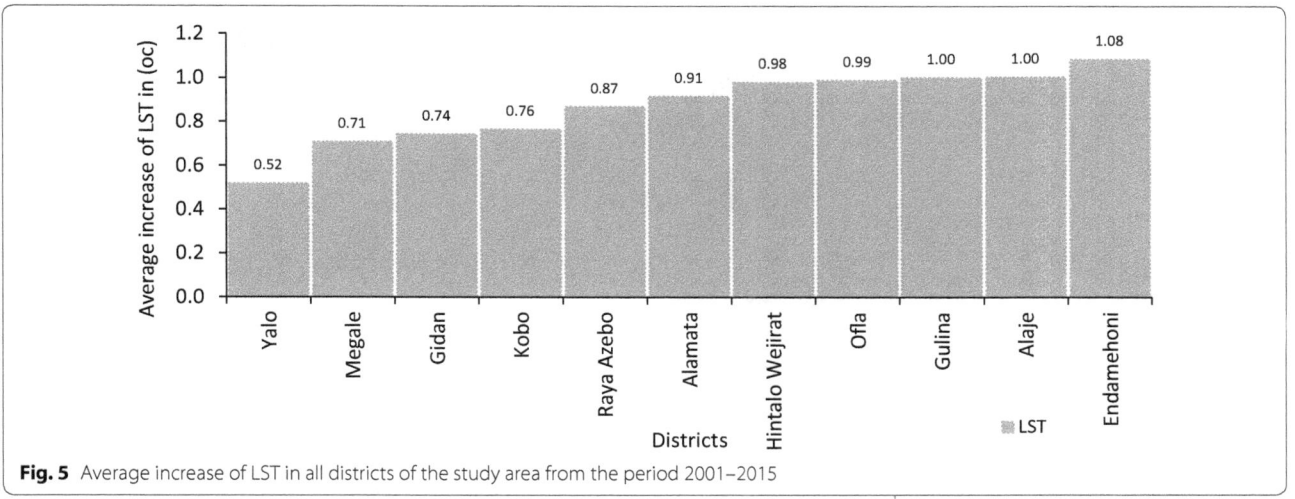

Fig. 5 Average increase of LST in all districts of the study area from the period 2001–2015

Table 2 Analysis of agricultural drought onset (O), cessation (C), duration (L), and recurrence interval (I) in the study area. Source: Gidey et al. (2017)

District	O	C	L (year)	I
Yalo and Gulina	2004	2009	6	1.36
	2011	2015	5	
Megale	2004	2006	3	1.5
	2008	2009	2	
	2011	2015	5	
Raya Azebo	2004	2004	1	2.5
	2008	2009	1	
	2013	2015	3	
Alamata	2009	2009	1	3
	2013	2015	3	
Hintalo Wejirat	2013	2015	3	5
Kobo, Endamehoni, Ofla, Alaje, Gidan	2013	2013	1	7.5
	2015	2015	1	

5 years from 2011 to 2015. During these periods, significant effects on the livestock and humans were observed because the livelihoods of the communities are largely relying on the rearing of animals. Furthermore, the recurrence of agricultural drought in these two districts was once in every 1.36 years. Therefore, drought is a regular event in the area. Likewise, the rest of the study area was extensively affected by the agricultural drought. However, the impacts both on the livestock and on humans were diminishing due to the support of the federal and local governments and other non-government or humanitarian organizations. For example, the drought-affected communities were supported and are still getting food aid (Cereals and Other) at monthly basis as per the FAO survival threshold. In the study area, the government is supporting about 2131 kilo calorie (kcal) per person per day and also supply pasture and drinking water in the highly drought affected areas.

Agro-ecological based frequency of agricultural drought incidence

In this analysis, the VHI was considered as a basic parameter to declare the regularity of drought. Besides, the analyses were done based on the thresholds stated in Table 1. Kogan and Guo (2016) reported that the Horn of Africa (including the study area) was affected by droughts yearly. This study found that there are no districts that were free from the incidence of agricultural drought in the last 15 years. The highest agricultural drought incidence, which covers about 4409.7 km^2, was observed in the lowland area. The frequency of agricultural drought event in these districts were 10–11 times in the last 15 years (Fig. 6). This means that drought is a regular event in the lowland area. The result is largely similar to what Kogan and Guo (2016) reported, but the return period is less in the highlands area (Fig. 6) and some parts of the midlands area. In the midland area (Raya Azebo, Alamata, Hintalo Wejirat, Kobo) the incidence is relatively lower and the area has been under the spell of drought for about 2–6 times covering about 6385 km^2. However, in the highlands area, agricultural drought was occurred for about two times covering 3738 km^2 in the last 15 years during the main rainy season. The return period of agricultural drought in this area is different due

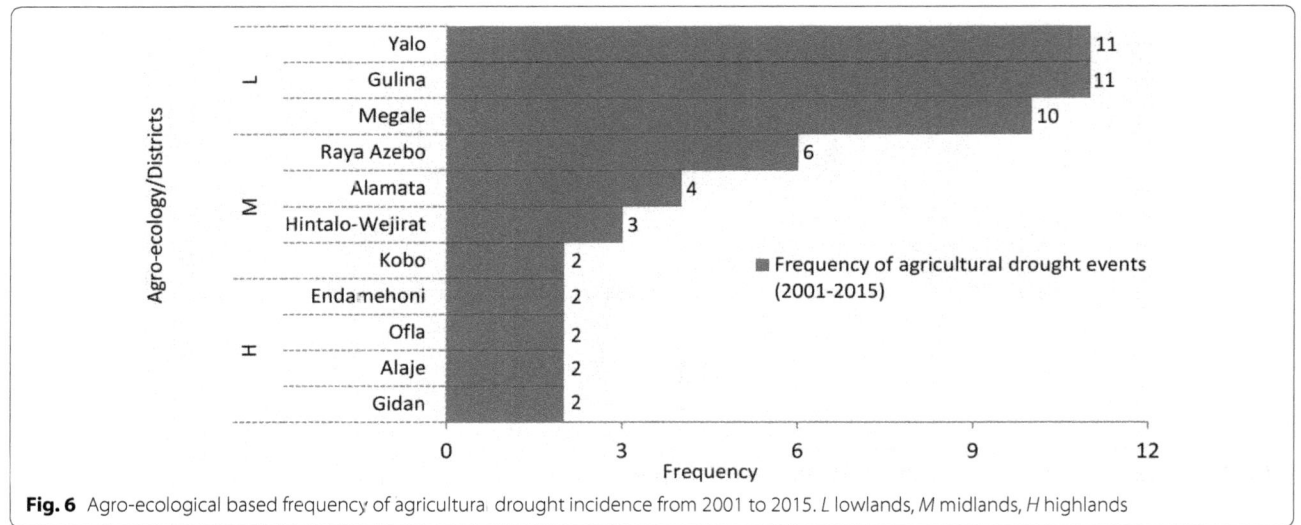

Fig. 6 Agro-ecological based frequency of agricultural drought incidence from 2001 to 2015. *L* lowlands, *M* midlands, *H* highlands

to the various levels of moisture stress, rainfall deficit, and Land Surface Temperature conditions.

Analysis of the spatio-temporal agricultural drought

Figure 7 shows that the study area was experiencing agricultural drought during the period 2001–2015. The year 2015 observed extreme drought period across the study area where the mean VHI value was less than 10 (Fig. 7). In this period, a catastrophic shortage of livestock forage, drinking water, and food occurred. Lei et al. (2016) suggested that exploring adaptation strategies to the expected increase in droughts incidence has become a critical issue of poverty reduction and agricultural sustainability. The impacts of drought can be reduced through involving the smallholder farmers and agro-pastoralists in a wide range of on- and off-farm practices.

Coefficient of variation (CV) analysis

Studies revealed that the coefficient of variation determined by the absolute dispersion of data relative to the mean and mainly expressed as a percentage. Analyzing the coefficient of variation is, therefore, useful to determine the statistical reliability and/or precision of estimation. The highest coefficient of variation depicting the greater level of dispersion, while the lowest value of the coefficient of variation corresponds to good precision. This study, therefore, found very high precision of estimation in all districts (Fig. 8). The overall coefficient of variation ranges from 6 to 20.7%. Hence, a higher (20.7%) degree of coefficient variation has reported in the districts of Hintalo Wejirat, and lower in Ofla (8.6%). One of

the possible reasons could be due to erratic rainfall distribution which increasing the seasonal rainfall variability among each district. This indicated that the coefficient of variation estimation was highly reliable as the maximum acceptable thresholds are below 29.9%.

Agricultural drought (VHI) response to the seasonal rainfall

This study found that the majority of the study area received below average seasonal rainfall, which can directly cause agricultural drought. The shortage of rainfall is thus the most important climatic constraint to the occurrence of agricultural drought. Figure 9 shows that how the agricultural drought (VHI) responded to the seasonal rainfall. Dutta et al. (2015) observed that a good agreement between the values of VCI and meteorological indices [e.g., Rainfall Anomaly Index (RAI)] and Yield Anomaly Index in India. Wan et al. (2004) found a linear correlation between Vegetation Temperature Condition Index (VTCI), and monthly precipitation in the southern Great Plains, USA. However, in this study, the relatively strong relationship between VHI and rainfall ($R^2 = 0.651$, $R^2 = 0.602$) at $p < 0.01$ significance level in the districts of Megale (Fig. 9b), and Hintalo Wejirat (Fig. 9g) observed. Similarly, in the lowland area presented in Fig. 9 Yalo (b) and Gulina (c), an $R^2 = 0.526$ and 0.463 was also observed. Likewise, in midlands area shown in Fig. 9d–f an R^2 of 0.596, $R^2 = 0.544$, and $R^2 = 0.516$ were observed in the districts of Raya Azebo, Alamata, and Kobo. Furthermore, in the highland area depicted under Fig. 9h–k an $R^2 = 0.411$, an $R^2 = 0.383$, $R^2 = 0.398$, and $R^2 = 0.357$ was observed. However, in these area, the slightly poor

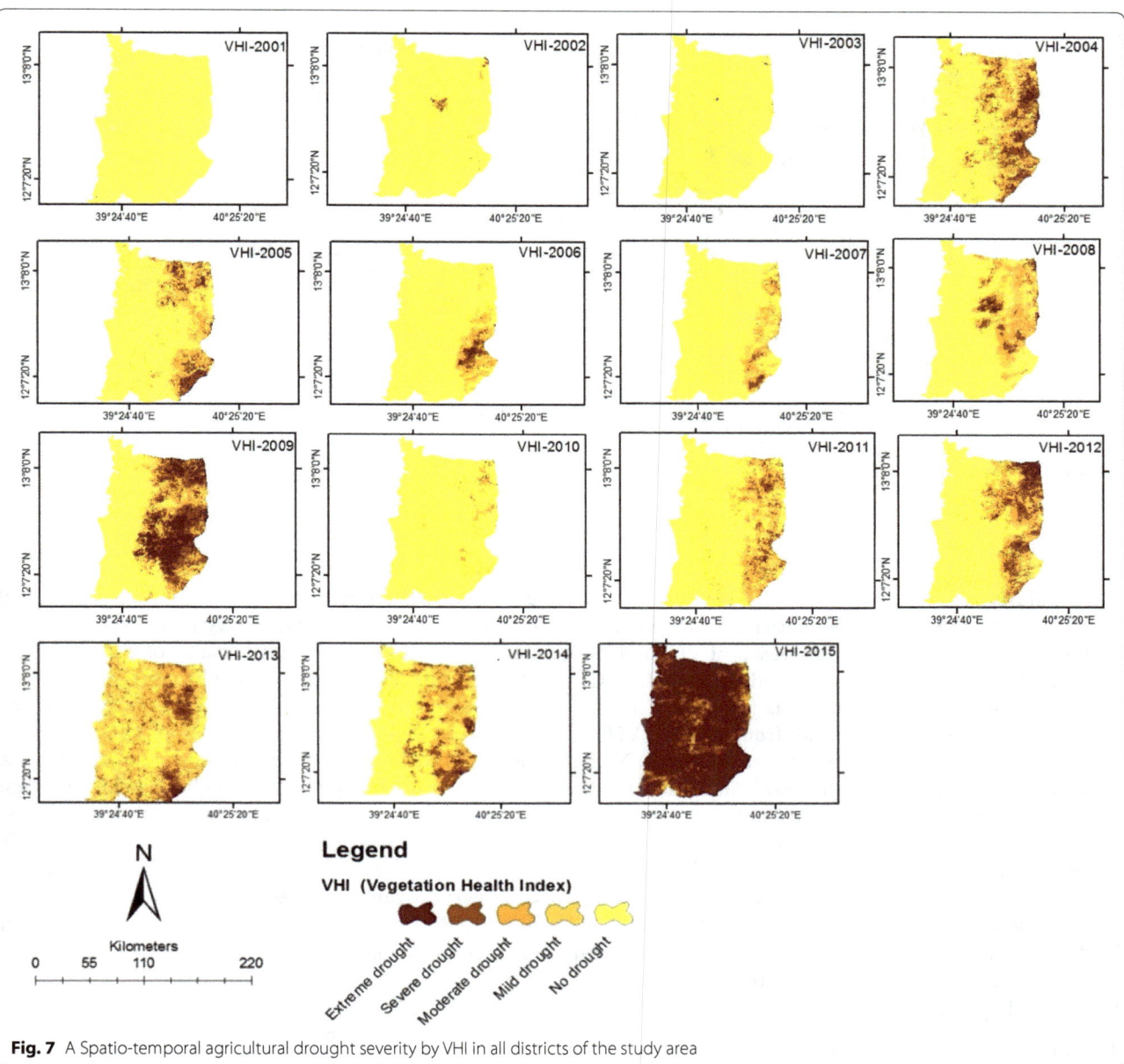

Fig. 7 A Spatio-temporal agricultural drought severity by VHI in all districts of the study area

regression result was associated with several factors such as topography. The relationship between VHI and rainfall is statistically significant at ($p < 0.01$ and $p < 0.05$) across all districts of the study area. Moreover, the regression analysis results of this study indicated that agricultural drought (VHI) positively responded to rainfall. This reveals that when rainfall increases, VHI also tends to increase. As a result, agricultural drought incidences significantly diminished. This study also demonstrated that the incident of agricultural drought was due to shortage of rainfall leading to high level of moisture stress.

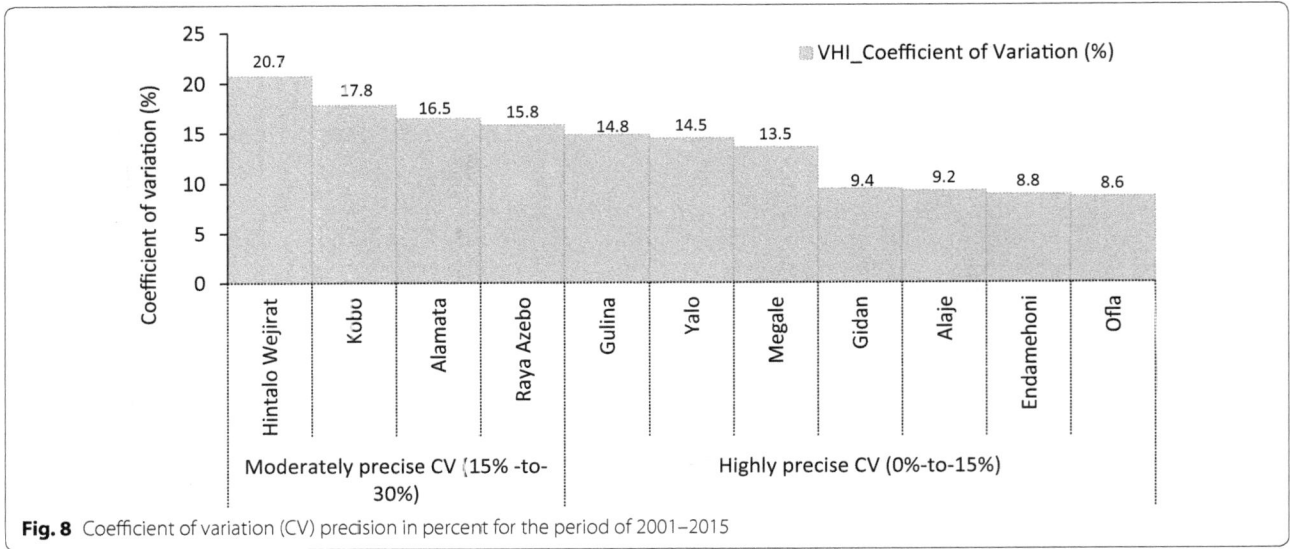

Fig. 8 Coefficient of variation (CV) precision in percent for the period of 2001–2015

Conclusions

Remote sensing and GIS-based agricultural drought can be better monitored by VHI composed of VCI and TCI drought indices. This study analyzed the onset, cessation, duration, recurrence interval, frequency, severity and spatial extent of agricultural drought using VHI at 3-month time-scale during the main rainy season. NDVI value was extremely low in the lowland area than the mid and highlands area. NDVI coverage during the main rainy season decreased by 3–4% in all districts of the study area. However, LST showed a significant increase by 0.52–1.08 °C across all agro-ecologies as well as districts in the last 15 years. LST was high both in the lowland and midlands area and it is an unfavorable condition for the vegetation because it causes stress, while the lowest LST is largely a favorable condition.

The increase in LST and the decrease in NDVI may contribute consid-erable moisture stress that can trigger the incidences of agricultural drought. Furthermore, the VHI and rainfall value diminished significantly during the main rainy season. This revealed that the incidence of agricultural drought became more frequent and severe, particularly in lowland and some parts of the mid and highlands area. There were no districts that were free from the incidence

of agricultural drought during the study periods. A high frequency of agricultural drought incidence (10–11 times) was observed in the lowland of the study area consisting of Yalo, Megale, and Gulina districts. The incidence is relatively lower (2–6 times) in the midland area (Raya Azebo, Alamata, Hintalo Wejirat, Kobo). Furthermore, the study noted that the frequency of drought was very low in the highlands (Endamehoni, Ofla, Alaje, and Gidan) of the area. Both the lowland and midlands area were more exposed to the agricultural drought than the highland area. VHI model showed that the year 2015 was extremely drought period across the study area where the mean VHI value was less than ten. The overall coefficient of variation ranged from 6 to 20%.

A higher (20.7%) coefficient variation was observed in Hintalo Wejirat, and lower in Ofla (8.6%). The relationship between rain-fall and VHI is positive ($R^2 = 0.357$ to $R^2 = 0.651$) and statistically significant at ($p < 0.01$ and $p < 0.05$) across all districts of the study area. This relationship reveals that when rainfall increases, VHI also tends to increase. As a result, agricultural drought incidences significantly reduced. This study suggests that the effect of drought could be reduced through involving the smallholder farmers in a wide range of on- and off-farm practices.

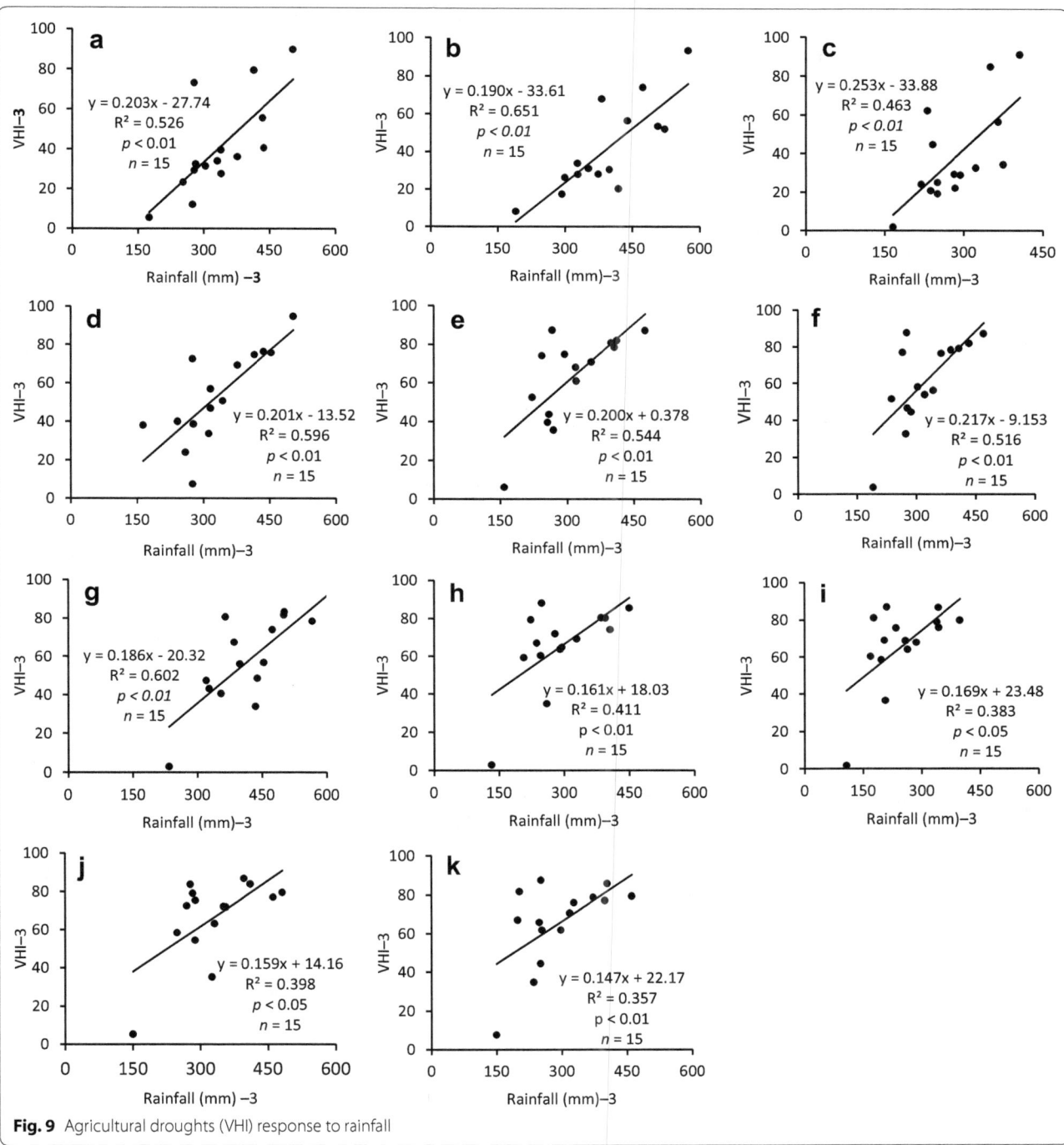

Fig. 9 Agricultural droughts (VHI) response to rainfall

The study may also support formulation and implementation of drought coping and mitigation programs in the study area.

Abbreviations

CV: Coefficient of Variation; LST: Land Surface Temperature; NDVI: Normalized Difference Vegetation Index; TCI: Temperature Condition Index; VCI: Vegetation Condition Index; VHI: Vegetation Health Index.

Authors' contributions
EG initiate the research idea, review relevant literature, design scientific methods, field data collection, data cleaning, data analysis and interpretation, prepare draft manuscripts for publication. OD, RS, ES, and AZ evaluate the research idea, supervise the overall research activities, and enrich the manuscript. All authors have contributed their well-grounded knowledge to the project. All authors read and approved the final manuscript.

Author details
[1] Department of Environmental Science, University of Botswana, Private Bag UB 0704, Gaborone, Botswana. [2] Land Resource Management and Environmental Protection, Mekelle University, P.O. Box 231, Mekelle, Ethiopia. [3] Institute of Climate and Society, Mekelle University, P.O. Box 231, Mekelle, Ethiopia.

Acknowledgements
The authors thank financial support of Mekelle University and Open Society Foundation-Africa Climate Change Adaptation Initiative (CSF-ACCAI) project of Mekelle University-Ethiopia. The lead author is grateful for the Ph.D. scholarship given by the Transdisciplinary Training for Resource Efficiency and Climate Change Adaptation in Africa (TreccAfrica II) project. The authors would also like to thank the National Aeronautics and Space Administration (NASA)—United States Geological Survey (USGS) and Famine Early Warning System Network (FEWS-NET) for the provision of satellite data. We are grateful for the constructive feedback of the two anonymous reviewers and the editor.

Competing interests
The authors declare that they have no competing interests.

Funding
This research was financially supported by Mekelle University under Grant Number CRPO/ICS/PhD/001/09 and the Open Society Foundation—Africa Climate Change Adaptation Initiative (OSF-ACCAI) project of Mekelle University.

References
Alemaw BF, Simalenga T (2015) Climate change impacts and adaptation in rainfed farming systems: a modeling framework for scaling-out climate smart agriculture in Sub-Saharan Africa. Am J Clim Change 4(04):313

Ayenew T, GebreEgziabher M, Kebede S, Mamo S (2013) Integrated assessment of hydrogeology and water quality for groundwater-based irrigation development in the Raya Valley, northern Ethiopia. Water Int 38(4):480–492

Barbosa HA, Huete AR, Baethgen WE (2006) A 20-year study of NDVI variability over the Northeast Region of Brazil. J Arid Environ 67(2):288–307

Bhuiyan C (2004) Various drought indices for monitoring drought condition in Aravalli terrain of India. In: Proceedings of the XXth ISPRS Congress, Istanbul, Turkey, pp 12–23

Bhuiyan C (2008) Desert vegetation during droughts: response and sensitivity. Int Arch Photogramm Remote Sens Spat Inf Sci 21:907–912

Bhuiyan C, Singh RP, Kogan FN (2006) Monitoring drought dynamics in the Aravalli region (India) using different indices based on ground and remote sensing data. Int J Appl Earth Obs Geoinf 8(4):289–302

Brown JF, Howard D, Wylie B, Frieze A, Ji L, Gacke C (2015) Application-ready expedited MODIS data for operational land surface monitoring of vegetation condition. Remote Sens 7(12):16226–16240

Choi M, Jacobs JM, Anderson MC, Bosch DD (2013) Evaluation of drought indices via remotely sensed data with hydrological variables. J Hydrol 476:265–273

Dutta D, Kundu A, Patel NR, Saha SK, Siddiqui AR (2015) Assessment of agricultural drought in Rajasthan (India) using remote sensing derived Vegetation Condition Index (VCI) and Standardized Precipitation Index (SPI). Egyptian J Remote Sens Space Sci 18(1):53–63

Frey CM, Kuenzer C, Dech S (2012) Quantitative comparison of the operational NOAA-AVHRR LST product of DLR and the MODIS LST product V005. Int J Remote Sens 33(22):7165–7183

Gidey E, Dikinya O, Sebego R, Segosebe E, Zenebe A (2017) Modeling the Spatio-temporal dynamics and evolution of land use and land cover (1984–2015) using remote sensing and GIS in Raya, Northern Ethiopia. Model Earth Syst Environ 3(4):1285–1301

Gu Y, Brown JF, Verdin JP, Wardlow B (2007) A five-year analysis of MODIS NDVI and NDWI for grassland drought assessment over the central Great Plains of the United States. Geophys Res Lett. https://doi.org/10.1029/2006G L029127

Jain SK, Keshri R, Goswami A, Sarkar A, Chaudhry A (2009) Identification of drought-vulnerable area using NOAA AVHRR data. Int J Remote Sens 30(10):2653–2668

Jenkerson CB, Schmidt GL (2008) eMODIS product access for large scale monitoring. In: Proceedings of the 17th William T. Pecora Memorial Symposium on Remote Sensing, Denver, CO

Jenkerson C, Maiersperger T, Schmidt G (2010) eMODIS: a user-friendly data source (No. 2010-1055). US Geological Survey

Karnieli A, Bayasgalan M, Bayarjargal Y, Agam N, Khudulmur S, Tucker CJ (2006) Comments on the use of the vegetation health index over Mongolia. Int J Remote Sens 27(10):2017–2024

Karnieli A, Agam N, Pinker RT, Anderson M, Imhoff ML, Gutman GG, Goldberg A (2010) Use of NDVI and land surface temperature for drought assessment: merits and limitations. J Clim 23(3):618–633

Kogan FN (1990) Remote sensing of weather impacts on vegetation in non-homogeneousarea. Int J Remote Sens 11(8):1405–1419

Kogan FN (1995) Application of vegetation index and brightness temperature for drought detection. Adv Space Res 15(11):91–100

Kogan FN (2000) Contribution of remote sensing to drought early warning. In: Wilhite DA, Sivakumar MVK, Wood DA (eds) Early warning systems for drought preparedness and drought management. Proceedings of an Expert Group Meeting held 5–7 September, 2000, in Lisbon, Portugal, vol 57. World Meteorological Organization, Geneva, pp 86–100

Kogan FN (2001) Operational space technology for global vegetation assessment. Bull Am Meteor Soc 82(9):1949

Kogan F, Guo W (2016) Early twenty-first-century droughts during the warmest climate. Geomatics Nat Hazards Risk 7(1):127–137

Kogan F, Adamenko T, Guo W (2013) Global and regional drought dynamics in the climate warming era. Remote Sens Lett 4(3):36472

Kogan F, Guo W, Strashnaia A, Kleshenko A, Chub O, Virchenko O (2016) Modelling and prediction of crop losses from NOAA polar-orbiting operational satellites. Geomatics Nat Hazards Risk 7(3):886–900

Lei Y, Liu C, Zhang L, Luo S (2016) How smallholder farmers adapt to agricultural drought in a changing climate: a case study in southern China. Land Use Policy 55:300–308

Liu WT, Kogan FN (1996) Monitoring regional drought using the vegetation condition index. Int J Remote Sens 17(14):2761–2782

Marufah U, Hidayat R, Prasasti I (2017) Analysis of relationship between meteorological and agricultural drought using standardized precipitation index and vegetation health index. In: IOP Conference Series: Earth and Environmental Science, vol 54, No. 1. IOP Publishing, Bristol. p 012008

Maybank J, Bonsai B, Jones K, Lawford R, O'brien EG, Ripley EA, Wheaton E (1995) Drought as a natural disaster. Atmos Ocean 33(2):195–222

Mishra AK, Singh VP (2011) Drought modeling—a review. J Hydrol 403(1):157–175

Owrangi MA, Adamowski J, Rahnemaei M, Mohammadzadeh A, Sharifan RA (2011) Drought monitoring methodology based on AVHRR images and SPOT vegetation maps. J Water Resour Protect 3(05):325

Parviz L (2016) Determination of effective indices in the drought monitoring through analysis of satellite images. Agric Forest Poljoprivreda i Sumarstvo 62(1):305–324

Peters AJ, Walter-Shea EA, Ji L, Vina A, Hayes M, Svoboda MD (2002) Drought monitoring with NDVI-based standardized vegetation index. Photogram Eng Remote Sens 68(1):71–75

Raya Valley Livelihood Zone (2007) Livelihood profile Tigray Region, Ethiopia. pp 1–5. http://www.dppc.gov.et/Livelihoods/Tigray/Pages/RVL_LZ.htm. Accessed 19 May 2017

Rhee J, Im J, Carbone GJ (2010) Monitoring agricultural drought for arid and humid regions using multi-sensor remote sensing data. Remote Sens Environ 114(12):2875–2887

Se AST, Dorosh P, Gemessa SA (2011) Crop production in Ethiopia: regional patterns and trends. Food Agric Ethiopia Prog Policy Chall 74:53

Segosebe EM (1990) Some reflections on the concept of critical herdsize and its relevance to drought-prone Botswana. Botswana Notes Records 22:105–112

Seiler RA, Kogan F, Sullivan J (1998) AVHRR-based vegetation and temperature condition indices for drought detection in Argentina. Adv Space Res 21(3):481–484

Sholihah RI, Trisasongko BH, Shiddiq D, La Ode SI, Kusdaryanto S, Panuju DR (2016) Identification of agricultural drought extent based on vegetation health indices of landsat data: case of Subang and Karawang, Indonesia. Proc Environ Sci 33:14–20

Singh RP, Roy S, Kogan F (2003) Vegetation and temperature condition indices from NOAA AVHRR data for drought monitoring over India. Int J Remote Sens 24(22):4393–4402

Song X, Saito G, Kodama M, Sawada H (2004) Early detection system of drought in East Asia using NDVI from NOAA/AVHRR data. Int J Remote Sens 25(16):3105111

Sruthi S, Aslam MM (2015) Agricultural drought analysis using the NDVI and landsurface temperature data; a case study of Raichur district. Aquat Proc 4:1258–1264

Swets DL (1999) A weighted least-squares approach to temporal smoothing of NDVI. In: Proceedings of the 1999 ASPRS annual conference, from image to information. Portland. American society for photogrammetry and remote sensing, Bethesda, 17–21 May 1999

Tsiros E, Domenikiotis C, Spiliotopoulos M, Dalezios NR (2004) Use of NOAA/AVHRR-based vegetation condition index (VCI) and temperature condition index (TCI) for drought monitoring in Thessaly, Greece. In: EWRA Symposium on water resources management: risks and challenges for the 21st century, Izmir, Turkey, pp 2–4

Umran Komuscu A (1999) Using the SPI to analyze spatial and temporal patterns of drought in Turkey. In: Drought Network News (1994–2001), vol 11, issue 1. Drought – National Drought Mitigation Center, DigitalCommons@University of Nebraska - Lincoln, University of Nebraska - Lincoln, pp 7–13

Van Hoolst R, Eerens H, Haesen D, Royer A, Bydekerke L, Rojas O, Racionzer P (2016) FAO's AVHRR–based Agricultural Stress Index System (ASIS) for global drought monitoring. Int J Remote Sens 37(2):418–439

Vicente-Serrano SM, Beguería S, Gimeno L, Eklundh L, Giuliani G, Weston D, Konte D (2012) Challenges for drought mitigation in Africa: the potential use of geospatial data and drought information systems. Appl Geogr 34:471–486

Wan Z (2006) MODIS land surface temperature products users'guide. Santa-Barbara, Institute forComputational Earth System Science, University of California

Wan Z, Wang P, Li X (2004) Using MODIS land surface temperature and normalized difference vegetation index products for monitoring drought in the southern Great Plains, USA. Int J Remote Sens 25(1):61–72

Wilhite DA, Glantz MH (1985) Understanding: the drought phenomenon: the role of definitions. Water Int 10(3):111–120

Zhang J, Mu Q, Huang J (2016) Assessing the remotely sensed Drought Severity Index for agricultural drought monitoring and impact analysis in North China. Ecol Ind 63:29609

Zhang L, Jiao W, Zhang H, Huang C, Tong Q (2017) Studying drought phenomena in the Continental United States in 2011 and 2012 using various drought indices. Remote Sens Environ 190:96–106

5

Deposition means storage and not loss

Rainer Schenk[1,2*] (iD)

Abstract

Background: By the authors Janicke (IBJparticle, Eine Implementierung des Ausbreitungsmodells. Bericht IBB Janicke, 2000; AUSTAL2000, Programmbeschreibung, Dunum, 2002), *Germany*, for the calculation of the dispersion of air pollutants under the designation AUSTAL2000 a "Model-based assessment system for the plant-related pollution control" is developed. This propagation model is declared binding in the Federal Republic of Germany for application. All other model developers have to validate their algorithms on the provided reference solutions. However, for example, Schenk (AUSTAL2000 ist nicht validiert. Immissionsschutz 01.15, pp 10–21, 2015a; Replik auf den Beitrag "Erwiderung der Kritik von Schenk an AUSTAL2000 in Immissionsschutz 01/2015". Immissionsschutz 04.15, pp 189–191, 2015b; Environ Ecol Res 5(1):45–58, 2017), *Germany*, demonstrates that these reference solutions are flawed. Main sentence and mass conservation theorem are violated. In various publications and other statements by Trukenmüller et al. (Erwiderung der Kritik von Schenk an AUSTAL2000 in Immissionsschutz 01/2015. Immissionsschutz 03/2015, S. 114–126, 2015); Trukenmüller (Äquivalenz der Referenzlösungen von Schenk und Janicke. Abhandlung Umweltbundesamt Dessau-Rosslau, IBS Archiv, S. 1, 2016; Stellungnahmen Umweltbundesamt vom 10.02.2017 und 23.03.2017. Dessau-Rosslau, IBS Archiv, S. 1–15, 2017), *Umweltbundesamt Deutschland*, the objections raised by Schenk (AUSTAL2000 ist nicht validiert. Immissionsschutz 01.15, pp 10–21, 2015a; Replik auf den Beitrag "Erwiderung der Kritik von Schenk an AUSTAL2000 in Immissionsschutz 01/2015". Immissionsschutz 04.15, pp 189–191, 2015b; Environ Ecol Res 5(1):45–58, 2017), *Germany*, are denied and it is vehemently disputed.

Results: In this paper, the identified contradictions are analyzed in depth and the results of all studies are summarized. The correct reference solution is specified. The authors of the AUSTAL invoke an allegedly universal agreement $F_c = c_0 \cdot v_d$, which was approved by Venkatram and Pleim (Atmos Environ 33:3075–3076, 1999), *USA;* should be justified. However, it turns out to be embarrassing that they can not be read there. Moreover, it is shown that in Venkatram and Pleim (Atmos Environ 33:3075–3076, 1999), *USA;* described relationship between deposition and sedimentation is not applicable. A correction is required. There is also evidence that because of a lack of physical basics soil concentrations must be calculated in a tricky way. Because of the undisputed importance of the agreement $F_c = c_0 \cdot v_d$, this relationship is referred to here as the *Janicke Convention*.

Conclusions: In summary, the author of this paper concludes that the propagation model AUSTAL is not validated and is not suitable for the calculation of pollutant propagation processes. Deposition, sedimentation and immission are calculated incorrectly. One understands by deposition loss and not storage. Statements of importance for health and safety, such as safety analyzes and hazard prevention plans are to be checked with physically justified model developments.

Keywords: AUSTAL2000, Dispersion calculations, Particle model, Sedimentation, Deposition, Air pollutants

*Correspondence: ibswettin@web.de
[1] Rosenberg 17, 06193 Wettin-Löbejün, Germany
Full list of author information is available at the end of the article

Background

In Axenfeld et al. (1984) describes a model for the calculation of dust precipitation, which was later developed into the LASAT propagation model and is explained extensively in Janicke (2001). This propagation model LASAT becomes Trukenmüller et al. (2015) promises to be a "mother model". It serves the authors of Janicke and Janicke (2002) for the development of a "model-based assessment system for plant-related pollution control" called AUSTAL2000. In this context, reference solutions are published for validation, which are already declared mandatory in VDI (2000) according to LASAT for all further model developments. The guidelines state that all further model developments in the Federal Republic of Germany must be based on this "mother model". However, Schenk (2015a) has shown that these reference solutions are all flawed and unsuitable for comparative calculations. The second law of thermodynamics and the law of mass conservation are violated. In Trukenmüller et al. (2015) is contradicted, and one refers to misunderstandings and lack of physics. At the same time, the derivation of the reference solutions will be published for the first time in the almost 30-year history of AUSTAL. In Schenk (2015b) the objections are rejected and it is proved by means of integral sentences that the violation of the main and conservation laws is universal and not limited to those in Janicke and Janicke (2002). In Trukenmüller (2016)[1] all criticism is ignored. With different defined deposition rates one wants to bring about equivalence of the reference solutions of the AUSTAL and according to Schenk (2015b), however one ignores the fact that the deposition rate is a material constant, which can not be defined at will. Finally, in Trukenmüller (2017) an even different definition of the deposition rate is given. This is now to be explained by the relation $K \cdot \partial c / \partial z + c \cdot v_s = c_0 \cdot v_d$, referring to a convention $F_c = c_0 \cdot v_d$ allegedly contained in Venkatram and Pleim (1999). It is true, however, that there can not be read about such a convention. However, this alleged convention has already been described in Axenfeld et al. (1984) used and explained there with incomprehensible ideas about the occurrence of deposition. The deposition current is to be formed there at a rate at which "... a column rising up on the earth's surface, which contains the material capable of being deposited, runs empty by deposition". It does not explain, however, where this material should be afterwards. Deposition obviously means loss and not storage. The study of the work of Venkatram and Pleim (1999), on the other hand, affirms the fact that the reference equations of the AUSTAL are trivial solutions because of the boundary condition chosen there $c_0 = 0$ and thus $F_c = 0$. The conclusiveness of this result is stubbornly denied in Trukenmüller (2017). Because of the recurrent importance of the relationship $F_c = v_d \cdot c_0$ in the development of the AUSTAL, it is henceforth referred to as the *Janicke Convention*. The relationship between deposition and sedimentation rates is described in Venkatram and Pleim (1999) not convincingly described.

In conclusion, in this work, finally, all opinions expressed by Trukenmüller (2017) are refuted. It is proven that the erroneous reference solutions of the AUSTAL not only violate the relevant integral sentences. Mass conservation is also differentially violated because of *Janicke's convention*. For two case studies, which are described in Janicke (2001), the inconsistency of the reference solutions of the AUSTAL is demonstrated using the example of calculated deposition and sedimentation flows. It also shows that soil concentrations have to be calculated speculatively because of the appearance of an indeterminate expression. The associated thought model describes strange rather a segregation than the spread of air pollutants. In addition, the correct reference solutions are derived and the concentration distributions thus calculated are compared with the faulty graphics of the AUSTAL. It is further shown that the method described in Venkatram and Pleim (1999), the relationship between deposition and sedimentation velocities is erroneous. This result is of considerable interest, since not only the authors of the AUSTAL use these incorrect connections. Also in Simpson et al. (2012) refers to these inadmissible contexts when developing a high-priority Western European transport model. In this work a correction is made. The use of the *Janicke Convention* justifies the triviality of all reference solutions. This proof is also kept. Finally, critical terms of the authors of the AUSTAL are analyzed. They show little physical understanding of deposition and sedimentation and explain all the differences with the theory of the spread of air pollutants.

Methods

Mathematics and mechanics are used alone as valid methods of incorruptible evidence to conduct subsequent investigations. It should not be forgotten that even in the field of air pollutants this sword alone is about credibility. Using differential and integral equations, analytical relationships are described and analyzed. As further methods fluidic and thermodynamic doctrines are used to the main and conservation theorems. In addition there are foundations of the theory of momentum, heat and mass transport as well as numerical mathematics.

The primary material is the publications by Axenfeld et al. (1984), Janicke (2000, 2001, 2002), Janicke and

---

[1] References Trukenmüller (2016, 2017) can be requested at ibswettin@web.de and alfred.trukenmueller@uba.de

Janicke (2002), Simpson et al. (2012), Trukenmüller et al. (2015) and Trukenmüller (2016, 2017). It describes the theoretical background of the AUSTAL, the algorithms used, the technical application and various case examples for validation.

Basic knowledge of fluid mechanics, numerical mathematics, modeling of the spread of air pollutants and the basics of coupled momentum, heat and mass transport can be found in the textbooks Albring (1961), Berljand (1982), Janenko (1969), Naue (1967), Schlichting (1964), and Truckenbrodt (1989) can be read as secondary knowledge.

Results
Calculation of deposition and sedimentation using the faulty *Janicke's convention*
Derivation of the faulty reference solution

In the more than 30-year history of the AUSTAL in Trukenmüller et al. (2015) described for the first time the analytical correlations for the derivation of the reference solution of the AUSTAL. Also in Janicke (2001), where the theoretical foundations of the "mother model" LASAT are explained, can not be read about.

The authors of the AUSTAL use the differential Eq. (1)

$$-v_s \cdot \frac{\partial c}{\partial z} = \frac{\partial}{\partial z}\left(K \cdot \frac{\partial c}{\partial z}\right) \tag{1}$$

out. This equation describes a one-dimensional and stationary mass transport. It is an ordinary differential equation of the second order. To solve this equation two physically justified boundary conditions are required. In this respect, $v_s(\mathrm{m/s})$ is the vertical downward sedimentation velocity, $c(\mu\mathrm{g/m^3})$ is the concentration in the control room, $K(\mathrm{m^2/s})$ is the diffusion coefficient, and $z(\mathrm{m})$ is the vertical coordinate. After a single integration you get the differential equation (2)

$$K\frac{\partial c}{\partial z} + v_S \cdot c = Fc, \tag{2}$$

where $Fc(\mu\mathrm{g/(m^2\,s)})$ is an integration constant. Instead of integrating the inhomogeneous first-order differential equation thus obtained by applying Lagrange's method, the authors of the Austal select a constant concentration distribution to determine Fc (3).

$$c(z) = const. = c_i \tag{3}$$

as a special solution. Thus, according to the relation (2) for the integration constants Fc, they obtain the expression (4).

$$F_c = v_S \cdot c_i, \tag{4}$$

The authors of the AUSTAL subsequently determine the homogeneous solution (5) for Eq. (2).

$$c_h = b \cdot \exp\left(-z \cdot \frac{v_s}{K}\right), \tag{5}$$

where the constant $b(\mu\mathrm{g/m^3})$ is to be determined from the first boundary condition of a constant soil concentration. Because of the linearity of Eq. (2), the superposition principle can be applied, after which the general integral is summed up as the sum of the special and the homogeneous solution (6).

$$c(z) = c_i + c_h(z) = \frac{F_c}{v_S} + b \cdot \exp\left(-z \cdot \frac{v_S}{K}\right) \tag{6}$$

receives. If one uses a constant soil concentration $c(0) = c_0$ as boundary condition, one can determine the integration constant b. Thus, the authors of the AUSTAL finally get Eq. (7)

$$c(z) = c_0 \cdot \exp\left(-z \cdot \frac{v_s}{K}\right) + \frac{F_c}{v_S} \cdot \left[1 - \exp\left(-z \cdot \frac{v_s}{K}\right)\right], \tag{7}$$

Equation (7) describes the solution of the differential equation (1). However, taking into account the already fulfilled boundary condition $c(0) = c_0$ as well as the chosen special solution (3) one already obtains the trivial expression (8)

$$c(z) = c(0) = const. = c_i = c_0, \tag{8}$$

Considering Eq. (4) and concluding with Eq. (8), this triviality is also confirmed by Eq. (7) with $F_c = v_S \cdot c_0$ (9)

$$c(z) = c_0 \cdot \exp\left(-z \cdot \frac{v_s}{K}\right) + c_0 \cdot \left[1 - \exp\left(-z \cdot \frac{v_s}{K}\right)\right] = c_0, \tag{9}$$

With this solution, no concentration distributions can be calculated and propagation models can be validated. Taking Eqs. (2) and (4) into consideration, the trivial solution, as expected, differentially satisfies the mass conservation rate (10)

$$0 + v_S \cdot \frac{F_c}{v_S} = Fc, \tag{10}$$

Although Eq. (1) is an ordinary differential equation of the second order, only one constraint is satisfied. The authors of the AUSTAL have to realize that with the trivial solution (7) or (9) no concentration distributions can be calculated. Due to the lack of deposition rate $v_d(\mathrm{m/s})$ as a free parameter, these equations do not allow the determination of deposition currents. The second integration constant Fc is not determined and still freely selectable.

Instead of determining these by means of a second physically justified boundary conditions, this is determined by the *Janicke convention*

$$F_c = c_0 \cdot v_d \tag{11}$$

replaced. A physical justification for the validity of this convention is not given. The authors of the AUSTAL thus obtain Eq. (12)

$$c(z) = c_0 \cdot \exp\left(-z \cdot \frac{v_S}{K}\right) + \frac{c_0 \cdot v_d}{v_S} \cdot \left[1 - \exp\left(-z \cdot \frac{v_S}{K}\right)\right], \tag{12}$$

with which finally all reference solutions are determined. However, the use of the *Janicke convention* according to Eq. (11) violates the differential mass conservation law. In analogy to Eq. (10) one obtains with the relation

$$0 + v_S \cdot \frac{F_c}{v_d} \neq Fc \tag{13}$$

contrary, also the confirmation.

In connection with the introduction of the *Janicke Convention*, it is called in Trukenmüller et al. (2015) to an alleged generality and refers to Venkatram and Pleim (1999). It is true, however, that one can not read about such a convention there. Later, Trukenmüller (2016) refers to two different definitions of the deposition rate and thus justifies the erroneous reference solution. The authors of the AUSTAL thus want to bring about an equivalence between their own non-correct reference solution (12) and the correct relationship (30) to be derived later. After that, all contradictions would have been cleared up. In addition, reference is again made to Venkatram and Pleim (1999) and again refers to the universality already questioned. In connection with further efforts to clarify all differences, a third definition of the deposition rate is given in Trukenmüller (2017). After that she would be through the relationship

$$K \cdot \frac{\partial c}{\partial z} + c \cdot v_S = c_0 \cdot v_d \tag{14}$$

established. The deposition rate alone would only be a proportionality factor that parametrizes the deposition and sedimentation flows. This would be the true definition. Again, it is ignored that the deposition rate alone can only be a material constant. Common to all three given definitions is the fact that Trukenmüller (2016, 2017) noticeably avoid describing a thought model that could physically justify the validity of the *Janicke convention*. One becomes entangled in contradictions and relies

on authority rather than mathematics and mechanics. Obviously, one has forgotten that already in Axenfeld et al. (1984) used the *Janicke Convention* and described a thought model for it. According to this model of thought, the deposition rate is the speed with which "-picturally speaking-… a pillar standing up on the earth's surface, containing the material capable of being deposited, runs empty by deposition". The inequality (13) conclusively confirms that this thought model is identical to the mathematical model used. It is understood by deposition loss and not storage. So it is understandable that the solution (12) violates the law of mass conservation and the II law of thermodynamics. Later, the Janicke models are adopted in VDI (1988), thus establishing a novel doctrine. However, as will be shown, deposition currents are calculated erroneously according to amount and direction. In Trukenmüller (2016, 2017), one persistently insists on the validity of the mathematical model. Mathematically and physically justified objections are ignored.

To derive the reference solutions of the AUSTAL, it should be noted in conclusion that the second boundary condition, which describes the penetration of the deposition-capable material into the soil, is not taken into account. It lacks the understanding that mass can not be lost. There is also a lack of insight that the model of thought used and the associated mathematical formulation for describing deposition and sedimentation are not permissible.

Calculation of sedimentation and deposition currents for the case examples 22a and 22b by means of the faulty reference solutions of the AUSTAL2000

In "Derivation of the faulty reference solution" section the erroneous reference solution of the AUSTAL was derived and discussed. In this section, for the case examples 22a and 22b described in Janicke (2000, 2002), the graphics obtained with the solution (12) are described. The results can be seen in Fig. 1. The deposition current is identical to the diffuse material flow at the bottom. In the case of the faulty reference solution according to Eq. (12), this results in the expression

$$\dot{m}_d = -K \cdot \frac{\partial c}{\partial z}(0) = v_S \cdot c_0 - F_c \neq -v_d \cdot c_0, \tag{15}$$

where $\dot{m}_d \left(\mu g/(m^2 s)\right)$ describes the deposition current. The negative sign takes into account that the conductive material flow is basically directed against the concentration gradient. A negative concentration gradient causes a positive material flow and vice versa. It can already be seen here that Eq. (15) explains the deposition current incorrectly, since it is formed from the product of deposition velocity and soil concentration $v_d \cdot c_0$

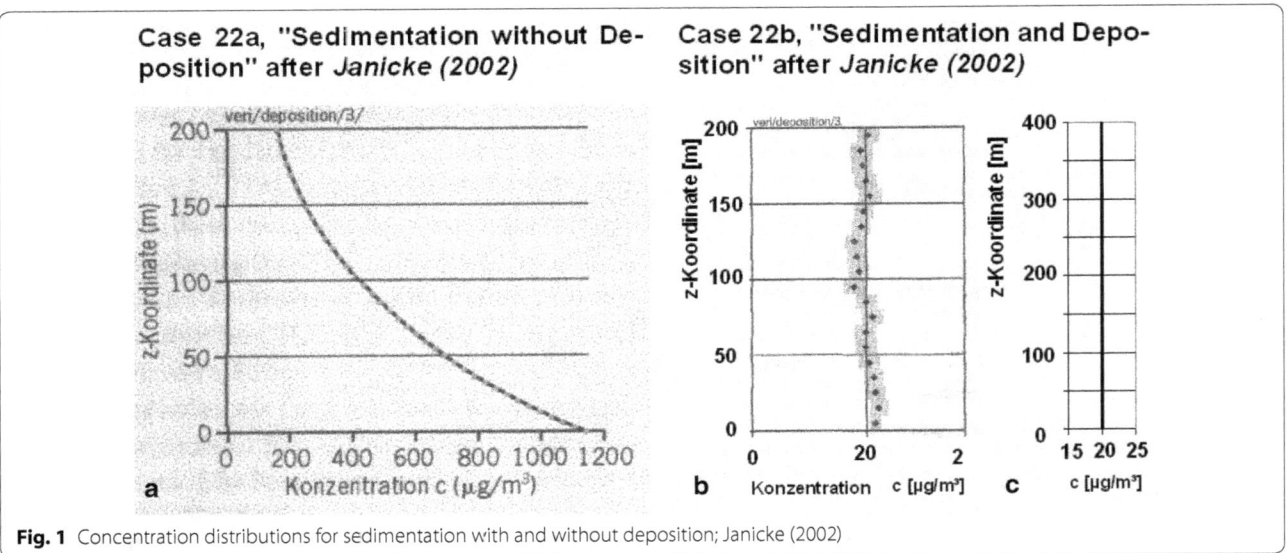

Fig. 1 Concentration distributions for sedimentation with and without deposition; Janicke (2002)

and not otherwise. The control room has a dimension of $L_x = 1000$ m, $L_y = 1000$ m and $L_z = 200$ m in both cases.

First, the case study 22a is of interest, in which, according to Janicke (2002), although sedimentation is allowed, but no deposition should take place. The results are described in Graph a of Fig. 1. With $L_z = 200$ m the sedimentation velocity is given. The diffusion coefficient is $K = 1$. The emissions are released by an evenly distributed volume source over the entire control area. Without further justification, the "specialization $F_c = 0$ "should continue to apply for this case, which implies according to Eq. (11) also $v_d = 0$, because one must assume a non-vanishing soil concentration c_0. If one compares the specification $F_c = 0$ with Eq. (2), this assumption implies that the propagation process should only be described by a homogeneous differential equation, for which only the homogeneous solution (5) with $b = c_0$ applies can be. Taking into account the definition of the deposition rate given in Trukenmüller (2017) according to Eq. (14), this assumption implies a vanishing conductive mass transport as well as a non-existent sedimentation stream. This leads to the conclusion that $F_c = 0$ can only be equated with a nonexistent emission. The other possibility that both transports cancel each other out also expresses a missing emission. This case is comparable to solving an eigenvalue problem according to Eq. (2) and is more theoretical than practical. In the case of practical applications, immissions can occur only in the presence of emissions, so that the barometric height distribution according to Eq. (5) can not be considered as a solution of the eigenvalue task (2) for propagation

calculations. Despite this inadmissibility one wants to make believe with the graphic a, one can compute concentration distributions with missing emissions. However, the authors of the AUSTAL have the difficulty that no calculation equation for the soil concentration is available. Because of the given propagation parameters with $F_c = 0$ and $v_d = 0$, the relation $F_c = c_0 \cdot v_d$ results in an indefinite expression $c_0 = 0/0$, which allows any numerical value. Without specifying a calculation rule, a soil concentration of $c_0 = 1100.6$ is given. One would have to know that a validation of propagation models without knowledge of the algorithm used is not possible. It is up to the reader to find out this calculation rule. It turns out that the soil concentration, contrary to all ideas for calculating the spread of air pollutants, is determined speculatively. But not only this contradiction is objectionable. According to the prerequisite, no deposition should take place for the case 22a. However, this would presuppose a vanishing concentration gradient on the ground for $z = 0$, but this can not be seen from Eq. (15). According to Eq. (15), a deposition current of $\dot{m}_d = v_s \cdot c_0 - F_c = 0.01 \cdot 1100.6 - 0 \approx 11$ is calculated, which, however, precludes the assumption without deposition. The negative concentration gradient calculated incorrectly on the ground according to Eq. (12) forces a material flow in the direction of the positive height coordinate. However, a deposition current is directed towards the ground and not vice versa in the free atmosphere. It is calculated incorrectly in terms of amount and direction, which is why the II. Law of thermodynamics is violated. The calculated concentration distribution also suggests that the source of the source should be located in the ground and not evenly distributed as a volume source throughout the control area. Already the consideration of the graphic a excites unrest.

In the case study 22b, one starts from the same control area again, but deposition and sedimentation should equally occur here. The results are described in Graph b of Fig. 1. The rate of deposition and the rate of sedimentation are identical and are $v_s = v_d = 0.05$. The diffusion coefficient is again $K = 1$. The emission with a strength of $F_c = 1$ is to be released at a level of 200 m. In contrast to case example 22a, a calculation equation is available here for determining the soil concentration. For this Eq. (11) $F_c = c_0 \cdot v_d$ is used again, and the soil value is given as $c_0 = F_c / \cdot v_d = 1/0.05 = 20$. According to the prerequisite, a deposition should take place for this case example. However, this would require a concentration gradient on the ground for $z = 0$ that is different from zero. However, such a concentration curve can not be seen in the graphic b. The deposition current is again calculated according to Eq. (15) and is $\dot{m}_d = v_s \cdot c_0 - F_c = 0.05 \cdot 20 - 1 = 0$. It turns out that no deposition can take place, which contradicts the assumption with deposition. The deposition current has to be different from zero and the depositable material must diffuse into the adjacent soil. Again, the deposit current is calculated incorrectly in terms of magnitude and direction, which also contradicts the II. Law of thermodynamics. As regards the situation of the source in 200 m, opposition is again to be reported. The concentration curve according to Graph b does not reveal a high-altitude source. The conditions also do not change when the control room is expanded to graphic c on. One does not understand how the differential equation (1) or the analytic solution (12) could even consider sources at different heights. The mathematical model that applies to this only describes the differential equation (16)

$$\frac{\partial c}{\partial t} + v_i \cdot \frac{\partial c}{\partial x_i} = \frac{\partial c}{\partial x_i}\left(K \cdot \frac{\partial c}{\partial x_i}\right) + \dot{q}, \quad (16)$$

where $t(s)$ denotes the time coordinate, $x_i(m)$ observing the summation convention, the different spatial directions, $v_i(m/s)$ the flow velocities, and $\dot{q}(\mu g/(m^3 \, s))$ a possible source term. The authors of the AUSTAL do not use this relationship but Eq. (1).

In the long history of AUSTAL by Axenfeld et al. (1984) on the "mother model" LASAT according to Janicke (2001) and VDI (1988) until the development of a model-based assessment system for occupational immission protection according to Janicke and Janicke (2002) is in Trukenmüller et al. (2015) published for the first time the derivation of the reference solutions of this dispersion model. In this section, the case examples 22a and 22b

were used as an example to show which distortions the use of the *Janicke Convention* leads to.

The soil concentration for case 22a must be calculated speculatively

For the case study 22a it is stated in "Calculation of sedimentation and deposition currents for the case examples 22a and 22b by means of the faulty reference solutions of the AUSTAL2000" section that the calculation of the soil concentration $c_0 = 0/0$ results in an indeterminate expression. The soil concentration can take any numerical value. It will be interesting to know how the authors of the AUSTAL proceed here.

First of all, it is noticeable that in contrast to the case 22b, in which the source is supposed to have been located in 200 m, the only exception is a "volume source over the entire computing area". A justification for this is not given. It is stated in this context that the control volume has a total mass pollutant of 100 kg. First one calculates with this information according to Eq. (17)

$$\bar{c} = \frac{E}{L_x \cdot L_y \cdot L_z} = \frac{100}{200 \cdot 1000 \cdot 1000} \cdot 10^9 = 500,$$
(17)

a mean concentration of $\bar{c}(\mu g/m^3) = 500$, which has settled after filling the control volume. In this equation, $E(kg) = 100$ also means the total emission. How the pollutant entry takes place, is not explained and is also of little concern. After adjusting the mean concentration in the entire control volume and, as here, no further emissions have to be taken into account, the stationary state with a complete concentration equalization has already been reached. Because of $\partial/\partial t = 0$, $\dot{q} = 0$ and $\partial/\partial x_i = 0$ the differential equations (1) and (16) are fulfilled identically. A further development of the already existing constant concentration profile is not possible because the solution has already been reached. It is therefore unnecessary any further propagation calculation. With $F_c = 0$, which is true for case 22a, the obtained constant concentration distribution precludes the faulty solution according to Eq. (7). According to Eq. (7), the concentration distribution should follow a barometric altitude distribution. It now has to be explained how a barometric concentration profile can be calculated in a static system and from an equally distributed pollutant admixture. The thought model developed for this purpose assumes that the pollutant particles inside the control room are redistributed by the authors of the AUSTAL so that they follow an exponential function. For this purpose the exponential distribution function according to Eq. (5) $c(z) = c_0 \cdot \exp(-z \cdot v_s/k)$ is used and the soil concentration c_0 is considered to be a free parameter, which has to be determined considering the already calculated mean concentration of $\bar{c} = 500$. It bypasses the indefinite expression. The calculation equation used for this is described by the expression (18).

$$c_0 = \frac{\bar{c} \cdot L_z}{\int_0^{L_z} \exp\left(-z \cdot \frac{v_s}{K}\right) \cdot dz} \tag{18}$$

After Trukenmüller et al. (2015), however, we do not use the soil concentration c_0 determined in this way for $z = 0$ but a concentration $c\,(5\text{m})$ at a reference level of $z = 5$. This finally gives Eqs. (19) and (20)

$$c(5\text{m}) = \bar{c} \cdot \frac{v_S \cdot L_z}{K} \cdot \frac{1}{\left[1 - \exp\left(-\frac{v_s \cdot L_z}{K}\right)\right]} \cdot \exp\left(-\frac{v_S}{K} \cdot 5\right) \tag{19}$$

$$c(5) = 500 \cdot \frac{0.01 \cdot 200}{1} \cdot \frac{1}{\left[1 - \exp\left(-\frac{0.01 \cdot 200}{1}\right)\right]} \cdot \exp\left(-\frac{0.01 \cdot 5}{1}\right)$$
$$= 1156.52 \cdot \exp\left(-0.05\right) = 1100.6 \tag{20}$$

Equation (20) is to be understood as a numerical value equation with which the soil concentration of $1100.6 \; \mu\text{g}/\text{m}^3$ specified in case 22a was determined. A comparison with Graph a of Fig. 1 confirms the agreement.

The question arises as to how the so-calculated redistribution of pollutant particles can be imagined. A concentration equalization always takes place in the direction of a potential gradient, and not the other way around, according to the second law of thermodynamics without external force. The thought model of the authors of the AUSTAL describes rather a procedural demixing than a method for the calculation of the dispersion of air pollutants with deposition and sedimentation. A procedural demixing requires an energy input, of which there can be no question here. The authors of AUTAL develop a very sophisticated algorithm to disguise the uselessness of their reference solutions in front of the public. This also includes the pollutant input distributed over a "volume source over the entire computing area", which is selected only for the case 22a alone. With this trick one wants to solve the problem that for this case for the calculation of the soil concentration only an indeterminate expression $c_0 = 0/0$ is available.

Calculation of deposition and sedimentation on the basis of physically justified boundary conditions (Schenk)

Derivation of the correct reference solutions

To derive the error-free relationship between deposition and sedimentation is again assumed by Eq. (1). After a single integration, Eq. (21)

$$K\frac{\partial c}{\partial z} + v_s \cdot c = A, \tag{21}$$

where the identity $A \equiv F_c$ has to be considered. With a further integration one obtains Eq. (22)

$$c(z) = \frac{1}{v_s} \cdot \left[A + D \cdot \exp\left(-\frac{v_s \cdot z}{K}\right)\right] \tag{22}$$

with the integration constants $A\left(\mu\text{g}/\left(\text{m}^2\,\text{s}\right)\right)$ and $D\left(\mu\text{g}/\left(\text{m}^2\,\text{s}\right)\right)$. For their determination, the boundary conditions (23) and (24) are used.

$$c(0) = c_0 \tag{23}$$

$$K \cdot \frac{\partial c}{\partial z}(0) - v_d \cdot c_0 = 0 \tag{24}$$

The boundary condition (23) takes into account a constant soil concentration. Equation (24) is derived from the equality of the conductive material flows $K \cdot \partial c/\partial z(0) = K_B \cdot \partial c/\partial z_*(T)$ at the lower boundary of the control space between the atmosphere and the ground with the relationship (25)

$$
\begin{aligned}
K \cdot \frac{\partial c}{\partial z}(0) &= K_B \cdot \frac{\partial c}{\partial z_*}(T) \\
&= \frac{K_B}{T} \cdot (c_0 - c_T) \approx \frac{K_B}{T} \cdot c_0 \\
&= v_d \cdot c_0
\end{aligned}
\tag{25}
$$

at a great depth $T(\text{m})$ of the soil, the mass concentration $c_T\left(\mu\text{g}/\text{m}^3\right)$ can be neglected in comparison to the soil concentration, $c_0 \gg c_T$. In Eq. (25), $K_B\left(\text{m}^2/\text{s}\right)$ is the effective mass transfer coefficient for the soil, and $z_*(\text{m})$ is the vertical soil coordinate in the interval $T \geq z_* \geq 0$. In comparison, the vertical height coordinate z is in the interval $H \geq z \geq 0$, where $H(\text{m})$ denotes the maximum height of the control room. Immediately at the boundary between the atmosphere and the ground, $z = 0$ and $z_* = T$ apply, as can be seen from Eq. (25). Equation (25) yields the boundary condition (24) and also the definition of the deposition velocity $v_d = K_B/T$. In Eq. (25) the derivative $\partial c/\partial z_* = (c_0 - c_T)/T$ of expression (27) is used.

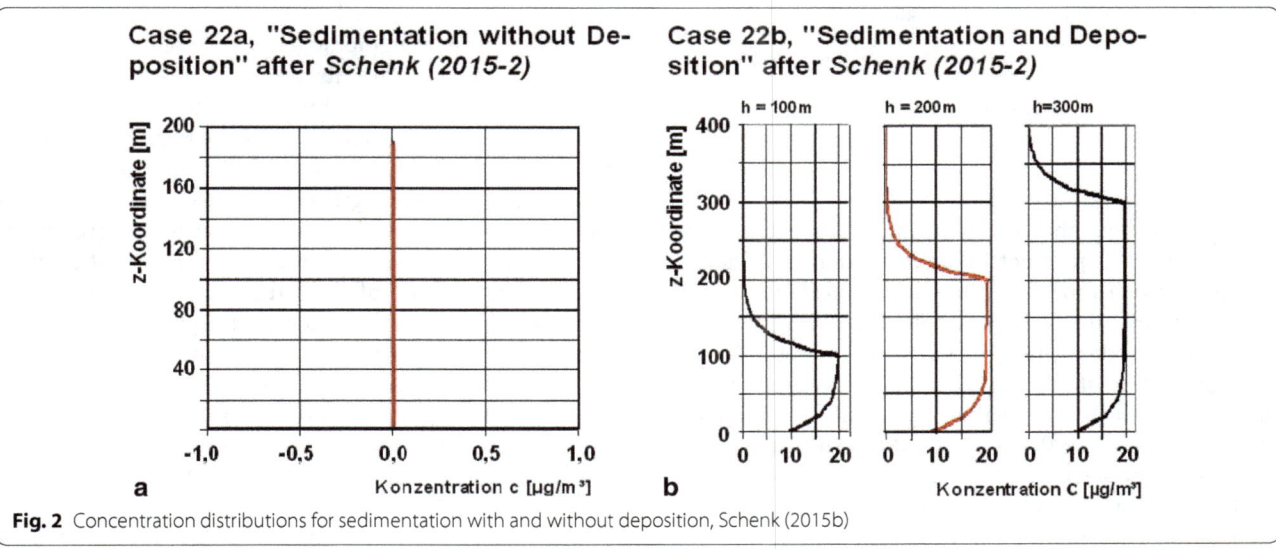

Fig. 2 Concentration distributions for sedimentation with and without deposition, Schenk (2015b)

Due to vanishing sedimentation velocity $v_s = 0$ and constant diffusion coefficient, the differential equation (25) with a linear concentration distribution according to the relation (27) applies for conductive transport in soil considering Eq. (1).

$$\frac{\partial^2 c}{\partial z_*^2} = 0 \qquad (26)$$

$$c(z_*) = \frac{c_0 - c_T}{T} \cdot z_* + c_T \qquad (27)$$

With the described boundary conditions (23) and (24), the integration constants A and D can be determined and Eqs. (28) and (29) result.

$$A = c_0(v_s + v_d) \qquad (28)$$

$$D = -c_0 \cdot v_d \qquad (29)$$

Equation (22) gives the relation (30)

$$c(z) = c_0 \cdot \frac{v_s + v_d}{v_s} \cdot \left[1 - \frac{v_d}{v_s + v_d} \cdot \exp\left(-\frac{v_s}{K} \cdot z\right) \right], \qquad (30)$$

The total mass flow $\dot{m}_g \left(\mu g/(m^2\,s) \right)$ is described by Eq. (31). For the deposition and sedimentation flows, $\dot{m}_d \left(\mu g/(m^2\,s) \right)$ and $\dot{m}_s \left(\mu g/(m^2\,s) \right)$, the expressions (32) and (33) are given.

$$\dot{m}_g = c_0 \cdot (v_s + v_d) \qquad (31)$$

$$\dot{m}_d = c_0 \cdot v_d \qquad (32)$$

$$\dot{m}_s = c_0 \cdot v_s \qquad (33)$$

The boundary condition (24) can, for example, also be read in Berljand (1982). It is older than the field of knowledge of the modeling of the spread of air pollutants, so that an author for this can no longer be specified. It simply results from the understanding of the equality of conductive flows and that mass can not be lost. Equation (30) describes the concentration distribution as a solution of the differential equation (1) and the function of deposition and sedimentation rates v_d and v_s as free parameters. The diffusion coefficient is also kept variable. With this relationship (30), deposition and sedimentation can be described free of defects.

Calculation of sedimentation and deposition for case studies 22a and 22b using the correct reference solution

"Derivation of the correct reference solutions" section described the correct solution of the differential equation (1). In this section, the concentration distributions obtained therewith are calculated for the already known case examples 22a and 22b and compared with the faulty representations of "Calculation of sedimentation and deposition currents for the case examples 22a and 22b by means of the faulty reference solutions of the AUSTAL2000" section. The results are explained in Graphs a and b of Fig. 2 of this section. They are to be compared with the graphics a and b of "Calculation of sedimentation and deposition currents for the case examples 22a and 22b by means of the faulty reference solutions of the AUSTAL2000" section. The geometrical dimensions of the control room and the flow parameters of the case examples 22a and 22b are already known. In contrast to

Eq. (15) one obtains the expression for the deposition current

$$\dot{m}_d = -K \cdot \frac{\partial c}{\partial z}(0) = -v_d \cdot c_0, \qquad (34)$$

It can be seen that the deposition current is formed solely from the product of deposition rate v_d and soil concentration c_0 and not differently according to Eq. (15).

First of all, the case example 22a with sedimentation without deposition is again of interest. For this, the graphic a of Fig. 2 is to be considered. Because of $A = F_c = 0$, a vanishing soil concentration $c_0 = A/(v_s + v_d) = 0/(0.01 + 0) = 0$ results in the entire control room according to Eq. (28), which is in contradiction to Graph a of "Calculation of sedimentation and deposition currents for the case examples 22a and 22b by means of the faulty reference solutions of the AUSTAL2000" section. The result follows the simple logic that no emissions can occur in the absence of emissions in the entire control volume. According to the assumption, no deposition should take place, which is why no deposition current can be observed even according to Eq. (34). The derivation of Eq. (30) for the soil is calculated as a vanishing concentration gradient $\partial c/\partial z(0) \equiv 0$ and thus $\dot{m}_d = -K \cdot \partial c/\partial z(0) = 1 \cdot 0 = 0$, which is confirmed by the concentration curve in Graph a of Fig. 2.

For the case 22b the results are described by the Graph b of Fig. 2. According to the authors of the AUSTAL, the source is said to have been at a height of $h(m) = 200$. However, it should be noted that the differential equation (1) is not suitable at all for considering sources at different heights. Also in the reference solution of the authors of the AUSTAL (12) no free parameter is recognizable, which allows to vary source heights. In fact, only the differential equation (16), which was simplified according to Eq. (35) for the present case, can be responsible for such tasks

$$\frac{\partial c}{\partial t} - v_S \cdot \frac{\partial c}{\partial z} = K \cdot \frac{\partial^2 c}{\partial z^2} - \dot{q} \qquad (35)$$

Due to a lack of analytical solution, this equation was solved numerically for the present task. The control room has been extended to show the complete solution history. Schenk (1980) describes the numerical algorithm used for this purpose. It is a general method for solving the three-dimensional non-stationary differential equations of momentum, heat and mass transport based on the interim step method according to Janenko (1969). In this respect, this simple task concerns a welcome application here. However, as the authors of the AUSTAL have

determined the graphic b of Fig. 1, is uncertain. For further discussion, the graphic b of Fig. 2 highlights the case $h = 200$. The soil concentration is $c_0 = 10$ and in 200 m the concentration is $c(200) \approx 20$. Because there is the source there, as expected, there is the maximum concentration. In the further course of concentration with the height, a considerably greater decrease of the concentrations can be observed in comparison to the area below the source. This can be explained by the fact that there the conductive part of mass transport changes its sign. Because of a positive concentration gradient for $z = 0$, the deposition current of $\dot{m}_d = -v_d \cdot c_0 = -0.05 \cdot 10 = -0.5$ is directed against the positive z-axis in the direction of the ground. In the interval up to $z = 200$ the numerical results agree very well with those of the analytic solution according to Eq. (30). This results in $c(0) = 10 \cdot (0.05 + 0.05)/0.05 \cdot [1 - 0.05/(0.05 + 0.05) \cdot 1] = 10$ and $c(200) = 10 \cdot (0.05 + 0.05)/0.05 \cdot [1 - 0.05/(0.05 + 0.05) \cdot \exp(-200 \cdot 0.05/1] \approx 20$, which again speaks for the quality of the numerical method. To prove its generality, the concentration distribution was still determined for the heights and. The location of the springs is very easy to recognize.

Concluding to this section, it can be estimated that the correct integral (30) of the differential equation (1), in contrast to Eq. (12), correctly describes the occurrence of deposition and sedimentation. The calculated concentration courses are plausible and physically justified.

Relationship between deposition and sedimentation
Incorrect calculation of deposition and sedimentation according to Venkatram and Pleim (1999)

The authors of the AUSTAL refer in all their publications to the reference Venkatram and Pleim (1999). The results there are intended to prove the validity of the *Janicke Convention*. An examination of this work, however, concluded that the authors of Venkatram and Pleim (1999) were more likely to derive an analytical link between deposition and sedimentation than to confirm any conventions. If one were able to calculate sedimentation velocities directly by sedimentation rates, then this would be of great use in the field of modeling the spread of air pollutants. However, the study brought the result that no convention can be read. The correctness of the relation between deposition and sedimentation can not be confirmed either. The following considerations are of interest here.

In deriving the relationship between deposition and sedimentation, the authors of Venkatram and Pleim (1999) of the differential equation already known from Eq. (2)

$$K \frac{\partial c}{\partial z} + v_S \cdot c = F, \tag{36}$$

again taking into account the indentities $F \equiv F_c \equiv A$. Equation (36) is an ordinary first-order inhomogeneous differential equation. In the solution process, the relationship (37) is obtained analogously to Eq. (22)

$$c(z) = \frac{1}{v_S} \cdot \{F + D \cdot \exp\left[-r(z) \cdot v_S\right]\}, \tag{37}$$

where D is to be understood as an integration constant which is to be determined by means of a boundary condition. F is given by the relation (36). For the determination of D the boundary condition (38) is used. It says that the soil concentration is identical to zero.

$$c(0) = 0, \tag{38}$$

Thus, the integration constant D is calculated to Eq. (39)

$$D = -F, \tag{39}$$

as a solution of Eq. (36) the relation (40) is given

$$c(z) = \frac{F}{v_S} \cdot \{1 - \exp\left[-r(z) \cdot v_S\right]\}, \tag{40}$$

In this equation, $r(z)$ according to Eq. (41) means the reciprocal conductive transport velocity.

$$r(z) = \int_0^z \frac{dz}{K(z)} = \frac{z}{K} \, f\ddot{u}r \, K = const. \tag{41}$$

The smaller the transport speed, the greater the resistance.

For the relationship between deposition and sedimentation, the authors of Venkatram and Pleim (1999) below describe the relationship (42)

$$v_d = \frac{v_S}{1 - \exp\left[-r(z) \cdot v_S\right]}, \tag{42}$$

which however can not be reconstructed. A comparison of coefficients between Eqs. (40) and (42) reveals that the expression (43)

$$F = c(z) \cdot v_d \tag{43}$$

has been used. However, the authors of Venkatram and Pleim (1999) overlook that Eq. (36) according to (2) is already the first integral of the differential equation (1). The constant F is also to be regarded as an integration

constant and can not be somehow determined according to Eq. (43). Likewise, the constant F can not be a variable $F(z)$ at the same time. Also, the use of the deposition rate v_d in Eq. (43) is not physically substantiated.

The constant F can not be determined arbitrarily. The constant F can not be a function of z. The deposition rate v_d is used inappropriately. Because of these three reasons, the correctness of Eq. (42) is not guaranteed. Their use is not allowed.

Nevertheless, the authors of the AUSTAL interpret Eq. (43) as their *Janicke convention* $F_c = c_0 \cdot v_d$, which is not recognizable and is not justified physically. Because of the boundary condition (38), the convention would result anyway $F_c = 0$, which calls into question the entire theory of reference solutions of the AUSTAL. But not only the founders of AUSTAL use these misleading descriptions of deposition and sedimentation to model the spread of air pollutants. Also the authors of Simpson et al. (2012) can be misleading in connection with the development of "The EMEP MSC-W chemical transport model".

Relevant calculation of deposition and sedimentation (Schenk)

The author of this article criticizes the results of Venkatram and Pleim (1999) regarding the description of deposition and sedimentation. A correction is required.

If one starts from Eq. (30), the relation (44) is obtained by a simple conversion.

$$v_d = v_S \cdot \frac{\frac{c(z)}{c_0} - 1}{1 - \exp\left[-r(z) \cdot v_S\right]} \tag{44}$$

If one uses the exact solution for the concentration distribution $c(z)$, the identities $v_d \equiv v_d$ and $v_s \equiv v_s$ result. In the case of different applications, however, it is possible to determine an average deposition rate $\bar{v}_d$ from a measured arbitrary concentration distribution. For this purpose, the integral (45) is formed

$$\bar{v}_d = \frac{v_S}{H} \cdot \int_0^H \frac{\frac{c(z)}{c_0} - 1}{1 - \exp\left(-r(z) \cdot v_S\right)} \cdot dz, \tag{45}$$

which can be determined approximately numerically. $H(m)$ means the height of the viewing area.

With the intention to use an analytic relationship, the estimate $c(z) \leq c_{max}$ can be used for all *zone persistently insists on*

and $\Delta c = c_{max} - c_0$ the integral (46)

$$\bar{v}_d^* = \frac{\Delta c \cdot v_s}{c_0 \cdot H} \cdot \int_{z_0}^H \frac{dz}{1 - \exp\left(-\frac{v_s \cdot z}{K}\right)} \tag{46}$$

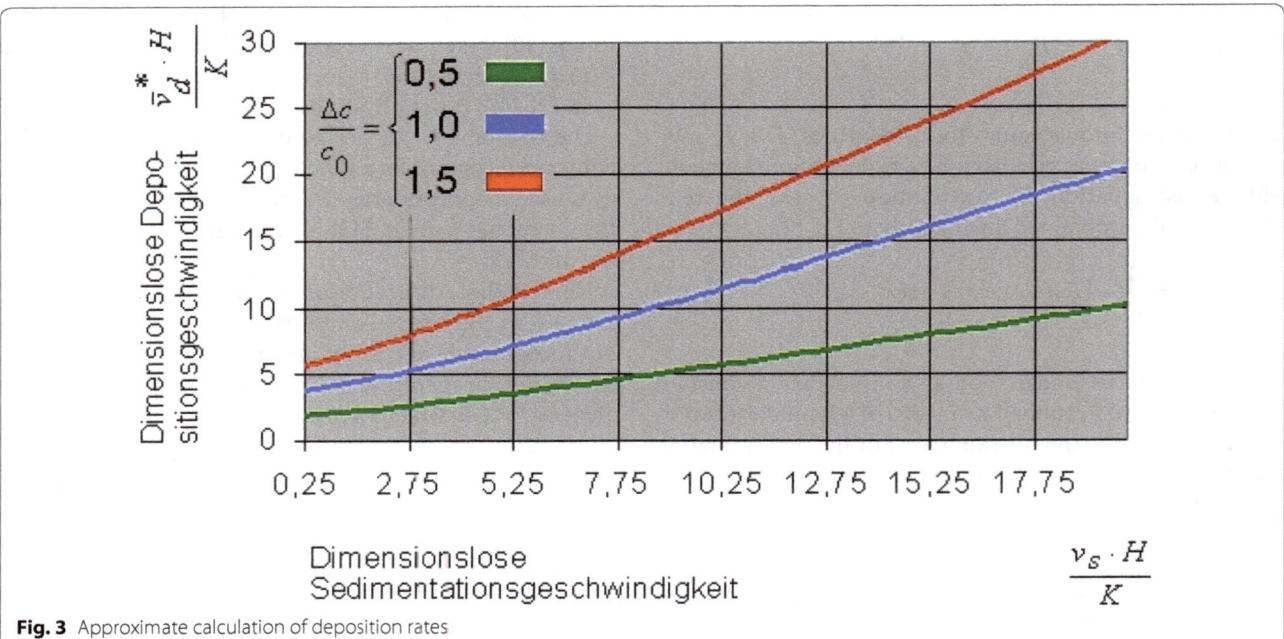

Fig. 3 Approximate calculation of deposition rates

receive. For this integral results after Gradstein and Ruschik (1963) a solution. This describes the relationship (47)

$$\bar{v}_d^* = \frac{\Delta c}{c_0} \cdot \frac{K}{H} \cdot \left\{ \frac{v_s}{K} \cdot (H - z_0) + \ln \left[\frac{1 - \exp\left(-\frac{v_s \cdot H}{K}\right)}{1 - \exp\left(-\frac{v_s \cdot z_0}{K}\right)} \right] \right\},$$

(47)

where z_0(m) is positively small enough in the vicinity of the singularity $z = 0$. $\bar{v}_d^*$(m) approximates the average deposition rate $\bar{v}_d^* \approx \bar{v}_d$.

In connection with practical applications, one uses a dimensionless notation according to Eq. (48)

$$\frac{\bar{v}_d^* \cdot H}{K} = \frac{\Delta c}{c_0} \cdot \left\{ \frac{v_s \cdot H}{K} \cdot \left(1 - \frac{z_0}{H}\right) + \ln \left[\frac{1 - \exp\left(-\frac{v_s \cdot H}{K}\right)}{1 - \exp\left(-\frac{v_s \cdot H}{K} \cdot \frac{z_0}{H}\right)} \right] \right\}$$

(48)

prefer. For example, for the intervals $0.25 \leq v_s \cdot H/K \leq 20$, $1.91 \leq \bar{v}_d^* \cdot H/K \leq 30.65$ and $z_0/H = 0.025$ and $\Delta c/c_0 = 0.5, 1.0, 1.5$, the following graph of Fig. 3 is obtained.

For the case study 22b described here, $H = 200$, $K = 1$, $v_s = 0.05$, $v_s = 0,05$, $\Delta c/c_0 = (c_{max} - c_0)/c_0 = (20 - 10)/10 = 1$ and $v_s \cdot H/K = 10$ give the dimensionless deposition rate $\bar{v}_d^* \cdot H/K = 11.256$. From this, the approximate average deposition rate is determined $\bar{v}_d^* = 0.056$. The exact value is $v_d = 0.05$. Improvements are achieved when the integral (45) is solved numerically.

The inadmissibility of the *Janicke Convention* justifies the triviality of all reference solutions of the AUSTAL2000

As already stated, the authors of the AUSTAL and Trukenmüller et al. (2015), Trukenmüller (2016, 2017) justifying the *Janicke Convention* $F_c = c_0 \cdot v_d$ on Venkatram and Pleim (1999). Because of the boundary condition (38) $c_0 = 0$, the banality (49)

$$F = F_c = 0,$$

(49)

The inadmissibility of *Janicke convention* is proved by Eq. (49). Their use for the development of models for the calculation of the spread of air pollutants is inadmissible. Deposition and sedimentation as well as immissions are calculated incorrectly.

Peculiar concepts

The object of the investigations was solely analytical and numerical considerations to solve a common differential equation for the description of a one-dimensional pollutant transport process. In the context of the model development of the AUSTAL and the associated reference solutions, strange terminology is used, which spreads confusion rather than education. So one speaks for example from a homogenizing and means diffusion. Homogenization, however, just like demixing and shredding, is one of the basic operations of process engineering. In your case, the mass transfer occurs through energy input, while in the case of diffusion for concentration equalization a potential gradient is responsible. When homogenizing, one notices strange oscillations at the region boundaries, which are not further explained. Whether numerical instabilities are responsible for this remains uncertain, because the authors of the AUSTAL themselves write that some solutions do not converge. Proofs for the numerical stability of their algorithms are not kept. It is calculated with so-called homogeneous and inhomogeneous turbulence approaches and actually means the effective diffusion, which, as is known, can also be spatially different. The belief is spread that one masters the modeling of turbulent flows, even though the equations of fluid mechanics are not the subject of all considerations. A description of the AUSTAL explains that calculations with so-called homogeneous turbulence and spatially variable step size can not be performed. In another report, however, such simulations are discussed. The reader is confused. Within closed buildings AUSTAL calculates concentrations. It is explained that Staupartikel vertical house walls can not "see" and therefore want to go through. In fact, it should be used to hide a careless way of programming. With mass / time, mass / (time * length^2) and "volume source over the entire computing area", three different source term definitions are given. However, a "volume source over the entire computing area" is unknown in the field of modeling the spread of air pollutants. It is chosen only to mask inefficiencies in the solution process. In fact, only point sources or mass flow rates are of interest. Deposition speeds are defined as desired. 3D wind fields are to be used for validation. In fact, one uses the rigid rotation of a solid in the plane.

The theoretical foundations of AUSTAL are also challenged by these strange views.

Discussion and summary of the results

As part of a non-university research in the field of propagation of air pollutants a model for the "Investment-related emission protection" called AUSTAL2000 is developed in the Federal Republic of Germany, Axenfeld et al. (1984). This model development begins with the model LASAT after Janicke (2001). This propagation model, including its theoretical foundations, is declared binding for all further model developments with reference to VDI (1988). AUSTAL provides reference solutions that other model developers use to validate their algorithms. Schenk (2015a) demonstrates that all reference solutions are flawed. Even the authors

of the AUSTAL could not have validated their propagation model. In Trukenmüller et al. (2015), the authors feel obliged to respond to the allegations. In this context, the derivation of the reference solutions is published for the first time in the more than 34-year history of AUSTAL. The errors and contradictions identified therein are explained in Schenk (2015b). The correct solution is given. It is also described that the convention used $F_c = c_0 \cdot v_d$ is inadmissible and useless. By unbelievable definitions of the deposition speed one wants to prove its validity in Trukenmüller (2016). However, this attempt fails because the deposition rate is a material constant and can not be defined as desired. One refers to Venkatram and Pleim (1999) and wants to make believe, there this connection would be described physically justified. Due to obvious contradictions, a further definition of the deposition rate is given below in Trukenmüller (2017) with $K \cdot \partial c / \partial z + c \cdot v_s = c_0 \cdot v_d$. Afterwards, this parameterizes the convective and conductive material flows. Thus one wants to prove the general validity of the convention $F_c = c_0 \cdot v_d$ and refers again to Venkatram and Pleim (1999) and other authors, e.g. Simpson et al. (2012). Again, this convention would be used, as it is claimed. It turns out, however, that these references are not justified. Neither in Venkatram and Pleim (1999) still in Simpson et al. (2012), this convention is the subject of consideration. In Venkatram and Pleim (1999) is more concerned with the derivation of an analytical relationship between deposition and sedimentation rates than with the justification of any convention. In Simpson et al. (2012) only the relationship between deposition and sedimentation according to Venkatram and Pleim (1999), but not the Convention $F_c = c_0 \cdot v_d$. Only the authors of the AUSTAL use this convention, which is why it has to be called the *Janicke Convention*. It can not be said that they are universally valid. Nevertheless, according to VDI (1988), this convention is adopted as a valid legal opinion in the "Scientific Handbook for Practices in Environmental Planning" of the Federal Environment Agency. With the doubtful research results Axenfeld et al. (1984) to develop a new theory of the spread of air pollutants. The authors of the AUSTAL and Trukenmüller (2016) as well as Trukenmüller (2017) ignore all substantiated objections. They insist on the validity of the *Janicke Convention* and authoritatively demand their recognition. The use of the *Janicke Convention* in all publications on air pollution control of the Federal Environmental Agency and in VDI (1988) has led to considerable distortions in the modeling of the spread of air pollutants. It is therefore justified to take a deeper look at the validity of the *Janicke Convention*, which is the subject of this work. Also the work of Venkatram and Pleim (1999) is critically examined.

The results of this article can be summarized.

It is shown that it is in the Trukenmüller et al. (2015) is a trivial relationship. Mathematics and mechanics as well as the fundamentals of the theory of ordinary differential equations have to be considered.

There is no free parameter in the solution to calculate deposition currents. Only the first boundary condition of a constant soil concentration is considered This difficulty is overcome by the arbitrary introduction of the *Janicke Convention*. It is explained that this differentially no mass conservation is guaranteed. One understands by deposition loss and not storage. The thought model for this purpose is defined according to Axenfeld et al. (1984), the deposition rate as the speed after the "-picturally speaking -... a rising on the surface of the earth, the column containing the depositionable material, by idling idle". The concentration distributions calculated with the faulty reference solution for the case examples 22a and 22b, which are described in Janicke and Janicke (2002) show that deposition and sedimentation are described inaccurately. The law of mass conservation and the II law of thermodynamics are violated. For the calculation of the soil concentration in case 22a, one obtains an indeterminate expression, which is why deviating from all other test cases, one must introduce a *"volume source over the entire computing area"* distributed there. The pollutant particles must be redistributed by the authors of the AUSTAL in the control volume so that they follow a barometric height distribution. However, this redistribution is more a result of procedural segregation rather than a calculation of the spread of air pollutants with physically based balance equations.

In contrast to the erroneous considerations of the authors of the AUSTAL the correct solutions are described. They assume that due to the equality of the conductive material flows at the lower limit of the control room on the ground mass can not be lost. The correct solutions thus obtained are used for the considered case studies for the erroneous concentration distributions according to Janicke and Janicke (2002). The law of mass conservation and the II law of thermodynamics are fulfilled.

In support of the *Janicke Convention*, the authors of the AUSTAL refer to Venkatram and Pleim (1999). For the purpose of enlightenment, the author of this work engrossed with the results. The authors of Venkatram and Pleim (1999) develop an analytical relationship between the deposition and sedimentation rates. An argument with any conventions does not take place. However, it should be noted that the authors of Venkatram and Pleim (1999) disregard the fact that their initial equation is already the first integral of the relevant balance equation for the propagation and the disturbance function

describes a constant quantity. The constant F can not be set as variable $F(z) = c(z) \cdot v_d$ at the same time. Again, the approach is chosen arbitrarily and physically not justified. The authors of the AUSTAL follow this erroneous view and see obviously inadmissible in a confirmation of the *Janicke Convention* $F_c = c_0 \cdot v_d$. They ignore the ones in Venkatram and Pleim (1999) used boundary condition $c_0 = 0$. The result is the expression $F_c = 0$. The *Janicke Convention* is not confirmed.

The author of this work doubts the validity of the in Venkatram and Pleim (1999) derived relationship between deposition and sedimentation. A correction is required. The result is an integral equation for which, according to Gradstein and Ruschik (1963) can be given an approximate solution. For the case example 22b, a numerical evaluation is made. The deposition rate can be calculated with 12% a deviation of about.

Finally, it is shown that for all reference cases $F_c = 0$ applies. The use of the *Janicke Convention* is inadmissible. This results in an eigenvalue task for all test examples. Their solution is out of the question for practical applications. It has no relation to the modeling of the spread of air pollutants.

Conclusions

Mathematics and mechanics proved that the propagation model AUSTAL2000 can not be validated. This applies to all further model developments that have been validated on the faulty reference solutions. Security plans and, for example, security analyzes prepared with AUSTAL are to be reviewed. These statements are also true for the "parent model" LASAT.

Abbreviations

A: constant of integration, $\mu g/(m^2 s)$; b: constant of integration, $\mu g/m^3$; c: concentration, $\mu g/m^3$; C_0: concentration, $\mu g/m^3$; $\overline{c}$: steady state concentration, $\mu g/m^3$; c_T: concentration in soil depth, $\mu g/m^3$; c_h: homogeneous solution, $\mu g/m^3$; C_i: special solution, $\mu g/m^3$; $\overline{c}$: medium concentration, $\mu g/m^3$; C_{max}: maximum concentration measured in the control room, $\mu g/m^3$; D: constant of integration, $\mu g/(m^2 s)$; E: total emission, kg; F: constant of integration, $\mu g/(m^2 s)$; F_c: source term Janicke, $\mu g/(m^2 s)$; H: height of the measuring cross section, m; K, K_B: diffusion coefficient, m^2/s; L_x, L_y, L_z: geometric dimensions, m; $\dot{m}_g$: total mass flow, $\mu g/(m^2 s)$; $\dot{m}_s$: sedimentation stream, $\mu g/(m^2 s)$; $\dot{m}_d$: deposition power, $\mu g/(m^2 s)$; $\dot{q}$: source term, $\mu g/(m^2 s)$; r: reciprocal transport speed, 1/(m/s); T: deep ground, m; t: time coordonate, s; v_d: deposition velocity, m/s; v_i: sedimentation speed, m/s; $\overline{v}_d$: mean deposition rate, m/s; $\overline{v}_d^*$: approximation mean deposition rate, m/s; v_i: flow coordinates, m/s; x_i: spatial coordinates, m; z: coordinates, m; z_0: nearby singularity, m; z_*: coordinate ground, m.

Authors' contributions
The author read and approved the final manuscript.

Authors' information
Prof. Dr.-Ing. habil. Rainer Schenk, Ordinarius für Strömungsmechanik, i.R, Rosenberg 17, 06193, WETTIN-LÖBEJÜM, OT Wettin, Germany, ibswettin@web.de.

Korr. Mitglied des Spiegelgremiums zu CENT/TC 264/WG 43 + 44 im VDI und DIN Normenausschuss KRdL.

Mitglied der Arbeitsgruppe VDI RL 3790, Blatt 3, Emissionen von Gasen, Gerüchen und Stäuben aus diffusen Quellen im VDI und DIN – Normenausschuss KRdL.

1968 doctorate to Dr.-Ing. at the Technical University of Merseburg. 1968–1970 additional studies in the field of "Computational Fluid Dynamics" at the Academy of Sciences of the former USSR in Novosibirsk, Akademgorodok. 1970 Lecturer in Theoretical Fluid Mechanics at the Technical University of Merseburg. Since 1972 active in the field of modeling of the spread of air pollutants at the Technical University of Merseburg and member of the main research area air pollution control at the Academy of Sciences of the former GDR. 1978 Calculation of transboundary pollutant flows and international cooperation with the Meteorological Institute of Leningrad University and with the NILU Institute Oslo. 1979 Calculation of long-distance transport Europe. 1979 Development of a 24 h forecast model and application by the Meteorological Service of the former GDR. 1980 Habilitation and scientific work in the field of numerical fluid mechanics and modeling of the spread of air pollutants under the direction of full members of the Academies of Sciences of the former USSR and former GDR Akademik JANENKO, Novosibirsk, and OM ALBRING, Dresden. 1980 participation in the construction of a data center east. 1980 Head of Environmental Monitoring at the Center for Environmental Design Wittenberg. 1982 Lecturer and University Professor of Fluid Mechanics at the Technical University of Zittau. 1985 Model for the calculation of the spread of radionuclides. 1994 honorary professor at the Technical University of Dresden, IHI Zittau. 1996 Research project model for the calculation of the spread of traffic emissions on behalf of the Ministry of the Environment Saxony-Anhalt. 2005 Research project model for the calculation of the expansion of heavy gases and vapors on behalf of the Ministry of the Environment of Saxony-Anhalt. 2007 Research project Mobile Environmental Data AVIS on behalf of the Arbeitsgemeinschaft für industrielle Forschung Berlin. 2008 Research Project Instruments Pollutant Prediction on behalf of the Arbeitsgemeinschaft für industrielle Forschung Berlin. 2010 Model for the calculation of the spread of traffic emissions taking into account moving point sources. 2015 Software developments for the evaluation of meteorological measurement series and for the development of cause analyzes.

Author details
[1] Rosenberg 17, 06193 Wettin-Löbejün, Germany. [2] Dresden University of Technology, International University Institute Zittau, Sachsen, Germany.

Competing interests
The Federal Environment Agency Germany will contradict and publish.

Funding
There is an equity financing.

References
Albring W (1961) Angewandte Strömungslehre. Akademie Verlag, Berlin
Axenfeld F, Janicke L, Münch J (1984) Entwicklung eines Modells zur Berechnung des Staubniederschlages. Umweltforschungsplan des Bundesministers des Innern Luftreinhaltung, Forschungsbericht 104 02 562, Dornier System GmbH Friedrichshafen, Im Auftrag des Umweltbundesamtes
Berljand ME (1982) Moderne Probleme der atmosphärischen Diffusion und Verschmutzung der Atmosphäre. Akademie-Verlag, Berlin
Gradstein IS, Ruschik IM (1963) Tabellen von Integralen, Summen und Reihen sowie ihre Anwendungen. Physikalisch-mathematische Literatur, Staatsverlag, Moskau, p 1963
Janenko NN (1969) Die Zwischenrittmethode zur Lösung mehrdimensionaler Probleme der mathematischen Physik. Springer-Verlag, Berlin
Janicke L (2000) IBJparticle, Eine Implementierung des Ausbreitungsmodells. Bericht IBB Janicke
Janicke (2001) Ausbreitungsmodell LASAT Referenzbuch zur Version 2.10, Dezember
Janicke (2002) AUSTAL2000, Programmbeschreibung, Dunum
Janicke U, Janicke L (2002) Entwicklung eines Modellgestützten Beurteilungssystems für den Anlagenbezogenen Immissionsschutz. IBJanicke
Naue G (1967) Einführung in die Strömungsmechanik. Reprocolor, Leipzig
Schenk R (1980) Numerische Behandlung instationärer Transportprobleme, Habilitation. TU Dresden, Dresden

Schenk R (2015a) AUSTAL2000 ist nicht validiert. Immissionsschutz 01.15, S. 10–21
Schenk R (2015b). Replik auf den Beitrag „Erwiderung der Kritik von Schenk an AUSTAL2000 in Immissionsschutz 01/2015". Immissionsschutz 04.15, S. 189–191
Schenk R (2017) The pollutant spreading model AUSTAL2000 is not validated. Environ Ecol Res 5(1):45–58
Schlichting H (1964) Grenzschichtheorie. Verlag G. Braun, Karlsruhe
Simpson D, Benedictow A, Berge H, Bergström R, Emberson LD, Fagerli H, Flechard CR, Hayman GD, Gauss M, Jonson JE, Jenkin ME, Hyiri A, Richter C, Semeena VS, Tsyro S, Tuovinen JP, Valdebenito A, Wind P (2012) The EMEP MSC-W chemical transport model—technical description. Atmos Chem Phys 12:7825–7865
Truckenbrodt E (1989) Fluidmechanik. Springer, Berlin
Trukenmüller A (2016) Äquivalenz der Referenzlösungen von Schenk und Janicke. Abhandlung Umweltbundesamt Dessau-Rosslau, IBS Archiv, S. 1
Trukenmüller A (2017) Stellungnahmen Umweltbundesamt vom 10.02.2017 und 23.03.2017. Dessau-Rosslau, IBS Archiv, S. 1–15
Trukenmüller A, Bächlin W, Bahmann W, Förster A, Hartmann U, Hebbinghaus H, Janicke U, Müller WJ, Nielinger J, Petrich R, Schmonsees N, Strotkötter U, Wohlfahrt T, Wurzler (2015) Erwiderung der Kritik von Schenk an AUSTAL2000 in Immissionsschutz 01/2015. Immissionsschutz 03/2015, S. 114–126
VDI 3945 Blatt3 (2000) Umweltmeteorologie—Atmosphärisches Ausbreitungsmodell—Partikelmodell. Beuth Verlag, Berlin
VDI Kommission Reinhaltung der Luft (1988) Stadtklima und Luftreinhaltung. Springer, Berlin
Venkatram A, Pleim J (1999) The electrical analogy does not apply to modeling dry deposition of particles. Atmos Environ 33:3075–3076

6

Minimizing environmental impacts of timber products through the production process "From Sawmill to Final Products"

Shankar Adhikari[1]*[iD] and Barbara Ozarska[2]

Abstract

As awareness of climate and environment issues increases and consumption habits change, new opportunities are opening up for the forest industry and wood construction to develop functional green solutions to meet consumers' needs. Wood is a versatile raw material and the only renewable construction material. The manufacture of wood products and structures consumes little energy in comparison to similar products and structures made of other materials. Unlike other materials, most of the energy needed to manufacture wood products is derived from renewable energy sources. The global timber sector currently faces the dual challenges of meeting the growing demand of quality timber products and minimising possible adverse impacts on the environment and human health. Major sources of environmental impacts occur throughout the wood supply chain from sawmills to final products. The major objective of this paper is to explore ways to reduce the environmental impacts of timber products, from sawmills to final products. The specific objectives include the identification of major sources and mechanisms of environmental impacts from timber products, the assessment of the status of energy consumption and GHG emission in wood products during timber processing and manufacturing as well as identifying the potential ways to minimize these environmental impacts.

Keywords: Assessment, Environmental impacts, Minimize, Sawmill, Timber products

Background

Amidst growing environmental consciousness and increasing demand for timber products, the importance of fulfilling growing demand for these products on the one hand, and at the same minimizing environmental impacts, is increasingly recognized. While FAO (2001) had predicted that by the end of 2020, global consumption of industrial timber products will increase by 45%, UK based sustainable real estate organization FIM, based on existing growth levels, has forecasted that global timber consumption in 2020 will be 2.3 billion cubic meters. This is an increase of 24% from the 2015 level and equivalent to a 4.4% increase per annum (FIM 2017). Moreover, The World Bank has also forecasted that global timber demand is set to quadruple by 2050 (FIM 2017). As a result, there is growing concern about fulfilling the need for increasing demand for timber products without deteriorating the world's forest resources. Hence, enhanced insight is required into ways of improving the efficiency of timber production process, reducing wood wastage and helping the timber sector to address growing environmental challenges (Eshun et al. 2012).

Timber products are regarded as products produced from renewable and sustainable environmental resources (Klein et al. 2016). However, as other products, timber products may create various kinds of environmental impacts at different stages of the timber product supply chain, from harvesting to their disposal (Fig. 1). A major source of the environmental impacts is the consumption of energy required to produce timber products and emission of greenhouse gases (GHG) during the manufacturing process from raw materials to the final products. Although production of timber products also involves

*Correspondence: adhikarishankar@gmail.com
[1] Department of Forests, Ministry of Forests and Environment, Kathmandu, Nepal

Minimizing environmental impacts of timber products through the production process...

77

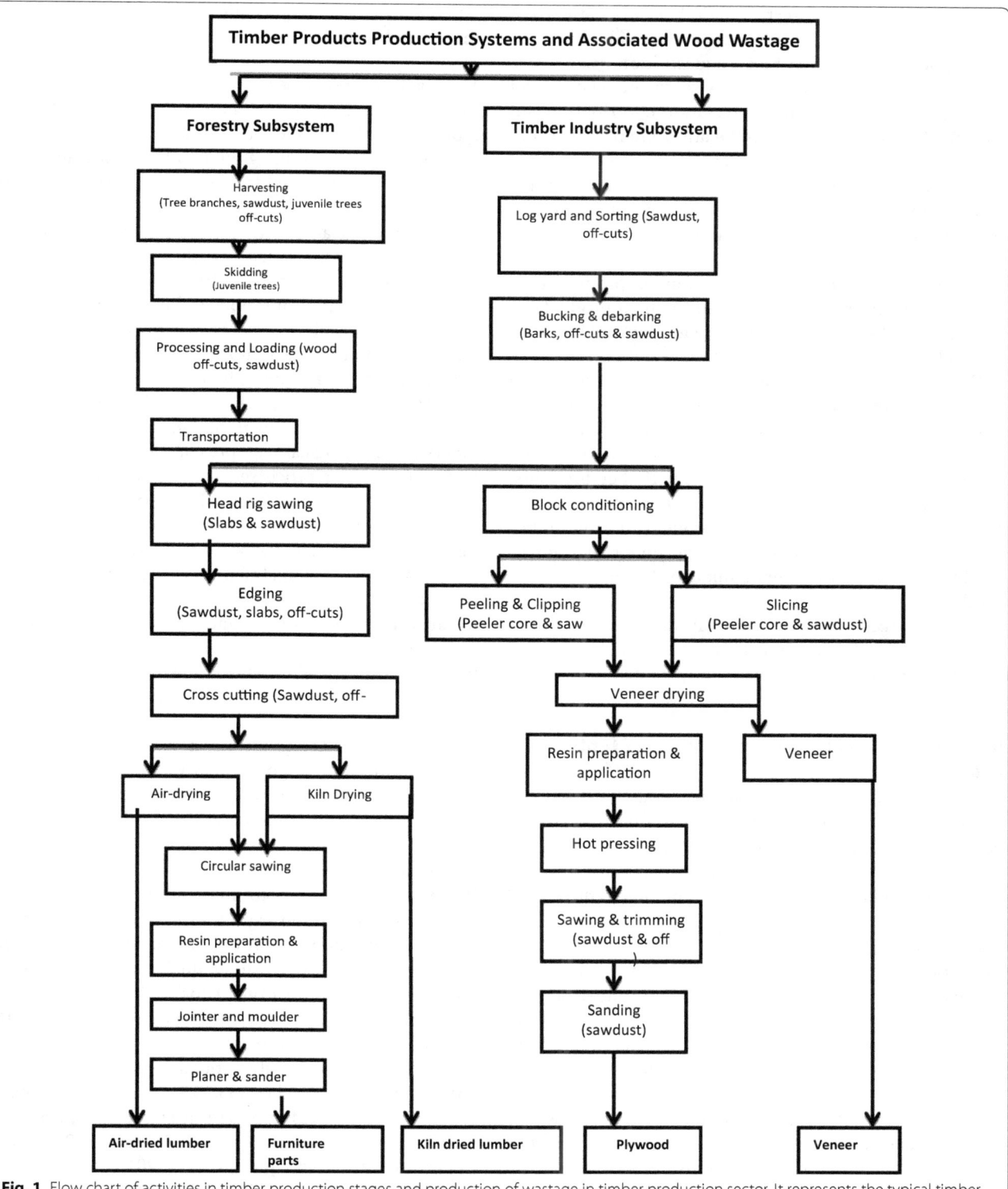

Fig. 1 Flow chart of activities in timber production stages and production of wastage in timber production sector. It represents the typical timber product production system, from harvesting to the final products through two subsystems viz the forestry and timber industry subsystems, within the timber production sector. Modified from Eshun et al. (2012)

emission of carbon, forest and timber provide carbon sinks because trees consume carbon dioxide from the atmosphere through carbon sequestration (Le Quéré et al. 2009). However, the forestry sector in general, and removal of trees through deforestation contribute to up to 17% of GHG emission into the atmosphere (Miles and Kapos 2008; Baccini et al. 2012). Other forms of environmental impact associated with timber products are due to the transportation of timber products (Lindholm and Berg 2005), use of chemicals, and wood wastage (Jurgensen et al. 1997; Wootton 2012).

Figure 1 represents the typical timber product production system, from harvesting to the final products. It shows two subsystems, the forestry and timber industry, within the timber production sector.

The major objective of this paper is to explore ways to reduce the environmental impacts of timber products, from sawmills to final products. The specific objectives include the identification of major sources and mechanisms of environmental impacts from timber products, the assessment of the current status of energy consumption and GHG emission in wood products during timber processing as well as identifying the potential ways to minimize these environmental impacts during the timber production process.

Main sources of environmental impacts of timber products can be categorised into physical impacts of timber processing, energy use and production of GHG emissions.

Sources attached to physical impacts of timber products

The production process for timber products, from log extraction to final products involves several stages, which can affect the surrounding environments in the form of land, air and water pollution. This paper looks into the impacts of timber products from sawmilling to disposal.

Sawmilling

The sawmilling process involves debarking and cutting of logs into sections, which are sawn into timber boards. Particulate environmental matter arises from log debarking, sawing into boards, wood residues and kiln drying as these processing stages create environmental hazards on the land. Similarly, heavy machinery is involved throughout the process with the impacts on land, water, and air quality. For example, sawmilling sector is the backbone of the wood based industry in Malaysia. A study by Ramasamy et al. (2015) on the environmental impact of sawmilling industry concluded that several gases such as CO_2, CH_4, NOx, N_2O, SO_2, and CO were found discharged into the environment and the impacts were found in the form of global warming, acidification,

human toxicity, eutrophication, and photo-oxidant formation in Malaysia.

Manufacturing processes

Timber processing and manufacturing involves different types of machines and processes such as sawing, drying, machining, jointing, gluing and finishing and so on, which can be connected to both environmental hazards, and workers occupational health and safety.

The major hazards with the machines could be classified into mechanical (e.g. crushing, cutting, trapping, shearing, abrasion, friction), structural (e.g. sharp edges, projections, obstructions, potential to fragment, collapse, overturn), physical (e.g. electricity, pressurized content, noise and vibration, heat, moisture or cold temperatures), chemical (e.g. gases, fumes, liquids, dusts, that can cause adverse health effects), ergonomic hazards (awkward working positions, manual handling, repetitive movements), and biological (e.g. present of bacteria, molds in materials used or processed in machinery) (Bluff 2014; Poisson and Chinniah 2016). Furthermore, as workers have to use machinery in all phases of its lifecycle from installation, through operation, maintenance, troubleshooting, repairs, adjustments, set-up, production disruptions, to cleaning and dismantling, they get exposed to various hazards (Poisson and Chinniah 2016; Rus et al. 2008). A study based on sixty-six Australian manufacturing firms which produce and supply machinery into local and international markets found that less than one in ten firms comprehensively recognized hazards, used safe place controls as the primary risk control measures, and provided substantial, good quality information to minimize environmental hazards, but the remaining firms did not consider the environmental impacts (Bluff 2014).

Wood waste and by-products

Preventing wood waste to improve the efficiency of primary wood utilization significantly helps to reduce the environmental impacts on the one hand, and fulfill timber product demands without further damage to world forest resources on the other. Dionco-Adetayo (2001) has found that out of 1 m^3 of tree cut and removed from the forest, about 50% goes to waste in the form of damaged residuals, followed by abandoned logs (3.75%), stumps (10%), tops and branches (33.75%), and butt trimmings (2.5%). Wood wastes comprise a significant portion of waste materials. For instance, in Germany, 401 million tonnes of wastes was produced in 2015, out of which waste wood accounts to 11.9 million tonnes. (Sommerhuber et al. 2015; Garcia and Hora 2017). The primary sources of waste wood were wood packaging (21%), demolition and construction (26.7%), wood processing industry (14%), municipal wastes (20.7%), imported wood

(9.7%) and others such as private households and railway construction (8%) (Sommerhuber et al. 2015; Garcia and Hora 2017). Similarly, around 1,781,000 tonnes of wood waste was being generated in Australia per annum until 2007 (Taylor and Warnken 2008).

This large amount of wasted wood is often used in the steam production boiler for drying wood products, or is dumped in a site (Eshun et al. 2012). These practices contribute to environmental impacts through wood waste and at the same time lead to depletion of timber resources. Eshun et al. (2012) have identified 19 wood waste sources in Ghana, out of them 3 related to the forestry sub system and 16 to the timber industry subsystem. Major sources of wood waste were low quality logs with large defects, bark, off-cuts, sawdust, slabs, and edged trimmings from sawn timber. There are new technologies for the utilization of low quality logs which can significantly reduce the wood wastage as well as specialized equipment which enables to maximize the wood recovery. However, in many enterprises, particularly in developing countries, these new production methods have not yet been used. Therefore, the major causes of wood wastage can be broadly classified into technology-based factors such as the use of obsolete equipment and inefficient procedures and production methods, management-based operational practices, and administrative and institutional issues.

Toxic chemicals

Different types of chemicals are used in the process of timber production, especially in preservative treatment, adhesive application and coating of final products. Though these chemicals have played the positive role of increasing the life span of timber products, they can also contribute to environmental impacts through the toxic elements they contain. For example, disposal of timber from demolition building sites still retaining high levels of preservatives is also another important environmental concern. Many countries have introduced policies, which prevent the use of toxic chemicals.

Adhesives

Even though adhesives are important materials made up of both natural and synthetic substances for bonding wood components into wood product they still might have some negative environmental impacts (Yang and Rosentrater 2015). Phenol–formaldehyde (PF) and urea–formaldehyde (UF) are the two commonly used adhesives in external environments due to high weather- and water-resistance properties (Cetin and Özmen 2002; Pizzi and Mittal 2011; Zhang et al. 2013). However,

even the completely cured adhesives regarded as non-toxic and safe, can produce hazardous materials for both humans and the environment (Yang and Rosentrater 2015). For example, some curing agents such as aliphatic amines, and cycloaliphatic amines might cause irritation or damage to the skin, eyes, lungs, and liver (Yang and Rosentrater 2015). Therefore, there is growing interest in the use of adhesives which are environmentally benign (McDevitt and Grigsby 2014).

Wood coatings

Wood coatings protect wood from environmental influences such as moisture radiation, mechanical and chemical damage, and biological deterioration. However, they contain liquid made up of either organic solvent or water, and have potential to emit volatile organic compounds (VOC). VOC such as those containing chlorofluorocarbon are considered a major environmental problem from both air pollution and human health and safety perspectives (de Meijer 2001).

Impacts associated with transportation

Environmental impacts associated with the transportation of timber from forest to sawmills, then sawn timber from sawmills to manufacturing companies, and finally to end-users, consume significant amounts of fossil fuel, and thereby emit greenhouse gas (GHG) to the environment. A study carried out in Swedish wood supply chain showed that transportation of timber from forests to industrial sites consumes more fossil fuels than any other part of the chain (Lindholm and Berg 2005). The energy used during the transportation system has impacts on the environment due to release of emission with likely effects on global warming, acidification and eutrophication. For example, organic compounds and phosphorus released to water, and emissions of nitrogenous compounds to both air and water, are the most serious environmental impacts. Similarly, road transport of timber account for almost half of the total GHG emissions. In East Norway, GHG emissions from the final felling, extraction and transport of timber, was found to have 17.893 kg CO_2-equivalents per m^3 of timber delivered to industry gate in 2010 (Timmermann and Dibdiakova 2014). As a result, transportation creates impacts on the atmosphere, land and water resources, and noise pollution.

Study by Timmermann and Dibdiakova (2014) assessed annual greenhouse gas effects from seedling, tree felling, transportation and processing of timber products. The study concluded that GHG emissions of forestry supply chain activities and found road transport of timber had the highest impact in climate change category.

Maintenance of timber products during use

Maintenance of timber products is carried out either in the form of their full or partial replacement, or by using chemicals to maintain or prolong their life. Therefore, proper care must be taken while maintaining timber products to produce minimum impact on the environment.

Disposal

Disposal of timber products creates various environmental impacts especially in urban area. Commercial and industrial wastes, construction and demolition activities, pallets and packaging; and utilities are the main sources of urban wood wastes (Taylor and Warnken 2008). When the products are disposed instead of being reused, recycled and refurbished they will create the outside pollution and GHG emissions in many ways due to transport from the source to a landfill site; disposal of synthetic materials contributes to toxic waste, which can leach from landfill, and finally, such materials take up a large amount of space in landfill sites and create the need for new waste disposal sites (ERDC 2001).

Although huge volume of waste wood is disposed of to landfill sites in major cities around the world, data on wood waste from the larger categories of waste is not differentiated in most cases. Data on wood waste from Landfills in Sydney and Melbourne, Australia, estimate that approximately 446,000 and 623,000 tonnes are annually disposed of respectively (Taylor et al. 2005). The figure of Melbourne city wood waste disposal is enough to fill the Melbourne Cricket Ground 1.5 time (Taylor et al. 2005).

Similarly, if disposal is carried out by burning of used products, it also produces smokes, contamination and emissions into the environment. For instances, solid contamination has disposal issues by reducing the efficiency of burning and producing waste, whereas excess chlorine in the burning also reduces the burning efficiency and can contribute to the production of dioxins (Taylor et al. 2005).

Sources of impacts due to the use of energy and emission of GHG

The energy involved in the process and stored in the product is called embodied energy (Ibn-Mohammed et al. 2013). Various types of energy source are used for different stages of timber production. Primarily energy is used for processing and materials handling, drying of raw materials, and associated utilities and services such as boiler steam, and condensation system, heating and lightning of premises (Bergman and Bowe 2008). As a result, there would be two phenomena involved together, energy consumption during the production process, and emission of greenhouse gas and other gasses as a consequence.

Sources of energy

The major sources of energy in sawmilling are either electrical energy or thermal energy. Electrical energy includes electricity supplied through the grid system, and is primarily used in sawing process, whereas thermal or heat energy is generated through biomass and used primarily for drying of sawn timber. Energy sources can also be classified based on the origin of the energy (Bergman and Bowe 2008). For example, if the energy is produced within the sawmill site, and used for drying or other purposes, it is called an onsite energy source. On the other hand, if energy requirements are fulfilled from outside of the sawmill site, they are referred as offsite energy sources.

On the other hand, sources of energy can also be classified based on the sources of carbon emission as the part of sawmilling procedures. For example, the energy produced as a result of the burning of wood biomass is called a biogenic energy, whereas energy derived from fossil fuel is called as the anthropogenic emission source (Gunn et al. 2012).

Sources of energy from fossil fuel have a significant impact on the environment and are non-renewable. If the sources of energy are renewable and have less impact on the environment such as hydroelectricity, wind energy, are known as renewable sources of energy. These have a lower environmental impacts and health hazards.

GHG emission of timber products

The energy sources and the ways they are used contribute to the production of GHG emissions and other environmental impacts.

Major environmental impacts associated with timber products include emission to air especially emission of GHG among others (Wilnhammer et al. 2015; Van et al. 2017). This kind of impact is called as carbon foot printing or the carbon impact of timber products (Box 1).

Box 1: The wood product carbon impact equation

$$A - B - C - D = E$$

A. *Manufacturing carbon*: Manufacturing uses energy and most energy production results in carbon dioxide release.

B. *Bio-fuel*: Wood residues are often burned for energy during the manufacture of wood products.

C. *Carbon storage*: Carbon dioxide is absorbed from the atmosphere during photosynthesis by the growing tree. This carbon is converted to wood, bark and other parts of the tree.

D. *Substitution*: There are alternatives to wood products for most applications. However, almost all of these non-wood alternatives require more energy for their manufacture, and the energy used is almost entirely fossil carbon.

E. *Total Carbon Footprint or Carbon Credit*: The bio-fuel (B), carbon storage (C) and substitution (D) effects reduce the carbon footprint of wood products. In fact, these effects together are almost always greater than the manufacturing carbon (A), so the overall carbon effect of using wood products is a negative carbon footprint (i.e. carbon credit or storage). Thus using wood products can help us to reduce contributions to climate change and conserve energy resources.

Source: Bergman et al. (2014)

The forest industry especially the timber production process contributes to global GHG emission in different ways from harvesting to end use and disposal. Manufacturing–related emissions dominate the GHG contribution from the sector by accounting for 55% of all emissions occurring throughout the value chain (Miner 2010) which is approximately, 490 million tonnes of CO_2 equivalent per year. This is mainly due to the fuel combustion at the manufacturing facilities. Similarly, a significant amount of emission of about 238 million tonnes, also occurs at the end of the life cycle, especially from methane emission (235 million tonnes) and emission associated with the burning of used products (3 million tonnes) (Miner 2010).

A study on life cycle impacts and benefits of wood along the value chain in Switzerland shows that high environmental benefits in construction and furniture are often achieved when replacing conventional heat production and energy-consuming materials. For instance, replacement of fossil fuels for energy or energy-intensive building materials, and taking appropriate measures to minimize negative effects such as particulate matter emissions could ensure high environmental benefits (Suter et al. 2017).

Methods of impact assessment

Major methods in vogue for the impact assessment of environmental sectors are life cycle assessment (LCA) (Gustavsson and Sathre 2006; Ramesh et al. 2010; Roy et al. 2009), input–output methods (Ivanova and Rolfe 2011), cost–benefit analysis (Atkinson and Mourato 2006), health hazard scoring (HHS) system, material input per service-unit (MIPS), Swiss eco-point (SEP) method, sustainable process index (SPI), Society of Environmental Toxicology and Chemistry's life-cycle impact assessment (SETAC LCA), and environmental priority system (EPS) (Hertwich et al. 1997).

Though most of these methods could be applied to examine the complex interaction among the timber production process from the sawmill to final product, and their impact on their corresponding environments, LCA can explain such a relationship in a more comprehensive way. This is because it is a procedure for evaluating the energy and environmental burdens related to a process or activity, which is carried out with the help of identifying the source of energy used or consumption, the materials used and their impact on the environment (Goedkoop et al. 2008).

So far, extensive studies on LCA and various aspect of timber production are well documented (Cabeza et al. 2014; Dodoo et al. 2014a, 2014b; Lippke et al. 2004; Mirabella et al. 2014; Puettmann and Wilson 2007). Among them, the Consortium for Research on Renewable Industrial Materials (CORRIM), has published a 22-module research plan and protocol to develop a LCA for residential structures and other wood uses while evaluating the life cycle inventory (LCI) databases for use in each stage of processing "from cradle to grave" (Lippke et al. 2004; NCASI 2006).

Benefits of using timber in various products

Wood competes with many other materials in various products and applications. The main competitors are: steel, concrete, aluminum, brick and plastic (Taylor 2003; George 2008). Many studies have been conducted which compared the environmental impacts of wood and its competing materials. Production of wood results in few greenhouse gas emissions, in which the main emission source is the energy used in wood processing. The energy saving requirements of the industry in wood processing can be met with the use of wood residue, which provides more energy savings compared to the use of fossil fuel based energy. On the other hand, production of most competitor materials results in high greenhouse

Table 1 Greenhouse gas emission profile of wood and key competitors. Adopted from George (2008)

Materials	Greenhouse gas emissions per tonne material (tCO_2-e/tonne)	Sources
Aluminum	22.4	Norgate et al. (2007); George (2008)
Steel (blast furnace production)	2.55	George (2008)
Steel (scrap-based electronic arc furnace production)	1.1	George (2008)
Cement	0.77	George (2008)
Hardwood (rough sawn kiln dried)	0.23	George (2003)
Softwood (rough sawn kiln dried)	0.234	George (2008)
Medium density fibreboard (MDF)	0.726	George (2008)
Particle board	0.982	George (2008)

emissions. The emission values of wood and competitor products are presented in Table 1.

Wood consumes less energy and emits less pollutant to the environment, thereby adds environmental values throughout the life of the structure. In contrast, steel and concrete use more energy, emit more greenhouse gases, and release more air and water pollutants during the manufacturing process than that of wood products (APA 2017). For example, wood is 105 times more efficient than concrete, and 400 times more efficient than steel. When it comes to energy consumption, steel and concrete consumes 12 and 20% more than wood products respectively. Similarly, steel emits 15% more GHG than wood and concrete emits 29% GHG more than wood. Likewise, steel and concrete significantly contribute in water pollution than that of wood products. For example, steel pollutes 300% more water resources, and concrete pollutes 225% more water than the wood products (APA 2017).

Unlike their competitors, wood products are part of the carbon cycle. Therefore, as tree absorb carbon dioxide and act as an important carbon sink, they contribute to carbon sequestration and climate change mitigation as well (George 2008). There are no environmentally perfect building and construction materials; however, wood is still an intelligent and informed choice especially for many commercial and residential buildings (APA 2017) mainly due to low energy use and CO_2 emission than that of steel and concrete products. For example, wood-based building construction consumes 3800 gigajouls (GJ) of total energy whereas steel and concrete based structure consumes 7350 and 5500 GJ energy, respectively.

Similarly, on carbon emission, wood-based construction emits 73000 kg carbon emission whereas steel and concrete based construction emits 105,000 and 132,000 kg carbon, respectively (APA 2017).

Most construction materials such as steel, concrete, aluminum and plastic require a high energy input during the manufacturing process while the manufacture of timber products uses much less energy than the competitive materials. (Figure 2).

Many studies have also confirmed that timber products have a net carbon storage value which means that they store more carbon than is required in their manufacture. Typical results for various materials are shown in Table 2.

Possible ways to reduce the environmental impacts of timber products

With the identification of potential sources of environmental impact and their mechanisms at different stages of the timber production process, the following methods can be applied to tackle the associated contemporary challenges.

Changes in energy sources and consumption pattern

As energy sources and consumption patterns are critical towards overall environmental impacts of energy consumption practices, environmentally friendly energy sources should be promoted. For example, fossil fuel based energy such as energy generated from coal, has more adverse environmental impacts than that of non-fossil based energy sources. Similarly, anthropogenic emissions due to fossil fuel have comparatively higher emission and negative environmental impacts, than that of biogenic emission from burning wood materials (Bergman and Bowe 2008). Therefore, while choosing energy sources for the timber production process, there needs to be proper care in the use of renewable energy instead of fossil fuel-based energy techniques. Even if fossil fuel based energy source are to be used, efforts must be made to use as little energy as possible.

Use of Sawmill by-products as a thermal energy

Instead of leaving the sawmill products within the premises of sawmills, and creating environmental hazards, they could be collected and used for producing thermal energy to reduce environmental impacts. This would help to minimize the reliance on offsite fossils fuel to some extent and promotes the production of bioenergy at the sawmill site. For example, the sawdust could be recycled into a bio-briquette. Such bio-briquettes have even higher heating value ranged from 14.88 up to 16.94 MJ/kg, than that of the briquette made from other substances (Lela et al. 2016).

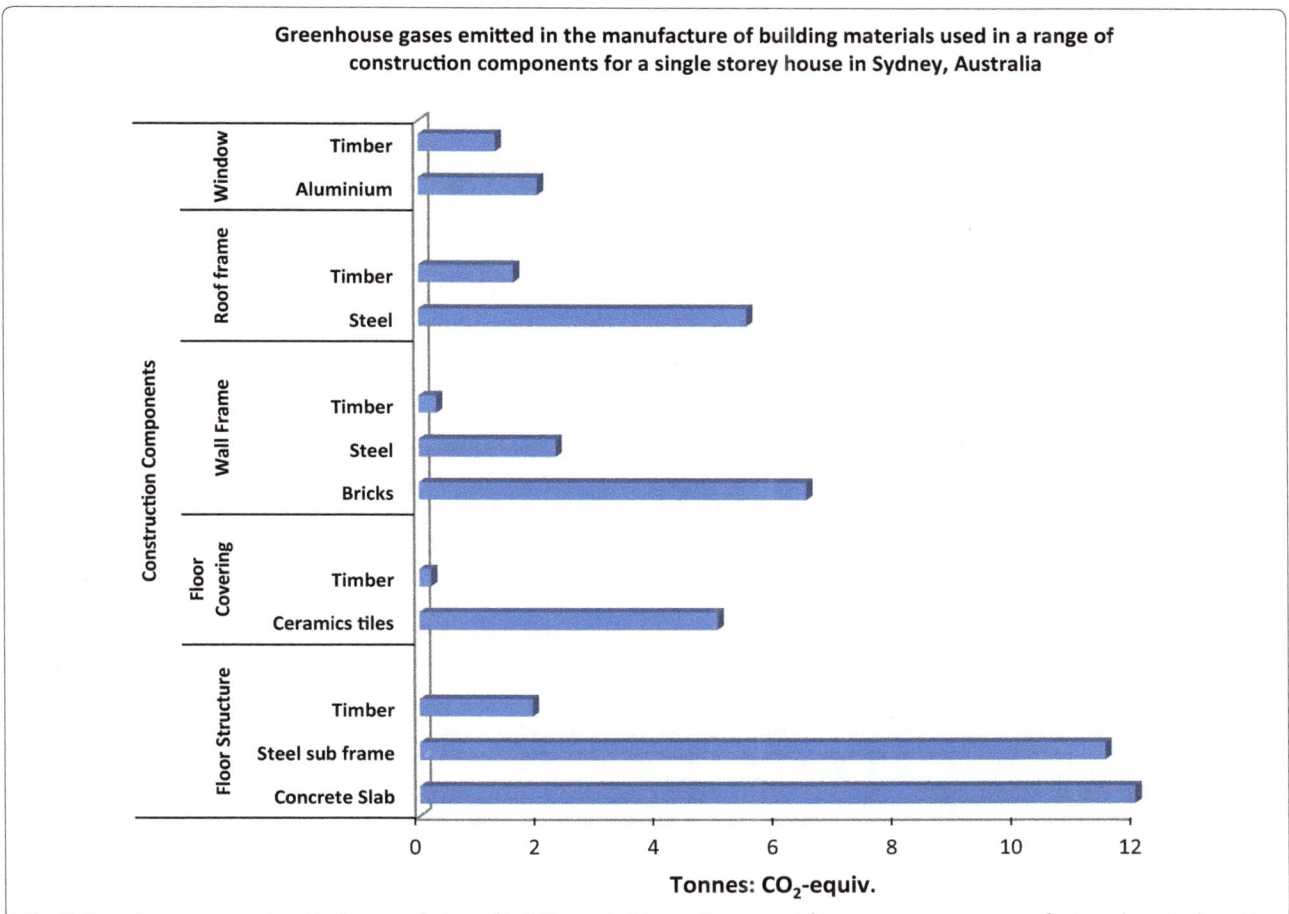

Fig. 2 Greenhouse gases emitted in the manufacture of building materials used in a range of construction components for a single storey house in Sydney, Australia. Modified from Australian Government, Forest and Wood Products Research and Development Corporation (2006)

Table 2 Carbon released in the manufacture of building materials compared with carbon stored in material. Adopted from Williamson et al. (2001)

Material	Sawn timber	Steel	Concrete	Aluminum
Carbon released (kg/m³)	15	5320	120	22,000
Carbon stored (kg/m³)	250	0	0	0

Improved sawmilling and sawing machinery

Improved sawmilling techniques, machinery and manufactured products help reducing the environmental hazards and human health problems (Harms-Ringdahl et al. 2000) and ultimately contribute to environmental sustainability in numerous ways (Gaussin et al. 2013). The use of recent technology and safety procedures could be helpful in these regards.

Laurent et al. (2016) conducted environmental assessment of a wood manufacturing industry and established environmental profile of the company so that company continue to maintain its environmental integrity as well as environmental profile of different wood products it manufactures.

First, improved and new varieties of machinery instead of old and obsolete one help reducing the wood waste, thereby reduce environmental impacts, while increasing the working efficiency in terms of time, energy and efforts. Second, hazardous energies related to machinery use can be minimised as safety and precautionary measures such as lockout system. The lockout measures is a step-by-step procedure, carried out by authorised employee to prevent inadvertent machine energization or the release of stored energy, which is in practice in Canada and the United States (Poisson and Chinniah 2016). Third, and most importantly, workers health and safety, and ergonomic measures have to be taken into account while planning and executing the sawmilling operation in the field (Jones and Kumar 2007, 2010).

Improved energy efficiency in drying system

Wood drying is the key to controlling wood quality of final products, and it consumes up to 90% of the processing time in hardwoods and more than 70% of primary processing cost, with the use of significant amounts of heat and energy (Goreshnev et al. 2013). The supplied heat is primarily used for the drying process, which is carried out in a drying kiln. Lead-time and wood quality are the major priority before energy consumption while producing the lumber (Anderson and Westerlund 2014). Therefore, the introduction of improved drying processes including simple yet environmentally friendly drying process would be beneficial to reduce the environmental impacts while ensuring the quality of final products. For example, solar drying provides opportunity as an alternative method of drying timber, while using renewable solar energy to address the shortcomings associated with fossil fuel based drying process. In addition, solar systems use the energy from sun, which is abundant, inexhaustible and nonpolluting (Akinola 1999; Akinola et al. 2006; Kumar and Kishankumar 2016), thereby has little environmental impact (Belessiotis and Delyannis 2011), unlike other forms of fossil fuel based drying methods. However, external factors such as air temperature, air velocity geographic locations, and relative humidity influence the potential drying rate. Yet, it has advantages over open-to-the-sun or air drying techniques, because the solar dryer traps solar energy to increase the temperature of circulating air and ensures the required equilibrium moisture content (EMC), enhanced shelf life, value addition, and quality enhancement (Helwa et al. 2004; LayThong 1999). These features can be further complemented by the controlled air humidity and other drying conditions, even with the use of water sprayers in some cases. However, there might still be chances that productivity is affected by weather condition such as rainfall, cloud cover, and less predictable outcomes than that of industrial kilns (Haque and Langrish 2005).

Solar kiln drying is usually affected by geographic and climatic conditions. For example, the temperature inside the kiln is affected by the ambient temperature and solar radiation (Hasan and Langrish 2014; Phonetip et al. 2017a). Areas with low humidity offer a productivity performance for solar kilns (Ong 1997). According to Phonetip et al. (2017b), decreasing the relative humidity (RH) level to 40% can dry boards faster than when the conditions are maintained at 60% RH. Taking advantage of a low ambient RH could result in several benefits, such as lowering the consumption of water and energy.

A study by Phonetip et al. (2018) described a method that used the combined tools of GIS and Fuzzy theory to identify the most suitable locations for solar kilns based on variables of geographical and climatic conditions and restricted areas, using an example location in Vientiane, Laos. This method can be applied to different geographical regions and local climatic seasons.

Therefore, in order to improve efficiency and reduce the environmental impacts, various kind of solar drying are in practice, such as integral, distributed and mixed type solar dryers based on the mode of utilization of solar heat, and greenhouse system, external collector, and mixed mode solar drying depending on greenhouse systems. Currently, enhanced solar timber kilns can also be used with characteristic features of solar energy storage with independent heating, integration of an air heater in the storage and in the drying chamber, and management of different drying cycles based on the quality control of the products (Ugwu et al. 2015).

Overall, solar drying has more environmental advantages due to shorter drying time, and better drying quality than that of air-drying. Similarly, it requires, low operating costs and lower training manpower, along with the chances of having EMC in broad range of climates, and ultimately constitutes an environmentally friendly technique due to its reliance on renewable resources and low environmental impact.

Studies on improving energy efficiencies have shown that if available state-of-the-art technologies are applied in drying kilns, it could reduce the heat consumption by about 60% (Anderson and Westerlund 2011, 2014; Johansson and Westerlund 2000). Moreover, a study by Anderson and Westerlund (2014) using the Torksim simulation program has further reported that energy recovery technologies in the sawmill industry could save considerable amounts of energy and biomass for the other purpose. According to the authors, use of a heat exchanger, mechanical heat pump, and open absorption system are the major energy recovery technologies. For instance, open absorption system is the most effective which will reduce energy consumption by 67.5%, whereas mechanical heat pump could also decrease a significant amount of energy usage and result in a large heat surplus in the drying system. However, the latter requires high consumption of electricity. In contrast, use of heat exchanger technology contributes only a marginal increase in energy efficiency of 4–10% depending upon the sawmill condition and drying scheme. Therefore, findings of such studies mainly related to the result of higher energy efficiency from open absorption system should be promoted to reduce the energy use and GHG emission, increase the efficiency, and minimize the environmental impacts.

Use of environmentally friendly chemicals
Preservatives

There is a growing trend towards environmentally friendly preservatives to reduce the environmental impacts while improving the durability of timber products. In this context, environmentally benign wood preservative systems can be developed with proper combination of an organic biocide with metal chelating and/or antioxidant additives (Schultz and Nicholas 2002). That will not only enhance protection of wood against fungi as compared to the biocide alone, but also consequently, help reduce the environmental impacts especially on land and water resources. Physical barriers have been accepted as alternative non-biocidal wood protection method in India as they reduce leaching and subsequent negative impacts of wood preservative components to the organisms in vicinity (Sreeja and Edwin 2013).

Policy and legislative measures to ban the use of toxic preservatives, and growing awareness on using less toxic and more environmentally friendly preservatives would be another way to reduce the environmental impacts (Lin et al. 2009). For example, a number of toxic preservatives such as CCA, cresote, and preservatives based on volatile organic solvent (VOC), are restricted in Europe and the USA. Instead, use of environmentally friendly preservatives such as copper-organic preservatives replacing CCA, CCB and CCP preservatives, microemulsion water-dilutable concentrates with organic fungicides and insecticides, and water and solvent-based coloured preservatives replacing creosote, have emerged to fill the gap (Coggins 2008; EU 2006). Therefore, stringent environmental policies will have to be practiced to reduce the use of harmful chemicals in wood preservatives, as practiced under Biocidal Products Directives within European Union (Hingston et al. 2001) and restricted pesticidal use of three primary heavy duty wood preservatives ("HDWPs") under Environmental Protection Agency, USA in 2008 (Tomasovic 2012).

Australian Government Department of Agriculture and Water Resources (2016) accepts certain permanent preservative treatments as biosecurity treatments for use on certain timber products and timber packaging. For a timber preservative treatment to sufficiently address biosecurity risks and be accepted as a biosecurity treatment by the department, it must meet the following requirements:

suitable treatment application methods, preservative penetration zone requirements, preservative retention requirements and accepted preservative formulations.

Adhesives

As biochemical adhesives have 22% fewer environmental impacts than that of petrochemical adhesives (Yang and Rosentrater 2015), use of biochemical adhesives should be encouraged. For example, Pizzi (2006) have identified bio-based adhesives such as tannin, protein, carbohydrate, lignin, and unsaturated oil to maintain both environmentally friendly alternatives and efficient traditional adhesives of the timber industries. Consistent with these findings, Navarrete et al. (2012) conducted a comparative study between the emission from particle board produced with UF and the natural adhesives and found that there was at least seven times higher emission of urea formaldehyde than that of biochemical based adhesive such as lignin and tannin. Yet, the impacts from these biochemical adhesive is quite significant, therefore various innovative measures have to be taken to reduce the impacts on the environment. For example, adhesive based on hexamine could be used to reduce the impact of formaldehyde. Similarly, environmentally-friendly products such as tannin-hexamine adhesive, and in case of lignin adhesive, adhesives pressed at high speed, in the presence of pre-methylated lignin could be used to reduce the environmental impacts (Yang and Rosentrater 2015). Furthermore, soy-based adhesive has also been effective in increasing the wet bond strength with the use of polyamidoamine–epichlorohydrin (PAE) resin as a co-reactant. That has led to resurgence in soy-based adhesive consumption with minimal environmental impacts (Frihart and Birkeland 2014).

In India, extensive research studies have been carried out since 1980 on extending the soya flour to synthetic resin (Sarkar et al. 1985; Zoolagud et al. 1997). Mamatha et al. (2011) developed phenol-soya adhesive for the manufacture of exterior grade plywood. About 40% substitution of phenol by soya was optimized for making exterior grade plywood having strength properties confirming to relevant standard requirements. The substitution not only helps to minimize the formaldehyde release from the products and disposal of waster for better utilization, but also reduces the air and water pollution along with minimization of production cost of the plywood products due to reduced cost in resin system (Mamatha et al. 2011).

A recently published book "Bio-based Wood Adhesives" by Zhongqi He (2017) provides the synthesis of the fundamental knowledge and latest research on bio-based adhesives from a remarkable range of natural products and byproducts, and identifies need areas and provides directions of future bio-based adhesive research.

Policy measures should be placed on restriction of VOCs to the atmosphere. Likewise, an interesting shift from using less environmentally harmful adhesive in joining wood components for furniture and interior joinery by wood welding technology without the use of adhesive has been also initiated. This could be explained by the polymerization and cross-linking of lignin and of carbohydrate-derived furfural (Gfeller et al. 2003). Many studies have been conducted on wood welding using high speed rotation welding (Pizzi et al. 2004; Belleville et al. 2016) and linear welding (Mansouri et al. 2010; Martins et al. 2013; Belleville et al. 2017). If this technique could be scaled up successfully, it would contribute to reduce the adhesive based emission and environmental hazard involved in the timber productions process.

While choosing the adhesive during the course of timber product manufacturing and production processes, proper attention has to be given to environmentally friendly either bio-based adhesive or techniques without using adhesive as far as possible to reduce the impact both on the environment, and the human health.

Wood coating

Over the past few years, regulation under the Clean Air Act (USA) and consumer demand for low-VOC finishes have led to the creation of a variety of new products. Many penetrating finishes, such as semi-transparent stains, have low solids content (pigment, oils, polymers) and are being reformulated to meet low-VOC regulations. To meet the VOC requirements, these reformulated finishes may contain higher solids content, reactive diluents (dilutants or thinners), new types of solvents and/or co-solvents, or other non-traditional substitutes. These low-VOC requirements favour film-forming formulations over products that penetrate the wood surface, since traditional wood stains were formulated to penetrate the wood, and the new formulations that meet the VOC requirements may not penetrate as well.

Another way to decrease air emissions from wood finishes is to change the formulation to a water-based coating. The new water-based products achieve a dramatic improvement over solvent-based finishes in terms of VOC emissions and human comfort and health. Companies that have successfully switched to water-based coatings have worked closely with their suppliers to determine the best water-based formula for their specific uses.

Wood waste management

Eshun et al. (2012) and EPA (2015) have listed ways to minimize wood waste and wood waste management. Main measures to wood waste management include, among others, good operating practices, technology changes, changes in input materials, waste recycling, and waste reuse/recover practices. Similarly, EPA (2015) has described the waste reduction opportunities via lumber receiving, drying and storage; rough end and gluing; machining and sanding; assembly; finishing;

packing, shipping and warehouse; building and equipment maintenance.

It is interesting to note that developed countries such as Australia and Sweden place more emphasis on waste recycling, and waste reuse/recover, whereas other countries such as Taiwan, South Africa, and India have put emphasis on improving almost all processing and manufacturing techniques identified above. This might be due to the fact that developed countries may already have good operating practices and required technology in the timber production sector. A study carried out by Daian and Ozarska (2009) in Australia has highlighted the need for using recovered and waste wood in the mulching and compost sector, bioenergy sector, animal product sector, and engineered wood product sector.

During 2013 and 2014, Italy re-used 95% of the waste wood to produce particleboard, while Germany and United Kingdom shared the account to 34 and 53% respectively (Garcia and Hora 2017).

In Europe, the Waste Framework Directive (2008/98/EC) provides a guideline of basic concepts and procedure related to the waste management. A concept called "end-of-waste criteria" has been introduced that is used as a guideline to determine when a waste ceases to be a waste and becomes a secondary raw material. In this concept, waste hierarchy is maintained from landfill through recovery, recycling, reuse to reduce from the least favoured to most favoured option (Garcia and Hora 2017). The values and ways to wood recovery and recycling, classified into direct and indirect recycling, have been well illustrated by Taylor and Warnken 2008 (Fig. 3). Indirect recycling of wood products results in compost or mulch which will decompose into carbon dioxide aerobically. Similarly, direct recycling and reuse of recovered wood into timber products prolong the service life of the timber and at the same time provides the opportunity of potential recovery at end-of-life. Degradable organic carbon contained in the wood disseminate into methane in the landfill site. Methane has 25 times higher global warming potential, so recovering wood will prevent the greenhouse gases (Taylor and Warnken 2008).

Integrated industrial sites

With due consideration of growing energy demand from the different industrial sectors, an essential strategy would be the development of highly integrated industrial sites. Such sites would serve to lower energy and resource consumption and, at the same time, complement one plant to another. For example, saw mills would supply huge biomass to other pellet plants, pulp and paper plant, and combined heat and power (CHP) plants, and some portion of such biomass would be used to fulfill the internal heat requirements as well (Anderson and Toffolo

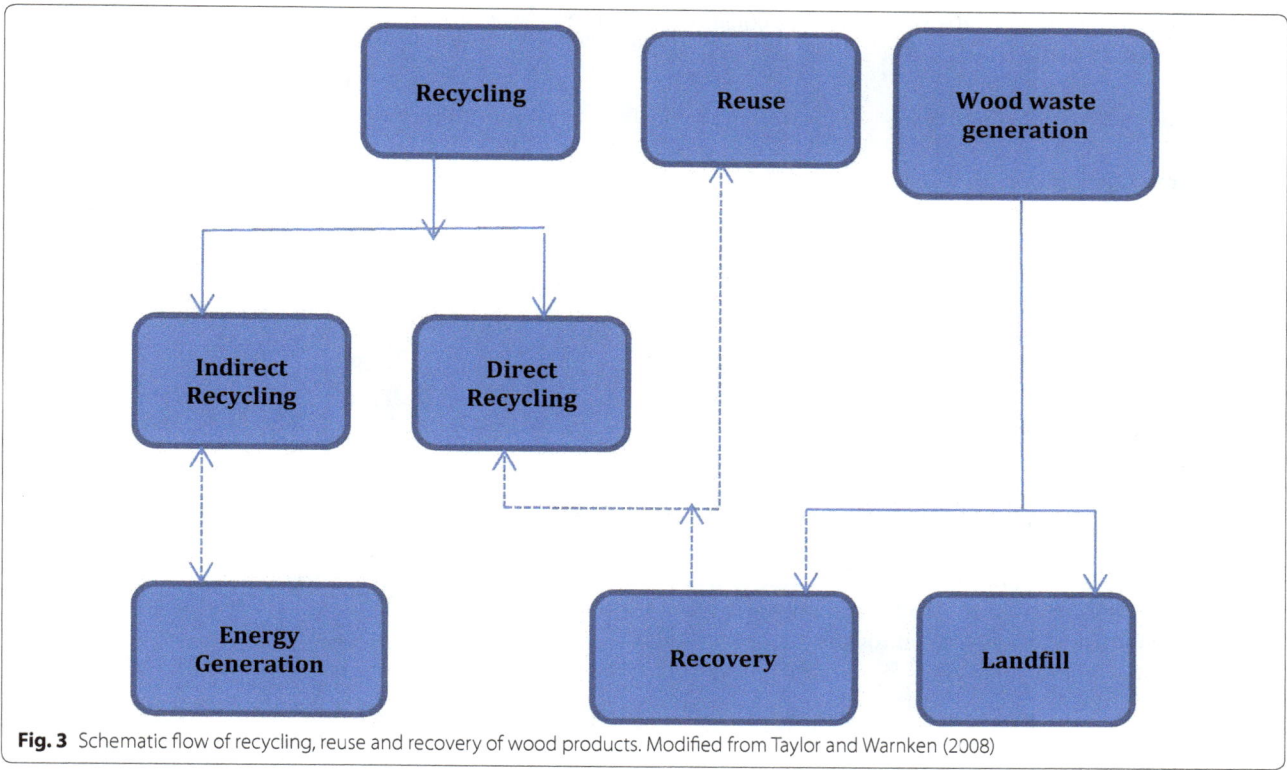

Fig. 3 Schematic flow of recycling, reuse and recovery of wood products. Modified from Taylor and Warnken (2008)

2013). Therefore, if these plants were combined it would reduce the energy and resource consumption and help reduce the environmental impacts.

Energy efficient biofuel and improved transportation system

Environmental impacts associated with transportation could be minimized by changing the source of energy and mode of transportation of the timber products. Use of renewable sources of energy such as electricity generated from hydropower, and biofuel, instead of fossil based energy would reduce emission during the transportation. Interestingly, in Sweden it was reported that transporting forest products via railway transport requires less process energy than by using road vehicles. Furthermore, use of biofuel instead of fossil fuel in a lorry could replace about 96% of fossil energy (Lindholm and Berg 2005).

By- or co-product or even wood waste can be a feed-stock for second generation biofuel (Cantrell et al. 2008; Havlík et al. 2011; Sklar 2008), or be supplied by dedicated plantations. The latter ones seems more promising and can be established on marginal lands (Tilman et al. 2006; Zomer et al. 2008, Havlík et al. 2011), or enter into direct competition with conventional agricultural production (Field et al. 2008; Gurgel et al. 2007) and other services. Therefore, improved transportation system for timber products with the use of energy efficient biofuel should be promoted.

Safe disposal

Environmental impacts related to disposal of wood wastage can be minimized by using a minimum amount of materials required for the production process, and renewable materials, and by avoiding materials that deplete natural resources while prompting recycle and recyclable material and waste by-products. Similarly, those left for disposal should be put into safe disposal landfill sites. Landfill sites represent a major disposal option for wood wastes in many countries. For example, in Australia, it is estimated that approximately 2.3 million tonnes of solid wood products are placed in all Australian landfills each year (Ximenes et al. 2008). There should be reliable landfill side for safe disposal of wood wastage.

Policy support

Overarching policy and institutional support should be in place in order to realize the improvements with regard to minimizing adverse environmental impacts as a result of the production process of timber products in general and sawmilling in particular. Similarly, it should encourage robust production planning (Zanjani et al. 2010), suitable policy measure of impacts minimization and quality enhancement (Loxton et al. 2013), and further collaboration with other stakeholders.

Others

Apart from aforementioned measures to minimize environmental impact as a result of timber production process, some other social, ecological and economic factors should also be taken into account. For example, in order to obtain sustained supply for raw timber from the forest, the timbers supplied from sustainably managed and certified forest is being encouraged (Päivinen et al. 2012). In addition, timber industry should incentivize and support the endeavors of both government and private sectors on plantation and management of forests, so that it would create harmony among them and help the regular supply of raw materials to the industry. Similarly, societal need, interest, and capacity should also be considered while designing, and operating the sawmill industry. Further, proper coordination and collaboration among different stakeholders are also crucial for the success of the industry.

Conclusion

Major sources of environmental impacts occur throughout the wood supply chain from sawmills to final products. Many studies have been conducted with the aim to identify environmental impacts of timber products. The studies, in particular the ones based on LCA methods have provided comprehensive coverage of different processes such as energy consumption, manufacturing process and their impacts on the environments. The impacts can be minimized in various ways: changes in energy consumption behavior, promotion of renewable energy, improved sawing and sawmilling practices, proper wood waste management, use of less toxic chemicals on the treatment of wood and timber products, and most importantly use of energy efficient and environment friendly drying techniques and energy sources such as effective air drying, improved solar and kiln drying, microwave modification and vacuum technology *inter alia*. Moreover, there needs to be proper policy support to promote the concept of integrated industrial site with effective coordination and collaboration among relevant stakeholders. That collaborative work not only helps produce quality forest products, but also reduces their concomitant environmental impacts. Moreover, it should help to ensure the broad goal of environmental sustainability, while recognizing the timber sector as a part of an integrated approach of sustainable development.

Authors' contributions
SA drafted the manuscript. BO provided conceptual guidance and polished and revised the manuscript. Both authors read and approved the final manuscript.

Author details
[1] Department of Forests, Ministry of Forests and Environment, Kathmandu, Nepal. [2] School of Ecosystem and Forests Science, University of Melbourne, Burnley Campus, Melbourne, Australia.

Competing interests
The authors declare that they have no competing interests.

Funding
Not applicable.

References

Akinola AO (1999) Development and performance evaluation of a mixed-mode solar food dryer. Federal University of Technology, Akure

Akinola O, Akinyemi A, Bolaji B (2006) Evaluation of traditional and solar fish drying systems towards enhancing fish storage and preservation in nigeria: Abeokuta local governments as case study. J Fisheries Int 1:44–49

Anderson J-O, Toffolo A (2013) Improving energy efficiency of sawmill industrial sites by integration with pellet and CHP plants. Appl Energy 111:791–800

Anderson J-O, Westerlund L (2011) Surplus biomass through energy efficient kilns. Appl Energy 88:4848–4853

Anderson J-O, Westerlund L (2014) Improved energy efficiency in sawmill drying system. Appl Energy 113:891–901

APA (2017) Sustainable building, sustainable future. APA, Washington D.C

Atkinson G, Mourato S (2006) Cost-benefit analysis and the environment: recent developments. Organisation for economic co-operation and development. OECD, Paris. ISBN 9264010041

Australian Government, Department of Agriculture and Water Resources 2016. Timber permanent preservative treatment requirements. http://www.agriculture.gov.au/import/goods/timber/approved-treatments-timber/permanent-preservative-treatment. Accessed 19 Mar 2018

Australian Government, Forest and Wood Products Research and Development Corporation (2006) Forest, wood and Australia's carbon balance. Australian Government, Forest and Wood Products Research and Development Corporation, Australia

Baccini AGSJ, Goetz SJ, Walker WS, Laporte NT, Sun M, Sulla-Menashe D, Hackler J, Beck PSA, Dubayah R, Friedl MA, Samanta S (2012) Estimated carbon dioxide emissions from tropical deforestation improved by carbon-density maps. Nat Climate Change 2:182

Belessiotis V, Delyannis E (2011) Solar drying. Sol Energy 85:1665–1691

Belleville B, Pizzi A, Ozarska B (2016) Assessing the potential of wood welding for Australian eucalypts and exotic species. Eur J Wood and Wood Prod 74:753–757

Belleville B, Amirou S, Plzzi A, Ozarska B (2017) Optimization of wood welding parameters for Australian hardwood species. BioResources 12:1007–1014

Bergman RD, Bowe SA (2008) Environmental impact of producing hardwood lumber using life-cycle inventory. Wood Fiber Sci 40:448–458

Bergman R, Puettmann M, Taylor A, Skog KE (2014) The carbon impacts of wood products. For Products J 64:220–231

Bluff E (2014) Safety in machinery design and construction: performance for substantive safety outcomes. Saf Sci 66:27–35

Cabeza LF, Rincón L, Vilariño V, Pérez G, Castell A (2014) Life cycle assessment (LCA) and life cycle energy analysis (LCEA) of buildings and the building sector: a review. Renew Sustain Energy Rev 29:394–416

Cantrell KB, Ducey T, Ro KS, Hunt PG (2008) Livestock waste-to-bioenergy generation opportunities. Biores Technol 99:7941–7953

Cetin NS, Özmen N (2002) Use of organosolv lignin in phenol–formaldehyde resins for particleboard production: I. Organosolv lignin modified resins. Int J Adhes Adhes 22:477–480

Coggins C (2008) Trends in timber preservation—a global perspective. J Tropical For Sci 20:264–272

Daian G, Ozarska B (2009) Wood waste management practices and strategies to increase sustainability standards in the Australian wooden furniture manufacturing sector. J Clean Prod 17:1594–1602

de Meijer M (2001) Review on the durability of exterior wood coatings with reduced VOC-content. Prog Org Coat 43:217–225

Dionco-Adetayo EA (2001) Utilization of wood wastes in Nigeria: a feasibility overview. Technovation 21:55–60

Dodoo A, Gustavsson L, Sathre R (2014a) Lifecycle carbon implications of conventional and low-energy multi-storey timber building systems. Energy Build 82:194–210

Dodoo A, Gustavsson L, Sathre R (2014b) Lifecycle primary energy analysis of low-energy timber building systems for multi-storey residential buildings. Energy Build 81:84–97

ERDC (2001) Aiming for sustainable product development furniture and building products. Energy Research and Development Corporation, Centre for Design at RMIT, Australia

Eshun JF, Potting J, Leemans R (2012) Wood waste minimization in the timber sector of Ghana: a systems approach to reduce environmental impact. J Clean Prod 26:67–78

EU (2006) Directive 2006/121/EC of the European Parliament and of the Council of 18 December 2006 amending Council Directive 67/548/EEC on the approximation of laws, regulations and administrative provisions relating to the classification, packaging and labelling of dangerous substances in order to adapt it to Regulation (EC) No 1907/2006 concerning the Registration, Evaluation, Authorisation and Restriction of Chemicals (REACH) and establishing a European Chemicals Agency

FAO (2001) Global forest resource assessment 2000. United Nations Food and Agriculture Organization (FAO), Rome

Field CB, Campbell JE, Lobell DB (2008) Biomass energy: the scale of the potential resource. Trends Ecol Evol 23:65–72

FIM (2017) Update on global timber demand. FIM Service Limited, Burford

Frihart CR, Birkeland MJ (2014) Soy properties and soy wood adhesives: soy-based chemicals and materials. American Chemical Society, Washington D.C

Garcia CA, Hora G (2017) State-of-the-art of waste wood supply chain in Germany and selected European countries. Waste Manag 70:189–197

Gaussin M, Hu G, Abolghasem S, Basu S, Shankar M, Bidanda B (2013) Assessing the environmental footprint of manufactured products: a survey of current literature. Int J Prod Econ 146:515–523

George A (2008) Impact of carbon trading on wood products, forest and wood products. Australia, Forest and Wood Products Australia Limited

Gfeller B, Zanetti M, Properzi M, Pizzi A, Pichelin F, Lehmann M, Delmotte L (2003) Wood bonding by vibrational welding. J Adhes Sci Tech 17:1573–1589

Goedkoop M, Heijungs R, Huijbregts M (2008) A life cycle impact assessment method which comprises harmonised category indicators at the midpoint and the endpoint level. Report I: characterisation. Den Haag, Impact world

Goreshnev M, Kazarin A, Lopatin V, Sekisov F, Smerdov O (2013) Combined timber drying method'. J Eng Phys Thermophys 86:336–339

Gunn JS, Ganz DJ, Keeton WS (2012) Biogenic vs. geologic carbon emissions and forest biomass energy production. GCB Bioenergy 4:239–242

Gurgel A, Reilly JM, Paltsev S (2007) Potential land use implications of a global biofuels industry. J Agric Food Ind Organ 5:2

Gustavsson L, Sathre R (2006) Variability in energy and carbon dioxide balances of wood and concrete building materials. Build Environ 41:940–951

Haque M, Langrish T (2005) Assessment of the actual performance of an industrial solar kiln for drying timber. Dry Technol 23:1541–1553

Harms-Ringdahl L, Jansson T, Malmén Y (2000) Safety, health and environment in small process plants—results from a European survey. J Saf Res 31:71–80

Hasan M, Langrish TA (2014) Numerical simulation of a solar kiln design for drying timber with different geographical and climatic conditions in Australia. Dry Technol 32:1632–1639

Havlík P, Schneider UA, Schmid E, Böttcher H, Fritz S, Skalský R, Aoki K, De Cara S, Kindermann G, Kraxner F, Leduc S (2011) Global land-use implications of first and second generation biofuel targets. Energy Policy 39:5690–5702

He Z (2017) Bio-based wood adhesives. CRC Press, Taylor and Francis Group, Boca Raton, p 366

Helwa N, Khater H, Enayet M, Hashish M (2004) Experimental evaluation of solar kiln for drying wood. Dry Technol 22:703–717

Hertwich EG, Pease WS, Koshland CP (1997) Evaluating the environmental impact of products and production processes: a comparison of six methods. Sci Total Environ 196:13–29

Hingston J, Collins C, Murphy R, Lester J (2001) Leaching of chromated copper arsenate wood preservatives: a review. Environ Pollut 111:53–66

Ibn-Mohammed T, Greenough R, Taylor S, Ozawa-Meida L, Acquaye A (2013) Operational vs. embodied emissions in buildings—a review of current trends. Energy Build 66:232–245

Ivanova G, Rolfe J (2011) Using input-output analysis to estimate the impact of a coal industry expansion on regional and local economies. Impact Assess Proj Apprais 29:277–288

Johansson L, Westerlund L (2000) An open absorption system installed at a sawmill: description of pilot plant used for timber and bio-fuel drying. Energy 25:1067–1079

Jones T, Kumar S (2007) Comparison of ergonomic risk assessments in a repetitive high-risk sawmill occupation: saw-filer. Int J Ind Ergon 37:744–753

Jones T, Kumar S (2010) Comparison of ergonomic risk assessment output in four sawmill jobs. Int J Occup Saf Ergon 16:105–111

Jurgensen M, Harvey A, Graham R, Page-Dumroese D, Tonn J, Larsen M, Jain T (1997) Review article: impacts of timber harvesting on soil organic matter, nitrogen, productivity, and health of inland northwest forests. Forest Sci 43:234–251

Klein D, Wolf C, Schulz C, Weber-Blaschke G (2016) Environmental impacts of various biomass supply chains for the provision of raw wood in Bavaria, Germany, with focus on climate change. Sci Total Environ 539:45–60

Kumar S, Kishankumar VS (2016) Thermal energy storage for a solar wood drying kiln: estimation of energy requirement. J Indian Acad Wood Sci 13:33–37

Laurent AB, Menard JF, Lesage P, Beauregard R (2016) Cradle-to-gate environmental life cycle assessment of the portfolio of an innovative forest products manufacturing unit. BioResources 11:8981–9001

LayThong H (1999) Rubberwood processing and utilization. Malayan forest records, 39th edn. Rubberwood processing and utilization, Malaysia

Le Quéré C, Raupach MR, Canadell JG, Marland G, Bopp L, Ciais P, Conway TJ, Doney SC, Feely RA, Foster P (2009) Trends in the sources and sinks of carbon dioxide. Nat Geosci 2:831–836

Lela B, Barišić M, Nižetić S (2016) Cardboard/sawdust briquettes as biomass fuel: physical–mechanical and thermal characteristics. Waste Manag 47:236–245

Lin L-D, Chen Y-F, Wang S-Y, Tsai M-J (2009) Leachability, metal corrosion, and termite resistance of wood treated with copper-based preservative. Int Biodeterior Biodegrad 63:533–538

Lindholm E-L, Berg S (2005) Energy requirement and environmental impact in timber transport. Scand J For Res 20:184–191

Lippke B, Wilson J, Perez-Garcia J, Bowyer J, Meil J (2004) CORRIM: life-cycle environmental performance of renewable building materials. For Prod J 54:8–19

Loxton EA, Schirmer J, Kanowski P (2013) Designing, implementing and monitoring social impact mitigation strategies: lessons from forest industry structural adjustment packages. Environ Impact Assess Rev 42:105–115

Mamatha BS, Sujatha D, Nath SK (2011) Soya based phenolic resins for plywood manufacture. J Indian Acad Wood Sci 8:112–116

Mansouri HR, Pizzi A, Leban J-M (2010) End-grain butt joints obtained by friction welding of high density eucalyptus wood. Wood Sci Technol 44:399–406

Martins SA, Ganier T, Pizzi A, Del Menezzi CHS (2013) Parameter scanning for linear welding of Brazilian Eucalyptus benthamii wood. Eur J Wood Prod. 71:525–527

McDevitt JE, Grigsby WJ (2014) Life cycle assessment of bio-and petrochemical adhesives used in fiberboard production. J Polym Environ 22:537–544

Miles L, Kapos V (2008) Reducing greenhouse gas emissions from deforestation and forest degradation: global land-use implications. Science 320:454–455

Miner R (2010) Impact of the global forest industry on atmospheric greenhouse gases. Food and Agriculture Organization of the United Nations, Rome

Mirabella N, Castellani V, Sala S (2014) LCA for assessing environmental benefit of eco- design strategies and forest wood short supply chain: a furniture case study. Intl J Life Cycl Assess 19:1536–1550

Navarrete P, Pizzi A, Tapin-Lingua S, Benjelloun-Mlayah B, Pasch H, Rode K, Delmotte L, Rigolet S (2012) Low formaldehyde emitting biobased wood adhesives manufactured from mixtures of tannin and glyoxylated lignin. J Adhes Sci Technol 26:1667–1684

NCASI (2006) Energy and greenhouse gas impacts of substituting wood products for non-wood alternatives in residential construction in the United States, 925th edn. National Council for Air and Stream Improvement, Inc., Canada

Norgate TE, Jahanshahi S, Rankin WJ (2007) Assessing the environmental impact of metal production processes. J Clean Prod 15:838–848

Ong KS (1997) Comparison between drying of timber in a solar dryer and in an electrically-heated kiln. Dry technol 15:1231–1237

Päivinen R, Lindner M, Rosén K, Lexer M (2012) A concept for assessing sustainability impacts of forestry-wood chains. Eur J For Res 131:7–19

Phonetip K, Brodie G, Ozarska B, Belleville B (2017a) Drying *Eucalyptus delegatensis* timber by solar kiln using an intermittent drying schedule of conventional laboratory kiln. Dry. Technol 17:1–2

Phonetip K, Ozarska B, Belleville B, Brodie G (2017b) Using a conventional laboratory kiln as a simulation of solar cyclic drying. In: *IUFRO* division 5 conference/SWST 60th international convention forest sector innovations for a greener future. Vancouver, Canada

Phonetip K, Ozarska B, Brodie G, Belleville B, Boupha L (2018) Applying a GIS-based fuzzy method to identify suitable locations for solar kilns. BioResources. 13:2785–2799

Pizzi A (2006) Recent developments in eco-efficient bio-based adhesives for wood bonding: opportunities and issues. J Adhes Sci Technol 20:829–846

Pizzi A, Mittal KL (2011) Wood adhesives CRC Press, Boston

Pizzi A, Leban J-M, Kanazawa F, Properzi M, Pichelin F (2004) Wood dowel bonding by high speed rotation welding. J Adhes Sci Technol 18:1263–1278

Poisson P, Chinniah Y (2016) Managing risks linked to machinery in sawmills by controlling hazardous energies: theory and practice in eight sawmills. Saf Sci 84:117–130

Puettmann ME, Wilson JB (2007) Life-cycle analysis of wood products: cradle-to-gate LCI of residential wood building materials. Wood Fiber Sci 37:18–29

Ramasamy G, Ratnasingam J, Bakar ES, Halis R, Muttiah N (2015) Assessment of environmental emissions from sawmilling activity in Malaysia. BioResources 10:6643–6662

Ramesh T, Prakash R, Shukla K (2010) Life cycle energy analysis of buildings: an overview. Energy Build 42:1592–1600

Roy P, Nei D, Orikasa T, Xu Q, Okadome H, Nakamura N, Shiina T (2009) A review of life cycle assessment (LCA) on some food products. J Food Eng 90:1–10

Rus RM, Daud A, Musa KI, Naing L (2008) Knowledge, attitude and practice of sawmill workers towards noise- nduced hearing loss in kota bharu, Kelantan. Malays J Med Sci 15:28

Sarkar A, Naha P, Nag J (1985) Development of suitable economic adhesives based on synthetic resin as well as proteinous materials and their application in the production of tea-chest plywood. IPIRTI research report No. 07

Schultz TP, Nicholas DD (2002) Development of environmentally-benign wood preservatives based on the combination of organic biocides with antioxidants and metal chelators. Phytochemistry 61:555–560

Sklar T (2008) Ethanol from wood waste an opportunity for refiners. Oil Gas J 106:54–59

Sommerhuber PF, Welling J, Krause A (2015) Substitution potentials of recycled HDPE and wood particles from post-consumer packaging waste in wood-plastic composites. Waste Manag. 46:76–85

Sreeja A, Edwin L (2013) Physical barriers: an alternative to prevent negative impacts of chemically treated wood. J Indian Acad Wood Sci 10:140–147

Suter F, Steubing B, Hellweg S (2017) Life cycle impacts and benefits of wood along the value chain: the case of Switzerland. J Ind Ecol 21:874–886

Taylor J (2003) Review of the environmental impact of wood compared with alternative products used in the production of furniture. Forest and Wood Products Research and Development Coorporation, Australia

Taylor J, Warnken M (2008) Wood recovery and recycling: a source book for Australia. Forest and Wood Products Research and Development Coorporation, Australia

Taylor J, Mann R, Reilly M, Warnken M, Pincic D, Death D (2005) Recycling and end-of-life disposal of timber products. Forest and Wood Products Research and Development Coorporation, Australia

Tilman D, Hill J, Lehman C (2006) Carbon-negative biofuels from low-input high-diversity grassland biomass. Science 314:1598–1600

Timmermann V, Dibdiakova J (2014) Greenhouse gas emissions from forestry in East Norway. Int J Life Cycle Assess 19:1593–1606

Tomasovic BS (2012) The fate of treated wood infrastructure. Va Environ Law J 30:28

Ugwu S, Ugwuishiwu B, Ekechukwu O, Njoku H, Ani A (2015) Design, construction, and evaluation of a mixed mode solar kiln with black-painted pebble bed for timber seasoning in a tropical setting. Renew Sustain Energy Rev 41:1404–1412

Van Hilst F, Hoefnagels R, Junginger M, Shen L, Wicke B (2017). Sustainable biomass for energy and materials: A greenhouse gas emission perspective. Working paper: Copernicus Institute of Sustainable Development, Utrecht, Utrecht University

Williamson TJ, Olweny M, Moosmayer V, Pullen SF, Bennetts H (2001) Environmentally friendly housing using timber-principles. Forest and Wood Products Research and Development Corporation, Australia

Wilnhammer M, Lubenau C, Wittkopf S, Richter K, Weber-Blaschke G (2015) Effects of increased wood energy consumption on global warming potential, primary energy demand and particulate matter emissions on regional level based on the case study area Bavaria (Southeast Germany). Biomass Bioenergy 81:190–201

Wootton JT (2012) Effects of timber harvest on river food webs: physical, chemical and biological responses. PLoS ONE 7:43561

Ximenes FA, Gardner WD, Cowie AL (2008) The decomposition of wood products in landfills in Sydney, Australia. Waste Manag 28(11):2344–2354

Yang M, Rosentrater K (2015) Environmental effects and economic analysis of adhesives: a review of life cycle assessment (LCA) and techno-economic analysis (TEA) in 2015 ASABE Annual International Meeting, p 1

Zanjani MK, Ait-Kadi D, Nourelfath M (2010) Robust production planning in a manufacturing environment with random yield: a case in sawmill production planning. Eur J Operational Res 201:882–891

Zhang W, Ma Y, Xu Y, Wang C, Chu F (2013) Lignocellulosic ethanol residue-based lignin–phenol–formaldehyde resin adhesive. Int J Adhes Adhes 40:11–18

Zomer RJ, Trabucco A, Bossio DA, Verchot LV (2008) Climate change mitigation: a spatial analysis of global land suitability for clean development mechanism afforestation and reforestation. Agric Ecosyst Environ 126:67–80

Zoolagud SS, Rangaraju TS, Narayana prasad TR, Mohandas KK (1997) Study on the suitability of deoiled soya bean flour and Jatropha seed cake flour as extenders for UF and PF resin adhesives for bonding MR and BWR grades of plywood. IPIRTI research report No. 97

7

Radon emanation and heavy-metals assessment of historical warm and cold springs in Nigeria using different matrices

E Babatope Faweya[1]*, O Gabriel Olowomofe[2], H Taiwo Akande[2] and T Adeniyi Adewumi[3]

Abstract

Background: In recent years, attention has been drawn to radon gas as main risk factor for lung cancer. Radon is colourless, odourless and tasteless radioactive noble gas. To mitigate radon effects, water consume by populace needs to be conserved. Radon concentration in water and heavy metals concentrations in sediment samples from historical cold and warm springs at Ikogosi were determined using Durridge RAD-7 analyzer with RAD H_2O accessory and atomic absorption spectrophotometer.

Results: The mean activity concentration of radon in water samples ranged from 0.07 to 0.36 with overall mean value 0.20 Bq L^{-1}, 35–210 with an overall mean value 75.9 Bq L^{-1} and 11.7–140.0 with an overall mean 79.4 Bq L^{-1} for bottled, cold and warm water samples respectively. The calculated total effective dose values were below 100 μSv $year^{-1}$ recommended by WHO. The result of elemental analysis showed that the mean values of metals concentrations were Pb (2.9–11.8 mg kg^{-1}), Cu (3.8–12.8 mg kg^{-1}), Fe (945.0–2010.0 mg kg^{-1}), Cd (0.6–1.7 mg kg^{-1}) and Ni (0.3–2.6 mg kg^{-1}).

Conclusions: The results revealed values not higher than recommended permissible limit and background values. The pollution load index, revealed that the overall contamination of metals indicated no significant pollution in all the studied samples.

Keywords: Dose, Enrichment factor, Geoaccumulation index, Heavy-metals, Pollution index, Radon

Background

The three naturally occurring radon isotopes ^{222}Rn, ^{220}Rn and ^{219}Rn are formed on the alpha decay of their radium parents ^{226}Rn, ^{224}Rn and ^{223}Rn respectively (IAEA 2013). The relevant physico-chemical properties such as half-lives, decay constants, average recoil energies on formation, diffusion coefficients in air (D_{MA}) and diffusion coefficients in water (D_{MW}) are (3.82 days, 2.10×10^{-6} s^{-1}, 86 keV, 1×10^{-5} m^2s^{-1}, 1 x10^{-9} m^2s^{-1}), (55.8 s, 1.2×10^{-6} s^{-1}, 103 keV) and (3.98 s, 1.74×10^{-1} s^{-1}, 104 keV) for ^{222}Rn, ^{220}Rn and ^{219}Rn respectively (IAEA 2013). Radon half-life and solubility

have allowed the use of radon gas as a natural groundwater tracer to identify and quantify groundwater discharge to surface water (Skeppstrom and Olofsson 2007; Schubert et al. 2011; Ortega et al. 2015). The short-lived decay products of radon are responsible for most of the hazard by inhalation and ingestion. If radon and its daughters are ingested through water or inhaled in the air and decay inside the human lungs, the radiation has the potential to split water molecules and produce free radicals such OH. The free radicals are very reactive and may damage the DNA of the cells in the lungs, thus causing cancer (Edsfeldt 2001). In addition, other organs, including the kidney and the bone marrow may receive certain amounts of doses if an individual drinks water in which radon was dissolved (Kendall and Smith 2002). Although the risk is very low when radon is in the open, in places such as caves, mines, volcanic soils, aluminous shale's, granite

*Correspondence: febdeprof@yahoo.co.uk;
faweya.ebenezer@lmu.edu.ng
[1] Radiation and Health Physics Division, Department of Physics, Faculty of Science, Ekiti State University, P.M.B 5363, Ado-Ekiti, Nigeria

and rocky area such as Ikogosi, it can build up to dangerous concentrations. This may cause substantial health effect after long-term exposure (Crawford-Brown 1991; USEPA 1999; Yu and Kim 2004). Radon is extracted from the volcanic deposits in which the aquifer resides (Hector et al. 2015), its release taking place via emanation, transport and exhalation through the fissures network in the fracture system or from mantle degassing. Typical example of fissures through which radon could be released are Ikogosi warm and cold springs. The quantity of radon dissolved in groundwater discharge to surface water depends on different factors such as the characteristics of the aquifer, water–rock interaction (as seen in Ikogosi), water residence time within aquifer and material content of radium (Gundersen et al. 1992; Choubey and Ramola 1997; Choubey et al. 1997). Ninety-five percent (95%) of exposure to radon is from indoor air; about one (1%) is from drinking water sources (Kendall and Smith 2002). Most of this 1% drinking water exposure is from inhalation of radon gas released from running water activities such as bathing, showering, cleaning and healing as the people of Ikogosi believed in the healing potentials of the two springs. In many countries, some home obtain drinking water from ground sources (springs, wells, adits and boreholes) (Greeman and Rose 1996; Marazio-tis 1996; De Martino et al. 1998). Ikogosi warm and cold springs are not an exception in this aspect. The water was bottled for consumption by UAC and was named GOSSY WATER. Ikogosi Ekiti people still consume the spring water untreated because it is believed that the water has a lot of therapeutic properties to cure hypertension, guinea worm, hook worm, kidney stone, rheumatism, body rashes and pimples by either drinking it or bathing with it (Hairul et al. 2013). Apart from bottled water, tourists do visit the place for various purposes such as swimming, bathing, health reasons, aesthetic appreciation and pleasure. Underground water often moves out in two places (hot and cold) through fissures in the rock. The water might be contaminated by radon because of its volatility. Many countries in the world have defined an action level of radon concentration to guide their program to control domestic exposure to radon but this is not readily available in Nigeria. Among the vast different contaminants affecting water resources, heavy metals receive particular concern considering their strong toxicities even at low concentrations due to their cumulative effects (Momodu and Anyakora 2010). The temperature of the springs at the meeting point was attributed to the circulation of the normal groundwater to a depth of one to several thousand feet (Rogers et al. 1969). The circulation of groundwater has a potential filtering effect and possibility of water pollution through weathering of basement rocks. Chemical species such as CO_3^{2-}, Ca, Mg, Na, K, Fe which

have some sanitary health effects as well as toxins such as Pb, Cd, SO_4^{2-} could easily be introduced into the water through leaching (Oladipo et al. 2005). Toxic elements could be transferred to human through ground elements, surface water (Rasheed 2010) and sediment obtained from bottom of rivers, streams and springs (Faweya and Farai 2006; Faweya 2007; Faweya et al. 2013). Heavy metals are discharged into the river from numerous sources. They may be transported as either dissolved species in water or as an integral part of sediments (Shuanxi 2014). Sediments have been integral part of river basin with the variation of habitats and environments (Morillo et al. 2004). Sediments are not only the integral part of river basin but carrier of contaminants and potential secondary source of contaminants in aquatic systems (Calmano et al. 1990). Sediments have been widely reported as environmental indicator for the assessment of metal pollution in natural water (Islam et al. 2015). Liao et al. reported poor quality of water along rivers Kaoping and Tungkang in many points due to sediments that act as both sinks and sources of heavy metals (Liao et al. 2006). Therefore, analysis of sediment is a useful method to assess regional environmental pollution (Lai et al. 2010). Consumption of water contaminated by heavy metals from sediments may results in spread of diseases and health challenges such as reduced mental and central nervous function, lower energy levels, damage to blood through accumulation of lead in the blood stream by ingestion of contaminated aquatic species, lungs, kidneys and other vital organs (Jarup 2003; Tukura et al. 2014). Ikogosi warm and cold springs being natural water are located in Ekiti West Local government of Ekiti State of Nigeria in a valley from surrounding hills. Both warm and cold springs in the area play important roles guaranteeing water supply for domestic, agricultural, tourists attraction and bottled GOSSY water for urban needs by UAC Nigeria. The interest of this study lies in the fact that (a) during the last two decades there has been increase in consumption of treated waters in Nigeria, (b) commercially bottled and sachet water has partially substituted the consumption of tap water from municipal supplies, (c) residents ingest untreated water from the springs (d) Ikogosi warm and cold springs are one of the most visited tourists centre in Nigeria. Therefore, the presents study was carried out to provide information on (i) level of radon ^{222}Rn concentration in both treated (bottled) and untreated (source) water consumed in urban and local areas (ii) radiological dose that could be accrued to infants, children and adults due to consumption of water (iii) health risks that could be accrued to the populace due to the presence of heavy metals and other contents. The results obtained would be compared with recommended values by UNSCEAR, WHO and USEPA drinking standards.

Location of study area

The two springs sprout out and flow with a constant temperature and volume up to 150 L s^{-1} from morning till night, at all seasons, all-year round (Kukoyi et al. 2013). The warm spring has a temperature of 70 °C at source and 36–37 °C after meeting the cold spring (Kukoyi et al. 2013; Oladipo et al. 2005). Ikogosi is a town in Ekiti West Local Government Area of Ekiti State. The warm spring lies on longitude 5°0′0″ East and latitude 7°40′0″ North

(Fig. 1 Schematic map of the study area). This was done using Arc Gis 10. The topographical elevation varies from less than 473 m in the valleys to 549 m on the hills (Ojo et al. 2011).

Materials and methods

Sampling procedure

Twenty water samples were collected in various points along the springs and five samples of GOSSY

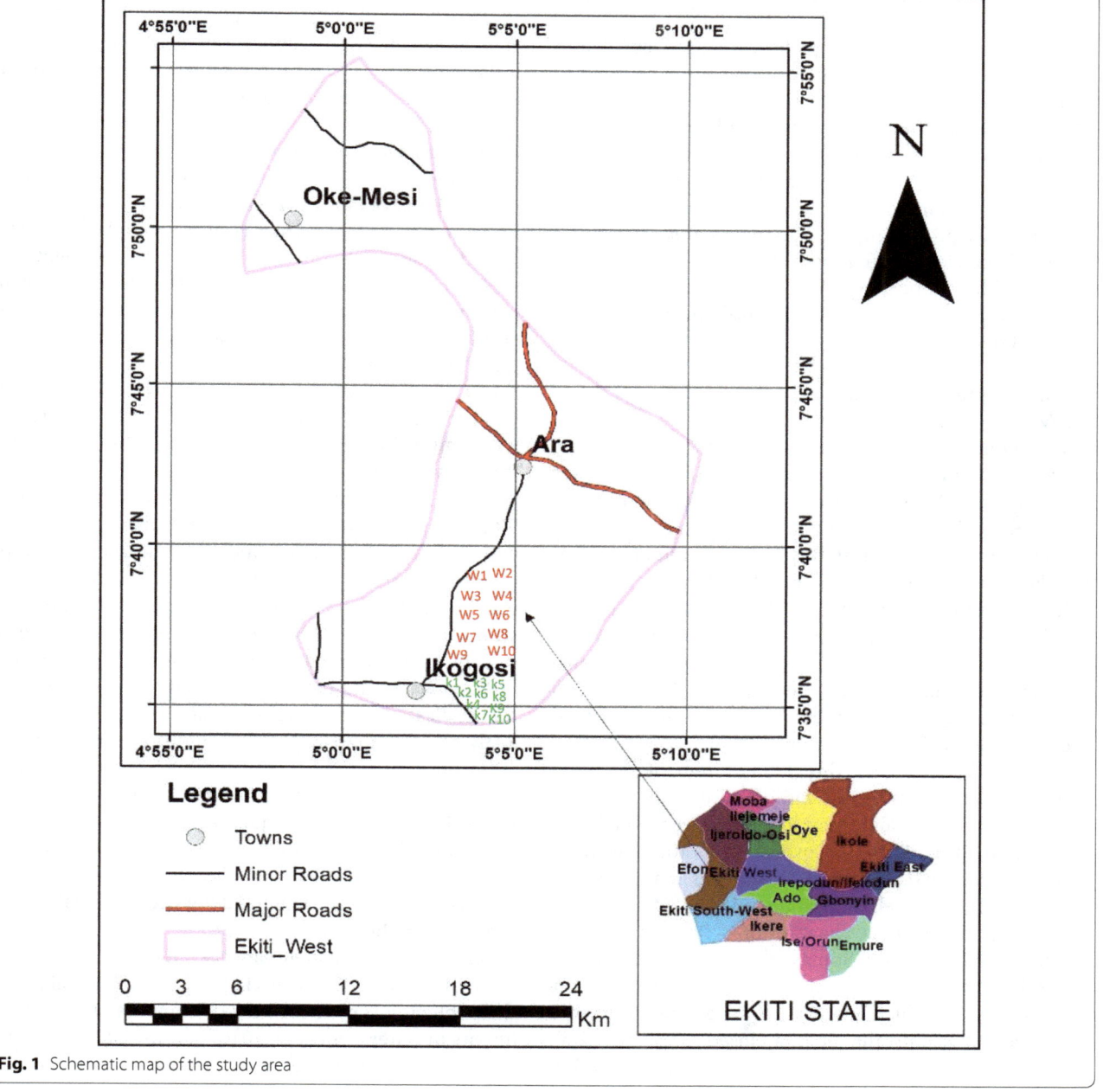

Fig. 1 Schematic map of the study area

bottled water. Ten water samples were collected along each spring. At each point, 250 mL vials designed for radon-in-water activity measurement were filled to the edge with the sampled water and then closed immediately (Stringer and Burnett 2004). Conscious effort was made to prevent bubbling of water, in order not allow escape of dissolved radon by degassing during collection and transportation to the laboratory (Oni et al. 2014, 2016). Samples were taken to the laboratory immediately. Radon levels were determined within 3–6 h after sample collection in order to minimize the effect of radioactive decay (Hector et al. 2015). This project was carried out between January and August 2017. Sediment samples were collected at the bottom of the point where water samples were taken in order to have a better representation. Samples were collected with plastic-made tools to avoid metallic contamination. This was done in triplicate at each sampling point. Samples were kept in polythene bags that are free from heavy metal and organic impurities (Faweya et al. 2013; Pravin et al. 2014). In the laboratory triplicate samples from each point were thoroughly mixed to give gross samples. The sediment samples were air-dried and sieved using mesh 0.5 mm for uniform particle size. The samples were also oven-dried at a temperature of 105 °C (Alan et al. 1997) until constant weights were obtained.

Laboratory measurements
Analysis of radon
Radon concentration in water samples was measured using an advanced radon-in-air RAD-7 radon analyzer (Durridge Co., USA) that uses alpha spectrometry technique (El-Taher 2012; Oni et al. 2014). The RAD7 radon detector was calibrated at the Durridge radon calibration facility at Billerica Massachusetts, United States. The calibration system was compared to a precision of better than 1%, with a secondary standard chamber, which was in turn calibrated by comparison with a National Institute of Standards and Technology (NIST) radon standard supplied through the U.S. Environmental Protection Agency. The calibration system's accuracy was also check by making a direct measurement of radon level from activity and emission of a European standard radon source. The calibration achieves a reproducibility of better than ±2% and an overall calibration accuracy of better than ±5%. The Rad-7 used was maintained at between 6 and 10% relative humidity for its efficiency not to decrease due to neutralization of ^{218}Po ions by water particle (Ravikumar and Somashekar 2014). In the setup, 250-mL sample bottle was connected to RAD-7 detector via bubbling kit which enables it to degas radon from a water sample in into the air in a closed loop (Oni et al. 2014). To achieve this, the equipment was set to wat-250

for 5 min. The equipment was allowed to rest for 5 min and then count each sample for 30 min in five cycles. Radon concentration was determined by RAD7 taking into consideration the calibration of RAD7, volume of the closed air loop of the set up and the size of the vial used. The counting time was shorter than 3.8 days the half-life of radon. This made RAD-7 better than other detectors for ^{222}Rn measurement in water. Five runs were done for each sample. At the end of the runs (after the start), the RAD-7 prints out automatically the summary, showing the average radon reading for the five cycles counted. The readings and typical alpha energy spectrum obtained from the capture software were shown in Figs. 5 and 6 of the Additional file 1. The samples were counted immediately after collection without any delay, therefore radon decay correction was not calculated (Ravikumar and Somashekar 2014; Hector et al. 2015).

Physico-chemical parameters study
The sediment samples were analyzed for pH, electrical conductivity, organic matter content, nitrogen and the heavy metal contents. The pH of the samples was determined using Jenway 3510 pH meter. The sediment samples were mixed in a ratio of 1:1 with distilled water in a beaker before inserting the probes. Readings were taken after the instrument had stabilized. Conductivity was taken by dipping the conductivity probe into a mixture 1:1 of the sediment samples using Jenway 4520 conductivity meter. Organic matter (OM) was determined by wet combustion method (APHA 1995). One (1) g of each sample was weighed into a Pyrex beaker and 20 mL of con HNO_3 was added to it. This was allowed to soak for 30 min and then transferred to a hot plate and heated at 400 °C until frothing stops and HNO_3 was almost evaporated. Five (5) mL of conc $HClO_4$ was added and watch glass placed of the beaker until sample became light strain in colour. The beaker was removed and allowed to cool, then the watch glass rinsed into the beaker with distilled water and the digest filtered into a 100 ml volumetric flask. Heavy metal contents were determined by analyzing the prepared sediment filtrate using Atomic Absorption Spectrophotometer (BUCK 210 VGP). Others physical and chemical properties were determined using the standard techniques and methods (Hem 1985; APHA, AWWA, WEF 1998).

Evaluation of doses in water and physico-chemical properties in the sediment
Evaluation of mean annual effective dose
Radon transports by water via ingestion and inhalation to the public is a very serious threat compared to other pollutants in water (Oni et al. 2016) because of dose accrued to the populace. Therefore, dose due to radon can be

divided into two parts: ingestion (through consumption of water) and inhalation (when radon is released from water to indoor air) (Manzoor et al. 2008; Ravikumar and Somashekar 2014). The mean annual effective dose rate for ingestion and inhalation were calculated according to parameters introduced by UNSCEAR (2000) and were calculated as:

$$EWI_{ing}\left(\mu \text{ Sv year}^{-1}\right) = CR_n W \cdot CW \cdot EDC \qquad (1)$$

$$EWI_{inh}\left(\mu \text{ Sv year}^{-1}\right) = C_{air} \cdot R \cdot T \cdot Đ \qquad (2)$$

From Eq. 1, EWI_{ing} is the effective dose from ingestion (μ Sv year^{-1}), $CR_n W$ is the radon concentration in Bq L^{-1}, CW are the estimated weight of used water found to be 100, 75 and 60 L year^{-1}) by infants, children and adults respectively and EDC is the effective dose coefficient for ingestion (3.5 n Sv Bq^{-1}).

From Eq. 2, EWI_{inh} is the effective dose of inhalation, R is the ratio of radon in air to radon in spring water (10^{-4}), C_{air} is the radon concentration in Bq L^{-1}, F is the equilibrium factor between radon and its decay products (0.4), T is the average indoor occupancy time per person (7000 h year^{-1}), and Đ is the dose conversion factor for radon exposure $\left[9 \text{ n Sv h}^{-1}\left(\text{Bq m}^{-3}\right)^{-1}\right]$. The contribution of the dose to the lungs and stomach is calculated by multiplying the inhalation and ingestion dose by a tissue weighting factor for lung (0.12) and stomach (0.12) (ICRP 2012).

Evaluation of physico-chemical properties
Heavy metals
The occurrence of heavy metals in soil and sediment could due to natural sources such as dissolution of naturally occurring minerals containing trace elements in the soil and sediments in the area (Faweya and Farai 2006; Faweya 2007; Faweya and Babalola 2010; Faweya et al. 2013). The heavy metals most frequently encountered in soil and sediment are Arsenic, Cadmium, Copper, Chromium, Zinc, Nickel, Iron, Cobalt and Manganese (Kumar et al. 2017a; Zhang et al. 2011; Song et al. 2015). Drinking water containing high levels of these harmful metals in bottom sediment and using the sediment for other purposes may be hazardous to health.

Enrichment factor (EF) and geo-accumulation index (I$_{geo}$) analysis
The sediments quality and metal contamination in the cold and warm springs were assessed using enrichment factor and geo-accumulation index. Variation in metal concentrations can be identified through EF by using geochemical normalization of the heavy metals data to conservative

elements such as Al, Si or Fe (Zhang et al. 2009; Ghrefat et al. 2011). In the present study the EF was determined based on Fe which was used as a conservative tracer to evaluate the anthropogenic impact in order to differentiate natural from anthropogenic components.

Mathematically, EF is expressed as follows

$$EF = \frac{\left(\frac{M}{Fe}\right)_{sample}}{\left(\frac{M}{Fe}\right)_{background}} \qquad (3)$$

where $\left(\frac{M}{Fe}\right)_{sample}$ is the ratio of metal and Fe concentrations of the sample and $\left(\frac{M}{Fe}\right)_{background}$ is the ratio of metal and Fe concentrations of a background, the background values used were 46,700 mg kg^{-1} for Fe, 0.3 mg kg^{-1} for Cd, 45 mg kg^{-1} for Cu, 20 mg kg^{-1} for Pb and 68 mg kg^{-1} for Ni respectively (Turekian and wedepohl 1961; Faweya et al. 2013). Degrees of enrichment are defined as; $1 \leq EF < 3$, minor enrichment; $3 \leq EF < 5$, moderate enrichment; $5 \leq EF < 10$, moderately severe enrichment; $10 \leq EF < 25$, severe enrichment $25 \leq EF < 50$, very severe enrichment; and EF > 50 extremely severe enrichment.

Another criterion commonly used to evaluate the heavy metal pollution in sediment is the geo-accumulation index. The index of geo-accumulation gives the assessment of contamination by comparing the current and pre-industrial concentrations (Muller 1969). The equation used for the calculation of I$_{geo}$ is expressed as follow:

$$I_{geo} = \log_2\left(C_n / 1.5B_n\right) \qquad (4)$$

where C_n is the measured concentration for the metal in the sediments and B_n is background value of the metal, and the factor 1.5 is used because of possible variations of the background data due to lithological variations. The geo-accumulation index has seven grades. The grades are as follows: $I_{geo} \leq 0$, uncontaminated; $0 < I_{geo} \leq 1$, uncontaminated/moderately contaminated; $2 < I_{geo} \leq 3$, moderately/strongly contaminated; $3 < I_{geo} \leq 4$, strongly contaminated $4 < I_{geo} \leq 5$ strongly/extremely contaminated $5 \leq I_{geo}$, extremely contaminated.

Contamination factor, degree of contamination and pollution load index
Contamination factor C_f^i is the ratio of toxicity of a heavy metal in the environment. It was calculated using the equation proposed by Hakanson (1980);

$$C_f^i = C^i / C_n^i \qquad (5)$$

where C^i is the mean concentration of metal i in the sediments and C_n^i is the background concentration of metal i. The following criteria are used to describe the values of the contamination factor; $C_f^i < 1$ low contamination factor $1 \leq C_f^i < 3$, moderate contamination factor; $3 \leq C_f^i < 6$, considerable contamination factor; and $C_f^i \geq 6$, very high contamination factor (Turekian and Wedepohl 1961; Hakanson 1980).

The pollution load index is a simple way of measuring the degree of metal pollution in a studied medium (Tomlinson et al. 1980). It is expressed as

$$PLI = \left(C_{f1}^i \cdot C_f^i \cdot C_f^i \right)^{\frac{1}{n}} \tag{6}$$

where n is the number of metals and C_f^i is the contamination factor. The pollution load Index can be classified as (PLI < 1), no pollution; (1 < PLI < 2), moderate pollution; (2 < PLI < 3), heavy pollution and (3 < PLI), extremely heavy pollution (Banerjee and Gupta 2012).

Quantification of contamination and quality of sediment

The index Q_oC as proposed by Asaah (Asaah et al. 2006) majorly defines the quantification of anthropogenic concentration of metal using the concentration in the background metal to represent the lithogenic material. It was calculated in the sediment samples using the following relation.

$$QoC(\%) = \left(\frac{C_X - C_n}{C_X} \right) \cdot 100 \tag{7}$$

Where C_X is the average concentration of metal and C_n is the average concentration of the metal in the background (Asaah et al. 2006), the value in percentage will determine if the impact is geogenic (negative values) or anthropogenic (positive values).

Since Nigeria has not established sediment quality guidelines at this time, the sediment quality criteria as used by (Zarei et al. 2014; Stephen et al. 2004; Orkun et al. 2011; Cevik et al. 2009) was used to classify sediment samples with regard to their potential toxicity.

In this study, sediment from cold and warm springs are compared with guidelines and global baseline values such as threshold effect level (TELs), effect range low values (ERLs), probable effect levels (PELs), effect range median values (ERMs), mean earth crust (MECs), mean world sediments (MWSs) and mean continental shale (MCSs).

Results and discussion

Radon concentration

The mean activity concentration of radon in bottled, cold and warm water samples as seen in the ninth and fourth columns of Tables 1 and 2 and ranged from 0.07 to 0.36 with an overall mean value 0.20 Bq L^{-1}, 35–210 with an overall mean value 75.9 Bq L^{-1} and 11.7–140. 0 with an overall mean value 79.4 Bq L^{-1} for bottled, cold and warm water samples respectively. The radon concentration was higher than 100 Bq L^{-1} recommended by WHO in C_5, C_6, W_3, W_4, W_5 and W_{10}. The higher values found in cold and warm spring were due to uranium content of the bed rocks which easily interact with water by the effect of lithostatic pressure (Toscani et al. 2001).

Among the samples, six samples (24%) showed radon concentration exceeding the maximum contamination level for radon in water for human consumption as suggested by EU (2001) and WHO (2011). The mean concentration of radon in bottled water was below 11 Bq L^{-1}, 100 Bq L^{-1} recommended by USEPA (1991), EU and WHO indicating the safety of bottled water for consumption. The maximum concentrations of radon in some of the samples in study such as C_3 (105), C_5 (210), C_6 (220), C_7 (105), C_8 (176) C_9 (105), C_{10} (140), W_1 (105) W_3 (175), W_4 (175), W_5 (140), W_7 (105), W_9 (140) and W_{10} (175) Bq L^{-1} were lower than maximum concentrations obtained at Mysore city India 435 Bq L^{-1} (Chandrashekara et al. 2012). Kumaun Himalayan region India 392 Bq L^{-1} (Bourai et al. 2012) Kamuan India 336 Bq L^{-1} (Yogesh et al. 2009), Virginia and Maryland US 296 Bq L^{-1} (Mose et al. 1990), Baoji China 127 Bq L^{-1}, (Xinwei 2006) and Sankey Tank area India 381.2 Bq L^{-1} (Ravikumar and somashekar 2014).

Annual effective dose rate

Table 1 shows annual effective dose rate to different age classification as recommended by ICRP using their average annual consumption rate. ICRP age classification of 0–1 years, 1–2 years, 2–7 years, 7–12 years, 12–17 years and 17 year-above in bottled water have annual effective dose rate which is 0.1% of the 1 mSv year^{-1} recommended by UNSCEAR and WHO for public. The calculated values were well below the reference level and hence bottled water does not pose any health problems from radon dose received from drinking bottled water. It suffices to say that radon with half-life 3.8 days must have decayed during the processing, bottling and storage of bottled water.

Annual effective dose rate values ranged from 0.04 to 0.20 mSv year^{-1} with a mean value 0.08 mSv year^{-1} and 0.02–0.14 with a mean 0.07 mSv year^{-1} for age classification 1. For age classification 2, its values varied from 0.05 to 0.27 mSv year^{-1} with an average value of 0.09 mSv year^{-1} and 0.02–0.182 mSv year^{-1} with average value 0.10 mSv year^{-1}. The corresponding annual effective dose rate for age classification 3 ranged from 0.053

Table 1 Mean radon activity in bottled water, cold and warm spring and annual effective dose rate

#	Water id	Sample	Temp (°C)	Rel Hum (%)	Radon conc. Min	Max	S.D	Mean (BqL⁻¹)	AEDR (µSv year⁻¹) Infant 1 <1 years	2 1–2 years	Children 3 2–7 years	4 7–12 years	5 12–17 years	Adult 6 >17 years	AEDR (mSv year⁻¹) Infants 1 <1 years	2 1–2 years	Children 3 2–7 years	4 7–12 years	5 12–17 years	Adult 6 >17 years
1	Bottled	B1	27.0	8.0	0.0	0.44	0.22	0.25	0.25	0.33	0.38	0.44	0.75	0.91	0.0003	0.0004	0.0004	0.0004	0.0008	0.0009
2		B2	26.4	6.0	0.0	0.44	0.19	0.22	0.22	0.29	0.33	0.39	0.66	0.80	0.0002	0.0003	0.0003	0.0004	0.0007	0.0008
3		B3	26.7	6.0	0.0	0.87	0.37	0.36	0.36	0.47	0.54	0.63	1.08	1.31	0.0004	0.0005	0.0005	0.0006	0.001	0.001
4		B4	27.7	7.0	0.0	0.29	0.14	0.11	0.11	0.14	0.17	0.19	0.33	0.40	0.0001	0.0002	0.0002	0.0002	0.0003	0.0004
5		B5	28.0	7.0	0.0	0.15	0.08	0.07	0.07	0.09	0.11	0.12	0.21	0.26	0.00007	0.00009	0.0001	0.0001	0.0002	0.0003
6	Cold	C1	32.6	8.0	35.0	35.0	0.02	35.0	35.0	45.50	52.50	61.25	105.00	127.75	0.035	0.046	0.053	0.061	0.105	0.128
7		C2	32.2	7.0	35.0	70.0	20.2	46.7	46.7	60.71	70.05	81.73	140.10	170.46	0.047	0.061	0.070	0.082	0.140	0.171
8		C3	32.2	7.0	0.0	105.0	53.5	58.4	58.4	75.92	87.60	102.20	175.20	213.17	0.058	0.076	0.088	0.102	0.175	0.213
9		C4	31.0	8.0	0.0	70.0	35.0	35.0	35.0	45.50	52.52	61.25	105.00	127.75	0.035	0.046	0.053	0.061	0.105	0.128
10		C5	31.0	7.0	35.0	210.0	92.7	105.0	105.0	136.50	157.50	183.75	315.00	383.25	0.105	0.137	0.158	0.184	0.315	0.383
11		C6	31.0	6.0	0.0	220.0	100.0	210.1	210.1	273.13	315.15	367.68	630.30	766.87	0.210	0.273	0.315	0.368	0.630	0.767
12		C7	30.7	6.0	0.0	105.0	60.5	35.0	35.0	45.50	52.50	61.25	105.00	127.75	0.035	0.046	0.053	0.061	0.105	0.128
13		C8	30.7	6.0	35.0	176.0	73.4	93.7	93.7	121.80	140.55	163.98	281.10	342.01	0.094	0.122	0.141	0.164	0.281	0.342
14		C9	30.7	5.0	0.0	105.0	53.5	46.7	46.7	60.71	70.05	81.73	140.10	170.46	0.047	0.061	0.070	0.082	0.140	0.171
15		C10	30.7	5.0	35.0	140.0	53.6	93.6	93.6	121.68	140.40	163.80	280.80	341.64	0.094	0.122	0.141	0.164	0.281	0.342
16	Warm	W1	30.4	6.0	70.0	105.0	20.2	93.4	93.4	121.42	140.10	163.45	280.20	340.91	0.093	0.121	0.140	0.163	0.280	0.341
17		W2	30.4	5.0	35.0	35.0	0.02	35.0	35.0	45.50	52.50	61.25	105.00	127.75	0.035	0.046	0.053	0.061	0.105	0.128
18		W3	30.4	5.0	105.0	175.0	35.0	140.0	140.0	182.00	210.00	245.00	420.00	511.00	0.140	0.182	0.210	0.245	0.420	0.511
19		W4	30.1	5.0	35.0	175.0	72.9	117.0	117.0	152.10	175.50	204.75	351.00	427.05	0.117	0.152	0.176	0.205	0.351	0.427
20		W5	29.2	7.0	70.0	140.0	35.0	105.0	105.0	136.50	157.50	183.75	315.00	383.25	0.105	0.137	0.158	0.184	0.315	0.383
21		W6	29.2	6.0	0.0	35.0	20.2	11.7	11.7	15.21	17.55	20.48	35.10	42.71	0.018	0.015	0.018	0.020	0.035	0.043
22		W7	29.2	6.0	70.0	105.0	20.2	81.7	81.7	106.21	122.55	142.98	245.10	298.21	0.082	0.106	0.123	0.143	0.245	0.298
23		W8	29.2	5.0	0.0	70.0	35.0	35.0	35.0	45.50	52.50	61.25	105.00	127.75	0.035	0.046	0.053	0.061	0.105	0.128
24		W9	29.2	6.0	35.0	140.0	60.7	70.0	70.0	91.00	105.00	122.50	210.00	255.50	0.070	0.091	0.105	0.123	0.210	0.256
25		W10	30.1	5.0	70.0	175.0	60.7	105.0	105.0	136.50	157.50	183.75	315.00	383.25	0.105	0.137	0.158	0.184	0.315	0.383

Table 2 Annual effective dose rate contribution to lungs and stomachs

	Water id	Sample	Mean (Bq L^{-1})	Ingestion (μSv year^{-1})			Inhalation (μSv year^{-1})	Lungs	Stomach (μSv year^{-1})			Total effective dose (μSv year^{-1})		
				Infant	Children	Adult			Infant	Children	Adult	Infant	Children	Adult
1	Bottled	B1	0.25	0.09	0.07	0.05	0.63	0.08	0.01	0.008	0.006	0.09	0.088	0.086
2		B2	0.22	0.08	0.06	0.05	0.55	0.07	0.01	0.007	0.006	0.08	0.077	0.076
3		B3	0.36	0.13	0.09	0.08	0.91	0.11	0.02	0.010	0.010	0.130	0.120	0.120
4		B4	0.11	0.04	0.03	0.02	0.28	0.03	0.005	0.004	0.003	0.035	0.034	0.033
5		B5	0.07	0.03	0.02	0.01	0.17	0.02	0.004	0.003	0.002	0.024	0.023	0.022
6	Cold	C1	35.0	12.25	9.19	7.35	88.20	10.58	1.47	1.10	0.88	12.05	11.68	11.46
7		C2	46.7	16.35	12.26	9.81	117.70	14.12	1.96	1.47	1.18	16.08	15.59	15.3
8		C3	58.4	20.44	15.33	12.26	147.20	17.66	2.45	1.84	1.47	20.11	19.50	19.13
9		C4	35.0	12.25	9.19	7.35	88.20	10.58	1.47	1.10	0.88	12.05	11.68	11.46
10		C5	105.0	36.75	27.56	22.05	264.60	31.75	4.41	3.31	2.65	36.16	35.06	34.40
11		C6	210.1	73.54	55.16	44.13	529.50	63.54	8.82	6.62	5.30	72.36	70.16	68.84
12		C7	35.0	12.25	9.19	7.35	88.20	10.58	1.47	1.10	0.88	12.05	11.68	11.46
13		C8	93.7	32.80	24.60	19.68	236.12	28.33	3.94	2.95	2.36	32.27	31.28	30.69
14		C9	46.7	16.35	12.26	9.81	117.70	14.12	1.96	1.47	1.18	16.08	15.59	15.30
15		C10	93.6	32.76	24.57	19.67	235.87	28.30	3.95	2.95	2.35	32.25	31.25	30.65
16	Warm	W1	93.4	32.69	24.52	19.61	235.37	28.24	3.92	2.94	2.34	32.16	31.18	30.58
17		W2	35.0	12.25	9.19	7.35	88.20	10.58	1.47	1.10	0.88	12.05	11.68	11.46
18		W3	140.0	49.00	36.75	29.40	352.80	42.34	5.88	4.41	3.53	48.22	46.75	45.87
19		W4	117.0	40.95	30.71	24.57	294.60	35.38	4.91	3.69	2.95	40.29	39.07	38.33
20		W5	105.0	36.75	27.56	22.05	264.60	31.75	4.41	3.31	2.65	36.16	35.06	34.40
21		W6	11.7	4.09	3.07	2.46	29.48	3.54	0.49	0.37	0.30	4.03	3.91	3.84
22		W7	81.7	28.60	21.45	17.16	205.88	24.71	3.43	2.57	2.06	28.14	27.28	26.77
23		W8	35.0	12.25	9.19	7.35	88.20	10.58	1.47	1.10	0.88	12.05	11.68	11.46
24		W9	70.0	24.50	18.38	14.70	176.40	21.17	2.94	2.21	1.76	24.11	23.38	22.93
25		W10	105.0	36.75	27.56	22.05	264.60	31.75	4.41	3.31	2.65	36.16	35.06	34.40

to 0.315 mSv year^{-1} and 0.053 to 0.21 mSv year^{-1} with average value 0.11 and 0.12 mSv year^{-1}. For age classification 4, it varied from 0.061 to 0.368 mSv year^{-1} with average value 0.133 mSv year^{-1} and 0.061–0.245 mSv year^{-1} with average value 0.139 mSv year^{-1}. The estimated values of annual effective dose rate for age classification 5 oscillated from 0.105 to 0.630 mSv year^{-1} with an average value of 0.238 mSv year^{-1} and 0.035–0.420 mSv year^{-1} with average value 0.238 mSv year^{-1} and varied from 0.128 to 0.767 mSv year^{-1} with average 0.277 mSv year^{-1}; 0.043–0.511 mSv year^{-1} with average value 0.290 mSv year^{-1} for age classification 6 for both cold and warm water samples respectively. The mean values were far below 1 mSv year^{-1} recommended by UNSCEAR and WHO for member of public.

The present study revealed that the annual effective dose rate values increased with respect to radon activity, age and water consumption rates. The annual effective dose rate received by ICRP age classifications of $1 < 2 < 3 < 4 < 5 < 6$. All the samples have annual effective dose rate values that were significantly lower than

1 mSv year^{-1} recommended by UNSCEAR and WHO for member of public.

Inhalation and ingestion dose and effect on stomach and lungs

The annual effective dose rate values received by stomach in columns 5, 6 and 7 due to ingestion from bottled water varied from 0.03 to 0.13 μSv year^{-1}, 0.02 to 0.09 μSv year^{-1}, 0.01 to 0.08 μSv year^{-1} with mean values 0.04, 0.03 and 0.02 μSv year^{-1} for infants, children and adults respectively. While annual effective dose rate values received by lungs due to inhalation of radon released from bottled water in column 8 ranged from 0.17 to 0.95 with a mean value 0.51 μSv year^{-1}. For cold water samples, the values received by stomach in columns 5, 6 and 7 of Table 2 due to consumption of cold water samples varied from 35.0 to 210.1 μSv year^{-1}, 12.25 to 73.54 μSv year^{-1}, 9.19 to 55.16 μSv year^{-1}, 7.35 to 44.13 μSv year^{-1} with mean values 26.57, 19.93 and 15.95 μSv year^{-1} for infant, children and adult respectively; while its values received by lungs due to inhalation of radon ranged from 88.20 to 529.5 µSv year^{-1} with mean value 191.33 µSv year^{-1}. For warm water samples, annual effective dose rate values received by stomach ranged from 4.09 to 49.00 µSv year^{-1}, 3.07 to 36.75 µSv year^{-1}, 2.46 to 29.40 µSv year^{-1} with mean values 27.78, 20.84 and 16.67 µSv year^{-1} for infant, children and adult respectively; while annual effective dose rate for lungs varied from 29.48 to 352.80 µSv year^{-1} with average value 200.1 µSv year^{-1}.

The contribution of the dose to the lungs and stomach was calculated by multiplying the inhalation and ingestion dose by a tissue weighing factor 0.12 for lung and stomach (ICRP 2012). The results obtained are shown in ninth, tenth, eleventh and twelfth columns of Table 2 for lungs and stomach respectively.

The annual effective dose rate values received by lungs due to inhalation from bottled, cold and hot spring water varied from 0.02 to 0.11 µSv year^{-1} with mean value 0.06 µSv year^{-1}; 10.58 to 63.54 µSv year^{-1} with mean value 22.96 µSv year^{-1} and 3.54 to 42.34 µSv year^{-1} with a mean 24.00 µSv year^{-1} respectively.

Annual effective dose rate values received by stomach due to consumption of water by infant, children and adult varied from 0.004 to 0.02 µSv year^{-1} with mean 0.01 µSv year^{-1}; 0.003 to 0.01 µSv year^{-1} with mean 0.01 µSv year^{-1}; 0.002 to 0.01 µSv year^{-1} with mean 0.01 µSv year^{-1} respectively for bottled water. The values for infant, children and adult ranged from 1.47 to 8.82 µSv year^{-1} with a mean 3.19 µSv year^{-1}; 0.49 to 5.88 µSv year^{-1} with mean 3.33 µSv year^{-1}; 1.10 to 6.62 µSv year^{-1} with mean 2.50 µSv year^{-1}; 0.88 to 5.30 µSv year^{-1} with mean 2.00 µSv year^{-1} for cold and warm spring respectively. The results show that dose contribution to lungs was higher than dose contributed to the stomach. The results agreed with that of radon found in drinking water in India, that indicate dose contribution to lungs higher than dose contribution to the stomach (Kumar et al. 2017b). The calculated effective dose (whole body) due to radon inhalation and ingestion for infant, children and adult ranged from 0.024 to 0.13 µSv year^{-1} with a mean value 0.07 µSv year^{-1}; 0.023 to 0.12 µSv year^{-1} with a mean value 0.07 µSv year^{-1}; 0.022 to 0.12 µSv year^{-1} with a mean value 0.07 µSv year^{-1} respectively for bottled water. It ranged from 12.05 to 72.36 µSv year^{-1} with a mean value 26.15 µSv year^{-1}; 4.03 to 48.22 µSv year^{-1} with a mean value 27.94 µSv year^{-1}; 11.68 to 70.16 µSv year^{-1} with a mean value 25.35 µSv year^{-1}; 3.91 to 46.75 µSv year^{-1} with a mean 24.87 µSv year^{-1} and 3.84 to 45.87 µSv year^{-1} with a mean value 26.00 µSv year^{-1}, for infant, children and adult in cold and warm spring respectively. The results for risk estimates indicate that inhalation of radon accounts for 88.89% of the individual risk associated

with the use of bottled, cold and warm water, while the remaining 11.11% resulting from the ingestion of radon gas. The results agreed with 89% inhalation and 11% ingestion revealed by USEPA (1999). The calculated total effective dose values were below 100 µSv year^{-1} which is safe limit recommended by WHO (2004) therefore, no radiological health problems is envisaged.

Physico-chemical evaluation

The concentration of metals in the cold and warm springs sediments, global baseline values and SQGs of the studied metals are presented in Tables 3 and 4. The variations in concentration values are depicted in Fig. 2a–d (Fig. 2a bar chart of readings and spectrum for Ni, Cd, Cu and Pb in selected points in cold spring Fig. 2b bar chart of readings and spectrum for Fe in selected points in cold spring Fig. 2c bar chart of readings and spectrum for Ni, Cd, Cu and Pb in selected points in warm spring Fig. 2c bar chart of readings and spectrum for Ni, Cd, Cu and Pb in selected points in warm spring Fig. 2d bar chart of readings and spectrum for Fe in selected points in warm spring) (Additional file 1). The results in Table 3 and Fig. 2b, d showed that Fe had the highest concentration in the sediments. The average concentration of Pb ranged from 2.9 mg kg^{-1} dw (W$_4$) to 11.80 mg kg^{-1} dw (C$_6$) respectively. A comparison of Pb highest concentration in sediments with the corresponding values of this metal in ERL, ERM, TEL, MEC, PEL, MWS and MCS showed that the levels of Pb were lower (3.9 8 times) than ERL, ERM (18.64 times) TEL (2.56 times), PEL (9.49 times), MEC(1.19 times), MWS (1.61 times) and MCS (1.69 times). The highest and lowest mean concentrations of Cu in sediments were found to be 3.80 and 12.80 mg kg^{-1} dw, respectively; The average concentrations of Fe in sediments in ranged from 945 (C$_5$) to 2010 mg kg^{-1} dw (W$_3$). The lowest and highest concentrations of Fe in the sediments were below the MCS. The highest and lowest concentrations of Cd in the sediment were 0.6 and 1.7 mg kg^{-1} dw.

The comparison of Cd highest concentration with the studied standard values showed that the levels of Cd were lower than (2.47 times) PEL, (5.67 times) MCS, (5.64 times) ERM, (1.42 times) ERL, (2.43 times) TEL. In the studied sediments the highest mean concentration of Ni 2.60 mg kg^{-1} dw was lower than ERL (88%), ERM (95%) and MCS (96%) respectively. The differences in the level of metals in the sediments of cold and warm springs of Ikogosi may be due to parameters such as pH, organic matter and environmental factors which control the solubility and availability of metals (Ebrahimpour and Mushrifah 2008). The resulting EF values in Table 4 showed that Pb, Cu and Cd are enriched in the sediment samples while there is no Ni enrichment in both

Table 3 Mean physicochemical parameters of sediments

Sample and parameter	No of sample	Metals (mg kg⁻¹)					Other parameters				Pollution indices									
											I_{geo}					EF				
		(Pb)	(Cu)	(Fe)	(Cd)	(Ni)	pH	Cond (μS)	OM (%)	N (%)	(Pb)	(Cu)	(Fe)	(Cd)	(Ni)	(Pb)	(Cu)	(Fe)	(Cd)	(Ni)
C_1	3	10.80	5.90	1510.0	1.60	2.00	6.51	93.5	0.22	0.045	0.108	0.027	0.0065	1.07	0.0060	1.67	0.41	–	16.50	0.091
C_2	3	9.10	4.80	1745.0	1.65	1.40	6.62	80.8	0.19	0.040	0.091	0.022	0.0057	1.10	0.0042	1.11	0.40	–	20.63	0.077
C_3	3	9.70	6.60	1840.0	1.60	2.60	6.51	93.2	0.17	0.030	0.097	0.030	0.0079	1.07	0.0078	1.23	0.37	–	13.54	0.097
C_4	3	8.30	7.10	1620.0	1.70	1.30	6.51	93.0	0.18	0.035	0.083	0.032	0.0070	1.13	0.0039	1.10	0.45	–	16.34	0.055
C_5	3	5.20	7.20	945.0	0.90	0.70	6.71	70.2	0.20	0.040	0.052	0.032	0.0041	0.60	0.0021	1.29	0.79	–	14.83	0.051
C_6	3	11.10	10.20	1015.0	0.90	0.90	6.71	72.0	0.20	0.040	0.111	0.046	0.0044	0.60	0.0027	2.55	1.04	–	13.80	0.061
C_7	3	9.20	3.80	1650.0	0.90	1.30	6.68	80.8	0.19	0.035	0.092	0.017	0.0071	0.60	0.0039	1.30	0.29	–	8.49	0.054
C_8	3	9.30	4.20	1245.0	1.10	1.40	6.68	80.9	0.19	0.035	0.093	0.019	0.0054	0.74	0.0042	1.74	0.35	–	13.75	0.077
C_9	3	6.70	12.80	1320.0	1.20	1.20	6.15	94.1	0.23	0.050	0.067	0.058	0.0057	0.80	0.0036	1.18	1.01	–	14.15	0.062
C_{10}	3	7.20	14.10	1445.0	1.10	1.10	6.50	94.6	0.16	0.040	0.072	0.063	0.0062	0.74	0.0033	1.16	1.01	–	11.85	0.052
W_1	3	8.50	6.60	1660.0	0.80	0.80	6.61	80.2	0.18	0.040	0.085	0.030	0.0071	0.54	0.0024	1.20	0.41	–	7.50	0.033
W_2	3	10.10	7.20	1825.0	0.60	0.70	6.68	80.9	0.20	0.035	0.101	0.032	0.0078	1.07	0.0021	1.29	0.41	–	5.12	0.026
W_3	3	10.20	6.80	2010.0	0.80	0.50	6.81	65.3	0.25	0.050	0.102	0.031	0.0086	0.54	0.0015	1.19	0.35	–	6.10	0.017
W_4	3	2.90	6.60	1920.0	0.80	0.30	6.82	64.9	0.23	0.050	0.029	0.037	0.0083	0.54	0.0009	0.35	0.36	–	6.49	0.011
W_5	3	4.60	6.10	1625.0	0.60	1.10	6.89	62.3	0.24	0.040	0.046	0.027	0.0070	0.40	0.0033	0.66	0.39	–	5.75	0.046
W_6	3	9.30	5.80	1770.0	1.50	1.10	6.91	54.2	0.20	0.050	0.093	0.026	0.0076	1.00	0.0033	1.23	0.34	–	13.19	0.043
W_7	3	10.60	6.60	1820.0	1.30	1.30	6.95	54.0	0.18	0.035	0.106	0.030	0.0078	0.87	0.0039	1.36	0.38	–	11.12	0.049
W_8	3	10.40	12.20	1990.0	1.40	1.20	6.51	93.4	0.18	0.030	0.104	0.055	0.0086	0.94	0.0036	1.22	0.64	–	10.95	0.041
W_9	3	9.10	10.20	1020.0	0.90	1.10	6.52	93.0	0.19	0.030	0.091	0.046	0.0044	0.60	0.0033	2.08	1.04	–	13.74	0.074
W_{10}	3	9.20	7.60	1250.0	0.80	0.90	6.75	69.0	0.19	0.030	0.092	0.034	0.0054	0.54	0.0027	1.71	0.63	–	9.96	0.049

Table 4 Contamination factor, pollution load index, quantification of contamination and sediment quality guidelines

Sample and parameter	C_f^i					PLI	Q_oC (%)				
	Pb	Cu	Fe	Cd	Ni	Ni	Pb	Cu	Fe	Cd	Ni
C_1	0.54	0.13	0.03	5.33	0.03	0.20	−85.19	−662.71	−2992.72	81.25	−330.00
C_2	0.46	0.11	0.03	5.50	0.02	0.18	−119.78	−837.50	−3651.00	81.82	−4757.1
C_3	0.49	0.15	0.04	5.33	0.04	0.23	−106.19	−581.82	−2438.04	81.25	−2515.38
C_4	0.42	0.16	0.03	5.67	0.02	0.19	−140.96	−533.80	−2787.70	82.35	−5130.77
C_5	0.26	0.16	0.02	3.00	0.01	0.12	−284.62	−525.00	−4841.70	66.67	−9614.29
C_6	0.56	0.23	0.02	3.00	0.01	0.15	−79.28	−341.18	−4500.99	66.67	−7455.56
C_7	0.46	0.08	0.04	3.00	0.02	0.15	−117.39	−1084.21	−2730.30	66.67	−5130.77
C_8	0.47	0.09	0.03	3.67	0.02	0.16	−115.05	−971.43	−3651.00	72.73	−4757.14
C_9	0.34	0.28	0.03	4.00	0.02	0.19	−198.51	−251.56	−3437.88	75.00	−5566.67
C_{10}	0.36	0.31	0.03	3.67	0.02	0.19	−177.78	−219.15	−3131.83	72.73	−6081.82
W_1	0.43	0.15	0.04	2.67	0.01	0.15	−135.29	−581.82	−2713.25	62.50	−8400.00
W_2	0.51	0.16	0.04	2.00	0.01	0.15	−98.02	−525.00	−2458.90	50.00	−9614.29
W_3	0.51	0.15	0.04	2.67	0.01	0.15	−96.08	−561.76	−2223.38	62.50	−13,500.00
W_4	0.15	0.15	0.04	2.67	0.00	0.00	−589.66	−581.82	−2332.29	62.50	−22,566.60
W_5	0.23	0.14	0.03	2.00	0.02	0.13	−334.78	−637.70	−2773.85	50.00	−6081.82
W_6	0.47	0.13	0.04	5.00	0.02	0.19	−115.05	−675.86	−2538.42	80.00	−6081.82
W_7	0.53	0.15	0.04	4.33	0.02	0.19	−88.68	−581.82	−2465.93	76.92	−5130.00
W_8	0.52	0.27	0.04	4.67	0.02	0.22	−92.31	−268.85	−2246.73	78.57	−5566.00
W_9	0.46	0.23	0.02	3.00	0.02	0.17	−119.78	−341.18	−4478.43	66.67	−6081.00
W_{10}	0.46	0.17	0.03	2.67	0.01	0.14	−117.39	−492.11	−3636.00	62.50	−7455.00

Global baseline values and sediment quality (mg kg^{-1})

	Pb	Cu	Fe	Cd	Ni
ERL	47	34	–	1.2	21
ERM	220	270	–	9.6	52
TEL	30.2	18.7	–	0.7	–
PEL	112	108	–	4.2	–
MEC	14	50	4.1	–	–
MWS	19	33	4.1	–	–
MCS	20	45	46,700	0.3	68
Present study	2.9–11.8	3.8–12.8	945.0–2010.0	0.6–1.7	0.3 Min–2.6 Max

cold and warm springs respectively. The EF values for Cd are the highest among the metals and it has a very severe to severe enrichment. This is similar to research carried out by Ghrefat et al. (2011) in the sediments of Kafrain Dam, Jordan. The EF values also indicate that Pb has a moderate enrichment to severe enrichment, Cu has minor enrichment to moderately severe enrichment, and Ni has no enrichment. The enrichment of metals in the sediments of the springs has been observed to be relatively high in the sediments. The fluctuations in EF values of different metals in the cold and warm springs may be due to the differences in the magnitude of input for each metal in sediment and or the removal rate of each metal from the sediment as reported by Ghrefat et al. (2011). The EF values of Pb, Cu and Cd that are greater than one suggest that the sources are more likely to be anthropogenic. The EF values in this study were compared with those available from other regions. The values obtained fell within results of Kafrain Dam; 10, 70, 37410, 140 and 100 (Ghrefat et al. 2011), Wadi Al-Arab Dam 9, 60, 11270 ND, ND (Ghrefat and Yusuf 2006), Seyhan Dam; 21, 198, 393500 ND, ND (Cevik et al. 2009). Ataturk Dam ND, 18.6, 15925, ND, 91.7 mg kg^{-1} (Karadede and Unlu 2000) for Cd, Cu, Fe, Pb and Ni respectively.

The EF values show that as the values of metals vary the classification of contamination levels vary. The classification of contamination level base on I_{geo} does not always vary as the content of metals vary (Ghrefat et al. 2011). Therefore, the calculations of I_{geo} are more reliable than those of EF for assessing metal pollution as seen in Table 3. The geoaccumulation index results in the Table 3 show that sediments are uncontaminated to

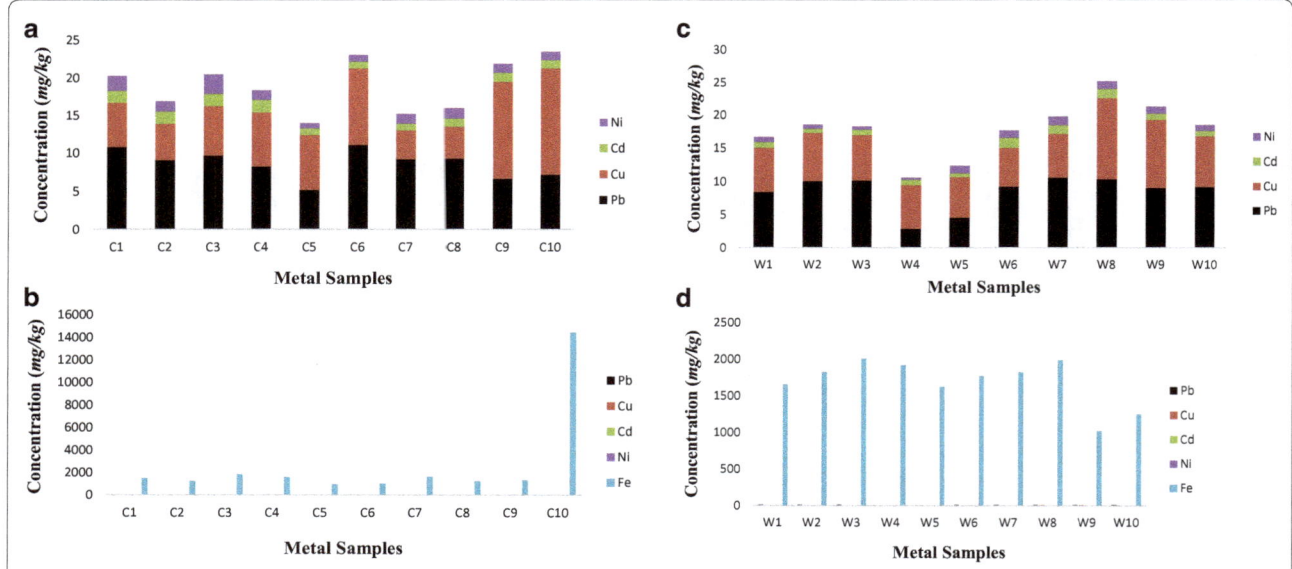

Fig. 2 **a** Bar chart of readings and spectrum for Ni Cd, Cu and Pb in selected points in cold spring. **b** Bar chart of readings and spectrum for Fe in selected points in cold spring. **c** Bar chart of readings and spectrum for Ni, Cd, Cu and Pb in selected points in warm spring. **d** Bar chart of readings and spectrum for Fe in selected points in warm spring

uncontaminated/moderately contaminated. The moderately contaminated values of Cd in few of the samples are probably a result of anthropogenic activities. The results of the analysis of the contamination factor C_f^i as proposed by Hakanson (1980) and pollution load index (PLI) (Tomlinson et al. 1980) for the studied metals are shown in Table 4. The values of C_f^i revealed low contamination levels for Pb (0.15–0.56), Cu (0.08–0.31), Fe (0.02–0.04), Ni (0.01–0.04) and indicate from moderate contamination levels to considerable contamination levels for Cd. The values of PLI indicated no pollution in all the studied samples at each sampling point and varied from 0.12 to 0.23. The analysis of Q_oC is normally used to describe the geogenic and anthropogenic sources of metal contamination in sediments samples (Zarei et al. 2014). Table 4 showed that the concentration of Pb, Cu, Fe, and Ni were mainly from geogenic sources because of the negative values while the values of Cd showed to have anthropogenic sources of contamination in all the study points. Q_oC for Cd values showed 50.00–82.35% magnitude for anthropogenic impacts that could be from tourist activities). From the result obtained, the pH lies between 6.15 and 6.95, 95% of the values obtained could be rounded up to 7, which indicates the neutrality of the sediments and pure water is neutral with a pH 7. The neutrality in sediments of both springs was attributed to factors such as CO_2 removal by photosynthesis through bicarbonate degradation and dilution of water with fresh water influx from both springs. Nitrogen in the sediments samples

varied from 0.03 to 0.05%. All the values obtained are almost the same; which can be attributed to the oxidation of organic matter that settled in the bottom sediment from the top layer. Positive correlation ($R^2 = 0.99$) obtained between organic matter (OM) and (N) % revealed the contribution of organic matter. OM content varied from 0.16 to 0.25%. The peak value 0.25% was obtained at point W_3. The calculated values of organic matter could be attributed to dead planktonic matter which settles at the bottom, oxidized and decomposed as reported by Martin et al. (2010). The results revealed conductivity values between 54.0 and 94.6, this is an indication that the two spring's sediments have conductivity not exceeding 150–500 $\mu S\ cm^{-1}$ ideally for freshwater as reported by Sharon and Montpelier (1997).

Statistical analysis

Pearson's correlation coefficients for Pb, Cu, Fe, Cd, Ni and pH values in the sediments samples of both springs are shown in Table 5. The matrix showed the strength of the linear correlation. The linear correlation coefficients showed that there is positive correlation between Pb and Ni (r = 0.66, P < 0.05), Fe and Ni (r = 0.76, P < 0.05), Cd and Ni (r = 0.66, P < 0.05) in cold spring. The positive correlation revealed the possibility of the same source of pollutants (Khuzestani and Souri 2013). Positive correlation (r = 0.57, P < 0.05) was obtained between Cd and Ni in the warm spring, while negative correlation (r = −0.81, P < 0.01) was obtained between Cu and

Table 5 Correlation analysis for metals in the warm and cold springs sediments *$P < 0.05$* **$P < 0.01$

Sediment Id and parameter		Pb	Cu	Fe	Cd	Ni	pH
Cold	Pb	1					
	Cu	− 0.37	1				
	Fe	0.28	− 0.18	1			
	Cd	0.27	− 0.23	0.10	1		
	Ni	0.66*	− 0.33	0.76*	0.66*	1	
	pH	0.27	− 0.37	− 0.31	0.33	− 0.21	1
Warm	Pb	1					
	Cu	0.34	1				
	Fe	− 0.01	− 0.18	1			
	Cd	0.43	0.28	0.08	1		
	Ni	0.43	0.33	− 0.30	0.57	1	1
	pH	− 0.27	− 0.81**	0.31	0.08	− 0.04	

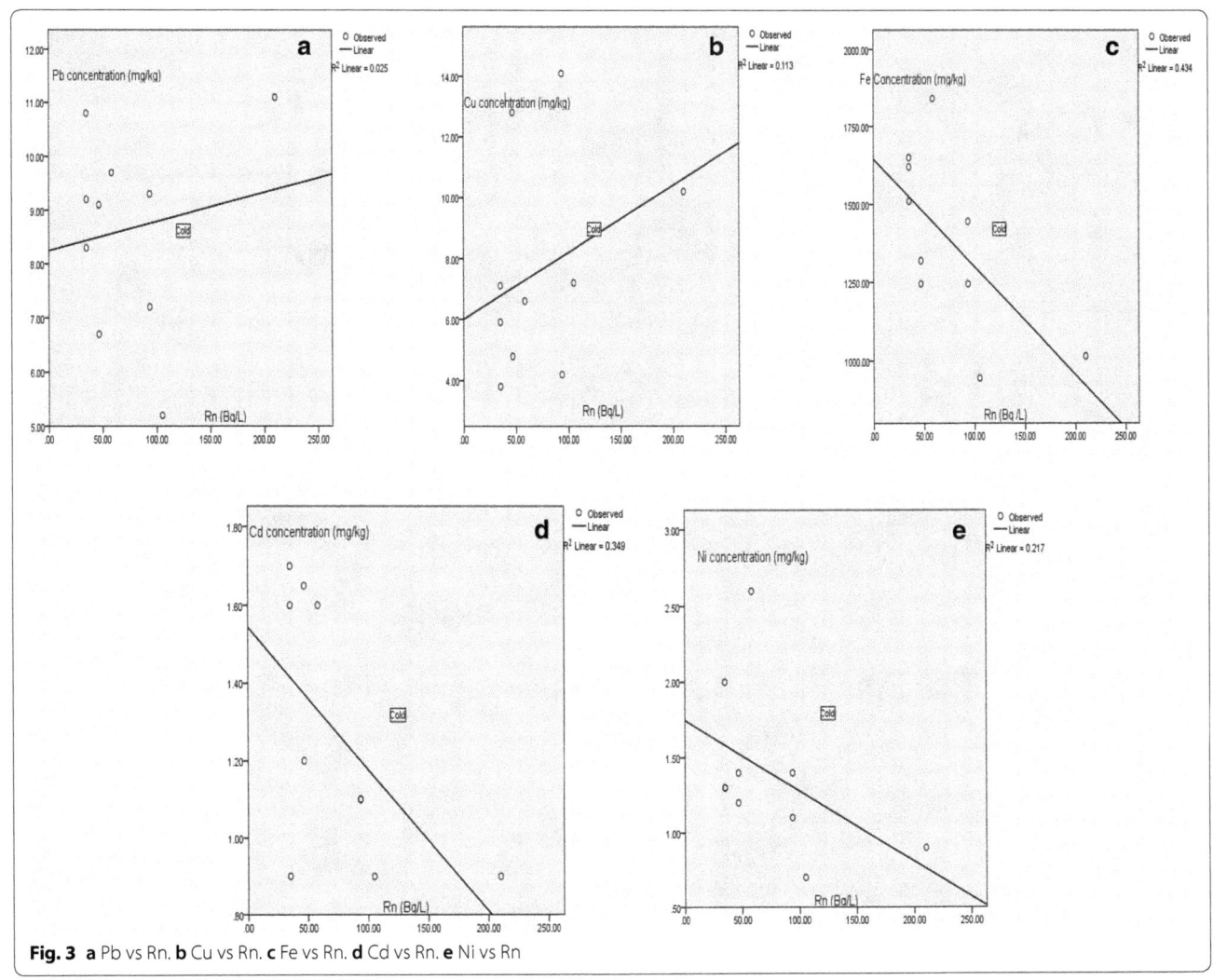

Fig. 3 **a** Pb vs Rn. **b** Cu vs Rn. **c** Fe vs Rn. **d** Cd vs Rn. **e** Ni vs Rn

pH). The pH values correlated with Pb, Cu, Fe, Cd and Ni showed no significant value in both cold and warm springs, an indication that the studied metals are immobile (Hamzeh et al. 2011). Both radon and heavy-metals are pollutants, Fig. 3a–e (Fig. 3a Pb vs Rn Fig. 3b Cu vs Rn Fig. 3c Fe vs Rn Fig. 3d Cd vs Rn Fig. 3e Ni vs Rn) and Fig. 4a–e (Fig. 4a Pb vs Rn Fig. 4b Cu vs Rn Fig. 4c Fe vs Rn Fig. 4d Cd vs Rn Fig. 4e Cu vs Rn) revealed the relationship between radon concentration and heavy-metals. The results indicated moderate positive correlation between Fe and Rn (Cold), Cd, Rn (cold), Cd and Rn (warm) and Ni and Rn (warm). Radon has poor negative correlations with all other elements which indicate different geochemical behaviour.

Conclusions

The results of the average radon concentration in bottled, cold and warm spring's water samples in Ikogosi area were within the reference range recommended by the USEPA and UNSCEAR. The water in the studied area is safe for the members of the public irrespective of age brackets. The variation in the radon concentration may be due to geological structure of the area. The effective dose due to inhalation and ingestion was found to be within the safe limit (100 µSv year^{-1}) recommended by WHO and EU. Dose due to inhalation of radon is higher as compared to ingestion. The mean concentration of metals increased according to this sequence Ni < Cd < Cu < Pb < Fe. The result obtained by the sediment quality guidelines classification revealed that most

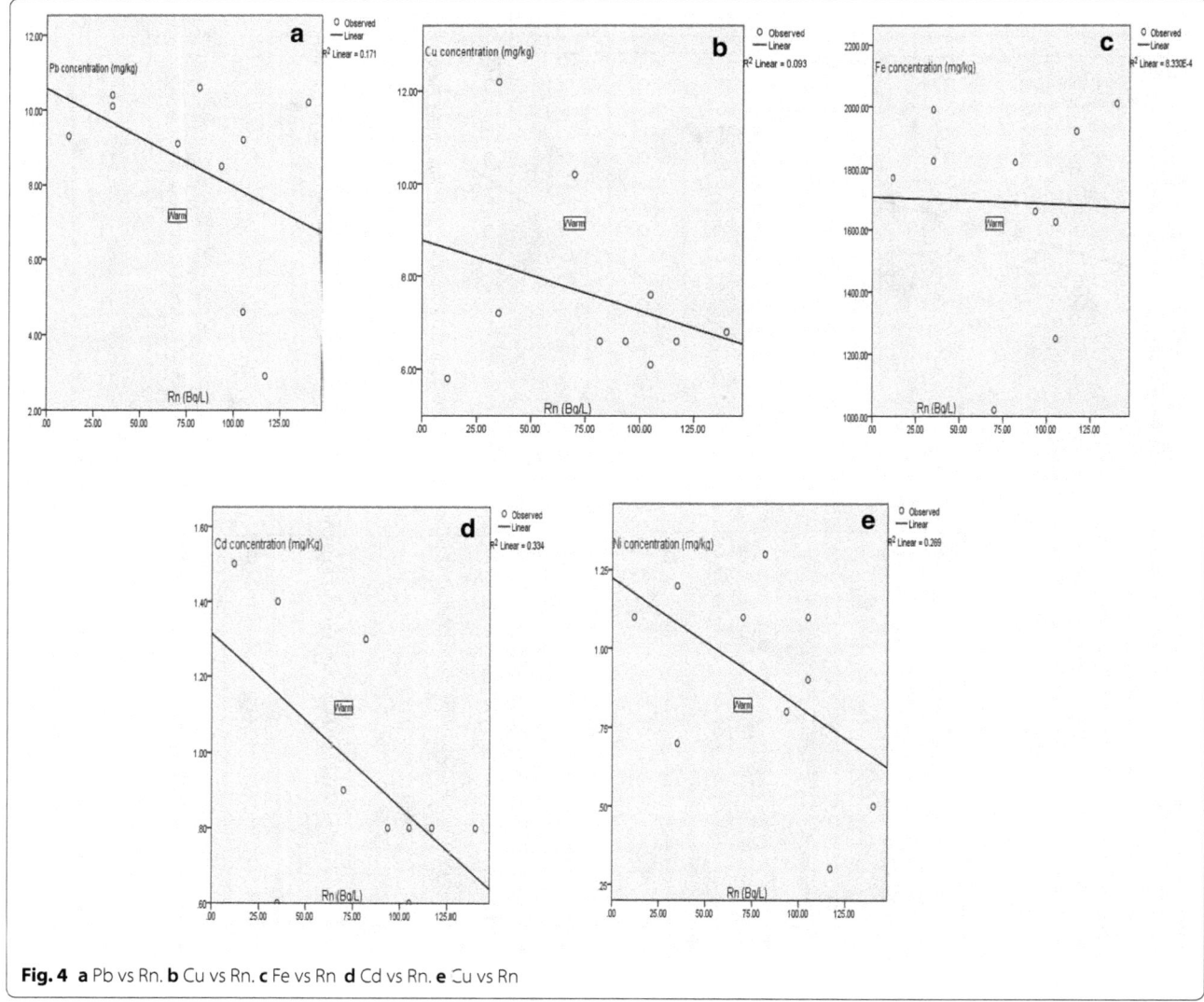

Fig. 4 **a** Pb vs Rn. **b** Cu vs Rn. **c** Fe vs Rn **d** Cd vs Rn. **e** Cu vs Rn

of the studied metals showed no negative biological effects such as reduced mental and central nervous function. Geoaccumulation index showed that all the samples are unpolluted with Pb, Cu, Fe and Ni, while the values of Cd demonstrated to have none to moderate contamination. The EF values of Cu and Ni were below 1 in 90% of the sampling points, indicating that these metals in the sediments of all sampling points were derived mainly from natural processes. EF values of Cd were enriched in the bottom sediments by anthropogenic activities. Analysis of QoC shows that Pb, Cu, Fe and Ni demonstrated a geogenic source with no evidence of anthropogenic impacts, while the values of Cd revealed anthropogenic source. The high values of Cd identified might be related to human activities such as wastes (Islam et al. 2017) from tourists' visitation, materials deposition from those that seek for healings and sacrificial materials by the two spring's worshippers. Similar results were also found for the analysis of contamination factor C_f^i. Contamination factor C_f^i demonstrated low contamination for all the studied metals except Cd. Contamination factor values for Cd were mostly evaluated to have moderate contamination to considerable contamination. The values of PLI, determining the overall metal pollution in sediments (Zarei et al. 2014), showed no pollution status in all the studied points. Therefore, no health hazard is envisaged when water and sediment samples from the two springs are used for various purposes.

Abbreviations
AAS: atomic absorption spectrophotometer; APHA: American Public Health Association; AWWA: America Water Works Association; Ci-n: cold; ICRP: International Commission on Radiological Protection; EU: European Union; UAC: United Africa Company; USEPA: United States Environmental Protection Agency; UNSCEAR: United Nations Scientific Committee on the Effects of Atomic Radiation; Wi-n: warm; WEF: water environment federation; WHO: World Health Organization; ND: not detectable.

Authors' contributions
EBF conceived designed and wrote the paper, OGO went for the samplings, HTA did the graphic and TAA conducted the data analysis. All authors read and approved the final manuscript.

Author details
[1] Radiation and Health Physics Division, Department of Physics, Faculty of Science, Ekiti State University, P.M.B 5363, Ado-Ekiti, Nigeria. [2] Department of Physics, Faculty of Science, Ekiti State University, Ado-Ekiti, Nigeria. [3] Department of Physics, Faculty of Science, Federal University, Lafia, Nigeria.

Acknowledgements
The authors appreciate the following people for samplings collection and laboratory analysis: they are Prof. Oni M.O, Mr. Yinka Ajiboye, Mr. Peter Elebonu, Mr. Jimoh Akeem and Mr. Adebayo Ayodeji.

Competing interests
The authors declare that they have no competing interests.

Funding
The research was funded by all the authors.

References
Alan MN, Chowdhury MI, Kamal M, Ghose S, Banu Hand-Chakrabarjy D (1997) Study of natural radionuclide concentrations in an area of elevated radiation background in the northern Districts of Bangladesh. Appl Radiat Isot 48:61–65

American Public Health Association (APHA) (1995) Standard methods of estimation of water and wastewater, 19th edn. America Water Works Association, Water Environment Federation, Washington

APHA, AWWA, WEF (1998) In: Clesceri LS, Greenberg AE, Eaton AD (eds) Standard methods for the examination of water and wastewater, 20th edn. American Water Work Asscciation, Water Environment Federation, Washington D.C, pp 1–541

Asaah VA, Abimbola AF, Suh CE (2006) Heavy metal concentrations and distribution in Surface soils of the Bassa industrial Zone 1, Douala, Cameroun. Arab J Sci Eng 31(2):147–151

Banerjee U, Gupta S (2012) Source and distribution of lead, cadmium, iron and manganese in the river Damodar near Asansol Industrial Area, West Bengal, India. Int J Environ Sci 2:1531–1542

Bourai AA, Gusain GS, Rautela BS, Joshi V, Prasad G, Ramola RG (2012) Variations in radon concentration in groundwater of Kumaon Himalaya, India. Radiat Prot Dosimetry. https://doi.org/10.1093/rpd/ncs186

Calmano W, Wolfgang A, Forstner U (1990) Exchange of heavy metals between sediment components and water. In: Broekaert JAC, Gucer S, Adams F (eds) Metal speciation in the environment. NATO ASI ser, ser G 23. Springer, Berlin, p 503

Cevik F, Ziya MLG, Derici OB, Findik O (2009) An assessment of metal pollution in surface sediments of Seyhandam by using enrichment factor, geoaccumulation index and statistical analyses. Environ Monit Assess 152:309–317

Chandrashekara MS, Veda SM, Paramesh L (2012) Studies on radiation dose due to radioactive elements present in ground water and soil samples around Mysore city, India. Radiat Prot Dosimetry 149(3):315–320. https://doi.org/10.1093/rpd/ncr231

Choubey VM, Ramola RC (1997) Correlation between geology and radon levels in groundwater, soil and indoor air in Bhilangana valley, Garhwal Himalaya, India. Environ Geol 32:258–262

Choubey VM, Sharma KK, Ramol RC (1997) Geology of radon occurrence around Jari in Parvati valley, Himachal Pradesh, India. J Environ Radioact 34:139–147

Crawford-Brown DJ (1991) Risk and uncertainty analysis for radon in drinking water final report. American Water Works Association, Chapel Hill

De Martino S, Sabbrase C, Monetti G (1998) Radon emanation and exhalation rates from soils measured with an electrostatic collector. Appl Radiat Isot 49(4):407–413

Ebrahimpour M, Mushrifah I (2008) Heavy metal concentration in water and sediments in Tasik chini, a freshwater lake, Malaysia. Environ Monit Assess 14(1–3):297–307

Edsfeldt C (2001) The radium distribution in some Swedish soils and its effect on radon emanation. Royal Institute of Technology, Stockholm, p 52

El-Taher A (2012) Measurement of radon concentrations and their annual effective dose exposure in groundwater from Qassim area, Saudi Arabia. J Environ Sci Technol 5:475–481. https://doi.org/10.3923/jest.2912-475.481

European Commission (2001) Commission recommendation of 20th December 2001 on the protection of the public against exposure to radon in drinking water. 2001/1982/Euratom.L344/85. Off J Eur Comm

Faweya EB (2007) Radiological implication of the natural radioactivity contents of sediments of rivers and streams in the Northern part of Ibadan city. Jurnal Fizik Malaysia 28(1&2):17–21

Faweya EB, Babalola AI (2010) Radiological safety assessment and occurrence of heavy metals in soil from designated waste dumpsites used for building and composting in Southwestern Nigeria. Arab J Sci Eng 35(2A):219–225

Faweya EB, Farai IP (2006) Natural radioactivity content of the sediment of rivers and streams in the Northern part of Ibadan city. J Appl Environ Sci 2(29):142–146

Faweya EB, Oniya EO, Ojo FO (2013) Assessment of radiological parameters and heavy metal contents of sediment samples from lower Niger River, Nigeria. Arab J Sci Eng 38:1903–1908

Ghrefat HA, Yusuf N (2006) Assessing Mn, Fe, Cu, Zn, and Cd pollution in bottom Sediments of Wadi Al-Arab Dam, Jordan. Chemosphere 65:2114–2121

Ghrefat HA, Yousef A, Marc AR (2011) Application of geoaccumulation index and enriched factor for assessing metal contamination in the Sediments of Kafrain Dam, Jordan. Environ Monit Assess 178:95–109

Greeman V, Rose AW (1996) Factors controlling the emanation of radon and thoron in soils of the eastern USA. Chem Geol 129:1–14

Gundersen LCS, Schumann RR, Otton JK, Owen DE, Dubiel RF, Dickinson KA (1992) Geology of radon in United States. Geol Soc Am Spec Pap 271:1

Hairul NBI, Ojo KA, Kasimu MA, Gafar OY, Okoloba V, Mohammed SA (2013) Ikogosi warm water resorts: what you don't know? Interdiscip J Contemp Res Bus 4(9):280–303

Hakanson L (1980) An ecological risk index for aquatic pollution control. A sedimentological approach. Water Res 14(8):975–1001

Hamzeh MA, Aftabi A, Mirzaee M (2011) Assessing geochemical influence of traffic and other vehicle-related activities on heavy metal contamination in urban soils of Kerman city using a GIS-based approach. Environ Geochem Health 33(6):577–594

Hem JD (1985) Study and interpretation of chemical characteristics of natural water, 3rd edn. U.S. Geological Survey, Washington D.C

Hector A, Tatiana C, Jesus GR, Jonay G, del Maria CC, Miguel AA, Alicia T, Alejandro R, Francisco JP, Pablo M (2015) Radon in groundwater of the Northeastern Gran Canaria Aquifer. Water 7(2):575–2590. https://doi.org/10.3390/w7062575

IAEA (2013) Measurement and calculation of radon releases from Norm residues. TRS 474 (Technical Report Series 474). IAEA, Vienna

International Commission on Radiological Protection (ICRP) (2012) A compendium of dose coefficients base on ICRP Publication 60. ICRP, Publication 119. Oxford. Ann. ICRP 41(Suppl.)

Islam MS, Ahmed MK, Raknuzzaman M, Habibullah-Al-Mamun M, Islam MK (2015) Heavy metal pollution in surface water and sediment: a preliminary assessment of an urban river in a developing country. Ecol Ind 48:282–291

Islam MS, Habibullah-Al-Mamun M, Feng Y, Tokumura M, Masunaga S (2017) Chemical speciation of trace metals in the industrial sludge of Dhaka city, Bangladesh. Water Sci Technol 76:256–267

Jarup L (2003) Hazards of heavy metal concentrations. Br Bull 68:167–182

Karadede H, Unlu E (2000) Concentrations of some metals in water, sediment and fish species from the Ataturk Dam Lake (Euphrates), Turkey. Chemosphere 41:1371–1376

Kendall GM, Smith TJ (2002) Dose to organs and tissues from radon and its decay products. J Radiol Prot 22:389–406

Khuzestani RB, Souri B (2013) Evaluation of heavy metal contamination hazards in nuisance dust particles in Kurdistan Province, Western Iran. J Environ Sci 25(7):1346–1354

Kukoyi IA, Tijani NO, Adedara MT (2013) Evaluation of Ikogosi warm spring: a potential Geotourist site in Ekiti State, Southwest, Nigeria. Eur J Hosp Tour Res 1(3):1–9

Kumar A, Kaur M, Mehra R, Kumar DK, Mishra R (2017a) Comparative study of radon concentration with two techniques and elemental analysis in drinking water samples of the Jamiu district, Jamiu and Kashmir, India. Health Phys 113(4):1–9

Kumar A, Kaur M, Sharma S, Mehra R (2017b) A study of radon concentration in drinking water samples of Amritsar city of Punjab (India). Radiat Prot Environ 39:13–19

Lai WL, Chen JJ, Chung CY, Lee CG, Liao SW (2010) The influence of lagoon on neighboring rivers by water and sediment quality. Water Sci Technol 61(10):2477–2489

Liao SW, Sheu JY, Chen JJ, Lee GG (2006) Water quality assessment and apportionment source of pollution from Neighborings Rivers in Tapeng Laggon (Taiwan) using multivariate analysis: a case study. Water Sci Technol 54(11–12):47–55

Manzoor F, Alaamer AS, Tahir SNA (2008) Exposure to ^{222}Rn from consumption of underground municipal water supplies in Pakistan. Radiat Prot Dosimetry 130:392–396

Marazio-tis EA (1996) Effects of intraparticle porosity on the radon emanation coefficient. Environ Sci Technol 30:2441–2448

Martin GD, Vijay TG, Laluraj CM, Madhu NV, Joseph T, Nair M, Gupla GV, Balachandran KK (2010) Fresh water influence on nutrient stoichiometry in a tropical estuary, Southwest coast of India. Appl Ecol Environ Res 6(1):57–64

Momodu MA, Anyakora CA (2010) Heavy metal contamination of ground water: the Surulere case study. Res J Environ Earth Sci 2:39–43

Morillo J, Usero J, Gracia I (2004) Heavy metal distribution in marine sediments from the southwest coast of Spain. Chemosphere 55:431–442

Mose DG, Mushrush GW, Chrosniak C (1990) Indoor radon and well water radon in Virginia and Maryland. Arch Environ Cotam Toxicol 19(6):952–956

Muller G (1969) Index of geoaccumulation in sediments of the Rhine River. GeoJournal 2(3):108–118

Ojo JS, Olorunfemi MO, Falebita DE (2011) An appraisal of the geologic structure beneath the Ikogosi warm spring in south western Nigeria. J Appl Sci 5(1):75–79

Oladipo AA, Oluyemi EA, Tubosun IA, Fasasi MK, Ibitoye FI (2005) Chemical examination of Ikogosi warm spring in south western Nigeria. J Appl Sci 5(1):75–79

Oni MO, Oladapo OO, Amuda DB, Oni EA, Olive-Adelodun AO, Adewale KY, Fasina MO (2014) Radon concentration in groundwater of areas of high background radiation level in southwestern Nigeria. Niger J Phys 25(1):64–67

Oni EA, Oni OM, Oladapo OO, Olatunde ID, Adediwura FE (2016) Measurement of radon concentration in drinking water of Ado-Ekiti, Nigeria. J Acad Ind Res 4(8):190–192

Orkun ID, Galip S, Cagatayhan BE, Turan Y, Bulent S (2011) Assessment of metal pollution in water and surface sediments of the Seyhan River, Turkey, using different indexes. Clean Soil Air Water 39(2):185–194

Ortega L, Manzano M, Custodio E, Hornero J, Rodriguez-Arevalo J (2015) Using ^{222}Rn to identify and quantify groundwater inflow to the Mundo River (SE Spain). Chem Geol 395:67–79

Pravin US, Ansari MVA, Dixit NN (2014) Assessment of physic-chemical properties of sediments collected along the Mahul creek near Mumbai, India. Int Lett Nat Sci 16:54–61. https://doi.org/10.18052/www.scipress.com/ILNS.16.54

Rasheed M (2010) Monitoring of contaminated toxic and heavy metals, from mine tailings through age accumulation, in soil and some wild plants in Southeast Egypt. J Hazard Mater 178:739–746

Ravikumar P, Somashekar RK (2014) Determination of the radiation dose dues to radon ingestion and inhalation. Int J Environ Sci Technol 11:493–508

Rogers AS, Imevbore AMA, Adegoke OS (1969) Physical and chemical properties of Ikogosi warm springs, Western Nigeria. J Min Geol 4:69–81

Schubert M, Brueggemman L, Knoeller K, Schirmer M (2011) Using radon as an environmental tracer for estimating groundwater flow velocities in single well tests. Water Resour Res. https://doi.org/10.1029/2010Wr009572

Sharon B, Montepelier VT (1997) Testing the waters: chemical and physical vital signs of a river: river watch network. ISBN: 0787234923. www.fosc.org/WQData/WQparameter.htm

Shuanxi F (2014) Assessment of heavy metal pollution in stream sediments for the Baoji city section of the Welhe River in Northwest China. Water Sci Technol 70(7):1279–1284. https://doi.org/10.2166/wst.2014.366

Skeppstrom K, Olofsson B (2007) Uranium and radon in groundwater. Eur Water 17:51–62

Song D, Zhuang D, Jiang D, Fu J, Wang Q (2015) Integrated health risk assessment of heavy metals in Suxian County, South China. Int J Environ Res Public Health 12:7100–7117

Stephen DM, Mohammad RS, Eric W, Sabine A, Roberto C (2004) An assessment of metal contamination in coastal sediments of the Caspian sea. Mar Pollut Bull 48:61–77

Stringer C, Burnett WC (2004) Sample bottle design improvements for radon emanation analysis of natural waters. Health Phys 87:642–646

Tomlinson D, Wilson J, Harris C, Jeffrey DD (1980) Problems in the assessment of heavy metal levels in estuaries and the formation of a pollution index. Helgolander Meeresuntersuchunger 33(1–4):566–575

Toscani L, Martinell G, Dalledonne C (2001) Proceedings 5th international conference on rare gas geochemistry, (EDS), Debrecen, Hungary, p 321

Tukura BW, Ayinya MIG, Ibrahim IG, Onche EU (2014) Assessment of heavy metals in groundwater from Nasarawa State, Middle Belt, and Nigeria. Am Chem Sci J 4:798–812

Turekian KK, Wedepohl KH (1961) Distribution of the elements in some major units of the earth's crust. Geol Soc Am Bull 72(2):175–192

United States Environmental Protection Agency (USEPA) (1991) Federal register 40 parts 141 and 142: national primary drinking water regulations; radionuclides: proposed rule. U.S. Government Printing Office, Washington, D.C; Federal Register 64:9559–9599

United States Environmental Protection Agency (USEPA) (1999) Radon in drinking water health risk reduction and cost analysis. USEPA, office of Radiation Programs, Washington, D.C; USEPA Federal Register 64

United Nations Scientific Committee on the Effects of Atomic Radiation (UNSCEAR) (2000) Sources and effects of ionizing radiation. In: Report to the general assembly with scientific annexes. United Nations, New York

World Health Organization (WHO) (2004) Guidelines for drinking water quality, vol 1. World Health Organization, Geneva

World Health Organization (WHO) (2011) Guidelines for drinking water quality. In: Radiological aspects. World Health Organization, Geneva, pp 214–217

Xinwei L (2006) Analysis of radon concentration in drinking water in Baoji China and the associated health effects. Radiat Prot Dosimetry 121(4):452–455

Yogesh PG, Choubry P, Ramola VC (2009) Geohydrological control on radon availability in groundwater. Radiat Meas 44(1):122–126

Yu D, Kim KJA (2004) A physiologically based assessment of human exposure to radon released from groundwater. Chemosphere 54:639–645

Zarei I, Pourkhabbaz A, Khuzestani RB (2014) An assessment of metal contaminant risk in sediments of Hara Biosphere Reserve Southern Iran with a focus on application of pollution indicators. Environ Monit Assess. https://doi.org/10.1007/s10661-014-3439-X

Zhang W, Feng H, Chang J, Qu J, Xie H, Yu L (2009) Heavy metal contamination in surface sediments of Yangtze River Intertidal Zone: an assessment from different indexes. Environ Pollut 157:1533–1543

Zhang C, Qiao Q, Piper JDA, Huang B (2011) Assessment of heavy metal from a Fe-smelting plant in urban river sediments using environmental magnetic and geochemical methods. Environ Pollut 159:3057–3070

8

Chemical forms of phosphorous and physicochemical properties of acid soils of Cheha and Dinsho districts, southern highlands of Ethiopia

Bereket Ayenew[1*], Abi M. Taddesse[2], Kibebew Kibret[3] and Asmare Melese[4]

Abstract

Background: Soil acidity and low availability of P are among the major problems limiting crop production in the highlands of Ethiopia. The current study was conducted to evaluate the distribution of chemical P forms and selected physical and chemical properties of soils in Cheha and Dinsho districts, southern highlands of Ethiopia. Twelve representative composite soil samples were collected from the surface layer (0–15 cm) and analyzed for selected soil properties. Analysis of available P and oxalate extractable P, Al and Fe, and P fractionation were done following standard procedures.

Results: The pH of soils in the study area ranged from 4.65 to 5.45. In the study area the Mehlich-3 extractable P ranged from 8.90 to 25.75 mg kg^{-1}, while the Bray-I extractable P varied from 8.54 to 22.81 mg kg^{-1}. The studied soils had a total P content that ranged from medium to very high. Mehlich-3 P was positively and significantly correlated with pH, CEC and OM, while it was negatively and significantly (P $\leq$ 0.01) correlated with exAc and exAl. Oxyhydroxides of iron and aluminum were dominating the exchange sites of the studied soils. In terms of the relative abundance of P forms, Res-P and HCl-P followed by NaOH-P$_o$ and NaOH-P$_i$ were the dominant P forms in Cheha and Dinsho districts respectively. The total sum of readily available P forms (NH$_4$Cl-P$_i$, NaHCO$_3$-P$_i$ and NaHCO$_3$-P$_o$) did not exceed 4% of the total P.

Conclusion: Soils of the study area were acidic in reaction and found to have low available P. Given the presence of considerable amount of total P, low P availability and high soil acidity in the study areas, further research has to be done on adsorption characteristics to determine the P requirements of the soils for better P management.

Keywords: Amendment, Fractionation, Oxyhydroxide, Phosphate, Acid soils, Ethiopian soils

Background

Soil acidity and low availability of phosphorous (P) are among the major problems limiting crop production in the highlands of Ethiopia where high rainfall, nutrient leaching, and soil erosion are more prevalent (Asmare et al. 2015). When soil pH drops below 5, active form of Al becomes soluble and results in reduced nutrient uptake (Achalu 2014). Reports from numerous

experiments in P-deficient soils in the highlands of east Africa indicate that the inorganic form of P assessed by the Olsen method is less than 5 mg P kg^{-1} soil (Ikerra 2004). Limited availability of P in many tropical soils can be attributed to severe P fixation or retention, which is particularly strong in soils with low P status. In acidic soils, crystalline and non-crystalline oxides of Fe and Al (sesquioxides) are the main adsorbing agents of phosphate (Jiang et al. 2015).

Despite the absence of specific system that offers an adequate solution for generalizing or comprehending the behavior of P in different soils, sequential extraction

*Correspondence: bersofsam12@gmail.com
[1] Department of Chemistry, School of Natural Sciences, Macda Walabu University, Bale-Robe, Ethiopia

has been used for characterization of complex mixtures that do not lend themselves to conventional chemical analysis (Herlihy and McCarthy 2006). In a review on the use of the Hedley method, contrasting agro-ecosystems of tropical and temperate climates, with different soil use and managements, Negassa and Leinwebe (2009) reported that inorganic and organic fractions of P could act as a source or sink of available P to plants. The Hedley sequential extraction or fractionation of P method has also been used to distinguish labile from non-labile fractions and to rank stable fractions and organic fractions in terms of their availability to plants (Cross and Schlesinger 1995). Based on this extraction method, P can be classified as available to plants or microorganisms (labile P), which includes the sum of inorganic P (Pi) and organic P (Po) extracted by ammonium chloride (NH_4Cl) and sodium bicarbonate ($NaHCO_3$). The non-labile P includes the sum of the remaining fractions (Pi and Po in hydroxide and in sonicate + hydroxide, Pi in HCl and residual P in sulfuric acid digestion). Accurately characterizing P forms has been proved to overcome the limited information that total P analysis can provide (Zhou et al. 2001). The quantity of P reserve to replenish solution P and the ability of the soil to maintain sufficient solution P concentration are the main factors governing P supply to plants (Buresh et al. 1997) and in the absence of fertilization, solution P is replenished from other soil P pools with different degree of availability.

Even though several works have been done on the relative distribution and forms of P (Duffera and Robarge 1996; Shiferaw 2004; Negassa and Leinwebe 2009; Fisseha et al. 2014; Achalu et al. 2014; Asmare et al. 2015), research related to the different P forms and distribution encompassing organic P in the soils of the study areas is scant. Besides, given the role played by phosphorous in agriculture, knowledge on the status of various P forms, their relative distribution and interaction with other soil attributes that influence P availability in soils of the study area is worth researching. Therefore, this work was initiated to characterize the forms and distributions of P and their relation with selected soil physical and chemical properties in acidic soils of Dinsho and Cheha districts, southern highlands of Ethiopia.

Methods

Description of the study area

The study was conducted in Ethiopian highlands of Dinsho district in Bale zone and Cheha district in Gurage zone, Ethiopia (Fig. 1). Dinsho district lies between 6°58′40″ and 7°20′0″ N, and 39°44′0″ and 40°26′40″ E. Physiographically, most of the land area of the district is situated above 2000 m above sea level (masl). The district is classified into three agro-climatic zones: highlands (2300–2600 masl), midlands (1500–2300), and lowlands (1200–1500). The district has a bimodal rainfall distribution with mean annual rainfall of about 1150 mm. The

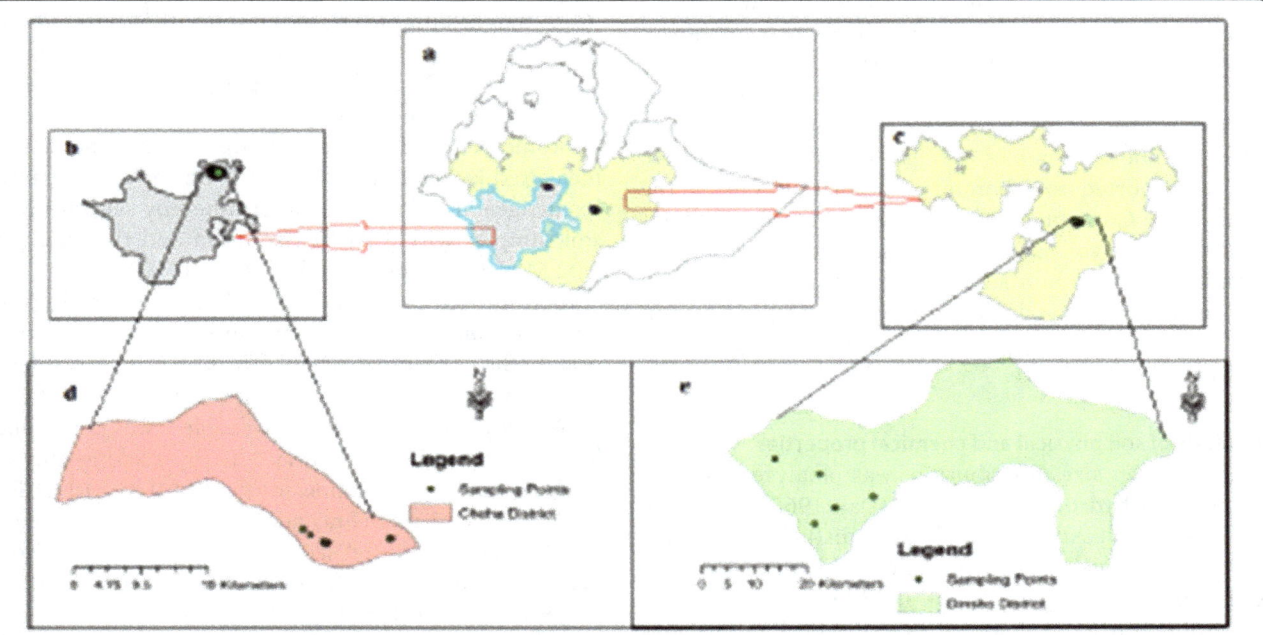

Fig. 1 Location map of the study areas, **a** Oromia and SNNP regions in Ethiopia, **b** Cheha district in SNNP region, **c** Dinsho district in Oromia region, **d** Cheha district, **e** Dinsho district

maximum and minimum mean annual temperatures of the District are 17.5 and 6 °C, respectively. Wheat and barley are some of the major cereal crops grown in the area. The major reference soil groups in the district are Pellic Vertisols, Eutric Cambisols, Eutric Nitosols (now Nitisols), and Chromic Luvisols (FAO Food and Agriculture Organization 1984).

Cheha district is situated between 8°32'0'' and 8°20'0''N, and 37°41'20'' and 38°2'40''E, at an elevation that ranges from 900 to 2812 masl. EIAR (Ethiopian Institute of Agricultural Research) (2011) classified the area into three agro-ecological zones i.e. highlands (2300–3200 masl), midlands (1500–2300 masl), and lowlands (500–1500 masl) based on the bimodal rain fall system. The 10 years mean annual rainfall of the district is about 1268 mm. The mean annual maximum and minimum temperatures are 24.97 and 10.69 °C, respectively. The dominant soil types are Eutric Nitosols, Leptosols, and Pellic Vertisols (FAO Food and Agriculture Organization 1984).

Site selection, soil sampling, and soil analysis

A preliminary soil survey and field observation was made using topographic map (1:50,000) of the study area. Soil pH (potentiometer), altitude (GPS), and slope (clinometer) were used as criteria for selection of soil sampling sites. Fifteen (15) sub-samples were collected from each sampling site to make one composite sample. Accordingly, twelve composite soil samples (0–15 cm), seven from Cheha district (Goha 1, Goha 2, Goha 3, Aftir, Abret, Kechot, and Moche), and five from Dinsho district (Doyomarufa 1, Doyomarufa 2, Tulu, Weni, and Ketasire) having pH values of less than 5.5 were obtained within an altitudinal range of 2000–3000 masl and slope less than 8% (Table 1). Soil pH was measured at field condition using portable pH meter to select soils having a pH of less than 5.5. The soil samples were put in plastic bag, tagged, and transported to laboratory for analysis during the 2014/2015. Consequently, adequate (about 1 kg) amount of composited soil samples were air dried and ground to pass through a 2 mm sieve for analysis of selected soil chemical and physical properties except organic carbon and total nitrogen in which case the samples were passed through a 0.5 mm sieve.

Analysis of soil physical and chemical properties

Soil particle size distribution was analyzed by the Bouyoucos hydrometer method (Day 1965). Soil bulk density was measured from three undisturbed soil samples collected using a core sampler following the procedure described by Jamison et al. (1950). The moisture retention at field capacity (FC) (− 0.33 bar) and permanent welting point (PWP) (− 15 bar) was determined using the pressure plate apparatus method (Gupta

Table 1 Sampling locations and site description

Sampling sites	Longitude (E)	Latitude (N)	Altitude (masl)	Slope (%)
SC1	037°93'891''	08°033'45''	2426	4
SC2	037°94'389''	08°032'35''	2508	3
SC3	037°94'134''	08°030'78''	2498	2
SC4	037°91'180''	08°052'36''	2302	5
SC5	037°92'114''	08°043'58''	2401	3
SC6	038°02'572''	08°038'63''	2603	5
SC7	038°02'377''	08°038'23''	2563	5
SD1	039°51'934''	07°075'65''	2819	4
SD2	039°51'900''	07°072'28''	2832	5
SD3	0390 52'387''	07°066'39''	2790	3
SD4	039°52'450''	07°071'01''	2768	2
SD5	039°52'714''	07°080'40''	2719	5

SC1 Goha 1, *SC2* Goha 2, *SC3* Goha 3, *SC4* Aftir, *SC5* Abret, *SC6* Kechot, *SC7* Moche, *SD1* Doyomarufa 1, *SD2* Doyomarufa 2, *SD3* Tulu, *SD4* Weni, *SD5* Ketasire

2004). Available water holding capacity (AWHC) was calculated from the difference between moisture content at FC and PWP. Soil pH measured potentiometrically in H_2O and 1 M KCl solution at the ratio of 1:2.5 for soil:water and soil:KCl solutions using a combined glass electrode pH meter (Van Reeuwijk 1992). The soil lime requirement (LR) was determined by Shoemaker, McLean and Pratt (SMP) single buffer procedure (Shoemaker et al. 1961). Cation exchange capacity (CEC) of the soils was determined by making use of the method suggested by Lavkulich (1981). Exchangeable calcium (Ca), magnesium (Mg), potassium (K), and sodium (Na) were determined by saturating the soil samples with 1 M NH_4OAc solution at pH 7.0. Then, Ca and Mg were determined by using atomic absorption spectrophotometer (AAS) while exchangeable Na and K were measured by flame photometer from the same extract. Exchangeable acidity was determined by saturating the soil samples with 1 M KCl solution and titrating with 0.02 M NaOH as described by Rowell (1994).

Total nitrogen was analyzed using the micro-Kjeldahl digestion followed by ammonium distillation and titration method (Bremner 1965). Organic carbon was determined by the dichromate oxidation method as described by Walkley and Black (1934). Available P was determined by different techniques: Bray-I P (by shaking the soil samples with an extracting solution of 0.03 M NH_4F in 0.1 M HCl for 1 min (Bray and Kurtz 1945) and Mehlich 3 P (by shaking the soil samples with an extracting solution of 0.2 M $CH_3COOH + 0.25$ M $NH_4NO_3 + 0.015$ M $NH_4F + 0.013$ M $HNO_3 + 0.001$ M EDTA for 5 min) (Mehlich 1984). Total soil P was determined using the method stated by Olsen and Sommers (1982).

Soil P fractionation

For P fractionation study, 30 ml of 1 M NH_4Cl was added in centrifuge tube containing 0.5 g of soil equilibrated for 16 h followed by centrifugation, and then the inorganic P was determined from the extract. To the remaining soil in the tube, 30 ml of 0.5 M $NaHCO_3$ was added and shaken for same period as before, in order to determine inorganic P and total P extracted by $NaHCO_3$. The soil in the tube was allowed to interact with 30 ml of 0.1 M NaOH. Then the solution was shaken for 16 h, and centrifuged for determination of inorganic P from the extract. In the fractionation scheme total P in the NaOH extract was determined by perchloric acid digestion of the aliquot sample. Subsequently, 30 ml of 0.1 M NaOH was added into the tube containing the soil and sonicated for 2 min at 75 W for the determination of inorganic and total P in the sonicated NaOH extract. The remaining soil was treated with 1 M HCl followed by shaking and centrifugation of the soil containing 30 ml of 1 N HCl in order to determine the inorganic P in the HCl extract. Finally, the soil was digested with a mixture of concentrated HNO_3 and $HClO_4$ acids in order to determine the residual P. Accordingly, the P fractions were successively extracted with 1 M NH_4Cl (available Pi), 0.5 $NaHCO_3$ (labile Pi and Po adsorbed on the soil surface), 0.1 M NaOH (moderately labile Pi and Po held more strongly by Chemisorption to surfaces of Al and Fe oxides), Sonicate + 0.1 M NaOH (Pi and Po adsorbed within surfaces of Al and Fe oxides of soil aggregates-occluded P), 1 M HCl (P associated to Ca, derived from primary mineral-apatite) and a mixture of HNO_3 and $HClO_4$ (residual P-non-labile, stable Po forms and relatively insoluble Pi forms) as described by Hedley et al. (1982) and modified by Chen et al. (2000).

The oxalate extractable P, Al and Fe (P_{ox}, Al_{ox} and Fe_{ox}) were determined with 0.05 M ammonium oxalate $[(NH_4)_2C_2O_4 \cdot 2H_2O$, pH 3.3] in the dark (Mckeague and Day 1966). Citrate bicarbonate dithionite-extractable Fe and Al (Fe_d and Al_d) were determined by the method of Mehra and Jackson (1960). Degree of P saturation was determined from total phosphorous sorption capacity (PSC_t) which is alienated into P occupied sites and remaining P sorption capacity (PSC_r). The amount of P sorbed on the PSC_t was evaluated from PSC_r and oxalate-extractable P (P_{ox}). Likewise, the PSC_r was calculated by multiplying an experimental P sorption capacity (F_r) by a factor of 2.75 (Maguire et al. 2001). Experimental P sorption capacity was determined by the standardized one-point short-term isotherm method (Bache and Williams 1971). Three gram of 2 mm soil was shaken with 60 ml of a 75 mg P l^{-1} as KH_2PO_4 equivalent to 1.5 g P kg^{-1} soil in 0.01 M $CaCl_2$ solution for 18 h with some drops of chloroform. The sorbed P (F_r) was calculated as the difference

between added P and P remaining in solution. Accordingly, the degree of P saturation (DPS) was determined as the percentage of the ratio of the different forms of P to the oxalate extractable Al and Fe (Al_{ox} and Fe_{ox}) as follows (Van der Zee et al. 1988):

DPS as the percentage of ratio of P_{ox} to PSC

$$DPS\ P_{ox} = (P_{ox}/PSC) \times 100$$

The P adsorption capacity (PSC, mmol kg^{-1}) of the soil was calculated and estimated by (Börling et al. 2001):

$$PSC = \alpha(Al_{ox} + Fe_{ox})$$

where Al_{ox} and Fe_{ox} are oxalate-extractable Al and Fe, and α is a scaling factor.

The α scale factor was estimated by:

$$PSC_t = PSC_r + P_{ox}$$
$$PSC_r = F_r * 2.75$$

$$\alpha = \frac{P_{ox} + PSC_r}{Fe_{ox} + Al_{ox}}$$

The F_r, PSC_r, PSC_t parameters and α scale factor were determined for the twelve soil samples and the mean α was used to calculate PSC and DPS.

Statistical analysis

Pearson's simple correlation coefficient was executed using Statistical Analysis System (SAS) version 9.1 (SAS (Statistical Analysis System) version 9.1. SAS Institute 2004) to evaluate the magnitudes and directions of relationship between the different P forms and other related soil properties.

Results and discussion

Characteristics of selected soil physical properties

The selected soil physical properties of the studied soils are indicated in Table 2. The textural class of the studied soils is predominantly clay. Generally, the clay content of the studied soils ranged from 36.6 to 49.8% and the highest clay content was recorded from SC3 soil and the lowest was obtained from SD3 soil. This difference in clay content could be attributed to the variability in the degree of weathering of the soils. The bulk density values ranged from 1.11 to 1.32 g cm^{-3} with a mean value of 1.21 g cm^{-3} and the bulk density values of the studied soils were found to be within the normal range suggested for clay soils, which ranges from 1.0 to 1.6 g cm^{-3} (Aubertin and Kardos 1965). The observed variability in bulk density among the studied soils might be due to differences in organic matter content of the soils. Moreover, potential root restriction of the soils is

Table 2 Selected physical properties of the soils studied

Sampling sites	Particle size distribution (%)			Textural class	Bulk density (g cm³)	Water content at (%)		AWHC (%)
	Sand	Silt	Clay			FC	PWP	
SC1	23.2	28.6	48.2	Clay	1.32	31.00	19.60	11.40
SC2	22.8	32.6	44.6	Clay	1.25	32.25	22.25	10.00
SC3	23.2	27.0	49.8	Clay	1.29	31.00	20.55	9.65
SC4	19.4	34.0	46.6	Clay	1.22	31.95	21.75	10.20
SC5	14.8	37.2	48.0	Clay	1.21	31.65	20.90	10.75
SC6	19.6	37.0	43.4	Clay	1.24	32.90	22.50	10.40
SC7	22.0	33.0	45.0	Clay	1.24	32.70	21.80	10.90
SD1	29.6	32.6	37.8	Clay	1.13	31.60	21.60	10.00
SD2	23.8	32.6	43.6	Clay	1.16	33.00	22.70	10.30
SD3	29.0	34.4	36.6	Clay loam	1.11	39.15	22.20	16.95
SD4	26.8	29.0	44.2	Clay	1.17	38.10	25.11	12.99
SD5	29.2	32.6	38.2	Clay loam	1.16	32.31	20.72	11.59

SC1 Goha 1, *SC2* Goha 2, *SC3* Goha 3, *SC4* Aftir, *SC5* Abret, *SC6* Kechot, *SC7* Moche, *SD1* Doyomarufa 1, *SD2* Doyomarufa 2, *SD3* Tulu, *SD4* Weni, *SD5* Ketasire

presumed to be less as the bulk densities are less than 1.4 g cm⁻³ (Hazelton and Murphy 2007). The soil water content at FC ranged from 30.2 to 39.2% with mean value of 33.1%. According to the result obtained from FC and PWP measurements, soil samples collected from Dinsho district showed relatively higher water retention compared with soil samples collected from Cheha. This difference could be attributed to the relatively higher organic matter content of the former. The PWP and AWHC ranged from 19.6 to 25.1 and 9.7 to 16.9% with mean values of 21.8 and 11.3%, respectively.

Soil pH, buffer pH and lime requirements

The pH (H₂O) values of the soils varied between 4.65 and 5.45 (Table 3). Based on the rating suggested by Jones (2003), 58% of the sampled soils were found to be very strongly acidic while 42% could be considered as strongly acidic. The pH (KCl) values of the studied soils varied from 3.98 to 4.46. The pH measured in potassium chloride solution was lower than pH measured in water for all the soil samples. The greater amount of exchangeable Al (0.21–1.54 cmol_c kg⁻¹) at the exchange sites might have contributed for the lower pH (KCl) value in solution of the studied soils. The finding of the present

Table 3 Soil pH, buffer pH and lime requirements

Sampling sites	pH			ΔpH	LR (tons CaCO₃ ha⁻¹)	
	H₂O	KCl	Buffer	KCl	pH (6.0)	pH (6.5)
SC1	4.65	3.98	5.11	0.66	9.20	11.50
SC2	4.79	4.05	5.41	0.74	7.80	9.60
SC3	4.73	4.04	5.20	0.68	8.70	10.70
SC4	4.81	4.10	5.31	0.71	8.20	10.00
SC5	5.13	4.22	5.61	0.90	6.70	8.20
SC6	4.76	4.12	5.30	0.64	8.20	10.00
SC7	4.91	4.09	5.31	0.82	8.20	10.00
SD1	5.30	4.38	6.20	0.92	3.70	4.60
SD2	5.01	4.21	6.11	0.79	4.10	5.00
SD3	5.12	4.29	6.51	0.83	2.10	2.60
SD4	5.36	4.42	6.31	0.94	3.30	4.00
SD5	5.45	4.46	6.41	0.99	2.80	3.40
Mean	5.00	4.20	5.73	0.79	6.08	7.47

LR lime requirements, *SC1* Goha 1, *SC2* Goha 2, *SC3* Goha 3, *SC4* Aftir, *SC5* Abret, *SC6* Kechot, *SC7* Moche, *SD1* Doyomarufa 1, *SD2* Doyomarufa 2, *SD3* Tulu, *SD4* Weni, *SD5* Ketasire

study concurs with previous reports which described that the pH measured in chloride solution is lower than pH measured in water because of the larger amount of H^+ ion forced to soil solution when Al^{3+} or Fe^{3+} are replaced by K^+ on the exchange sites (Kissel and Vendrell PF Isaac 2004; Asmare et al. 2015). According to USDANRCS (2004) report, the numerical difference in the values of pH measured in KCl and H_2O is referred to as the delta pH (ΔpH). When this difference is negative, the colloid has a net negative charge, and when positive, it has a net positive charge. The ΔpH value obtained for the studied soil samples indicates that the colloid has a net positive charge. The highest difference between pH-H_2O and pH-KCl was found for soil sample SD5 from Dinsho (0.99 pH units) and the lowest (0.66 pH units) was for the soil sample SC1 from Cheha district. The variation in abundance of exchangeable Al which ranged from 0.21 to 1.54 $cmol_c$ kg^{-1} might be a possible rationale for the disparity of ΔpH among soils of the study area.

The hydrogen ion (H^+) concentration that exists on the soil colloids (buffer pH) was evaluated for soil samples and varied between 5.11 and 6.51 pH units. The LR values, as determined by the SMP single buffer method to raise the pH of the soils to target pH values of 6.0 and 6.5 are presented in Table 3. Consequently, the amount of lime required to raise the respective pH values of the studied soils to the target pH of 6.0 and 6.5 ranged from 2.10 to 9.20 t $CaCO_3$ ha^{-1} and 2.60–11.50 t $CaCO_3$ ha^{-1} respectively. Therefore, the soils of the study area need substantial amount of lime to alleviate the acidity problem and increase the productivity of acid sensitive crops. Likewise, the output of the correlation matrix in

Table 8 shows a strong significant and negative correlation between pH and LR (r = − 0.88, P ≤ 0.01). Hence, the lower the pH value of the soil, the higher is the amount of liming material needed to increase the pH to a suitable range for sustainable production of acid sensitive crops.

Exchangeable cations and exchange properties

The cation exchange capacity of soils of Cheha district (SC1–SC5) showed a relatively lower range (21.15–31.54 $cmol_c$ kg^{-1}) than soils from Dinsho district (SD1–SD5) (31.14–36.03 $cmol_c$ kg^{-1}) (Table 4). According to Hazelton and Murphy (2007) the CEC of the soils were in the range of moderate to high values. The variation in CEC among the soils might be due to the difference in OM content, clay mineralogy between some soil forms and the pH range in which the soils exist. In agreement with the current study, Peinemann et al. (2000) declared that clay and organic matter are the main sources of CEC and the more clay and organic matter (humus) a soil contains, the higher is its CEC.

As per the ratings suggested by Hazelton and Murphy (2007), the exchangeable Na was in the range of very low to moderate in the study area. Whereas the exchangeable Ca and Mg were in the range of low to moderate for soils from Cheha and moderate to high for Dinsho district. In Cheha district, the variation in concentration levels of exchangeable K among the soils followed the same trend with that of Ca and Mg whereas it was moderate in Dinsho district. Barber (1984) indicated that the critical level of exchangeable K for optimum crop production is 0.38 $cmol_c$ kg^{-1}. The status of exchangeable bases in soil from Dinsho was slightly higher than soils from Cheha. The

Table 4 Exchangeable cations and exchange properties

Soil	Exchangeable cations and CEC ($cmol_c kg^{-1}$)									Percent saturation (%)			K/Mg
	Ca	Mg	K	Na	TEB	exAc	exAl	CEC	ECEC	PAcS	PAIS	PBS	Ratio
SC1	4.21	0.98	0.20	0.04	5.43	2.61	1.54	19.15	10.56	13.63	8.04	28.36	0.20
SC2	8.45	2.23	0.26	0.09	11.03	2.18	0.82	23.88	17.77	9.13	3.43	46.19	0.12
SC3	4.91	1.11	0.25	0.09	6.36	2.38	1.29	20.88	11.12	11.40	6.18	30.46	0.23
SC4	7.86	1.69	0.28	0.11	9.96	1.57	0.35	26.98	14.09	5.82	1.30	36.92	0.17
SC5	8.95	2.38	0.30	0.11	11.74	1.09	0.42	22.81	15.39	4.78	1.84	51.47	0.13
SC6	6.40	1.42	0.29	0.12	8.25	2.11	0.43	29.54	12.93	7.14	1.46	27.93	0.20
SC7	9.42	2.05	0.32	0.16	11.92	2.29	0.49	25.03	16.77	9.15	1.96	47.62	0.16
SD1	13.30	2.87	0.34	0.25	16.76	0.59	0.28	33.70	18.73	1.75	0.83	49.73	0.12
SD2	9.29	2.14	0.32	0.21	11.97	0.87	0.32	29.80	14.44	2.92	1.07	40.17	0.15
SD3	14.78	4.39	0.41	0.38	19.96	0.62	0.24	34.03	25.95	1.82	0.71	58.65	0.09
SD4	10.87	3.52	0.34	0.20	14.93	0.56	0.22	29.14	17.87	1.92	0.75	51.24	0.10
SD5	12.90	3.77	0.37	0.23	17.27	0.59	0.21	30.66	21.25	1.92	0.68	56.33	0.10
Mean	9.28	2.38	0.31	0.17	12.13	1.46	0.55	27.13	16.41	5.95	2.35	43.75	0.13

TEB total exchangeable bases, *exAc* exchangeable acidity, *exAl* exchangeable Aluminum, *CEC* cation exchange capacity of soil, *ECEC* Effective cation exchange capacity, *PAcS* percent acid saturation, *PAIS* percent aluminium saturation, *PBS* percent base saturation

likely reason for the observed variation in exchangeable bases could be the difference in soil pH and organic carbon content. It has been reported that soil pH and soil OC are regarded as the main factors influencing the variation in exchangeable cations (YuGe et al. 2013). Based on this critical level, except soil from SD3 the entire soil samples from both districts had an exchangeable K concentration of below the optimum level. Other study results have also reported about the insufficiency of K in acidic soils of Ethiopia (Mesfin et al. 2014; Asmare et al. 2015). The persuasive reason for the scarcity of exchangeable K could be attributed to the higher intensity of weathering and intensive cultivation.

The distributions of these basic cations in the exchange complex followed the order of $Ca^{2+} > Mg^{2+} > K^+ > Na^+$. The variations in the order of distribution of these cations could be related to the charge density in which the divalent cations (Ca^{2+} and Mg^{2+}) have higher affinity towards the colloidal surface than monovalent ions (K^+ and Na^+) (Alemu et al. 2016). Similar order of distribution of these cations was reported by several studies (Teshome et al. 2013; Asmare et al. 2015; Okubay et al. 2015). Antagonistic effects are known to exist between Mg and K (FAO 2006) when disproportionate quantities of the cations are present in soil. In this study, the ratio was evaluated and ranged from 0.09 to 0.23 (Table 4). According to the recommendation noted by Abayneh and Ashenafi (2006), potassium is deficient for field crops (5:1), vegetables (3:1) and fruits (2:1) in soils from Dinsho and Cheha districts. On the other hand, according to the ratio (0.7:1) suggested by Loide (2004), the clay textured soils of the study area showed Mg induced K deficiency. The observed order ($Ca^{2+} > Mg^{2+} > K^+ > Na^+$) of cations concentration in the exchange complex could also substantiate the existence of Mg induced K deficiency.

The percentage base saturation (PBS) of the studied soils varied from 25.66 to 55.39% (Table 4). Based on the rating suggested by Hazelton and Murphy (2007), about 50% of the soils had a PBS value that is within the range of low (25.66–34.35%) to moderate (42.64–47.3%).The possible reason for the low PBS value of soils might be the loss of soluble basic cations through leaching and erosion. Research results made on other Ethioipian soils also revealed low values of PBS in acidic soils possibly due to intensive cultivation, enhanced loss of basic cations through leaching and erosion (Getachew and Heluf 2007; Mohhammed et al. 2016).

The exchangeable acidity (exAc) and exchangeable aluminium (exAl) of the studied soils were found to be in the range of 0.56–2.61 and 0.21–1.54 $cmol_c\ kg^{-1}$, respectively (Table 4). Meanwhile, the acid saturation (PAcs) and Al saturation (PAls) ranged from 1.75 to 13.63 and 0.68 to 8.04%, respectively. Variability was observed among soils

of the two districts in terms of exAc and exAl. The disparity in these values between the two locations might be due to the difference in soil CEC. In agreement with the present study, Al Baquy et al. (2017) attributed the difference in exAc and exAl to the difference in soil type, organic matter content and CEC of soils.

The correlation result in Table 8 shows that exAc and exAl were negatively correlated with pH, whereas significant positive correlation was observed between pH and extractable P (Mehlich 3-P). This indicates that the lower the pH of the soil, the higher will be the concentrations of exAl and exAc. Strong and positive correlation was also observed between LR of soils and exAl (r = 0.71, P ≤ 0.01) and exAc (r = 0.94, P ≤ 0.01). The result of this correlation study is concomitant with what was reported by Asmare et al. (2015) and indicates that higher dose of lime is required for soils with higher exchangeable Al and acidity. Therefore, pH, amount of exchangeable acidity and Al can provide preliminary information on LR of soils. Cation exchange capacity (CEC) was negatively correlated with exAc and exAl [r = − 0.78, P ≤ 0.01 and r = − 0.83, P ≤ 0.01]. This correlation indicates that soils with high exAc and exAl have lower CEC. This could be an indication of weaker desorption of P associated with oxyhydroxides of Fe and Al for soils having higher exchangeable Al and exchangeable acidity. Furthermore, the strong negative correlation between CEC of soil and clay content [r = − 0.67, P ≤ 0.05] reveals that the difference in CEC can be explained by the difference in clay contents among some soils (De Kimpe et al. 1979).

Organic carbon, total nitrogen, available and total phosphorus

As the result of soil analysis for OC, total N, C:N ratio, available and total P contents of the studied soils illustrated the percentage of organic carbon and total nitrogen in the soils of Cheha district varied from 1.58 to 2.11 and 0.17–0.19%, respectively and were appeared to be higher values compared to soils of Dinsho district (2.15–2.58 and 0.19–0.22% respectively) (Table 5). In general, as per the rating suggested by Hazelton and Murphy (2007), OC content of soils of the study area were within the range of moderate (1.00–1.80%) to high (1.80–3.00%). Even though soils of the study area have been intensively tilled, the crop residue that is often left after crop harvest, a common practice in the study area, might have contributed to this level of organic carbon in the soils. Furthermore, high clay content, which controls organic matter enrichment ability of soils, could also be another reason for the moderate to high organic carbon content in soils of the study area. The variation in percent clay fraction, management practices and intensity of cultivation might have contributed for the slight disparity in OC matter

Table 5 Organic carbon (OC), total nitrogen (TN), available and total phosphorus (P) OC = organic carbon, TN = total nitrogen, C:N = carbon to nitrogen ratio

Sampling sites	OC (%)	TN	C:N	Mehlich 3 P (mg kg^{-1})	Bray 1 P	Total P
SC1	1.58	0.17	9.55	8.90	8.54	648.62
SC2	1.86	0.17	11.24	10.14	9.65	806.55
SC3	1.93	0.19	9.89	9.73	8.95	708.11
SC4	1.97	0.18	11.23	9.15	7.78	728.20
SC5	2.11	0.19	11.38	10.79	7.61	693.18
SC6	2.00	0.18	11.43	13.94	7.90	789.43
SC7	1.91	0.19	10.32	10.38	7.61	736.77
SD1	2.39	0.19	12.89	23.26	19.36	1094.25
SD2	2.37	0.20	11.85	18.33	16.15	1022.80
SD3	2.58	0.22	11.98	24.35	21.41	1175.80
SD4	2.56	0.21	12.17	22.29	20.24	1045.60
SD5	2.52	0.21	12.29	21.34	20.12	1130.15
Mean	2.15	0.19	11.35	14.84	12.59	881.62

SC1 Goha 1, *SC2* Goha 2, *SC3* Goha 3, *SC4* Aftir, *SC5* Abret, *SC6* Kechot, *SC7* Moche, *SD1* Doyomarufa 1, *SD2* Doyomarufa 2, *SD3* Tulu, *SD4* Weni, *SD5* Ketasire

content of the soils. Many studies confirm that carbon retention in soil is influenced by crop management systems, such as crop rotation (Saljnikov et al. 2004), tillage (Saljnikov et al. 2009), residue management (Rasmussen et al. 1980) and fertilization and fertility (Saljnikov et al. 2005).

In addition to the above-mentioned reasons, the high silt plus clay content of the soils might have contributed to the moderate to high organic carbon content. Numerous studies have described the positive relationship between clay (or clay + silt) content and soil organic matter (OM) in soils from different sites in the tropics (Xiea and Steinbergera 2001). These studies have shown that clay-silt content is a relatively important determinant of soil OM or total OC level in soils. As a case in point, studies in Spanish semi-arid Mediterranean bare soils have shown different soil OM mineralization rates for different soil textures (GarcõÂa and HernaÂndez 1996) and indicated the importance of the physical and chemical properties, which might have played an important role in soil water availability. It has also been suggested that fine particles with high surface activity may physically and chemically protect soil OM from decomposition by binding strongly and create physical barrier to protect microbial access (Hassink et al. 1993).

As per the rating suggested by Hazelton and Murphy (2007), the total nitrogen (TN) content of soils in the study area was within the range of medium (0.17–0.22%). However, medium TN level in soils of the study area could not be taken as a guaranty for adequate availability of N to plants. Therefore, in order to maintain or enhance the nitrogen status of the soils imbalanced use of

fertilizers, nutrient mining through intensive cultivation and crop residue removal should be avoided. The slight variation within sites in particular might be associated to the variability of soils OC content.

The C:N ratio of the studied soils varied between 9.54 and 12.89. According to the recommendation made by Newey (2006), 17 and 83% of the studied soils were categorized under soils having very low (< 10) and low (10–15) C:N ratio respectively indicating the fast decomposition of OM in the studied soils. Abera and Belachew (2011) investigated soils from cultivated lands of Bale Zone including Dinsho and reported a C:N ratio ranging from 12.57 to 13.53, which is comparable with the findings of the current study.

Total P content in the surface soil samples ranged from 648.62 mg kg^{-1} in the SC1 soil collected from Cheha farmers' field to 1175.8 mg kg^{-1} in the SD3 soil collected from the Dinsho district (Table 5).

The amount of P extracted with Mehlich-3 and Bray and Kurtz-1 ranged from 8.90 to 25.75 and 8.54 to 22.81 mg kg^{-1} respectively (Table 5). However, there was a considerable variation among soils of the districts in terms of available P (Table 5). This may be ascribed to the presence of higher concentrations of oxyhydroxides of Al and Fe which are responsible for fixation of P at the exchange sites for soils of Cheha district compared to soils of Dinsho. As per the rating established by EthioSIS (2014), the Mehlich-3 extractable P of the studied soils in Cheha fall in the very low range (0–15 mg kg^{-1}), whereas soils of Dinsho exhibited low (15–30 mg kg^{-1}) Melich-3 P. Similarly, the rating proposed by Horneck et al. (2011) also indicated the predominantly low available P for

Cheha (< 20 mg kg^{-1} for Bray-1 P) and low to medium P for Dinsho soils (< 20 mg kg^{-1} Bray 1 P 20–40 mg kg^{-1}). The insufficiency may be due to acidic reaction in which free oxides of Fe and Al tend to fix P on the exchange sites. Reports of low available P in many parts of Ethiopian soils have been documented possibly due to the impacts of fixation, abundant loss by crop harvest and erosion (Asmare et al. 2015; Wolde et al. 2015).

The amounts of P extracted from studied soils decreased in the following order: Bray and Kurtz-1 < Mehlich 3. The differences among the P extraction methods probably arose from the fact that plant available P in the soil is not from a prudent fraction but from a continuum of fractions. Thus extracting agents preferentially extract P from different fractions depending on their reactions with soil components involved in P sorption. Furthermore, each extracting solution has a different ability to extract varying segment of soil P because they were embattled at different pools of soil P (Zhang et al. 2004). Extractable P by the Mehlich-3, and Bray and Kurtz-1 in this study were correlated among themselves (Table 8). The correlation coefficient (r) was appeared to be 0.96 indicating that any of these extractants could be used to estimate P extractability (available P) in soils. In spite of the strong positive correlations among these extractants, Mehlich-3 extraction method could be used as a substitute over the former conventional extraction method Bray and Kurtz-1 (1945) because of its capability to generate opportunities for optimization of laboratory management and cheaper procedure (Mesfin et al. 2014).

As per the ratings of Murphy (1968), the studied soils had a total P content that ranges from medium (501–750 mg kg^{-1}) to very high (> 1000 mg kg^{-1}). A gamut of results have been produced by several scholars on the total P contents of several soils of Ethiopia from various localities: 200–800 mg kg^{-1} (Eylachew 1987), 185 to 1981 mg kg^{-1} (Tekalign and Haque 1991), 226–1570 mg kg^{-1} (Duffera and Robarge 1996), 553–976 mg kg^{-1} (Achalu et al. 2012) and 685–1432 mg kg^{-1} (Asmare et al. 2015). The result also indicates that the total P content of the studied soils was much higher than 200 mg kg^{-1} which was the value indicated by Olsen and Engelstad (1972) as the maximum total P value for highly weathered tropical soils. The total P content determined in this study was very far from a report made on soils from northwestern highlands of Ethiopian for which the total P content ranged from 6900 to 24,000 mg kg^{-1} with a mean value of 12,275 mg kg^{-1} (Birru et al. 2003). However, the normal range of total P in soils is 100–3000 mg kg^{-1} (Frossar et al. 2000). Viewing the total P contents of some tropical soils, Asmare et al. (2015) observed that the total P contents of Ethiopian soils were not as poor as the other tropical soils. Surprisingly, despite the medium to high contents of total P, the available P contents of the studied soils were in low to very low range (Table 5).

Inorganic and organic phosphorus fractions

The amount, distribution and sum of various P fractions of the soils studied are presented in Table 6. The relative abundance of P forms was in the following order: Res-P > NaOH-Po > NaOH-Pi > HCl-P > (NaOH-Pi)sn > (NaOH-Po)sn > NaHCO$_3$-Po > NaHCO$_3$-Pi > NH$_4$Cl in soils from Cheha district, whereas

Table 6 Inorganic and organic P fractions and their distribution

Sites	NH$_4$Cl-Pi	NaHCO$_3$-Pi	NaHCO$_3$-Po	NaOH-Pi (mg kg^{-1})	NaOH-Po	(NaOH-Pi)sn	(NaOH-Po)sn	HCl-P	Resl-P	Sum $\sum$P
SC1	1.93	2.43	2.74	135.31	160.41	31.74	35.31	71.32	164.15	606.34
SC2	3.23	3.72	3.91	139.71	196.87	40.93	85.17	86.89	216.25	776.68
SC3	1.69	2.45	2.78	130.45	167.59	33.9	36.28	77.01	185.18	637.33
SC4	2.53	2.06	3.63	129.66	180.82	37.76	76.68	77.88	187.30	698.32
SC5	2.32	2.59	2.97	124.13	168.69	36.33	72.85	75.57	177.55	663.00
SC6	2.59	2.72	4.04	138.82	190.94	44.24	79.32	83.55	213.41	759.63
SC7	2.06	2.11	3.85	131.76	182.48	38.10	77.6	77.95	190.94	706.85
SD1	5.90	7.61	9.35	89.65	154.49	28.28	72.29	379.32	317.24	1064.13
SD2	3.90	5.12	5.63	82.76	144.32	23.42	70.50	354.86	301.16	992.67
SD3	7.40	7.75	8.95	97.37	168.11	30.40	77.76	407.26	340.73	1145.73
SD4	5.90	6.51	6.53	86.63	147.56	27.01	69.03	362.64	303.60	1015.41
SD5	6.90	7.14	7.97	96.12	161.64	31.22	72.71	391.69	324.87	1100.26
Mean	3.86	4.35	5.19	115.20	168.66	33.61	68.79	203.83	243.53	847.20

NH$_4$Cl-P ammonium extractable P, *NaHCO$_3$-Pi* sodium bicarbonate extractable inorganic P, *NaHCO$_3$-Po* sodium bicarbonate extractable organic P, *NaOH-Pi* sodium hydroxide extractable inorganic P, *NaOH-po* sodium hydroxide extractable organic P, *(NaOH-Pi)sn* sonicated sodium hydroxide extractable inorganic P, *(NaOH-po)sn* sonicated sodium hydroxide extractable organic P, *HCl-P* hydrochloric acid extractable P, $\sum$P sum of P fractions

HCl-P > Res-P > NaOH-Po > NaOH-Pi > (NaOH-Po) sn > (NaOH-Pi)sn > NaHCO$_3$-Po > NaHCO$_3$-Pi > NH$_4$Cl was the order for soils of Dinsho district. The labile P (Pi in NH$_4$Cl+Pi and Po in NaHCO$_3$) varied between 1.1 and 2.2%, the moderately labile P (Pi+Po in NaOH), varied between 22.9 and 48.8%, the occluded P (Pi+Po in Sonicate+NaOH) varied between 9.4 and 16.5%, the P associated to Ca (HCl-Pi) varied between 11.0 and 35.8% and the residual-P varied between 26.8 and 30.3%. It could be concluded that soils of Cheha district are dominated by Res-P (occluded Pi associated with the remaining inorganic minerals and non extractable Po) followed by NaOH extractable P (NaOH-Po and NaOH-Pi). On the other hand, it was HCl-P that dominates P fractions in soils of Dinsho district indicating (Ca)-Pi mineral were very prominent in these soils. Similar work has been conducted by Duffera and Robarge (1996) who characterized organic and inorganic P in the highland Plateau Alfisols of Ethiopia and reported that the relative abundance in cultivated farmers' fields was in the ensuing order: Residual P > NaOH-Po > NaOH-Pi > Sonic-Po > NaHCO$_3$-Po > Resin-P = Sonic-Pi > NaHCO$_3$-Pi = HCl-P.

In the studied soils considerable proportions of P (NaHCO$_3$-P$_o$) that may be easily mineralized were noticed and ranged from 36.00 to 48.00% indicating the significance of organically bound P as a source of labile, plant available P (Duffera and Robarge 1996). In this regard Cross and Schlesinger (1995) suggested that the bicarbonate Po as a percent of the total labile forms of P (resin Pi, bicarbonate Pi, and bicarbonate Po) represents a minimum index of the proportion of P that may be easily mineralized through biological processes. In clayey soils of Dinsho district, the sum of NH$_4$Cl-P, NaHCO$_3$-P

and NaOH-P was smaller than oxalate-P, indicating that the oxalate method extracted some residual P that contained insoluble inorganic P associated with Ca minerals and a highly resistant pool of inorganic occluded P and organic P (Table 6).

The result of the present study also revealed that the studied soils from Cheha district were dominated by moderately labile P fractions, accounting for 43.3–48.8% of the total P, whereas NH$_4$Cl-P was the lowest in concentration, 0.3–0.4% of the total P. On the other hand, the studied soils of Dinsho district were dominated by HCl-P, 35.6–35.8% of the total P, while NH$_4$Cl-P was the least in proportion (0.4–0.7%). Residual-P, which accounted for about 29.5–30.3% (Dinsho) and 26.8–28.1% (Cheha) of the total P, was the second higher fraction followed by moderately labile-P (22.9–23.4%) and occluded-P (11.0–16.5%) fractions, respectively (Table 6). The relatively higher abundance of Al and Fe bound P could be associated with the presence of variable Al and Fe contents in soils at various stages of relative development and their reaction with soil P (Kiflu et al. 2017). On the other hand, the P associated to Ca (HCl-Pi) varied between 11.0 and 35.8%, which could be attributed to the difference in Ca content of the studied soils and the residual-P varied between 26.8 and 30.3% (Table 7).

A noticeable variability in proportion of P fractions in the order of their relative abundance in the studied soils of the two districts was observed. The management practices being used by farmers and the extent of weathering of soils might have contributed for the variations in concentrations and distributions of different forms of P. As an illustration, NaOH extractable P in soils of Cheha district, which accounted for about

Table 7 Oxalate and dithionite citrate bicarbonate extractable Al, Fe and P

Sites	Al$_{ox}$ (mmol kg^{-1})	Fe$_{ox}$	P$_{ox}$	PSC	P$_{ox}^a$ DPS (%)	Al$_d$	Fe$_d$ (mmol kg^{-1})	Al$_d$-Al$_{ox}$	Fe$_{ox}$/Fe$_d$	Scaling factor (α)
SC1	152.37	97.70	5.51	133.54	4.13	173.33	633.66	20.6	0.24	0.62
SC2	139.07	78.54	6.17	116.20	5.31	151.85	478.04	12.78	0.29	0.62
SC3	145.96	88.18	5.82	125.03	4.65	164.07	551.61	18.11	0.27	0.57
SC4	130.07	79.96	6.07	112.16	5.41	144.07	456.25	14.00	0.29	0.59
SC5	126.37	77.46	7.07	108.85	6.49	149.26	491.61	22.89	0.26	0.61
SC6	131.85	76.39	7.32	111.20	6.58	161.11	476.61	29.26	0.28	0.62
SC7	134.37	81.75	6.45	115.41	5.59	150.74	498.75	16.37	0.27	0.56
SD1	90.00	66.21	9.74	83.42	11.68	119.26	313.93	29.26	0.29	0.42
SD2	91.85	82.29	8.26	92.99	8.88	112.19	355.54	20.33	0.26	0.45
SD3	71.89	51.75	12.29	66.02	18.62	134.07	281.07	62.19	0.26	0.45
SD4	90.37	59.43	8.39	79.99	10.49	128.52	307.14	38.15	0.29	0.44
SD5	78.15	63.36	10.35	75.56	13.69	124.07	291.43	45.93	0.27	0.44
Mean	115.19	75.25	7.79	101.70	8.46	142.71	427.97	27.52	0.27	0.53

Al$_{ox}$ oxalate extractable Al, Fe$_{ox}$ oxalate extractable P$_{ox}$ oxlate extractable P, PSC phosphorous sorption capacity, DPS P saturation with respect to Pox, Al$_d$ dithionite extractable Al, Fe$_d$ dithionite extractable Fe

59.9% of the total P, was in higher proportion compared to the 32.6% in soils of Dinsho district. Likewise, HCl extractable P, which was 11.4% of the total P in Cheha soils, was much lower than the 35.7% in Dinsho soils. This could be due to the dominance of free Fe and Al oxides and higher extent of weathering in soils of Cheha district compared to Dinsho soils as indicated in Table 7.

The dependence of different forms of P upon the physical and chemical properties of the soil was tested using correlation analysis (Table 9). Thus, NH_4Cl-P_i, $NaHCO_3-Po$ and $NaHCO_3-Pi$ were positively and significantly correlated with pH, CEC, OC and available P but negatively and significantly correlated with exAc and exAl. The other forms of P, such as $NaOH-Po$ and $NaOH-Pi$, were negatively and significantly correlated with pH, CEC, OC, and available P (Mehlich 3 P). Furthermore, the present study revealed that NH_4Cl-P_i (r=-0.87, P≤ 0.01), $NaHCO_3-Pi$ (r=-0.85, P≤ 0.01), $NaHCO_3-Po$ (r=-0.89, P≤ 0.01) and HCl-P (r=-0.82, P≤ 0.01) were negatively and significantly correlated with clay content. It may be inferred that the concentrations and distributions of these P fractions are smaller in soils where there is high clay content. The outcome of the present study was found to be in agreement with the correlation result obtained by Duffera and Robarge (1996) whose result exhibited negative correlation between clay content and the readily available P fractions, the bicarbonate Pi and HCl extractable P for highland soils from Ethiopia.

The P contents in terms of the sum of P fractions in the surface soil samples ranged from 606.3 mg kg^{-1} in the SC1 soil collected from Cheha district to 1145.7 mg kg^{-1} in the SD3 soil collected from the Dinsho district indicating a closer similarity between total P determined through direct digestion technique and summation of the P fractions. In general, the sum of the various P fractions was within 4% of the total P content determined by the Olsen and Sommers (1982) approach, indicating that the fractionation procedure does account for almost all of the P present in the soils.

Oxalate and dithionite citrate bicarbonate extractable Al, Fe and P

The distribution of oxalate and dithionite citrate bicarbonate extractable Al and Fe oxides values of the studied soils are presented in Table 7. Accordingly, considerable variations among the studied soils with respect to DCB and oxalate extractable Al and Fe were observed. The disparity in relative abundance of aluminum and iron oxides in studied soils, which are the most important phosphate adsorbents in weathered acidic soils, might have contributed to the variations observed in extractable Al and Fe.

Substantial differences were observed between dithionite extractable Fe and Al in the soils. This might be due to the fact that the dithionite in the dithionite citrate bicarbonate procedure is a reducing agent and it might have reduced Fe^{3+} to Fe^{2+} and thereby increased the solubility of Fe. However, Al is not a redox sensitive element and hence treatment with reducing agents such as dithionite is not likely to cause much extraction of Al.

The differences between DCB-extractable Fe and Al oxides and oxalate-extractable Fe and Al oxides represent the amounts of extracted crystalline Fe (c-Fe) and Al (c-Al) oxides. There was a large amount of c-Fe (76.9–84.6% of total Fe). In contrast, c-Al content was low (8.4-46.4% of total Al) (Table 7). The observed variation in the amount of crystalline Fe and Al might be ascribed to the amount of organic matter present in the soil. In agreement with the results of the existing study, Kang et al. (2009) reported that organic matter in soil inhibits Al oxides crystallization. The oxalate extractable Al, and Fe values ranged from 1941.03 to 4113.99 mg Al kg^{-1} and from 2898 to 5471.2 mg Fe kg^{-1} were comparable with acid soils of northwestern highlands of Ethiopia that contained 3528–5432 mg Fe kg^{-1} 3699–4779 mg Al kg^{-1} (Asmare et al. 2015). However, these extractable forms were higher than acidic soils of European countries that are characterized by an average of 1835.2 Fe_{ox} kg^{-1} to 2349.8 mg Al_{ox} kg^{-1} (Roberto and Giampietro 2005). The ratio of Fe_{ox}/Fe_d has been taken as an indicator of the maturity or crystallinity of free Fe oxides in soils (Blume and Schwertmann 1969). The value of the Fe_{ox}/Fe_d for the studied soils ranged from 0.24 to 0.29. All the values of Fe_{ox}/Fe_d ratio for studied soil samples were > 0.1, which indicates that poorly crystalline iron oxides are dominant in these soils (Fitzpatrick and Schwertmann 1982).

As it may also be seen in Table 7 the P_{ox} of the studied soils was ranged from 5.51 to 12.89 mmol kg^{-1} and were higher than acidic soils of West African soils which held 0.35 to 3.42 mmol P_{ox} kg^{-1} (Narteh and Sahrawat 1999). The observed difference in P_{ox} of the soils might be attributed to the variation in their poorly crystalline Fe and Al which are the most reactive fractions of soil iron and aluminum oxides.

The correlation matrix (Tables 8 and 9), Table 9 indicates that readily available P fractions (NH_4Cl, $NaHCO_3-Pi$ and $NaHCO_3-Po$) had a significantly negative relationship with Fe_{ox} [(r=-0.89, P≤ 0.01), r=-0.82, P≤ 0.01, r=-0.84, P≤ 0.01)], Al_{ox} [(r=-0.94, P≤ 0.01), r=-0.93, P≤ 0.01, r=-0.92, P≤ 0.01)] respectively. This result indicates that soils with high concentrations of free oxyhydroxides of aluminum and iron are deprived of NH_4Cl and $NaHCO_3$ extractable P. The negative relationship among oxalate and dithionite extractable metal oxides and readily available fractions of P is in agreement

Table 8 Correlation table for some soil properties

	Clay	pH	exAc	exAl	OC	Mehl 3P	Bra P	Fe_{ox}	Al_{ox}	Fe_d	PSC	DPS	P_{ox}
pH	−0.66*												
exAc	0.51	−0.90**											
exAl	0.65*	−0.72**	0.79**										
LR	0.81**	−0.88**	0.94**	0.71**									
OC	0.83**	0.88**	−0.94**	−0.79**									
M 3P	−0.82**	0.81**	−0.86**	−0.86**	0.93**								
Bra P	−0.80**	0.81**	−0.84**	−0.54	0.90**	0.96**							
Fe_{ox}	0.80**	−0.79**	0.81**	0.81**	−0.88**	−0.85**	−0.80**						
Al_{ox}	0.83**	−0.85**	0.93**	0.93**	−0.97**	−0.95**	−0.94**	0.86**					
F_{ed}	0.99**	−0.85**	0.92**	0.92**	−0.96**	−0.92**	−0.90**	0.90**	0.97**				
PSC	0.85**	−0.86**	0.92**	0.92**	−0.93**	−0.95**	−0.93**	0.93**	0.98**	0.98**			
DPS	−0.88**	0.73**	0.81**	−0.81**	0.87**	0.91**	0.89**	−0.89**	−0.93**	−0.87**	−0.95**		
P_{ox}	−0.91**	0.75**	−0.82**	−0.82**	0.88**	0.92**	0.88**	−0.88**	−0.94**	−0.88**	−0.95**	0.99**	
CEC	−0.67*	0.66*	−0.78**	−0.78**	0.84**	0.87**	0.77**	−0.87**	−0.83**	−0.91**	0.86**	0.83**	0.86**

LR lime requirements, *Mehl 3P* Mehlich 3P, *Bra P* Bray 1 P, *Ols P* Olsen P, *LR* lime requirement, *exAl* exchangeable Al, *exAc* exchangeable Acidity, *OC* organic carbon, *Alox, Feox, and Pox* oxalate extractable Al, Fe, and P, respectively, *PSC* phosphorous sorption capacity, *DPS* degree of P saturation, *Fed* citrate dithionite bicarbonate extractable Fe

*Significant at 0.05 and **Significant at 0.01 probability levels, respectively

Table 9 Correlation table for fractions of P and some soil properties

	Clay	pH	exAc	exAl	OC	M 3P	Fe_{ox}	Al_{ox}	Fe_d	PSC	DPS	P_{ox}	CEC
NH_4Cl	−0.87**	0.82**	−0.84**	−0.84**	0.89**	0.94**	−0.89**	−0.94**	−0.93**	−0.96**	0.94**	0.94**	0.82**
$NaHCO_3$ pi	−0.85**	0.81**	−0.84**	−0.84**	0.88**	0.96**	−0.82**	−0.93**	−0.90**	−0.92**	0.91**	0.91**	0.80**
$NaHCO_3$ Po	−0.89**	0.78**	−0.82**	−0.82**	0.85**	0.95**	−0.84**	−0.92**	−0.91**	−0.92**	0.92**	0.93 **	0.89**
NaOH Pi	0.05	−0.81**	0.89**	0.89 **	−0.89**	−0.89**	0.64*	0.91**	0.86**	0.85**	−0.74**	−0.75**	−0.72**
NaOH po	−0.23	−0.55	0.60	0.60	−0.55	−0.57	0.19	0.56	0.43	0.46	−0.39	−0.39	−0.26
HCl p	−0.82**	0.81**	−0.87**	−0.87**	0.92**	0.97**	−0.77**	−0.96**	−0.92**	−0.93**	0.88**	0.89**	0.81**

LR lime requirements, *Mehl 3P* Mehlich 3P, *Bra P* Bray 1 P, *Ols P* Olsen P, *LR* lime requirement, *exAl* exchangeable Al, *exAc* exchangeable Acidity, *OC* organic carbon, *Alox, Feox, and Pox* oxalate extractable Al, Fe, and P, respectively, *PSC* phosphorous sorption capacity, *DPS* degree of P saturation, *Fed* citrate dithionite bicarbonate extractable Fe

*Significant at 0.05 and **Significant at 0.01 probability levels, respectively

with the general assumption that the higher the value of free oxides of aluminum and iron, the stronger is the binding capacity of the soil for P. Consequently, the amount of P desorbed (NH_4Cl-P, $NaHCO_3$-Pi and $NaHCO_3$-Po) from the exchange sites with increasing levels of these free oxides could be decreased. The bicarbonate Po had strong negative [(r = −0.91, P ≤ 0.01)] correlation with F_{ed}, as it had positive and significant association [(r = 0.85, P ≤ 0.01) and (r = 0.89, P ≤ 0.01)] with OC and CEC, respectively. The correlation result of the present study is in agreement with what has been found by Tiessen et al. (1984) where organic P fractions were positively related to organic C despite the contrasting correlation result observed from bicarbonate Po and dithionite extractable Fe (Fe_d).

Extent of phosphorus saturation

As shown in Table 7, the degree of phosphorous saturation of the studied soils varied considerably ranging from 4.1 to 6.6 and 8.9–18.6% in Cheha and Dinsho districts respectively. The portions of soil exchangeable sites that were bound with P were greater for soils collected from different sampling sites of Dinsho as compared to soils collected from Cheha which had higher free metallic oxides. The extent at which the potential sites for P adsorption already occupied was designated by DPS (%). Accordingly, the highest DPS (18.6%) was recorded from Dinsho (SD3) and the lowest from Cheha (SC1) sampling site. Recent studies have shown that the degree of P saturation (DPS), as a function of the portion of the soil exchangeable sites that are bound with P (P sorbed) in relation to the number of sites available for P binding capacity (PBC), is a good indicator of the soil's potential to release P (Hooda et al. 2000). Therefore, the variation in DPS of the studied soils may be attributed to the disparity among these soils in terms of the number of sites available for binding P. In view of the cut off point for loss of P due to runoff (Ige et al. 2005), all the samples showed DPS values lower than 20% indicating no risk of P loss from soil (Table 7). These low values might be attributed to the higher adsorption capacity of these soils.

Extractable P (NH_4Cl, $NaHCO_3$-P_i and $NaHCO_3$-Po) was positively correlated with DPS (r=0.94, P≤0.01; r=0.91, P≤0.01; r=0.92, P≤0.01, respectively) and P_{ox} (r=0.94, P≤0.01; r=0.91, P≤0.01; r=0.93, P≤0.01, respectively). This confirms that these soils are characterized by higher amounts of readily available P fractions, larger concentrations of oxalate extractable P (P_{ox}) and greater degree of phosphorous saturation.

Phosphorous determined from NaOH extracts (NaOH-P_i), which is thought to be Fe and Al bound P, was positively correlated with Fe_{ox} (r=0.64, P≤0.05) and Al_{ox} (r=0.91, P≤0.01) despite the fact that it was negatively correlated with DPS (r=−0.74, P≤0.01) and P_{ox} (r=−0.75, P≤0.01). The correlation result substantiate that in the case where higher concentrations of oxalate and dithionite extractable iron and aluminum oxides are observed, the amount of P determined from NaOH extracts is higher. The same result has been found by Adhami et al. (2013) who reported that oxalate extractable iron had positive association with NaOH extractable P. This could be ascribed to the higher Al and Fe oxides on the exchange sites that might have resulted in lower pH at which the concentration of NaOH extractable P would become higher. Furthermore, P fraction determined from the HCl extract had a strong and negative association with Fe_{ox} (r=−0.77, P≤0.01) and Al_{ox} (r=−0.96, P≤0.01) although it was positively correlated with DPS (r=0.88, P≤0.01). Likewise, the correlation result proved that the higher the concentrations of extractable iron and aluminum in soil, the lower the concentrations of P determined from HCl extract. It has been reported by several studies that it is oxyhydroxydes of aluminum and iron in weathered acidic soils rather than Ca that supersede the active surface of soils on which P is adsorbed (Agbenin 2003; Yan et al. 2013; Campos et al. 2016). Therefore, with increasing concentrations of these free oxides, the amount of HCl extractable P decreases. Besides, NaOH and HCl extractable P (NaOH-Pi and HCl-P) could be best predicted from Al_{ox} and therefore put forth considerable information on the abundances and distributions of NaOH-Pi and HCl-P in the studied soils.

Mehlich 3 extractable P was negatively correlated with Fe_{ox}, Al_{ox} and Fe_d (r=−0.85, P≤0.01; r=−0.95, P≤0.01; r=−0.92, P≤0.01) respectively. It is evident that the higher the concentrations of free oxides of Fe and Al, the lower the pH, which triggers adsorption of P at the exchange sites; in effect Mehlich 3 extractable P (available P) would be decreased. Moreover, the Mehlich 3 extractable P could best be predicted from Fe_{ox}, Al_{ox} and F_{ed} and therefore put forth thought full information on the availability of P in the soils is still lacking.

Conclusions

The soils of the study area were strongly to very strongly acidic in reaction that triggers the fixation of P at the exchange sites in the presence of oxyhydroxides of Fe and Al. Despite the medium to high contents of total P, the Mehlich-3 extractable available P contents of the studied soils is in the low to very low range indicating P deficiency in the study area. Among the different fractions of P, HC1 extractable P fraction dominates the soils P pool for soils of Dinsho district, whereas soils' P pool of Cheha district was dominated by res-P. The sum of the labile P fractions (NH_4Cl-Pi, $NaHCO_3$-Pi, and $NaHCO_3$-Po) constitutes less than 4% of total P in soils from both districts reflecting very low reserves of crystalline Fe and Al-P. The sum of P fractions, which ranged from 606.3 to 1145.7 mg kg^{-1}, is almost equivalent to the total P determined through fusion method and reveals that the Hedley fractionation procedure could be used for determination of total P. The high content of Al and Fe oxides and hydroxides present in the soil might be responsible for the strong fixation of the native P as well as the applied P. Given the presence of considerable amount of total P, low P availability and high soil acidity in the study areas further research has to be done on adsorption characteristics to determine the P requirements of the soils for better P management.

Authors' contributions

BA: collected, analyzed, interpreted the data and made the final write up which was part of his Doctoral thesis in Soil Science at Haramaya University, Ethiopia. AMT, KK and AM, as co-authors. All authors read and approved the final manuscript.

Author details

[1] Department of Chemistry, School of Natural Sciences, Madda Walabu University, Bale-Robe, Ethiopia. [2] Department of Chemistry, College of Natural and Computational Sciences, Haramaya University, Dire Dawa, Ethiopia. [3] School of Natural Resources Management and Environmental Sciences, Haramaya University, Dire Dawa, Ethiopia. [4] Department of Plant Science, College of Agriculture and Natural Resource Sciences, Debre Berhan University, Debre Birhan, Ethiopia.

Acknowledgements

The project was funded by SIDA (Swedish International Development Cooperation Agency) and Madawelabu University Thus, we would like to thank these institutions. We would also like to thank the anonymous reviewers who contributed significantly to the improvement of the article.

Competing interests
The authors declare that they have no competing interests.

Funding
Ministry of education and SIDA provided fund for carrying out this study.

References

Abayneh E, Ashenafi A (2006) Soils of Sinnana Agricultural Research Center. National Soil Research Center. Soil Survey and Land Evaluation Section, Addis Ababa

Abera Y, Belachew T (2011) Effects of land use on soil carbon and nitrogen in soils of Bale, Southeastern Ethiopia. Trop Subtrop Agroecosyst 14:229–235

Achalu C (2014) Assessment of the severity of acid saturations on soils collected from cultivated lands of east Wollega Zone, Ethiopia. Sci Technol Arts Res J 3(4):42–48

Achalu C, Heluf G, Kibebew K, Abi T (2012) Response of barely to liming of acid soils collected from different land use systems on Western Oromia, Ethiopia. J Biodivers Environ Sci 2:1–13

Achalu C, Martti E, Kari Y (2014) Sequential fractionation patterns of soil phosphorus collected from different land use systems of Dire Inchine District, West Shawa Zone, Ethiopia. Am Eur J Sci Res 9(3):51–57

Adhami E, Owliaie HR, Molavi R, Rezaei RM, Esfandbod M (2013) Effects of soil properties on phosphorous fractions in subtropical soils of Iran. J Soil Sci Plant Nutr 13(1):11–21

Agbenin JO (2003) Extractable iron and aluminum effects on phosphate sorption in a savanna alfisol. Soil Sci Soc Am J 67:589–595

Al Baquy MA, Jiu YL, Chen YX, Khalid M, Ren KX (2017) Determination of critical pH and Al concentration of acidic Ultisols for wheat and canola crops. Solid Earth 8: 149–159. http://www.solid-earth.net/8/149/2017/ https://doi.org/10.5194/se-8-149-2017

Alemu L, Tekalign M, Wassie H, Hailu S (2016) Assessment and mapping of status and spatial distribution of soil macronutrients in Kambata Tembaro Zone, Southern Ethiopia. Adv Plants Agric Res 4(4):00144. https://doi.org/10.15406/apar.2016.04.00144

Asmare M, Heluf G, Markku YH, Birru Y (2015) Phosphorus status, inorganic phosphorus forms, and other physicochemical properties of acid soils of Farta District, Northwestern Highlands of Ethiopia. Hindawi Publish Corp Appl Environ Soil Sci. https://doi.org/10.1155/2015/748390

Aubertin GM, Kardos LT (1965) Root growth through porous media under controlled conditions. Soil Sci Am Proc 29:290–294

Bache BW, Williams EG (1971) A phosphate sorption index for soils. J Soil Sci 22:289–301

Barber SA (1984) Liming materials and practices. In: Adams F (ed) Soil acidity and liming. American Society of Agronomy Inc, Madison, pp 171–209

Birru Y, Heluf G, Gupta VP (2003) Sorption characteristics of soils of the northwestern highlands of Ethiopia. Ethiop J Nat Resour 5:1–16

Blume HP, Schwertmann U (1969) Genetic evaluation of profile distribution of aluminum, iron, and manganese oxides. Soil Sci Soc Am J 33(3):438–444

Börling K, Otabbong E, Barberis E (2001) Phosphorus sorption in relation to soil properties in some cultivated Swedish soils. Nutr Cycl Agroecosyst 59:39–46

Bray RH, Kurtz LT (1945) Determination of total, organic, and available forms of phosphorus in soils. Soil Sci 59:39–45

Bremner JM (1965) Methods of soil analysis. Part 2. American Society of Agronomy, Madison American Society of Agronomy, pp 1149–1178

Buresh RJ, Smithson P, Hellums DT (1997) Building soil phosphorus capital in Africa. In: Buresh RJ, Sanchez PA, Calhoum F (eds) Replenishing Soil Fertility in Africa. SSSA Special Publication No. 51. Soil Science Society of America, Madison, pp 111–149

Campos MD, Antonangelo JA, Alleoni LRF (2016) Phosphorus sorption index in humid tropical soils. Soil Tillage Res 156:110–118

Chen GC, He ZL, Huang CY (2000) Microbial biomass phosphorus and its significance in predicting phosphorus availability in red soils. Commun Soil Sci Plant Anal 31:655–667

Cross AF, Schlesinger WH (1995) A literature review and evaluation of the Hedley fractionation: applications to the biogeochemical cycle of soil phosphorus in natural ecosystems. Geoderma 64:197–214

Day PR (1965) Hydrometer method of particle size analysis. In: Black CA (ed)

Methods of Soil Analysis. Agronomy Part I, No. 9. American Society of Agronomy, Madison, pp 562–563

De Kimpe CR, Laverdiere MR, Martel YA (1979) Surface area and exchange capacity of clay in relation to mineralogical composition of gleysolic soils. Can Soil Sci 59:341–347

Duffera M, Robarge WP (1996) Characterization of organic and inorganic phosphorus in the highland plateau soils of Ethiopia. Commun Soil Sci Plant Anal 27(15):2799–2814

EIAR (Ethiopian Institute of Agricultural Research) (2011) Coordination of national agricultural research system, Ethiopia. English and Amharic Version, Addis Ababa

EthioSIS (Ethiopian Soil Information System) (2014) Soil fertility status and fertility recommendation atlas for Tigray regional state, Ethiopia, July 2014. EthioSIS, Addis Ababa

Eylachew Z (1987) Study on the phosphorous status of different soil types of Charcher Highlands, South Eastern Ethiopia (Ph.D. dissertation). University of Jestus Liebig, Giessen

FAO (Food and Agriculture Organization) (1984) Provisional soil map of Ethiopia. Land Use Planning Project, Addis Ababa

FAO (Food and Agriculture Organization) (2006) Plant nutrition for food security: a guide for integrated nutrient management, fertilizer and plant nutrition bulletin 16. FAO, Rome

Fisseha H, Heluf G, Kibebew K, Birru Y (2014) Study of phosphorus adsorption and its relationship with soiproperties, analyzed with Langmuir and Freundlich models. Agric For Fisheries 3(1):40–51. https://doi.org/10.11648/j.aff.20140301.18

Fitzpatrick RW, Schwertmann U (1982) Al-sabstitution goethite an indicator of pedogenic and other weathering environments in South Africa. Geoderma 27:335–347. https://doi.org/10.1016/0016-7061(82)90022-2

Frossar E, Condron LM, Oberson A, Sinaj S, Fardeau JC (2000) Processes governing phosphorous availability in temperate soils. J Environ Qual 29:15–23

GarcôÂa G, HernaÂndez T (1996) Organic matter in bare soils of the Mediterranean region with a semi-arid climate. Arid Soil Res Rehabil 10:31–41

Getachew F, Heluf G (2007) Characterization and fertility status of the soils of Ayehu Research Substation, northwestern highlands of Ethiopia. East Afr J Sci 1(2):160–169

Gupta PK (2004) Soil, plant, water and fertilizer analysis. AGROBIOS Publisher, India

Hassink J, Bouwman LA, Zwart KB, Bloem J, Brussard L (1993) Relationship between soil texture, physical protection of organic matter, soil biota, and C and N mineralization in grassland soils. Geoderma 57:105–128

Hazelton P, Murphy B (2007) Interpreting soil test results: what do all the numbers mean?, 2nd edn. CSIRO Publishing, Clayton

Hedley MJ, Stewart JWB, Chauhan BS (1982) Changes in inorganic and organic soil phosphorus fractions induced by cultivation practices and by laboratory incubations. Soil Sci Soc Am J 46:970–976

Herlihy M, McCarthy J (2006) Association of soil-test phosphorus with phosphorus fractions and adsorption characteristics. Nutr Cycl Agroecosyst 75:79. https://doi.org/10.1007/s10705-006-9013-2

Hooda PS, Rendell AR, Edwards AC, Withers PJA, Aitken MN, Trusedale VW (2000) Relating soil phosphorus indices to potential phosphorus release to water. J Environ Qual 29:1166–1171

Horneck DA, Sullivan DM, Owen JS, Hart JM (2011) Soil test interpretation guide EC1478. Oregon State University Extension Service, Corvallis

Ige DV, Akinremi OO, Flaten DN (2005) Environmental index for estimating therisk of phosphorus loss in calcareous soils of Manitoba. J Environ Qual 34:1944–1951

Ikerra S (2004) Use of Minjingu phosphate rock combined with different organic inputs in improving phosphorus availability and maize yields on a Chromic Acrisol in Morogoro, Tanzania. PhD thesis, Sokoine University of Agriculture, Morogoro

Jamison VC, Weaver HH, Reed IF (1950) A hammer-driven soil core sampler. Soil Sci 69:487–496

Jiang X, Bol R, Willbold S, Vereecken H, Klumpp E (2015) Speciation and distribution of P associated with Fe and Al oxides in aggregate-sized fraction of an arable soil. Biogeosciences 12:6443–6452. https://doi.org/10.5194/bg-12-6443-2015

Jones JB (2003) Agronomic handbook: management of crops, soils, and their fertility. Crcpress LLC, Boca raton

Kang JH, Hesterberg D, Osmond DL (2009) Soil organic matter effects on phosphorus Sorption: a path analysis. Soil Sci Soc Am J 73:360–366

Kiflu A, Sheleme B, Schoenau J (2017) Fractionation and availability of phosphorus in acid soils of Hagereselam, Southern Ethiopia under different rates of lime Chem. Biol Technol Agric 4:21. https://doi.org/10.1186/s40538-017-0105-9

Kissel DE, Vendrell PF Isaac B (2004) Third quarter NAPT report, 2004. http://www.naptprogram.org/files/napt/publications/methodpapers/2004salt-concentration-and-measurement-of-soil-ph.pdf

Lavkulich LM (1981) Methods manual, pedology laboratory. Department of Soil Science, University of British Columbia, Vancouver, British Columbia

Loide V (2004) About the effect of the contents and ratios of soil's available calcium, potassium and magnesium in liming of acid soils. Agron Res 2(1):71–82

Maguire RO, Foy RH, Bailey JS, Sims JT (2001) Estimation of the phosphorus sorption capacity of acidic soils in Ireland. Eur J Soil Sci 52:479–487

McKeague JA, Day JH (1966) Dithionite and oxalate extractable Fe and Al as acids in differentiating various classes of soils. Can J Soil Sci 46:13–22

Mehlich A (1984) Mehlich-III soil test extractant: a modification of Mehlich 2. Commun Soil Sci Plant Anal 15:1409–1416

Mehra OP, Jackson ML (1960) Iron oxide removal from soils and clays by a dthionite–citrate system buffered with sodium bicarbonate. Clays Clay Miner 7:317–327

Mesfin B, Abi T, Heluf G, Asmare M (2014) Relation between Universal Extractants and Soybeans (Glycine max L.) response to P and K application under greenhouse conditions. Am J Plant Nutr Fertil Technol 4(2):57–67

Mohhammed M, Kibebew K, Tekalign M (2016) Fertility mapping of some micronutrients in soils of Cheha District, Gurage Zone, Southern Ethiopia. Afr J Soil Sci 4(3):313–320

Murphy HF (1968) A report on fertility status and other data on some soils of Ethiopia. Experimental station bulletin, no. 44. College of Agriculture HSIU, Ethiopia.

Narteh LT, Sahrawat KL (1999) Oxalate and EDTA extractable soil phosphorous and iron in relation to P availability in lowland rice soils of West Africa. Ghana Jnl Agric Sci 32:189–198

Negassa W, Leinwebe P (2009) How does the Hedley sequential phosphorus fractionation reflect impacts of land use and management on soil phosphorus: a review. J Plant Nutr Soil Sci 172:305–325. https://doi.org/10.1002/jpln.200800223305

Newey A (2006) Litter carbon turnover with depth. Ph.D. thesis, Australian National University, Canberra

Okubay G, Heluf G, Tareke B (2015) Soil fertility characterization in vertisols of southern Tigray, Ethiopia. Adv Plants Agric Res 2(1):1–7

Olsen RA, Engelstad OP (1972) Soil phosphorus and sulfur. Soils of the humid tropics. Natural Academy of Sciences, Washington, DC, pp 88–101

Olsen SR, Sommers LE (1982) Phosphorus. In: Page AL et al (ed) Methods of soil analysis. Part 2. 2nd ed. Agron Monogr 9. ASA and SSSA, Madison, p. 403–430

Peinemann N, Nilda MA, Pablo Z, Maria BV (2000) Effect of clay minerals and organic matter on the cation exchange capacity of silt fractions. J Plant Nutr Soil Sci 163(1):47–52. https://doi.org/10.1002/(sici)1522-2624(200002)163:1<47::aid-jpln47>3.0.co,2-a

Rasmussen PE, Allmaras RR, Rohde RR, Roager NC (1980) Crop residue influences on soil carbon and nitrogen in a wheat-fallow system. Soil Sci Soc Am J 44:596–600

Roberto I, Giampietro D (2005) Evaluating phosphorus sorption capacity of acidic soilsby short-term and long-term equilibration procedures. Commun Soil Sci Plant Anal 35(15–16):2269–2282

Rowell DL (1994) Soil Sience: method and applications. Addison Wesley Longman, London

Saljnikov KE, Funakawa S, Akhmetov K, Kosaki T (2004) Soil organic matter status of Mollisols soil in North Kazakhstan: effects of summer fallow. Soil Biol Biochem 36:1373–1381

Saljnikov E, Hospodarenko H, Funakawa S, Kosaki T (2005) Effect of fertilization and manure application on nitrogen mineralization potentials in Ukraine. Zemljiste I Biljka 54(3):221–230

Saljnikov E, Cakmak D, Kostic L, Maksimov c S (2009) Labile fractions of soil organic carbon in Mollisols from different climatic regions. Agrochimica LIII:6

SAS (Statistical Analysis System) version 9.1 (2004) SAS Institute

Shiferaw B (2004) Soil phosphorous fractions influenced by different cropping system in andosols and nitisols in Kambata–Tenbaro and Wolaita Zones, SNNPRS, Ethiopia, M.S. Thesis, Alemaya University, Dire Dawa, Ethiopia

Shoemaker HE, McLean EO, Pratt PF (1961) Buffer methods of determining lime requirements of soils with appreciable amounts of extractable aluminum. Soil Sci Soc Am J 25(4):274–277

Tekalign M, Haque I (1991) Phosphorous status of some Ethiopian soils, II. Forms and distribution of inorganic phosphates and their relation to available phosphorus. Trop Agric 68(1):2–8

Teshome Y, Heluf G, Kibebew K, Sheleme B (2013) Impacts of land use on selected physicochemical properties of soils of Abobo Area, Western Ethiopia. Agric For Fisheries 2(5):177–183. https://doi.org/10.11648/j.aff.20130205.11

Tiessen H, Stewart JWB, Cole CV (1984) Pathways of P transformations in soils of differing pedogenesis. Soil Sci Soc Am J 48:853–858

USDANRCS (2004) United States Department of Agriculture, Natural Resources Conservation Service Soil survey laboratory methods manual. Version No. 4.0. Soil Survey Investigations Report No. 42

Van der Zee SEATM, Nederlof MM, Van Riemsdijk WH, de Haan FAM (1988) Spatial variability of phosphate adsorption parameters. J Environ Qual 17:682–688

Van Reeuwijk LP (1992) Procedures for soil analysis (3rd Ed) International Soil Reference and Information Center (ISRIC), Wageningen, the Netherlands, p 34

Walkley A, Black IA (1934) An examination of the Degtjareff method for determining soil organic matter and a proposed modification of the chromic acid titration method. Soil Sci 37:29–38

Wolde Z, Wassie H, Dhyna S (2015) Phosphorus sorption characteristics and external phosphorus requirement of Bulle and Wonago Woreda, Southern Ethiopia. Adv Crop Sci Tech 23(2):89–99

Xiea G, Steinbergera Y (2001) Temporal patterns of C and N under shrub canopy in a loessial soil desert ecosystem. Soil Biol Biochem 33:1371–1379

Yan X, Wang D, Zhang H, Zhang G, Wei Z (2013) Organic amendments affect phosphorus sorption characteristics in a paddy soil. Agric Ecosyst Environ 175:47–53

YuGe Z, ZhuWen X, DeMing J, Yong J (2013) Soil exchangeable base cations along a chronosequence of Caragana microphylla plantation in a semi-arid sandy land, China. J Arid Land 5(1):42–50

Zhang M, Wright R, Heaney D, Vanderwel D (2004) Comparison of different phosphorus extraction and determination methods using manured soils. Can J Soil Sci 84:469–475

Zhou Q, Gibson CE, Zhu Y (2001) Evaluation of phosphorus bioavailability in sediments of three contrasting lakes in China and the UK. Chemosphere 42:221–225

9

Distribution and ecological risk assessment of trace metals in surface sediments from Akaki River catchment and Aba Samuel reservoir, Central Ethiopia

Alemnew Berhanu Kassegne[1,2*], Tarekegn Berhanu Esho[3], Jonathan O. Okonkwo[4] and Seyoum Leta Asfaw[1]

Abstract

Background: Due to fast urban expansion and increased industrial activities, large quantities of solid and liquid wastes contaminated by trace metals are released into the environment of the Addis Ababa city, most often untreated. This study was conducted to investigate spatial distribution, seasonal variations and ecological risk assessment of selected trace metals (Cd, Cr, Cu, Fe, Mn, Pb, Ni and Zn) in the surface sediments from Akaki River catchment and Aba Samuel reservoir, Central Ethiopia.

Methods: Twenty-two surface sediment samples were collected, digested using the Mehlich-3 procedure and analyzed quantitatively using inductively coupled plasma optical emission spectrometer.

Results: The trace metals occurred in varying concentrations along the course of the sampling stations. The decreasing order of trace metal concentrations in the dry season was: Mn > Fe > Pb > Cr > Zn > Ni > Cu > Cd and in the rainy season was Mn > Fe > Pb > Cr > Ni > Zn > Cu > Cd. Little Akaki River contained a higher load of trace metals than the other regions, which is due to the existence of most of the industrial establishments and commercial activities. Relatively lower levels of trace metals were recorded at Aba Samuel reservoir due to the lower residence time of the sediment (reservoir rehabilitated recently). Ecological risk assessment using USEPA sediment guidelines, geo-accumulation index, contamination factor and pollution load index revealed the widespread pollution by Cd and Pb. These were followed by Mn, Ni and Zn.

Conclusion: The concentrations of Pb, Cd, Mn, Ni and Zn in sediments were relatively greater and at levels that may have adverse biological effects to the surrounding biota. Therefore, regular monitoring of these pollutants in water, sediment and biota would be required.

Keywords: Addis Ababa, Greater Akaki River, Little Akaki River, Sediment, Trace metals

Background

Contamination of the aquatic environment by trace metals in excess of the natural loads has become a problem of increasing concern. Large quantities of trace metals are discharged into the environment due to anthropogenic activities such as urbanization, industrialization and extension of irrigation and other agricultural practices. The situation is particularly alarming in developing countries, where most rivers, lakes and reservoirs are receiving untreated wastes due to poor setup of environmental sustainability (Mwanamoki et al. 2014; Awoke et al. 2016). The most vulnerable river-reservoir systems are those crossing large cities and densely populated areas, as well as near the industrial establishments (Mwanamoki et al. 2014; Yousaf et al. 2016). Trace metals are among the conservative pollutants that are not subject to degradation process and are permanent additions to

*Correspondence: alexbk2010@gmail.com
[1] Centre for Environmental Science, Addis Ababa University, P. O. Box 1176, Addis Ababa, Ethiopia

aquatic ecosystems (Igwe and Abia 2006; El Nemr et al. 2016). As a result, higher levels of trace metals are found in soil, sediment and biota. Most of these trace metals are persistent, toxic, bioaccumulative and that they exert a risk for humans and ecosystems even when the exposure is low (Gao and Chen 2012; Diop et al. 2015; Tang et al. 2016).

Monitoring of trace metals in sediment is extremely important as it can serve as sources of information about the long term trends in geological and ecological conditions of the aquatic ecosystem and the corresponding catchment area (Mekonnen et al. 2012; Dhanakumar et al. 2015). The occurrence of high levels of these pollutants in sediments can be a good indicator of anthropogenic pollution, rather than natural enrichment of the sediment by geological weathering.

Trace metals, originating from man-induced pollution and geological sources, have low solubility in water, and thus they get adsorbed on suspended particles and strongly accumulate in the sediments (Li et al. 2017). Therefore, sediments are reservoirs for trace metal contaminants and help to characterize the degree of environmental contamination and thus are suitable targets for pollution studies (Iqbal and Shah 2014; Liang et al. 2015).

Addis Ababa, the capital city of Ethiopia and the headquarters of the African Union, with approximately 5 million population, is one of the fast expanding cities in the country. There are two major rivers draining the city from North to South. These are Greater Akaki River (GAR) and Little Akai River (LAR). Greater Akaki River and Little Akai River meet at Aba Samuel reservoir, 37 km South-West of Addis Ababa. Aba Samuel reservoir was built in 1939, for hydropower production. The reservoir was, however, abandoned for several years due huge pollution issues and siltation (Gizaw et al. 2004). It was rehabilitated recently and have come back to life in 2016. Since Addis Ababa is the country's commercial, manufacturing and cultural center, large quantities of solid, liquid and gaseous wastes are released into the environment of the city, primarily nearby water bodies most often untreated (Alemayehu 2001, 2006; Awoke et al. 2016; Aschale et al. 2017). The water bodies around Addis Ababa receive increasing amounts of unlicensed discharge of effluents from industrial and domestic wastes and the water quality is deteriorating (Akele et al. 2016). The primary sources of trace metals pollution in the river system include metal finishing industries, tannery operations, textile industries, domestic sources, agrochemicals and leachates from landfills and contaminated sites (Melaku et al. 2007). The two Akaki Rivers and Aba Samuel reservoir, which are the main focuses of this study, serve as dumping grounds and pollutant sinks from upstream Addis Ababa and surrounding catchment areas.

Previous studies on trace metals in sediment are hardly representative and sufficient in the catchment area. There exists an information gap regarding the systematic study of occurrence, distribution, ecological risk and seasonal distribution of trace elements in sediment. The available literature has focused on trace metal levels in water, soil/sediment and vegetables on LAR and not including GAR, Aba Samuel reservoir and downstream areas (Itanna 2002; Arficho 2009; Prasse et al. 2012; Akele et al. 2016; Aschale et al. 2017; Woldetsadik et al. 2017). To the best of our knowledge, no such comprehensive work has been done on the level of trace metals in sediment from Akaki River catchment and Aba Samuel reservoir. Moreover, this study presented one of the earliest set of environmental monitoring data for the reservoir from the feeder Rivers following the restoration of the reservoir in 2016.

Therefore, the objective of this work was to determine the occurrence, distribution, ecological risk and seasonal variation of trace metals (Cd, Cr, Cu, Fe, Mn, Ni, Pb and Zn) in surface sediments from Akaki River catchment and Aba Samuel reservoir.

Materials and methods
Study area
The Akaki catchment is located in central Ethiopia along the western margin of the main Ethiopian Rift Valley. The catchment is geographically bounded between $8°46'–9°14'N$ and $38°34'–39°04'E$, covering an area of about 1500 km^2 (Demlie and Wohnlich 2006). Addis Ababa, which lies within Akaki catchment, has a fast population growth, uncontrolled urbanization and industrialization, poor sanitation, uncontrolled waste disposal, which results in a serious deterioration of surface and ground water quality. As it is the country's commercial, manufacturing and cultural center, large quantities of solid, liquid and gaseous wastes are generated and released into the environment of the city, most often untreated (Alemayehu 2006). There are two major rivers draining into the city from North to South, namely Greater Akaki River (GAR) (locally known as Tiliku Akaki River) and Little Akai River (LAR) (locally known as Tinishu Akaki River). GAR and LAR meet at Aba Samuel reservoir, 37 km South-West of Addis Ababa. Aba Samuel reservoir was built in 1939. It was the first hydropower station in Ethiopia, but it was abandoned in 1970s, because of many years of lack of maintenance, siltation and pollution issues (Gizaw et al. 2004). It was rebuilt and revived in 2016. The local people in the Akaki River catchment and Aba Samuel reservoir use the water for irrigation, drinking water for cattle, washing clothes, waste disposal site and other domestic needs without information on the level of water quality parameters (Melaku et al. 2007). Therefore, this study has been

conducted in some parts of GAR, LAR, Aba Samuel reservoir and downstream to the reservoir (Fig. 1).

Sampling sites and sample collection

Twenty-two (22) sediment samples were collected in August, 2016 and January, 2017 representing the rainy and dry seasons respectively. Composite samples were collected at the following sampling stations: GAR at Entoto Kidanemihiret Monastery (S1, control site 1), GAR at Tirunesh Beijing hospital (S2), GAR below Akaki town (S3), LAR above Geferesa reservoir (S4, control site 2), LAR at Lafto bridge (S5), LAR at Jugan Kebele, boundary between Addis Ababa and Oromia Special zone (S6), Aba Samuel reservoir below the confluence point of GAR and LAR (S7), Aba Samuel reservoir at the midpoint (S8), Aba Samuel reservoir above the Dam (S9), downstream about 50 m from the reservoir (S10) and downstream about 1000 m from the reservoir (S11). The distribution of the sampling points was chosen based on

topography, the purpose of the study and anthropogenic interference. Approximately 500 g of the top few centimeters of the sediment were collected using a stainless steel Ekman bottom Grab sampler. Each sample was obtained by mixing four randomly collected sediment samples. Samples were placed in clean polyethylene bags, labeled, stored in cooler box and transported into the laboratory. In the laboratory, coarse particles, leaves or large material was removed. Subsequently, sediment samples were air dried at ambient temperature and powdered using ceramic coated grinder. The dried and powdered samples were then sub-sampled and passed through a stainless steel sieve (45 μm mesh size) and transferred to labeled double-cap polyethylene bottles until further treatment.

Sample digestion and instrumental analysis

For determination of trace metals, 2 g sediment samples (< 45 μm) were digested using 20 ml of Mehlich 3 extractant [0.2 M CH_3COOH, 0.25 M ammonium

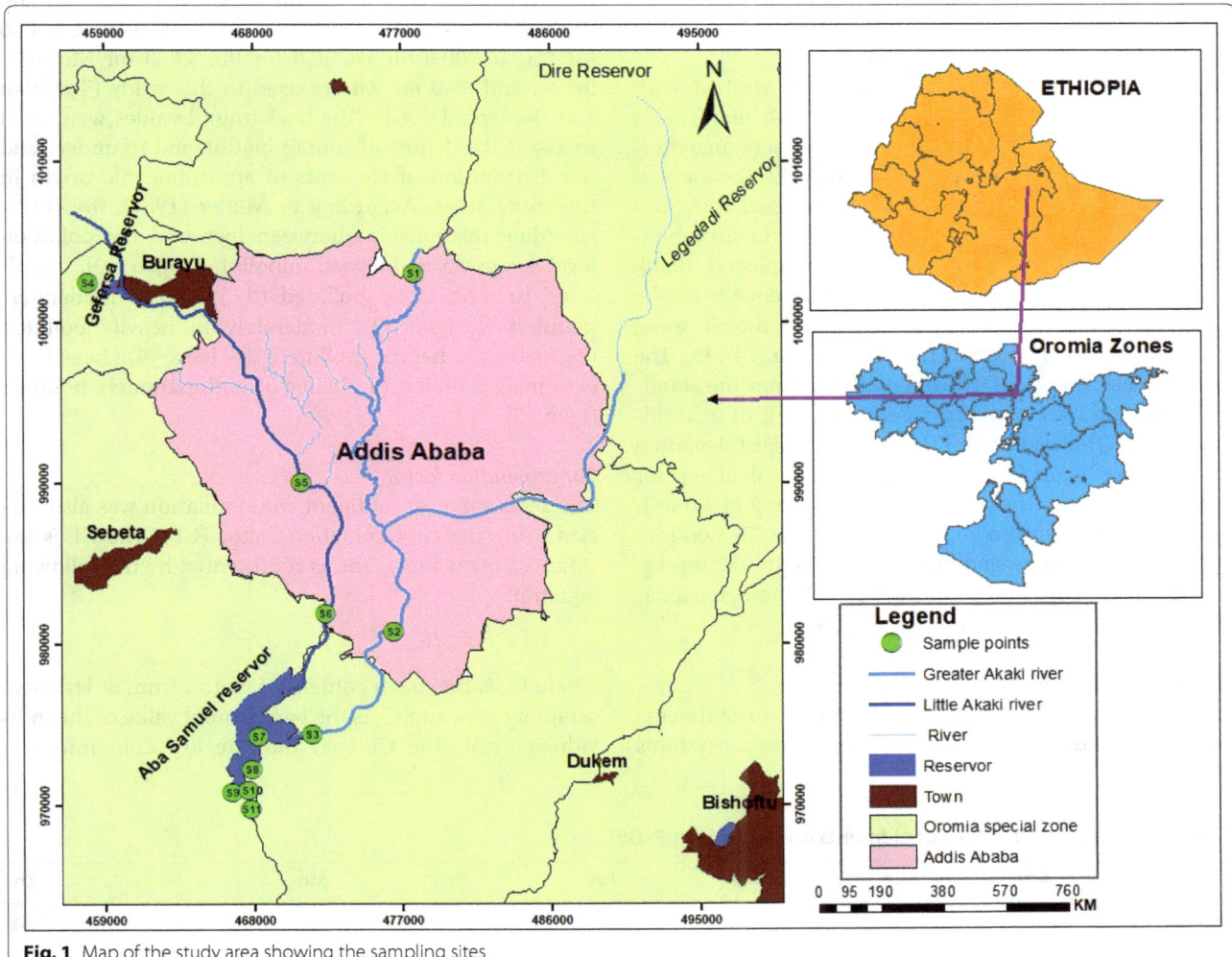

Fig. 1 Map of the study area showing the sampling sites

nitrate (NH_4NO_3), 0.015 M ammonium fluoride (NH_4F), 0.013 M HNO_3, and 0.001 M ethylene diamminetet-raacetic acid (EDTA)] (Mehlich 1978). An inductively coupled plasma optical emission spectrometer, ICP-OES, (Arcos FHS2, Germany) was used for the determination of trace metal concentrations in sediment samples. Argon gas (99.99%) was used as a plasma with a flow rate of 81 l/min. Calibration curves were prepared using 10, 20, 30, 40 and 50 mg/l of Fe and Mn; 0.04, 0.08, 0.12, 0.20, 0.40, 0.80, 1.20, 1.60, and 2.00 for Cu and Zn; 0.5, 1, 2, 3, 4, 5 for Cd, Ni and Pb and 1, 2, 4, 6, 8, 10 for Cr. In all cases, standard purity was ≥99.8%. Quantification of the elements were recorded at 214.438, 267.716, 324.757, 262.567, 220.353, 257.611, 231.604 and 213.856 nm, which correspond to the most sensitive emission wavelengths of Cd, Cr, Cu, Fe, Pb, Mn, Ni, and Zn respectively. The calibration curve showed linearity (r > 0.995) by the detector response for the quantified elements. This indicates good correlation between concentration and emission intensities of the detected elements and thus proper calibration of the instrument.

Quality control and quality assurance
All the glassware used were thoroughly washed with detergent, soaked in 10% HNO_3 for 24 h and rinsed with de-ionized water. All reagents used were analytical grade. In order to validate and evaluate the accuracy of the method used, certified reference material (ISE-952) obtained from Wageningen University Environmental Sciences section, Netherlands, was employed. Blank analyses were carried out to check interference from the laboratory. Mean recovery rates of the 4 metals were: Zn, 100.52%; Fe, 106.69%; Mn, 118.52%; Cu, 74.14%. The limit of detection (LOD) based on three times the standard deviation (3σ) of the blank and the limit of quantification (LOQ) based on ten times the standard deviation (10σ) of the blank for the ICP-OES were calculated for each analyte ions. The results are summarized in Table 1. The LOD was found to be in the range of 0.07–1.06 mg/kg, whereas LOQ ranged between 0.23 and 3.52 mg/kg. These ranges were found satisfactory for the determination of analyte ions in sediment samples.

Assessment of sediment contamination
The excessive accumulation of trace metals in sediments posed a potential ecological risk to freshwater ecosystems (Olivares-Rieumont et al. 2005; Chen et al. 2007). Different pollution assessment methods of trace metals were applied to evaluate the pollution degree and potential ecological risk posed by trace metals in sediment of Akaki River catchment and Aba Samuel reservoir. To this end, USEPA sediment guidelines, geo-accumulation index (Igeo), contamination factor and pollution load index were used (Wang et al. 2014).

Geo-accumulation index (Igeo): the geo-accumulation index (Igeo) is a geochemical criterion used to assess heavy metal accumulation in surface sediment studies (Muller 1981; Singh 2001; Aschale et al. 2017). It is expressed as

$$Igeo = \log_2 [C_n/1.5B_n].$$

where C_n is the measured total concentration of the element n in the sediment and B_n is the average concentration of element n in shale (background) value. The constant 1.5 is introduced to include possible variations of the background values due to lithogenic effects in sediments (Loska et al. 2004). Thus, the background concentrations (mg/kg) of 0.3 for Cd, 90.0 for Cr, 45.0 for Cu, 46,700.0 for Fe, 20.0 for Pb, 850.0 for Mn, 68.0 for Ni and 95.0 for Zn are used in this study (Turekian and Wedepohl 1961). The background values were used to assess the degree of contamination and to understand the distribution of elements of anthropogenic origin in the study areas. According to Muller (1981), the corresponding relationships between Igeo and the pollution level are given as follows: unpolluted (Igeo ≤ 0), unpolluted to moderately polluted (0 < Igeo ≤ 1), moderately polluted (1 < Igeo ≤ 2), moderately to heavily polluted (2 < Igeo ≤ 3), heavily polluted (3 < Igeo ≤ 4), heavily to extremely polluted (4 < Igeo ≤ 5) and extremely polluted (Igeo > 5).

Contamination factor
The assessment of sediment contamination was also carried using the contamination factor (CF). The CF is the single element index and is represented by the following equation

$$CF = C_o/C_n$$

where C_o is the mean content of metals from at least five sampling sites and C_n is the background value of the individual metal. The CF may indicate low contamination

Table 1 LOD and LOQ values of elements analyzed by ICP-OES

Element	Cd	Cr	Cu	Fe	Pb	Mn	Ni	Zn
LOD	0.11	0.30	0.07	0.78	0.23	1.06	0.55	0.08
LOQ	0.35	1.00	0.23	2.60	0.77	3.52	1.83	0.27

(CF < 1), moderate contamination (1 < CF < 3), considerable contamination (3 < CF < 6) and very high contamination (CF > 6) (Hakanson 1980).

Pollution load index (PLI)

Pollution load index (PLI) was examined to assess the overall pollution status of a sampling site. The index was determined by calculating the geometrical mean of the concentrations of all the trace elements in the particular sampling site (Usero et al. 1997; Chakravarty and Patgiri 2009).

The PLI is computed by the formula:

$PLI = (CF_1 \times CF_2 \times CF_3 \times \cdots \times CF_n)^{1/n}$, where n is the number of metals investigated and CF is the contamination factor. The PLI value > 1 is polluted whereas PLI value < 1 indicates no pollution (Chakravarty and Patgiri 2009).

Statistical analysis

Analysis of variance (ANOVA) was applied to assess significant differences in trace element concentrations at the various sampling sites. Multivariate analysis of the element concentration was performed through cluster analysis technique. It was performed to classify elements of different sources on the basis of their concentration similarities using dendrograms, and identify relatively homogeneous groups of variables with similar properties. Pearson's correlation coefficient was used to determine the association and possible sources of trace metals. Statistical analyses of the results were carried out using Origin Pro (version 9.4, 2017) and Microsoft Excel 2007.

Results and discussion
Concentrations of trace metals in sediment samples

The average concentrations along with standard deviations of 8 selected trace metals in sediment samples from the Akaki River catchment and Aba Samuel reservoir, Ethiopia in two seasons are presented in Table 2. Based on the elemental concentrations, the pattern in sediment was: Mn > Fe > Pb > Cr > Zn > Ni > Cu > Cd in the dry season and Mn > Fe > Pb > Cr > Ni > Zn > Cu > Cd in the rainy season. In both seasons, a similar pattern was observed. The concentration (mg/kg) ranges of trace metals in the dry season were 2.1–2.9 for Cd, 16.2–43.7 for Cr, 1.6–15.3 for Cu, 406.4–844.8 for Fe, 124.4–256.4 for Pb, 335.5–1319.2 for Mn, 15.6–36.2 for Ni and 4–110 for Zn. Similarly, in the rainy season the concentrations (mg/kg) were in the range of 2.5–3.1 for Cd, 18.3–29.4 for Cr, 2.1–6.2 for Cu, 415.2–1442 for Fe, 101.4–133.7 for Pb, 385.1–1833.4 for Mn, 14.6–24 for Ni and 4.8–38.2 for Zn. In both seasons, the minimum concentration was observed for the known toxic elements (such as Cd and Cu) while

the highest concentration is observed for Mn and Fe. Iron and manganese pollution of the catchment area, possibly arises from effluents from iron and steel manufacturing industries established within the catchment area of Akaki River (Melaku et al. 2007). The highest concentrations of Mn and Fe observed could also be related to the geological sources in addition to anthropogenic inputs (Alemayehu 2006). The geology of Addis the Ababa area is characterized by basaltic volcanic rocks with minor amounts of Quaternary alluvial sediments (Demlie et al. 2008). The rocks underlying the city and its environs were altered by intensive hydrothermal activity resulting in the characteristics reddish color of the residual soils (Gizaw 2002). Kaolin deposits found in many parts of the city are particularly good evidence of hydrothermal activity on lava flows. Alemayehu (2006) indicated that rock and soil outcrops of the Addis Ababa area are anomalously rich in trace metals derived from hydrothermal activity, which are related to geologic sources. From the hydrogeological point of view, the major rock types forming a reservoir of groundwater in the Addis Ababa area are considered to be the volcanic rocks consisting of basalts, trachytes, rhyolites, scoriaes and trachy-basalts. Studies indicated that the main aquifers in the Addis Ababa area include: shallow aquifers, deep aquifers and thermal aquifers (located at depths greater than 300 m) (Alemayehu 2006; Demlie et al. 2007).

Table 3 compares the results obtained from Akaki River Catchment and Aba Samuel reservoir and those from other freshwater ecosystems to understand the extent of trace metal pollution of the study area. A comparison between a study in LAR (Aschale et al. 2016) and this study indicates that samples from the latter showed higher concentrations of Cd and Pb and lower average concentrations of Cr, Cu, Fe, Mn and Zn. The Pb contamination in the sediments of this study was relatively higher than values from all the other studies. The concentrations of Cd, Cr, and Ni were within the ranges observed in the other polluted sediments. The levels of Cu and Zn were generally lower than values for other sediments. Overall, this comparison indicated that there was high accumulation of trace metals in the sediments of Akaki River and Aba Samuel reservoir and that it requires special care and management interventions such as proper waste collection, treatment and disposal.

Spatial distribution and seasonal variations of Trace metals in sediment

The spatial distribution of trace elements in the river and reservoir sediment depends on many factors including distance of the element sources to the reservoir, the chemical characteristics of the element and the hydrological conditions of the river and reservoir system

Table 2 Mean concentrations of trace metals (mean ± SD, mg/kg dry weight) in sediment samples

No.	Area	Cd	Cr	Cu	Fe
Dry season					
S1	GAR at Entoto Kidanemihiret Monastery	2.5 ± 0.12	22.0 ± 1.11	5.7 ± 0.03	436.4 ± 0.37
S2	GAR at Tirunesh Beijing Hospital	2.6 ± 0.16	28.4 ± 1.19	3.6 ± 0.01	844.8 ± 2.93
S3	GAR below Akaki town	2.6 ± 0.14	23.1 ± 1.23	3.2 ± 0.01	557.0 ± 1.21
S4	LAR above Gefersa reservoir	2.9 ± 0.16	16.2 ± 1.38	1.6 ± 0.01	406.4 ± 2.66
S5	LAR at Lafto Bridge	2.7 ± 0.16	26.9 ± 1.24	15.3 ± 0.09	579.3 ± 1.19
S6	LAR at Jugan kebele	2.1 ± 0.17	43.7 ± 1.66	2.2 ± 0.03	539.1 ± 2.57
S7	Below the confluence point of GAR and LAR	2.6 ± 0.12	26.3 ± 1.29	5.7 ± 0.05	478.5 ± 4.22
S8	Aba Samuel reservoir at the midpoint	2.6 ± 0.17	27.7 ± 1.45	5.2 ± 0.01	492.1 ± 2.48
S9	Aba Samuel reservoir above the Dam	2.7 ± 0.15	21.0 ± 1.20	2.6 ± 0.01	437.3 ± 3.26
S10	Aba Samuel reservoir below the Dam	2.5 ± 0.17	31.4 ± 1.35	4.0 ± 0.19	818.4 ± 3.59
S11	Aba Samuel reservoir 1000 m downstream	2.8 ± 0.13	32.1 ± 1.23	7.4 ± 0.02	539.9 ± 3.83
		Pb	**Mn**	**Ni**	**Zn**
S1	GAR at Entoto Kidanemihiret Monastery	145.6 ± 0.34	423.5 ± 0.68	18.4 ± 1.15	17.6 ± 0.12
S2	GAR at Tirunesh Beijing Hospital	134.1 ± 5.34	1103.9 ± 8.55	26.3 ± 0.90	18.6 ± 0.10
S3	GAR below Akaki town	133.3 ± 4.68	576.9 ± 2.47	20.7 ± 0.98	29.1 ± 0.17
S4	LAR above Gefersa reservoir	124.4 ± 3.57	373.3 ± 2.79	15.6 ± 1.64	7.0 ± 0.05
S5	LAR at Lafto Bridge	170.8 ± 5.04	669.6 ± 3.66	26.2 ± 1.18	59.4 ± 0.25
S6	LAR at Jugan kebele	256.4 ± 7.33	1319.2 ± 10.51	36.2 ± 1.37	5.2 ± 0.04
S7	Below the confluence point of GAR and LAR	152.7 ± 3.50	464.8 ± 0.90	21.7 ± 0.97	13.7 ± 0.08
S8	Aba Samuel reservoir at the midpoint	128.5 ± 2.80	591.7 ± 3.17	22.9 ± 1.62	4.0 ± 0.02
S9	Aba Samuel reservoir above the Dam	129.1 ± 4.48	335.5 ± 1.62	20.4 ± 1.45	6.2 ± 0.02
S10	Aba Samuel reservoir below the Dam	145.4 ± 1.41	1061.5 ± 0.90	25.4 ± 1.20	12.5 ± 0.06
S11	Aba Samuel reservoir 1000 m downstream	171.2 ± 1.31	522.0 ± 3.11	26.3 ± 1.40	110.0 ± 0.55
No.	Area	Cd	Cr	Cu	Fe
Rainy season					
S1	GAR at Entoto Kidanemihiret Monastery	2.6 ± 0.20	18.3 ± 1.02	2.1 ± 0.01	472.0 ± 2.38
S2	GAR at Tirunesh Beijing Hospital	2.8 ± 0.13	29.4 ± 1.17	2.4 ± 0.02	1094.9 ± 2.99
S3	GAR below Akaki town	2.8 ± 0.15	21.9 ± 0.80	3.3 ± 0.03	536.9 ± 1.52
S4	LAR above Gefersa reservoir	3.1 ± 0.17	28.7 ± 1.39	4.5 ± 0.01	1442.0 ± 10.23
S5	LAR at Lafto Bridge	2.6 ± 0.15	22.6 ± 0.91	5.3 ± 0.01	415.2 ± 3.35
S6	LAR at ugan kebele	2.5 ± 0.14	22.8 ± 1.38	6.2 ± 0.03	442.5 ± 3.12
S7	Below the confluence point of GAR and LAR	2.9 ± 0.16	23.4 ± 1.22	2.2 ± 0.01	524.2 ± 1.44
S8	Aba Samuel reservoir at the midpoint	2.6 ± 0.18	26.0 ± 1.18	2.5 ± 0.02	551.8 ± 1.63
S9	Aba Samuel reservoir above the Dam	2.7 ± 0.13	22.9 ± 0.96	2.5 ± 0.02	481.2 ± 3.93
S10	Aba Samuel reservoir below the Dam	2.7 ± 0.14	23.8 ± 1.07	3.6 ± 0.01	499.1 ± 1.16
S11	Aba Samuel reservoir 1000 m downstream	2.5 ± 0.13	29.1 ± 1.14	2.4 ± 0.02	541.2 ± 1.81
		Pb	**Mn**	**Ni**	**Zn**
S1	GAR at Entoto Kidanemihiret Monastery	111.5 ± 1.09	552.4 ± 1.94	14.6 ± 0.98	10.3 ± 0.02
S2	GAR at Tirunesh Beijing Hospital	114.4 ± 1.77	1531.4 ± 3.77	24.0 ± 1.16	19.8 ± 0.02
S3	GAR below Akaki town	122.5 ± 1.29	467.1 ± 3.44	20.7 ± 1.34	22.2 ± 0.18
S4	LAR above Gefersa reservoir	101.4 ± 3.10	1833.4 ± 11.63	23.1 ± 1.09	25.7 ± 0.11
S5	LAR at Lafto Bridge	125.9 ± 2.96	385.1 ± 2.84	19.9 ± 1.84	20.1 ± 0.09
S6	LAR at Jugan kebele	133.7 ± 0.17	422.5 ± 1.58	19.9 ± 1.54	31.1 ± 0.05
S7	Below the confluence point of GAR and LAR	132.6 ± 3.05	392.1 ± 1.56	21.2 ± 1.63	4.8 ± 0.04
S8	Aba Samuel reservoir at the midpoint	127.4 ± 2.10	541.4 ± 2.20	20.9 ± 1.13	38.2 ± 0.21
S9	Aba Samuel reservoir above the Dam	133.0 ± 2.10	393.4 ± 2.13	19.7 ± 1.19	17.2 ± 0.03
S10	Aba Samuel reservoir below the Dam	132.7 ± 3.33	431.8 ± 1.42	21.7 ± 1.02	18.2 ± 0.02
S11	Aba Samuel reservoir 1000 m downstream	131.0 ± 4.75	546.4 ± 2.37	20.6 ± 0.92	14.4 ± 0.04

Table 3 Average trace metals contents (mg/kg) in sediment from Akaki River catchment and Aba Samuel reservoir compared with aquatic environments from Ethiopia and other parts of the world

Location	Cd	Cr	Cu	Fe	Pb	Mn	Ni	Zn	References
Aba Samuel reservoir, Ethiopia[a]	2.6	25	4.5	469.3	136.8	464	20.6	10	This study
GAR, Ethiopia[a]	2.6	24.5	4.2	612.7	137.7	701.4	19.8	21.8	This study
LAR, Ethiopia[a]	2.6	28.9	6.4	508.3	183.9	787.4	21	23.9	This study
LAR, Ethiopia	0.2	262.1	32.0	30,475.2	45.4	1089.1	31.2	108.6	Aschale et al. (2016)
Lake Awassa, Ethiopia	0.2	8.3	8.7	–	15.7	–	20.2	93.8	Yohannes et al. (2013)
Lake Ziway, Ethiopia	–	32.8	31.1	–	39.9	–	14.2	30.3	Mekonnen et al. (2015)
Awash River Basin, Ethiopia	2.6	120.6	79.4	–	13.5	–	89.5	382.7	Dirbaba et al. (2018)
Lake Victoria, Tanzania	2.5	11.0	21.6	–	29.6	–	–	36.4	Kishe and Machiwa (2003)
Buriganga River, Bangladesh	1.5	173.4	344.2	–	31.4	–	153.3	481.8	Mohiuddin et al. (2015)
Tembi River, Iran	14.5	42	51.5	232	182	409	87.8	35.0	Shanbehzadeh et al. (2014)
Lijiang River, China	1.72	56.4	38.1	–	51.5	–	–	142.2	Xu et al. (2016)
Ergene River, Turkey	–	160.0	65.0	26,935	99.0	356.0	64.0	177.0	Halli et al. (2014)

[a] Average of three sampling stations in the dry season was taken

(Zhang et al. 2013). The level of trace elements in surface sediments of Akaki River catchment and Aba Samuel reservoir is shown in Figs. 2 and 3. In order to better understand the distribution and seasonal variation of trace metals, the study area was divided into four regions: GAR, LAR, reservoir and Downstream to the reservoir. Overall, LAR is more contaminated than GAR (Melaku et al. 2007; Akele et al. 2016). Statistical analyses of the results (p < 0.05) indicated that there were no significant spatial variations of the trace metal among the sampling stations in both seasons except Pb. The concentration of Pb varied significantly from upstream to downstream area in both seasons (p = 0.02). Dry season trace metal concentrations were slightly higher than the rainy season for Pb, Cr, Zn, Ni, and Cu. This might be attributed to the lower dilution process in the dry season than in the rainy season. During the dry season, the highest concentrations of Cr, Pb, Mn, and Ni were observed at S6 and Cu at S5 (Figs. 2 and 3). Both sampling sites lie along the Little Akaki River (LAR), which is in agreement with the previous results (Melaku et al. 2007; Akele et al. 2016). The average levels of Mn and Fe were higher in the rainy than in the dry season, which might be attributed to the anthropogenic and geologic inputs (Alemayehu 2006). It is presumed that pollutants from GAR and LAR finally ended up at the Aba Samuel reservoir. However, the levels of most trace metals investigated were lower at the Aba Samuel reservoir (S7, S8 and S9) than the upstream areas (Figs. 2 and 3). The relatively lower concentration of sediment-bound trace metals in the reservoir might have been due to less accumulated metals in the sediment because reservoir was rehabilitated in 2016. Furthermore, the natural processes that can attenuate the concentration of the chemicals/pollutants on their pathway (mixing, dilution, volatilization and biological degradation) might have contributed to the attenuation.

Correlation analysis

Pearson's correlation coefficients were computed to see if the elements were interrelated with each other in the sediment samples from the different sampling sites in both dry and rainy seasons. Examination of correlations also provides clues on the source(s) of pollution, distribution and similarity of behaviors of trace metals (Zhang et al. 2013; Diop et al. 2015). Table 4 shows the correlation matrix of the determined elements. A significant positive correlation was observed for Pb with Cr ($r = 0.85$), Mn with Cr ($r = 0.81$), Mn with Fe ($r = 0.73$), Mn with Pb ($r = 0.60$), Ni with Cr ($r = 0.97$), Ni with Pb ($r = 0.85$) and Ni with Mn ($r = 0.83$) in the dry season. Similarly, a significant positive correlation was observed between Fe with Cd ($r = 0.75$), Fe with Cr ($r = 0.67$), Mn with Cd ($r = 0.66$), Mn with Cr ($r = 0.66$), Mn with Fe ($r = 0.99$), Ni with Cr (0.81), Ni with Fe ($r = 0.61$) in the rainy season. This significant positive correlation suggests that the elements might have a common origin. Concentration of Zn was not significantly correlated with any of the studied trace metals. Significant negative correlations were also found between Cd with Cr ($r = -0.75$), Cd with Pb ($r = -0.72$), Cd with Mn ($r = -0.72$), Cd with Ni ($r = -0.71$) in the dry season and Pb with Fe ($r = -0.78$), Mn with Pb ($r = -0.82$) in the rainy season.

Cluster analysis

The hierarchical clustering by applying group average method and Euclidean distances for similarities in the variables was performed on the dataset. Altogether,

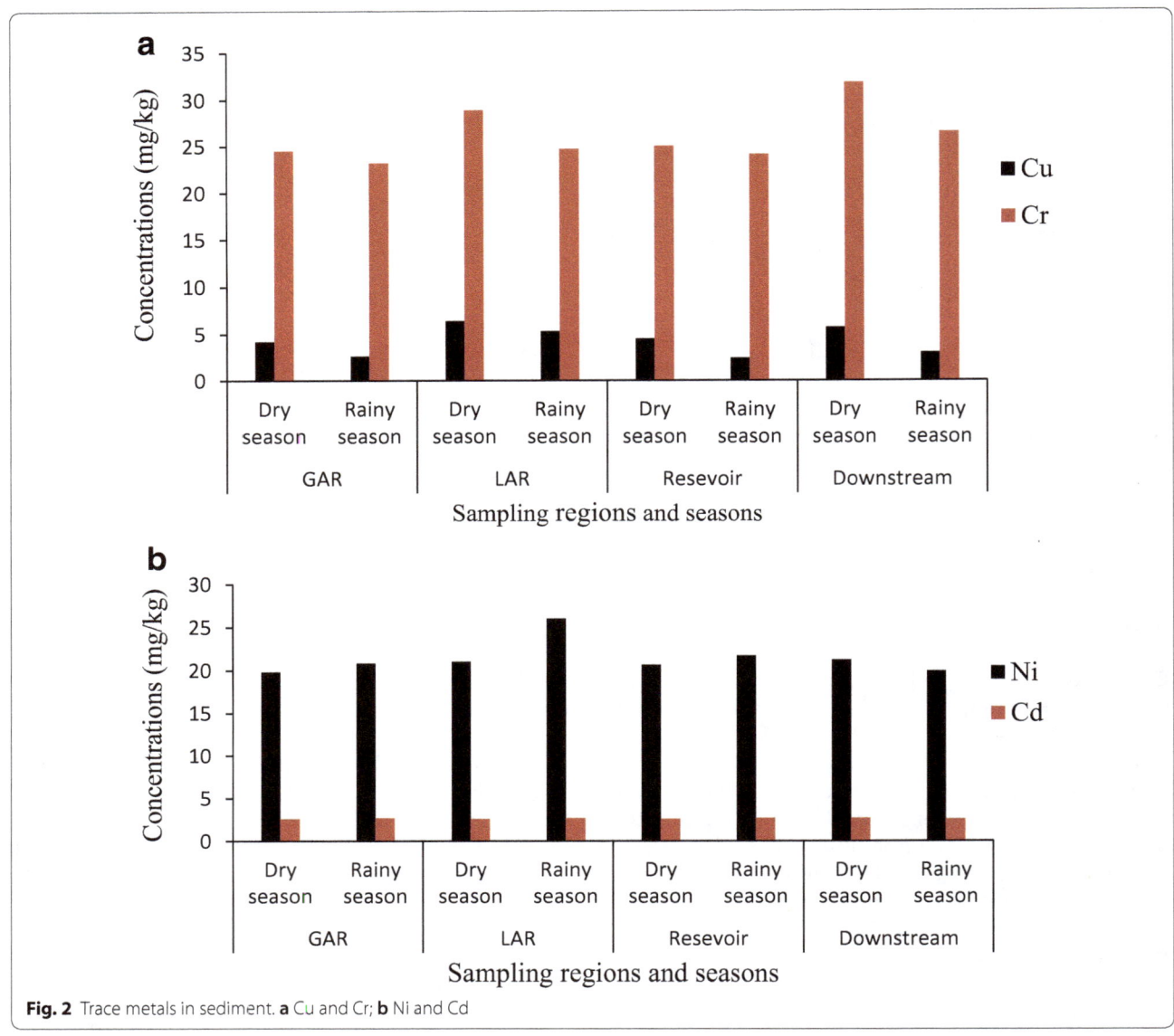

Fig. 2 Trace metals in sediment. **a** Cu and Cr; **b** Ni and Cd

8 variables (Cd, Cr, Cu, Fe, Pb, Mn, Ni and Zn) from 11 sampling sites in dry and rainy seasons were subjected to the cluster analysis. The dendrogram derived from the cluster analysis is shown in Fig. 4. In both seasons, similar types of clusters of the elements were observed. When we cut the dendrogram at an imaginary distance between 1000 and 2000 cm, it leaves two major clusters. Cluster 1 (Cd, Cu, Cr, Ni, Zn and Pb) and cluster 2 (Fe and Mn). From the dendrogram, there are two distinct source factors; one that relates to the soil/geologic inputs (which may introduce Fe and Mn) and another that relates to a variety of activities in the catchment including anthropogenic sources, which collectively contributed to the remaining metals in the sediment. Elements belonging to the same clusters or groups are likely to have originated from common sources (Faisal et al. 2014). The source factors for the cluster analysis of some of the trace metals are presented as follows. Chromium (Cr) contamination of the study area might have originated from one or some of the industries including electroplating and tannery industries, paints and inks, wood preservatives, textile and refractoriness. The highest concentration of Cr was observed at S6 where the majority of Tannery industries were located on the bank of Little Akaki River. The reason for the elevated concentration of Cr at S4 (control site), is not clear. Ni pollution in the study area might arise from sources like domestic

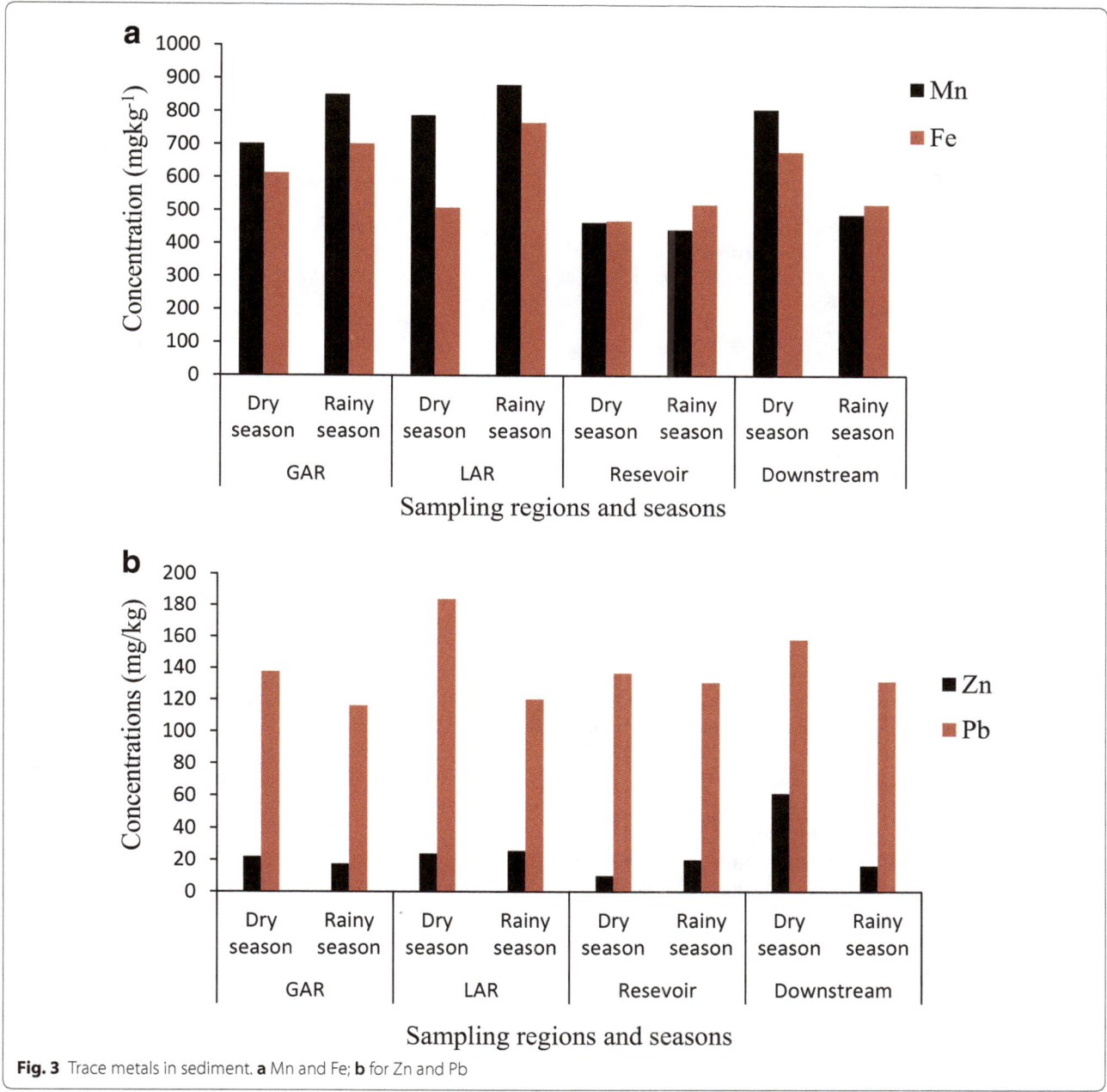

Fig. 3 Trace metals in sediment. **a** Mn and Fe; **b** for Zn and Pb

wastes, municipal sewage, electroplating coal, oil combustion, pigments and batteries (Aschale et al. 2017). Zn pollution in the study area might arise from the expected sources such as textile and metal works/iron and steel works. In addition to the geological sources, anthropogenic Pb pollution in the study area may arise as a result of activities such as industrial discharge from smelters, paints and ceramics, through vehicular emissions, runoff from contaminated land areas and sewage effluent.

Assessment of trace metals pollution
Assessment of sediment pollution using sediment contamination guidelines

After generating reliable data on the level of trace metals in sediment, interpretive tools are required to relate sediment chemistry information to the risk. To this end, numerical sediment quality guidelines (SQGs) established based on biological tests can be used (Macdonald et al. 1996). Most recent SQGs are derived from matching chemistry and toxicity data. The average concentration of

Environmental Science: A Global Outlook

Table 4 Correlation matrix among the different trace metals in sediment from Akaki River catchment and Aba Samuel reservoir in dry and rainy seasons

	Cd	Cr	Cu	Fe	Pb	Mn	Ni	Zn
Dry season								
Cd	1							
Cr	−0.75*	1						
Cu	0.22	C.04	1					
Fe	−0.21	C.39	0.04	1				
Pb	−0.72*	C.85*	0.11	0.01	1			
Mn	−0.72*	C.81*	−0.12	0.73*	0.60*	1		
Ni	−0.71*	0.97*	0.13	0.43	0.85*	0.83*	1	
Zn	0.38	0.15	0.59	0.06	0.15	−0.15	0.17	1
Rainy season								
Cd	1							
Cr	0.28	1						
Cu	−0.09*	−0.08	1					
Fe	0.75*	0.67*	0.02	1				
Pb	−0.57	−0.19	0.03	−0.78*	1			
Mn	0.66*	0.66*	0.02	0.99*	−0.82*	1		
Ni	0.52	0.81*	0.12	0.61*	−0.06	0.55	1	
Zn	−0.15	0.26	0.46	0.17	−0.05	0.17	0.28	1

*Correlation is significant at p < 0.05 level (two-tailed)

trace metals in surface sediments and guideline values is presented in Table 5. Based on the guideline, the river system and reservoir were non-polluted with Cd and Cu, but non-to moderately polluted with Cr, Ni and Zn. All sampling sites were heavily polluted with Pb.

The concentration of Ni in the analyzed samples were within the same range or slightly higher than the background values for sediment quality guidelines. Only one site, S11 in the dry season is moderately polluted with Zn. In both seasons, all the sampling stations were heavily polluted with Pb. The Federal Democratic Republic of Ethiopia (FDRE) has formulated three proclamations that are directly and/or indirectly related to the environment and pollution (Mekonnen et al. 2015). However, based on the results obtained in this study, the proclamations seem to have not properly implemented. The river system and reservoir need immediate attention of those trace metals having higher concentrations. Unless control measures are made possible, the situation could be worsened and affect biota in Akaki River system and Aba Samuel reservoir and downstream Awash River, which is the most productive inland river in Ethiopia.

"Effects range low" (ERL) and "effects range median" (ERM) developed by the National Oceanic and Atmospheric Administration are another sediment toxicity guidelines for trace metals and other contaminants (Long et al. 1995; Macdonald et al. 1996). ERL and ERM values identify threshold concentrations that, if exceeded, are expected to have adverse ecological or biological effects (Mekonnen et al. 2015). Based on the ERL- ERM range the level of Pb at S6 in LAR in the dry season could be toxic to bottom dwelling aquatic organisms (Table 6), while for Cr, Cu and Ni are less than the ERL range. Some of the results from the sampling sites lie between ERL-ERM range for Cd, Pb and Ni.

Geo-accumulation index, contamination factor and pollution load index
Geo-accumulation index
In this study, the calculated value of the geo-accumulation index (Igeo) is presented in Table 7. According to the Muller scale, the calculated results of Igeo values (Table 7) indicated that the sediments from the 11 sampling sites were found to be in class 0, thus are uncontaminated with Cr, Cu, Fe, Ni, and Zn. Mn concentrations represent unpolluted conditions at all stations except S6 (Igeo = 0.05) in the dry season and S2 (Igeo = 0.26) and S4 (Igeo = 0.52) in the rainy seasons. However, all the sediment samples were moderately to strongly contaminated with Cd. Similarly, sediment samples were moderately to strongly contaminated with Pb.

Contamination factor (CF) and pollution load index (PLI)
Pollution severity and its variation along the sites were determined with the use of pollution load index (PLI). This index is a quick tool to compare the pollution

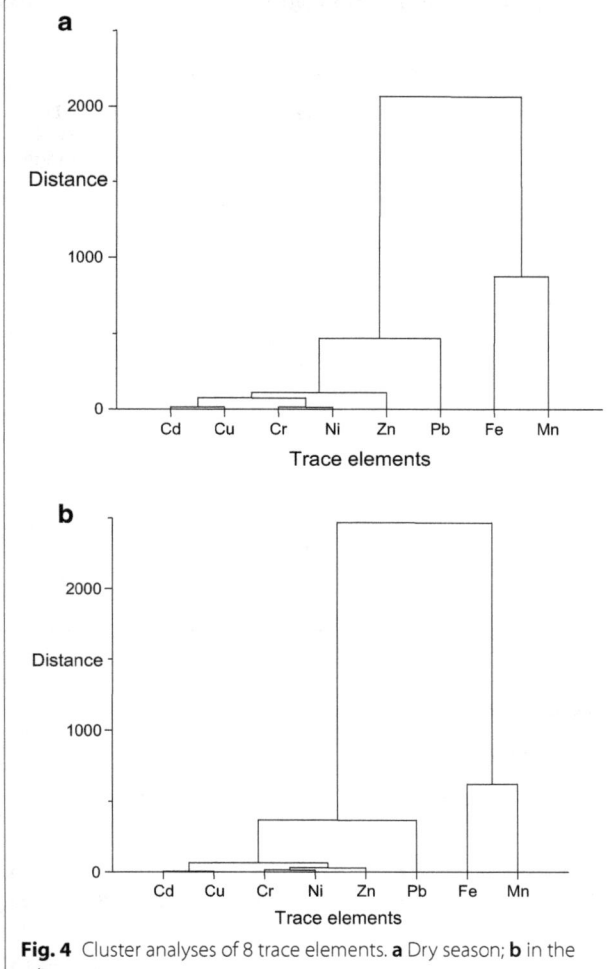

Fig. 4 Cluster analyses of 8 trace elements. **a** Dry season; **b** in the rainy season

(CF = 1.16) in the dry season. CF values for Cd in both dry and rainy seasons were > 6 at all the sampling sites, indicating very high contamination. Cadmium is one of the highly toxic, non-essential elements. Therefore, even at low concentrations, Cd could be harmful to living organisms. This amount of Cd from the study area may be attributed to the release of chemicals from sewage and industrial wastes from the nearby Addis Ababa city. The CF value of Pb > 6 in all the sampling sites except S1 (CF = 5.58), S2 (CF = 5.72) and S4 (CF = 5.07) in the rainy season, indicating very high contamination. The study area might be exposed to Pb pollution from various activities such as industrial discharge from smelters, paints and ceramics, through vehicular emissions, runoff from contaminated land areas and sewage effluent. CF values for Mn at S2 (CF = 1.30) and S10 (CF = 1.25) in the dry season and S2 (CF = 1.80) and S4 (CF = 2.16) in the rainy season suggesting moderate contamination, while values of CF < 1 in all other sites indicate low contamination. The relatively higher CF value of Mn in the rainy season is in agreement with the cluster analysis result that may in turn suggest soil runoff/geologic inputs as the potential source of the trace metal. The values of PLI (Table 7) were found to be generally low (< 1) in all the studied stations. The higher value of PLI (0.59 at S5 and 0.58 at S11 in the dry season) implies appreciable input of trace metals from anthropogenic sources (Table 8).

Conclusions

This study was targeted at generating up-to-date data on the spatial and seasonal variation and contamination levels of trace metals in surface sediments from Akaki River catchment and Aba Samuel reservoir, Central Ethiopia. The decreasing order of trace metal concentrations in the dry season was: Mn > Fe > Pb > Cr > Zn > Ni > Cu > Cd and in the rainy season was Mn > Fe > Pb > Cr > Ni > Zn > Cu > Cd. When comparing the sampling regions in Akaki River Catchment and Aba

status of different sampling locations. The contamination factor values (Table 8) for Cr, Cu, Fe and Ni were (< 1) at all the sampling sites in both seasons indicating low contamination. The CF of Zn represents low contamination at all the sampling sites except S11

Table 5 Concentration of trace metals in sediments and its comparison with sediment quality guidelines (SQGs) in mg/kg dry wt

Element	This study mean		This study range		SQG non polluted	SQG moderately polluted	SQG heavily polluted
	Dry season	Rainy season	Dry season	Rainy season			
Cd	2.6	2.7	2.1–2.9	2.5–3.1	–	–	>6
Cr	27.2	24.4	16.2–43.7	18.3–29.4	<25	25–75	>75
Cu	5.1	3.4	1.6–15.3	2.1–6.2	<25	25–50	>50
Pb	153.8	124.2	124.4–256.4	101.4–133.7	<40	40–60	>60
Ni	23.6	20.6	15.6–36.2	14.6–24	<20	20–50	>50
Zn	25.8	20.2	4.0–110.0	4.8–38.2	<90	90–200	>200

Table 6 Concentrations of trace metals in River and reservoir sediment and its comparison with ERL-ERM ranges of SQGs (mg/kg, dry wt)

Element	ERL-ERM range	This study range	Percentage of sampling sites		
			< ERL-ERM range	Between ERL-ERM range	> ERL-ERM range
Dry season					
Cd	1.2–9.6	2.1–2.9	–	100	–
Cr	81–370	16.2–43.7	100	–	–
Cu	34–270	1.6–15.3	100	–	–
Pb	46.7–218	124.4–256.4	–	91	9
Ni	20.9–51.6	15.6–36.2	36	64	–
Zn	150–410	4.0–110.0	100	–	–
Rainy season					
Cd	1.2–9.6	2.5–3.1	–	100	–
Cr	81–370	18.3–29.4	100	–	–
Cu	34–270	2.1–6.2	100	–	–
Pb	46.7–218	101.4–133.7	–	100	–
Ni	20.9–51.6	14.6–24.0	55	45	–
Zn	150–410	4.8–38.2	100	–	–

Table 7 Geoaccumulation index (Igeo) values of trace elements in the sediment

	Dry season								Rainy season							
Sites	Cd	Cr	Cu	Fe	Pb	Mn	Ni	Zn	Cd	Cr	Cu	Fe	Pb	Mn	Ni	Zn
S1	2.47	−2.62	−3.57	−7.33	2.28	−1.59	−2.47	−3.02	2.53	−2.88	−5.01	−7.22	1.89	−1.21	−2.80	−3.79
S2	2.53	−2.25	−4.23	−5.38	2.16	−0.21	−1.96	−2.94	2.64	−2.20	−4.81	−6.00	1.93	0.26	−2.09	−2.85
S3	2.53	−2.55	−4.40	−5.98	2.15	−1.14	−2.30	−2.29	2.64	−2.62	−4.35	−7.02	2.03	−1.45	−2.30	−2.68
S4	2.69	−3.06	−5.40	−7.43	2.05	−1.77	−2.71	−4.35	2.78	−2.23	−3.91	−7.31	1.76	0.52	−2.14	−2.47
S5	2.58	−2.33	−2.14	−6.91	2.51	−0.93	−1.96	−1.26	2.53	−2.58	−3.67	−7.40	2.07	−1.73	−2.36	−2.83
S6	2.22	−1.63	−4.94	−7.02	3.09	0.05	−1.49	−4.78	2.47	−2.57	−3.44	−7.31	2.16	−1.59	−2.36	−2.20
S7	2.53	−2.36	−3.57	−7.20	2.35	−1.46	−2.23	−3.38	2.69	−2.53	−4.94	−7.06	2.14	−1.70	−2.27	−4.89
S8	2.53	−2.28	−3.70	−7.16	2.10	−1.11	−2.16	−5.15	2.53	−2.38	−4.76	−6.98	2.09	−1.24	−2.29	−1.90
S9	2.58	−2.68	−4.70	−7.33	2.11	−1.93	−2.33	−4.52	2.58	−2.56	−4.76	−7.18	2.15	−1.69	−2.37	−3.05
S10	2.47	−2.10	−4.06	−6.38	2.28	−0.26	−2.01	−3.51	2.58	−2.50	−4.06	−7.14	2.14	−1.56	−2.23	−2.97
S11	2.64	−2.07	−3.19	−7.02	2.51	−1.29	−1.96	−0.37	2.47	−2.21	4.81	−7.02	2.13	−1.22	−2.31	−3.31

Samuel reservoir, Little Akaki River contained a higher trace metal load than the other regions. From the result, it can be concluded that the catchment area have high influx of trace metals as a result of uncontrolled urbanization, industrialization, poor sanitation and uncontrolled waste disposal from municipal, industrial and agricultural sources in the upstream Addis Ababa city. However, relatively lower concentration of sediment-bound trace metals were recorded in the reservoir which might be due to less accumulated metals in the sediment because the reservoir was rehabilitated in 2016. Ecological risk assessment using the USEPA guideline, Igeo, CF and PLI revealed the widespread pollution by Cd and Pb. These were followed by Mn, Ni and Zn. Hence, high level of trace metals in sediments probably have adverse effects to the bottom dwelling aquatic organisms as well as to the health of the people who depend on the water for various activities. Therefore, strict policy measures are required to decrease the degree of contamination since some of the elements are known to be toxic to biota. Furthermore, regular monitoring of these pollutants in water, sediment and biota is recommended.

Table 8 Contamination factor (CF) and pollution load index (PLI) for trace elements in sediments

Sites	Contamination factor (CF)								PLI
	Cd	Cr	Cu	Fe	Pb	Mn	Ni	Zn	
Dry season									
S1	8.33	0.24	0.13	0.01	7.28	0.50	0.27	0.19	0.39
S2	8.67	0.32	0.08	0.02	6.71	1.30	0.39	0.20	0.48
S3	8.67	0.26	0.07	0.01	6.67	0.68	0.30	0.31	0.40
S4	9.67	0.18	0.04	0.01	6.22	0.44	0.23	0.07	0.27
S5	9.00	0.30	0.34	0.01	8.54	0.79	0.38	0.63	0.59
S6	7.00	0.49	0.05	0.01	12.82	1.55	0.53	0.05	0.41
S7	8.67	0.29	0.13	0.01	7.64	0.55	0.32	0.14	0.40
S8	8.67	0.31	0.12	0.01	6.43	0.70	0.34	0.04	0.34
S9	9.00	0.23	0.06	0.01	6.46	0.39	0.30	0.06	0.29
S10	8.33	0.35	0.09	0.02	7.27	1.25	0.37	0.13	0.47
S11	9.33	0.36	0.16	0.01	8.56	0.61	0.39	1.16	0.58
Rainy season									
S1	8.67	0.20	0.05	0.01	5.58	0.61	0.21	0.11	0.30
S2	9.33	0.33	0.05	0.02	5.72	1.80	0.35	0.21	0.47
S3	9.33	0.24	0.07	0.01	6.13	0.55	0.30	0.23	0.37
S4	10.33	0.32	0.10	0.03	5.07	2.16	0.34	0.27	0.56
S5	8.67	0.25	0.12	0.01	6.29	0.45	0.29	0.21	0.38
S6	8.33	0.25	0.14	0.01	6.69	0.50	0.29	0.33	0.42
S7	9.67	0.26	0.05	0.01	6.63	0.46	0.31	0.05	0.30
S8	8.67	0.29	0.06	0.01	6.37	0.64	0.31	0.40	0.41
S9	9.00	0.25	0.06	0.01	6.65	0.46	0.29	0.18	0.35
S10	9.00	0.26	0.08	0.01	6.64	0.51	0.32	0.19	0.37
S11	8.33	0.32	0.05	0.01	6.55	0.64	0.30	0.15	0.34

Abbreviations
ANOVA: analysis of variance; CF: contamination factor; CRM: certified reference material; ERL: effects range low; ERM: effects range median; GAR: Greater Akaki River; ICP-OES: inductively coupled plasma optical emission spectrometer; Igeo: geo-accumulation index; LAR: Little Akai River; LOD: limit of detection; LOQ: limit of quantification; PLI: pollution load index; SD: standard deviation; SQGs: sediment quality guidelines.

Authors' contributions
All authors have contributed at different stages of this study. ABK designed the study, collected and analyzed samples and interpreted the data. He also wrote the draft manuscript. SLA and TBE involved on the design of the study, supervised the progress and provide comments on the manuscript. JOO supervised the work and provided comments on the manuscript. All authors read and approved the final manuscript.

Author details
[1] Centre for Environmental Science, Addis Ababa University, P. O. Box 1176, Addis Ababa, Ethiopia. [2] Department of Chemistry, Debre Berhan University, P. O. Box 445, Debre Berhan, Ethiopia. [3] Central Research Laboratories, Addis Ababa Science and Technology University, P. O. Box 16417, Addis Ababa, Ethiopia. [4] Department of Environmental, Water & Earth Sciences, Tshwane University of Technology, 175 Nelson Mandela Drive, Arcadia, Private Bag X680, Pretoria, South Africa.

Acknowledgements
The first author would like to thank Addis Ababa University for financial support.

Competing interests
The authors declare that they have no competing interests.

Funding
The authors greatly acknowledge Addis Ababa University Vice President office for Research and Technology Transfer for Financial support via thematic research project.

References
Akele M, Kelderman P, Koning C, Irvine K (2016) Trace metal distributions in the sediments of the Little Akaki River, Addis Ababa, Ethiopia. Environ Monit Assess 188:389

Alemayehu T (2001) The impact of uncontrolled waste disposal on surface water quality in Addis Ababa, Ethiopia. SINET 24:93–104

Alemayehu T (2006) Heavy metal concentration in the urban environment of Addis Ababa, Ethiopia. Soil Sediment Contam 15:591–602

Arficho DF (2009) Status, Distribution, and Phytoavailability of Heavy Metals and Metalloids in Soils Irrigated with Wastewater from Akaki River, Ethiopia: Implications for environmental management of heavy metal/metalloid affected soils. Addis Ababa University, Addis Ababa

Aschale M, Sileshi Y, Kelly-Quinn M, Hailu D (2016) Evaluation of potentially toxic element pollution in the benthic sediments of the water bodies of the city of Addis Ababa, Ethiopia. J Environ Chem Eng 4:4173–4183

Aschale M, Sileshi Y, Kelly-Quinn M, Hailu D (2017) Pollution assessment of toxic and potentially toxic elements in agricultural soils of the city Addis Ababa, Ethiopia. Bull Environ Contam Toxicol 98:234–243

Awoke A, Beyene A, Kloos H, Goethals PL, Triest L (2016) River water pollution status and water policy scenario in Ethiopia: raising awareness for better implementation in developing countries. Environ Manag 58:694–706

Chakravarty M, Patgiri AD (2009) Metal pollution assessment in sediments of the Dikrong River, NE India. J Hum Ecol 27:63–67

Chen CW, Kao CM, Chen CF, Dong CD (2007) Distribution and accumulation of heavy metals in the sediments of Kaohsiung Harbor, Taiwan. Chemosphere 66:1431–1440

Demlie M, Wohnlich S (2006) Soil and groundwater pollution of an urban catchment by trace metals: case study of the Addis Ababa region, central Ethiopia. Environ Geol 51:421–431

Demlie M, Wohnlich S, Wisotzky F, Gizaw B (2007) Groundwater recharge, flow and hydrogeochemical evolution in a complex volcanic aquifer system, central Ethiopia. Hydrogeol J 15:1169–1181

Demlie M, Wohnlich S, Ayenew T (2008) Major ion hydrochemistry and environmental isotope signatures as a tool in assessing groundwater occurrence and its dynamics in a fractured volcanic aquifer system located within a heavily urbanized catchment, central Ethiopia. J Hydrol 353:175–188

Dhanakumar S, Solaraj G, Mohanraj R (2015) Heavy metal partitioning in sediments and bioaccumulation in commercial fish species of three major reservoirs of river Cauvery delta region, India. Ecotoxicol Environ Saf 113:145–151

Diop C, Dewaelé D, Cazier F, Diouf A, Ouddane B (2015) Assessment of trace metals contamination level, bioavailability and toxicity in sediments from Dakar coast and Saint Louis estuary in Senegal, West Africa. Chemosphere 138:980–987

Dirbaba NB, Xue Y, Wu H, Wang J (2018) Occurrences and Ecotoxicological Risk Assessment of Heavy Metals in Surface Sediments from Awash River Basin, Ethiopia. Water 10:535

El Nemr A, El-Said GF, Ragab S, Khaled A, El-Sikaily A (2016) The distribution, contamination and risk assessment of heavy metals in sediment and shellfish from the Red Sea coast, Egypt. Chemosphere 165:369–380

Faisal B, Majumder RK, Uddin MJ, Abdul M (2014) Studies on heavy metals in industrial effluent, river and groundwater of Savar industrial area, Bangladesh by principal component analysis. Int J Geomatics Geosci 5:182–191

Gao X, Chen CTA (2012) Heavy metal pollution status in surface sediments of the coastal Bohai Bay. Water Res 46:1901–1911

Gizaw B (2002) Hydrochemical and environmental investigation of the Addis Ababa region, Ethiopia. Unpublished PhD Thesis, Ludwig Macmillan, University of Munich, Munich

Gizaw E, Legesse W, Haddis A, Deboch B, Birke W (2004) Assessment of factors contributing to eutrophication of Aba Samuel Water reservoir in Addis Ababa, Ethiopia. Ethiop J Health Sci 14:112–223

Hakanson L (1980) An ecological risk index for aquatic pollution control. A sedimentological approach. Water Res 14:975–1001

Halli M, Sari E, Kurt MA (2014) Assessment of arsenic and heavy metal pollution in surface sediments of the Ergene River, Turkey. Pol J Environ Stud 23:1581

Igwe J, Abia A (2006) A bioseparation process for removing heavy metals from waste water using biosorbents. Afr J Biotechnol 5:11

Iqbal J, Shah MH (2014) Occurrence, risk assessment, and source apportionment of heavy metals in surface sediments from Khanpur Lake, Pakistan. J Anal Sci Technol 5:28

Itanna F (2002) Metals in leafy vegetables grown in Addis Ababa and toxicological implications. Ethiop J Health Dev 16:295–302

Kishe M, Machiwa J (2003) Distribution of heavy metals in sediments of Mwanza Gulf of Lake Victoria, Tanzania. Environ Int 28:619–625

Li N, Tian Y, Zhang J, Zuo W, Zhan W, Zhang J (2017) Heavy metal contamination status and source apportionment in sediments of Songhua River Harbin region, Northeast China. Environ Sci Poll Res 24:3214–3225

Liang J, Liu J, Yuan X, Zeng G, Lai X, Li X, Wu H, Yuan Y, Li F (2015) Spatial and temporal variation of heavy metal risk and source in sediments of Dongting Lake wetland, mid-south China. J Environ Sci Health 50:100–108

Long ER, Macdonald DD, Smith SL, Calder FD (1995) Incidence of adverse biological effects within ranges of chemical concentrations in marine and estuarine sediments. Environ Manag 19:81–97

Loska K, Wiechuła D, Korus I (2004) Metal contamination of farming soils affected by industry. Environ Int 30:159–165

Macdonald DD, Carr RS, Calder FD, Long ER, Ingersoll CG (1996) Development and evaluation of sediment quality guidelines for Florida coastal waters. Ecotoxicology 5:253–278

Mehlich A (1978) New extractant for soil test evaluation of phosphorus, potassium, magnesium, calcium, sodium, manganese and zinc. Commun Soil Sci Plant Anal 9:477–492

Mekonnen KN, Ambushe AA, Chandravanshi BS, Abshiro MR, McCrindle RI, Panichev N (2012) Distribution of mercury in the sediments of some freshwater bodies in Ethiopia. Toxicol Environ Chem 94:1678–1687

Mekonnen KN, Ambushe AA, Chandravanshi BS, Redi-Abshiro M, McCrindle RI (2015) Occurrence, distribution, and ecological risk assessment of potentially toxic elements in surface sediments of Lake Awassa and Lake Ziway, Ethiopia. J Environ Sci Health 50:90–99

Melaku S, Wondimu T, Dams R, Moens L (2007) Pollution status of Tinishu Akaki River and its tributaries (Ethiopia) evaluated using physico-chemical parameters, major ions, and nutrients. Bull Chem Soc Ethiop 21:13–22

Mohiuddin K, Alam M, Ahmed I, Chowdhury A (2015) Heavy metal pollution load in sediment samples of the Buriganga river in Bangladesh. J Bangladesh Agril Univ 13:229–238

Muller G (1981) The heavy metal pollution of the sediments of Neckars and its tributary: a stocktaking. Chem Ztg 105:157–164

Mwanamoki PM, Devarajan N, Thevenon F, Birane N, de Alencastro LF, Grandjean D, Mpiana PT, Prabakar K, Mubedi JI, Kabele CG (2014) Trace metals and persistent organic pollutants in sediments from river-reservoir systems in Democratic Republic of Congo (DRC): spatial distribution and potential ecotoxicological effects. Chemosphere 111:485–492

Olivares-Rieumont S, de la Rosa D, Lima L, Graham DW, Katia D, Borroto J, Martínez F, Sánchez J (2005) Assessment of heavy metal levels in Almendares River sediments—Havana City, Cuba. Water Res 39:3945–3953

Prasse C, Zech W, Itanna F, Glaser B (2012) Contamination and source assessment of metals, polychlorinated biphenyls, and polycyclic aromatic hydrocarbons in urban soils from Addis Ababa, Ethiopia. Toxicol Environ Chem 94:1954–1979

Shanbehzadeh S, Vahid Dastjerdi M, Hassanzadeh A, Kiyanizadeh T (2014) Heavy metals in water and sediment: a case study of Tembi River. J Environ Public Health 2014:858720

Singh M (2001) Heavy metal pollution in freshly deposited sediments of the Yamuna River (the Ganges River tributary): a case study from Delhi and Agra urban centres, India. Environ Geol 40:664–671

Tang W, Shan B, Zhang H, Zhu X, Li S (2016) Heavy metal speciation, risk, and bioavailability in the sediments of rivers with different pollution sources and intensity. Environ Sci Poll Res 23:23630–23637

Turekian KK, Wedepohl KH (1961) Distribution of the elements in some major units of the earth's crust. Geol Soc Am Bull 72:175–192

Usero J, Gonzalez-Regalado E, Gracia I (1997) Trace metals in the bivalve molluscs Ruditapes decussatus and Ruditapes philippinarum from the Atlantic Coast of Southern Spain. Environ Int 23:291–298

Wang L, Wang Y, Zhang W, Xu C, An Z (2014) Multivariate statistical techniques for evaluating and identifying the environmental significance of heavy metal contamination in sediments of the Yangtze River, China. Environ Earth Sci 71:1183–1193

Woldetsadik D, Drechsel P, Keraita B, Itanna F, Gebrekidan H (2017) Heavy metal accumulation and health risk assessment in wastewater-irrigated urban vegetable farming sites of Addis Ababa, Ethiopia. Int J Food Contam 4:9

Xu D, Wang Y, Zhang R, Guo J, Zhang W, Yu K (2016) Distribution, speciation, environmental risk, and source identification of heavy metals in surface

Distribution and ecological risk assessment of trace metals in surface sediments from Akaki River...

137

sediments from the karst aquatic environment of the Lijiang River, Southwest China. Environ Sci Poll Res 23:9122–9133

Yohannes YB, Ikenaka Y, Saengtienchai A, Watanabe KP, Nakayama SM, Ishizuka M (2013) Occurrence, distribution, and ecological risk assessment of DDTs and heavy metals in surface sediments from Lake Awassa—Ethiopian Rift Valley Lake. Environ Sci Poll Res 20:8663–8671

Yousaf B, Liu G, Wang R, Imtiaz M, Zia-Ur-rehman M, Munir MA, Niu Z (2016) Bioavailability evaluation, uptake of heavy metals and potential health risks via dietary exposure in urban-industrial areas. Environ Sci Poll Res 23:22443–22453

Zhang D, Zhang X, Tian L, Ye F, Huang X, Zeng Y, Fan M (2013) Seasonal and spatial dynamics of trace elements in water and sediment from Pearl River Estuary, South China. Environ Earth Sci 68:1053–1063

10

Identification of two low-cost and locally available filter media (pumice and scoria) for removal of hazardous pollutants from tannery wastewater

Mekonnen Birhanie Aregu[*], Seyoum Leta Asfaw and Mohammed Mazharuddin Khan

Abstract

Background: Tannery wastewater contains the most hazardous pollutants. Therefore, identifying potentially efficient, low-cost and locally available filter media as an adsorbent for the treatment of tannery wastewater is critical. The aim of this study is to identify and assess the ability of identified adsorbents and compare their efficiency. The volcanic rocks of pumice and scoria were collected from the rift valley area of Oromia region, Ethiopia and their chemical characteristics were determined using X-ray fluorescence analysis. Batch mode experimental study design was carried out. The rocks were crushed and effective size was determined by using a standard sieve. The composite tannery wastewater was collected from Dire tannery, Addis Ababa, Ethiopia and treated with pumice and scoria. Two adsorption kinetics and isotherm models were conducted to predict the removal mechanism and capacity of the adsorbents on the reduction of NO_3–N, PO_4–P and $[Cr]_T$ from tannery wastewater. Analysis of wastewater samples was done before and after different retention time. R statistical software and Originlab pro 2017 was run for data analysis and graphing.

Results: The untreated tannery wastewater revealed that the mean concentration of BOD_5, COD, TSS, orthophosphate, ammonium, nitrite, nitrate, sulfide, sulfate and chromium were beyond the permissible limits. Nitrate removal efficiency of scoria and pumice were 99 and 95% respectively at retention time of 72 h. Phosphate removal was better by scoria on the first 24 and 48 h. The efficiency of pumice to remove sulfate was 83–84%, whereas scoria shows 75–77%. In the first 24 and 48 h retention time, pumice and scoria achieved 76 and 71% in chromium reduction respectively.

Conclusion: This study revealed that both scoria and pumice have a potential capacity to treat tannery wastewater. Conversely comparing the average efficiency to reduce hazardous pollutants scoria showed better results than pumice.

Keywords: Tannery wastewater, Pumice, Scoria, Filter media, Removal efficiency

Background

Industrial wastes are usually generated from different industrial processes, as a result the amount and toxicity of waste released from industrial activities varies with the type of industrial processes. Among all the industrial wastewater tannery wastewaters are the most source of

hazardous pollutants (Shen 1999). Tannery wastewater contains the most hazardous pollutants of industry. Major problems caused by tannery wastewater containing heavy metals, nutrients, toxic chemicals, chloride, lime with high dissolved and suspended salts, and other pollutants. In developing countries, many industrial units are operating in a small and medium scale. These industrial units can generate a considerable pollution load by discharging untreated or partially treated effluents directly into the nearby environment (Asfaw 2014).

*Correspondence: mekonnen.birhanie@aau.edu.et
Centre for Environmental Sciences, College of Natural Science, Addis Ababa University, Addis Ababa, Ethiopia

In Ethiopia currently, there are more than 30 tannery industries in operation. Among them the majority found in the Oromia region, especially Modjo town and around six established in the capital city Addis Ababa. These tanneries have 153,650 sheep and goat skin soaking capacity and 9725 cowhides soaking capacity per day together they also employ 4577 persons (UNIDO 2012).

The total wastewater discharge estimation from tanneries is about 400 million m³/year in Ethiopia. About 90% of world leather production use chrome-tanning processes rather than vegetable tanning. In Chrome tanning process tanneries utilize chromium in the form of basic chromium-sulphate for hide stabilization against microbial degradation and provision of flexibility of the leather. In chrome tanning process about 60–80% of chromium reacts with the hides and about 20–40% of the chromium amount remaining in the solid and liquid wastes (Rezic and Zeiner 2008).

Tanneries generate wastewater in the range of 30–35 L/ kg of skin or hide processed with variable pH and high concentrations of suspended solids, BOD and COD. Major problems are due to wastewater containing heavy metals, toxic chemicals, chloride, lime with high dissolved and suspended salts and other pollutants (Durai et al. 2011). Hexavalent chromium from tannery wastewater is one of the major concerns of environmental pollution. This is due to discharge of tannery wastewater in large quantities without or with partial treatment (Lofrano et al. 2008).

Developing countries face numerous challenges related to preserving the environment from industrial wastewater pollution. Like many other developing countries, Ethiopia also grieves from environmental pollution problems of wastewater particularly industrial wastewater. This issue seems to be a subject which has not yet received adequate attention during the development of industries. Therefore, there is a need to develop efficient and low-cost wastewater treatment technologies for the removal of heavy metals and other pollutants. Among these technologies, adsorption and filtration is a user-friendly technique for this purpose.

In filtration technique, wastewater containing suspended matter is added to the top of the filter medium as the wastewater filters through the porous medium, the suspended matter in the wastewater is removed by a different of mechanisms. These mechanisms are straining, sedimentation, impaction, interception, adhesion, adsorption, flocculation and biological degradation especially for organic removals on the top of the filter medium. A natural characteristics of filter medium are important in the pollutant removal performance. Some of them are effective size, size distribution, slope, density and porosity (Boller and Kavanaugh 1995).

Adsorption on the other hand is the process of accumulation of a substance on the surface of another substance. When a solid surface is exposed to a water or wastewater, molecules from the solution phase accumulate or concentrate at the solid surface. It is recognized as one of the most effective wastewater purification technique used in several industries. The basic principle of adsorption is mass transfer and adsorption of a molecule from a liquid into a solid surface. It will happen if the pollutant has low solubility in the wastewater, greater affinity for the substrate than wastewater and the combination of the two (Rao et al. 2007).

Depending on the type of force of attractions between the pollutant and adsorbent, the adsorption process can be divided into two types. Physical adsorption or chemical adsorption. If the force of attraction between pollutant and adsorbent is weak that is a Vanderwaal force of attraction, the process is called physical adsorption or commonly known as physisorption. Physical adsorption takes place with the formation of multilayer of pollutant on the adsorbent. But if the force of attraction between pollutant and adsorbent is chemical forces of attraction or chemical bond, the process is called chemical adsorption or chemisorption. Chemisorption takes place with the formation of a single layer of pollutant on the adsorbent. In general, there are factors influencing adsorption such as surface area, nature of the pollutant, hydrogen ion concentration (pH) of the wastewater, temperature, mixed solutes and nature of adsorbent (Dabrowski 2001).

Adsorption has been identified as one of the most promising mechanism for removal of dissolved heavy metal fractions and nutrients from wastewater. Although commercial adsorbents are available for use in adsorption, they are very expensive, resulting in various new low-cost adsorbents being studied by researchers. (Babel and Kurniawan 2003), reviewed the technical feasibility of various low-cost adsorbents for heavy metals removal from wastewater and concluded that the use of low-cost adsorbents may contribute to the sustainability of the surrounding environment and offer promising benefits for commercial purpose in the future.

Therefore, identifying potentially efficient, low-cost and locally available filter media as an adsorbent is critical for proper practice of environmental management by tanning industries. On the other hand ordinary sand for filter media is costly because of construction-expansion in the country, not available readily and not efficient in the removal of hazardous pollutants by adsorption hence there is a need to substitute pumice and scoria instead of sand filtration.

Pumice is a light, porous volcanic rock that forms during explosive eruptions (Fig. 1). It resembles a sponge as it consists of a network of gas bubbles frozen amidst

Fig. 1 Typical pumice: photograph Mekonnen Birhanie March/2016, Ethiopia

fragile volcanic glass and minerals. All types of magma (basalt, andesite, dacite, and rhyolite) will form pumice, however it is most commonly formed from rhyolite. During an explosive eruption, volcanic gases dissolved in the liquid portion of magma also expand rapidly to create a foam or froth; in the case of pumice, the liquid part of the froth quickly solidifies to glass around the glass bubbles. Pumice is considered a glass because it has no crystal structure. Like many of the materials considered in this report, it is an aluminosilicate (Akbal et al. 2000).

Studies have shown using substrate rich in iron (Fe), aluminum (Al) or calcium (Ca) concentrations enhances phosphate removal in experimental subsurface wetlands beyond that which can be achieved by using native soils (Arias et al. 2001). It was believed that sedimentation of particulate phosphorus and sorption of soluble phosphorus (onto the pumice) were responsible, and that the high concentrations of iron (18.2%), aluminum (13.7%), calcium (12.7%) and magnesium (7.3%) in the pumice were the source of this high sorption ability. The low specific gravity and high porosity of pumice make it important for a number of applications in water and wastewater

treatment processes. Pumice was used as a filter medium and as a support material for microbial growth in water and wastewater treatment (Farizoglu et al. 2003).

The other volcanic ash is scoria generally denser than the pumice. Scoria is somewhat porous material with high surface area and strength with density larger than one. Scoria is an excellent medium which holds water in its pores and allow air circulation to the root zone of the plant. Both pumice and scoria are widely available in the Rift valley area of Ethiopia.

Scoria is bomb-sized, generally vesicular pyroclastic rock with basaltic composition, which is reddish brown to black in color and is of low density (Fig. 2). It has been used in several industrial applications, such as the manufacturing of a lightweight concrete mixture, a heating-insulating material, low-cost fillers in paints, and Sorbents (Moufti et al. 2000; Alemayehu and Lennartz 2009). Scoria is abundant in many places worldwide including Central America, Southeast Asia (Vietnam, etc.), East Africa (Ethiopia, Kenya, etc.), and Europe (Greece, Italy, Spain, Turkey, etc.) (Kwon et al. 2005; Alemayehu and Lennartz 2009).

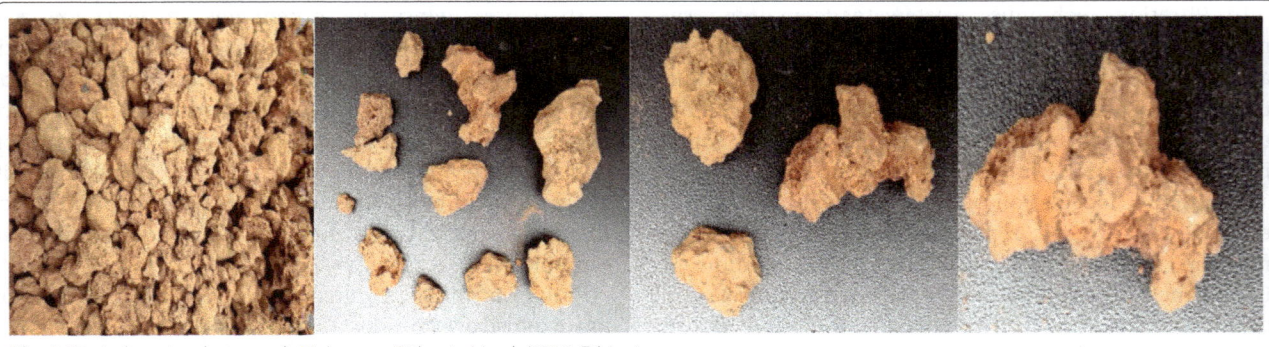

Fig. 2 Typical scoria: photograph Mekonnen Birhanie March/2016, Ethiopia

Sorption of contaminants onto scoria mainly takes place at the outside surface at the initial stage. Changes of ionic composition during sorption experiments suggest that cation exchange is likely the dominant mechanism of heavy metals sorption onto scoria, while considerable As(III) removal by scoria is explained by specific sorption of the neutral As(III) species and electrical adsorption of negatively charged As(V) species via As oxidation onto hematite (Kwon et al. 2010). The experimental investigation conducted demonstrates that the scoria is able to concurrently reduce concentrations of heavy metals and arsenic in aqueous solutions. Kwon et al. (2010), recommend that scoria can be used as an economic and efficient Sorbent to treat contaminated water with heavy metals.

Taking into account the growth of industrialization in Ethiopia and the expected demand for industrial wastewater management, low-cost, appropriate and eco-friendly approaches will play a critical role in the development of future wastewater treatment technology in the country. In this practical approach, this work deal with the principles of adsorption and filtration for the removal of hazardous pollutants from tannery wastewater by identifying these two volcanic ashes (scoria and pumice) as a filter media instead of conventional sand.

Methods

Study area and period
This study has been conducted in Addis Ababa University by transporting sample wastewater from Dire tanning industry located at Kolfe Keranio sub city Addis Ababa Ethiopia from May to August 2016.

Study design
Batch mode comparative experimental study design has been carried out to assess the ability and determine the efficiency of scoria and pumice filter media on removal of hazardous pollutants by the treatment of industrial wastewater, the case of tannery wastewater filtration.

Materials, experimental, design and set-up establishments
The volcanic rocks were collected from volcanic cones in the rift valley area of Oromia region, East Shewa Zone, Ethiopia, (pumice collected from the area lies in between: 8°28′36″N and 39°14′29″E; scoria collected from the area lies in between: 8°35′47″N and 39°08′45″E) approximately 50–100 km East of Addis Ababa (Fig. 3).

The rocks are local volcanic rocks with various chemical and mineralogical structure and transported to Addis Ababa University. The chemical characteristics of both filter media were determined by XRF analysis (Table 1).

These filter media were crushed and graded. The effective size was determined by using a standard sieve. Based on the analysis the effective size (ES) (d_{10}) of each media was 1.5–4.5 and the uniformity coefficient (UC) (d_{60}/d_{10}) is 3.5–4. After grading the filter materials were washed with tap water and dry in sunlight for 1 week.

Two filtration tanks were made of metal sheet, each with the following dimension, 60 cm height and 28 cm diameter and also were fitted with a half-inch an outlet tap (faucet) 5 cm above from the bottom of each tank. The filtration tanks were installed at College of natural and computational science, Addis Ababa University. After installation the filter media were filled in the filtration tank 10 cm depth with 10–25 mm grain size, drainage layer at the bottom, 30 cm depth filter layer with a grain size of 1.5–4.5 mm in the middle and the distribution layer (flat coarse gravel) was added 5 cm depth at the top of the filter media to protect erosion of filter's top layers, then it is ready for sample tannery wastewater filtration (Fig. 4).

Wastewater sample collection and filtration
The composite sample tannery wastewater was collected from Dire tannery and transported to Addis Ababa University in 40 L plastic 'Jerican'for each filter tank. The onsite measurement of the physicochemical parameters was undertaken. The collected raw tannery wastewater was added to the two filtration tank at the time and a sample also transported to Addis Ababa EPA water and wastewater analysis laboratory and Ethiopian Construction Design and Supervision Works Corporation Research Laboratory (ECDSWCRL) for the raw tannery wastewater characterization.

Filtrated sample collection and laboratory analysis
The physicochemical analysis of wastewater samples has been done before and after the treatment with the two filter media, using standard methods (APHA 2005). Optimum operating treatment time was determined for maximum removal of these impurities by running the experiment for 24, 48 and 72 h, respectively. Filtrated samples were taken by 2 L plastic bottle after each fixed retention time that is over 24, 48, and 72 h and transported to Addis Ababa EPA water and wastewater analysis laboratory and ECDSWCRL after taking each sample.

The analytical parameters were pH, DO, BOD_5, COD, TSS, ammonium N, nitrite N, nitrate N, phosphate, sulfide, sulfate and chromium. On-site measurement of the wastewater like temperature, pH and DO were carried out at the site in the tannery environmental quality control laboratory using a portable pH meter (Wagtech International N374, M128/03IM, USA) and DO meter (Hach P/N HQ30d, Loveland. CO, USA).

COD, ammonium–nitrogen, nitrite–nitrogen, nitrate–nitrogen, phosphate, sulfide and sulfate were measured

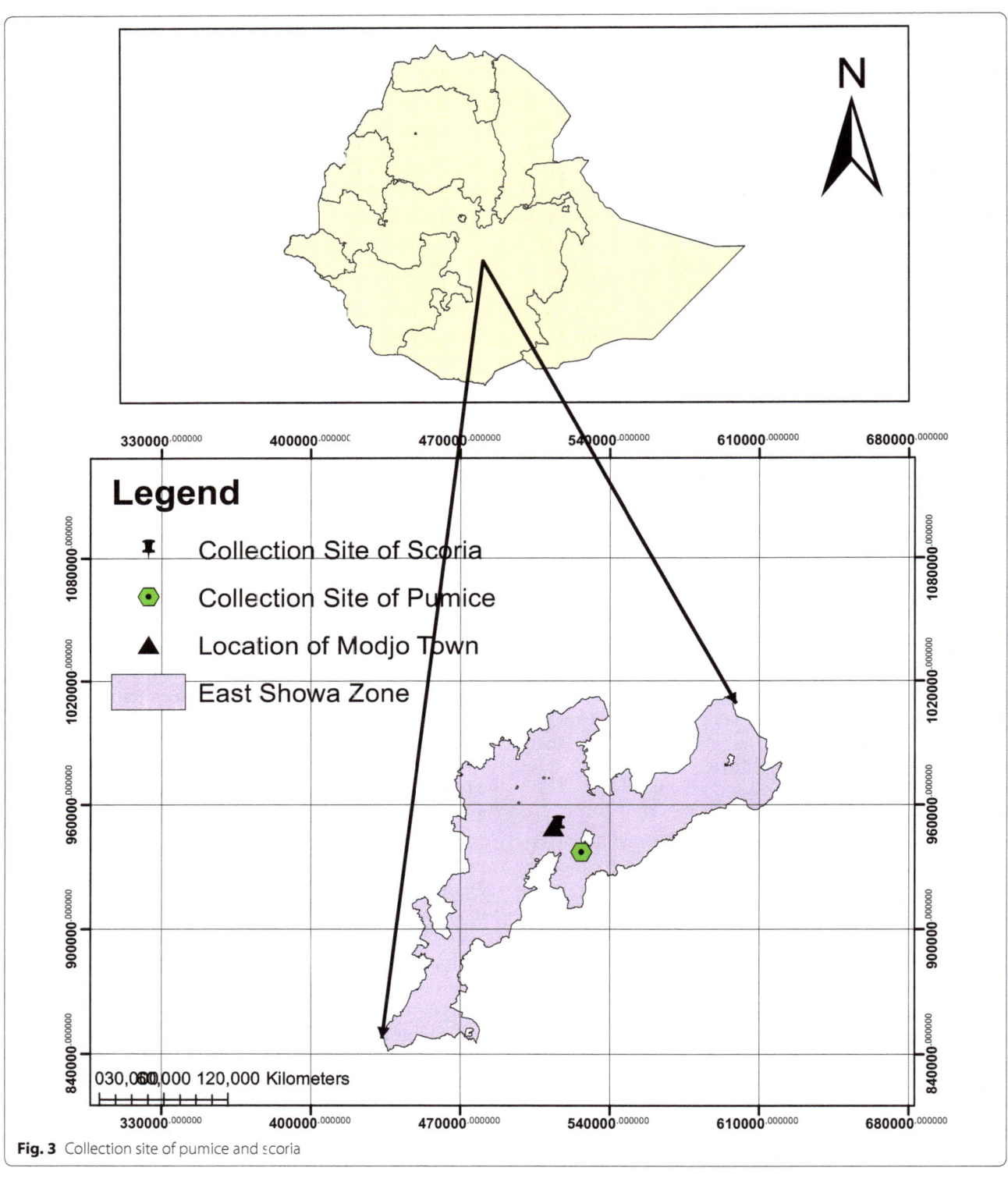

Fig. 3 Collection site of pumice and scoria

by using a spectrophotometer (Hach model DR/3900 portable spectrophotometer, Germany) according to Hach instructions. BOD_5 and total Cr were analyzed using BOD sensor and inductive stirring system AQUA LYTIC model type ET618-4 and flame atomic absorption spectrophotometer (AAS), (model AAS NOUA-400,

Table 1 Physical and chemical characteristics of scoria and pumice

Chemical composition	Percent weight	
	Scoria	Pumice
SiO_2	52.46	64.92
Al_2O_3	18.14	10.82
Fe_2O_3	5.40	4.62
CaO	9.40	5.74
K_2O	0.20	4.26
Na_2O	3.28	4.92
MgO	7.44	0.92
MnO	0.12	0.20
P_2O_5	0.36	0.14
TiO_2	0.41	0.15
H_2O	1.12	0.85
LOI	2.08	3.67
pH	7.81	7.53
[a]Physical properties, particle size $= 0.075$–0.425 mm		
Porosity (%)	36	73
Particle density (gcm^{-3})	2.96	2.33
Specific surface area (BET) (m^2 g^{-1})	2.49	3.5
Cation exchange capacity (CEC), mequiv. 100 g^{-1}	0.09	0.84

LOI loss on ignition

[a] Alemayehu et al. (2011)

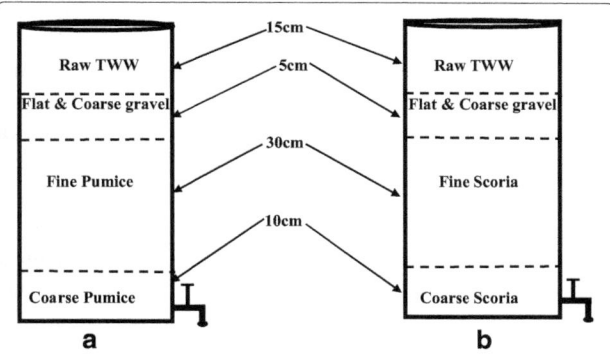

Fig. 4 Schematic layout of the filter tanks **a** components of pumice, **b** components of scoria

Germany) respectively. Total suspended solids (TSS) were determined according to the standard methods for the examination of water and wastewater gravimetric method (APHA 2005).

The amount of chromium from the wastewater adsorbed onto the filter medium at any time, q_t, was calculated as Eq. (1).

$$q_t = \frac{V(C_0 - C_t)}{W} \qquad (1)$$

At equilibrium, $q_t = q_e$ and $C_t = C_e$; therefore the amount of adsorbed chromium, q_e, was calculated by Eq. (2).

$$q_e = \frac{V(C_0 - C_e)}{W} \qquad (2)$$

where C_0, C_t and C_e are the initial concentration, concentration at any time and equilibrium concentrations of chromium in the wastewater (g/L), respectively, V is the volume of the wastewater (L), and W is the mass of the filer medium (kg) (Elmorsi 2011).

The removal efficiency of the filter medium for the selected parameters including chromium were calculated as Eq. (3).

$$\text{Removal\%} = \frac{(C_0 - C_t)}{C_0} \times 100 \qquad (3)$$

where C_0 is the parameter concentration in the untreated wastewater and C_t is the parameter concentration in the treated wastewater at the final hydraulic retention time t.

Statistical data analysis

Mean and standard deviations were calculated to estimate the concentration of each parameter of the samples. The hypothesis has been tested by student t test using R statistical software: R version 3.2.2 (2015-08-14), Platform: x86_64-w64-mingw32/x64 (64-bit) to determine whether an observed difference between the means of the groups is statistically significant or not, based on the treatment efficiency of the filter materials. OriginPro 2017 64-bit was used also for graphing and plotting for adsorption kinetics and isotherms model analysis.

Data quality management

To assure quality of the data by minimizing the errors the following measures had been undertaken: Apparatuses were calibrated; expiry date of reagents had been checked before starting the real analysis and standard control also prepared. Each test had been triplicated.

Result and discussion

Physicochemical characteristics of Dire tannery wastewater

The raw wastewater was taken from Dire tannery around Kolfe Keranio Sub city Asko area, Addis Ababa, Ethiopia and transported to the Addis Ababa EPA laboratory and ECDSWCRL for physicochemical analysis. Based on this investigation the mean concentrations of selected physicochemical parameters were presented in Table 2.

This study revealed that the mean concentration of BOD_5, COD and TSS were 1081 ± 160, $12{,}913 \pm 6875$ and 2426 ± 515.20 mg/L respectively (Table 2). This result is basically similar to different studies in Ethiopia with slight differences for different parameters, for example a study done at Modjo tannery indicated that

Table 2 Characteristics of Dire tannery wastewater

SNo	Parameter	Concentration (mg/L) except pH and T°	Range
1	pH	9.1 ± 3.10	6.5–12.50
2	T° (°C)	20.6 ± 2.30	19–22
3	BOD$_5$	1081 ± 160	924–1243
4	COD	12,913 ± 6875	8046–21,025
5	TSS	2426 ± 515.20	1849–2840
6	NH$_4$–N	314 ± 60	259–378
7	NO$_2$–N	1.7 ± 0.30	1.4–1.99
8	NO$_3$–N	124 ± 13	110–135
9	PO$_4$–P	168 ± 74	112–252
10	Sulfide	417 ± 131	334–568
11	Sulfate	1307 ± 224	1118–1555
12	Total chromium	35.7 ± 90	28–45

the mean concentration of COD was laid between 7950 and 15,240 mg/L with the mean of 11,123 ± 563.90 mg/L (Leta et al. 2003). Another study also undertaken with same tannery wastewater showed that the mean concentration of BOD$_5$ was 1054 ± 448 mg/L (Tadesse and Seyoum 2015). But the concentration of total suspended solid was found to range from 1849 to 2840 mg/L (Table 2) this result is a bit greater than some studies, for instance a study done in India indicated 1244 mg/L (Mandal et al. 2010).

Nutrients like orthophosphate, ammonium, nitrite and nitrate concentration of Dire tannery were characterized in this study, the result revealed that 168 ± 74, 314 ± 60, 1.7 ± 0.30, 124 ± 13 mg/L respectively. This result is comparable to a study done by Sivakumar et al. (2015) which indicates the concentration of nitrate in untreated tannery effluent as 116 mg/L. The result of ammonium is in the range of the results done at Bahir Dar tannery wastewater characterization (96–420 mg/L) (Wosnie and Wondie 2014). According to Sugasini and Rajagopal (2015), the nitrite concentration of untreated tannery wastewater was 1.3 mg/L almost parallel to this study finding which accounts 1.7 ± 0.30 mg/L (Table 2). Whereas the concentration of orthophosphate in this study was 168 ± 74 mg/L, this result shows that the concentration of phosphate in Dire tannery wastewater is higher than other study results done previously to characterize another tannery wastewaters. The variation may be due to the utilization of phosphorus containing chemicals for different purposes and tanning activities in Dire tannery.

The total suspended solids in Dire tannery found to be 2426 mg/L this result is more or less similar with the results of tannery wastewater analyzed by Banuraman and Meikandaan (2013). The concentration level of both

sulfide and sulfate were 417 ± 131 and 1307 ± 224 mg/L respectively. In this case the amount of sulfide found in this study wastewater was more or less equivalent to study done by Islam et al. (2014) that is 380 ± 50 mg/L. Sugasini and Rajagopal (2015), also characterize the tannery wastewater based on their result the concentration of sulfate was 1517 mg/L which is almost parallel to this investigation. In terms of chromium concentration, Dire tannery comprised 35.7 ± 90 mg/L is similar to other different results presented from various tannery wastewaters in Ethiopia for example a study done by Leta et al. (2003) indicates 32.2 ± 5.70 mg/L. On the other hand two more study results showed, the chromium concentration lied to be in the ranges of this investigation 28–45 mg/L as in Table 2 (Tadesse and Seyoum 2015; Asaye 2009).

Even though Wastewater of each tannery process consists of varying pH and temperature values, this study results (9.1 ± 3.10 and 20.6 ± 2.34 °C) respectively were analogous to different studies. Likewise a large variation exists in values of physicochemical parameters in general like BOD$_5$, COD, TSS, phosphate, sulfide, sulfate, etc. in every tannery wastewater characteristics, this may be because of different tanning process, methods, technology and raw material utilization by various tanning industries.

Adsorption kinetics for the pollutant removal process of tannery wastewater treatment using pumice and scoria filter medium

From the experimental data three selected pollutant parameters were tested to fit into different kinetic models for the adsorption rate, the removal mechanism and process and predict information about the interaction between these two naval adsorbents (filter medium) and those selected three pollutants (NO$_3$–N, PO$_4$–P and [Cr]$_T$) in the tannery wastewater (Martins et al. 2013). In this study, two models were used, these are the pseudo-first-order (Lagergren 1898) and the pseudo-second-order kinetics model (Ho and McKay 1999).

Pseudo-first-order equation

Pseudo-first-order equation was given first by Lagergren (1898) to determine the rate constant of adsorption process as Eq. (4).

$$\log (q_e - q_t) = \log q_e - \left(\frac{k_1}{2.303}\right) t \qquad (4)$$

where q_e and q_t are the amounts of the pollutants adsorbed (g/kg) at equilibrium and at time t (h), respectively and k_1 is the rate constant of adsorption (h^{-1}). Values of k_1 and q_e were calculated from the slope and the intercept of the plots of $\log (q_e - q_t)$ versus t for NO$_3$–N, PO$_4$–P and [Cr]$_T$ adsorption from the tannery wastewater onto pumice and scoria shown in Figs. 5a and 6a respectively.

The results in Tables 3 and 4 show that the values of correlation coefficients (R^2) were high for all parameters adsorption on to both adsorbents except chromium adsorption onto scoria (Table 4) and the experimental q_e value agree with the calculated value with very slight difference for all parameters except PO_4–P and $[Cr]_T$ adsorption onto scoria (Tables 3, 4). Therefore, this study revealed that the adsorption of the three pollutants onto both adsorbents fit with the pseudo-first-order

kinetic model except $[Cr]_T$ does not fit only onto scoria substrate.

Pseudo-second-order equation

Pseudo-second-order equation formulated from the adsorption equilibrium (Ho and McKay 1999). The equation can be expressed in Eq. (5):

$$\frac{d_q}{d_t} = k_2(q_e - q_t)^2 \tag{5}$$

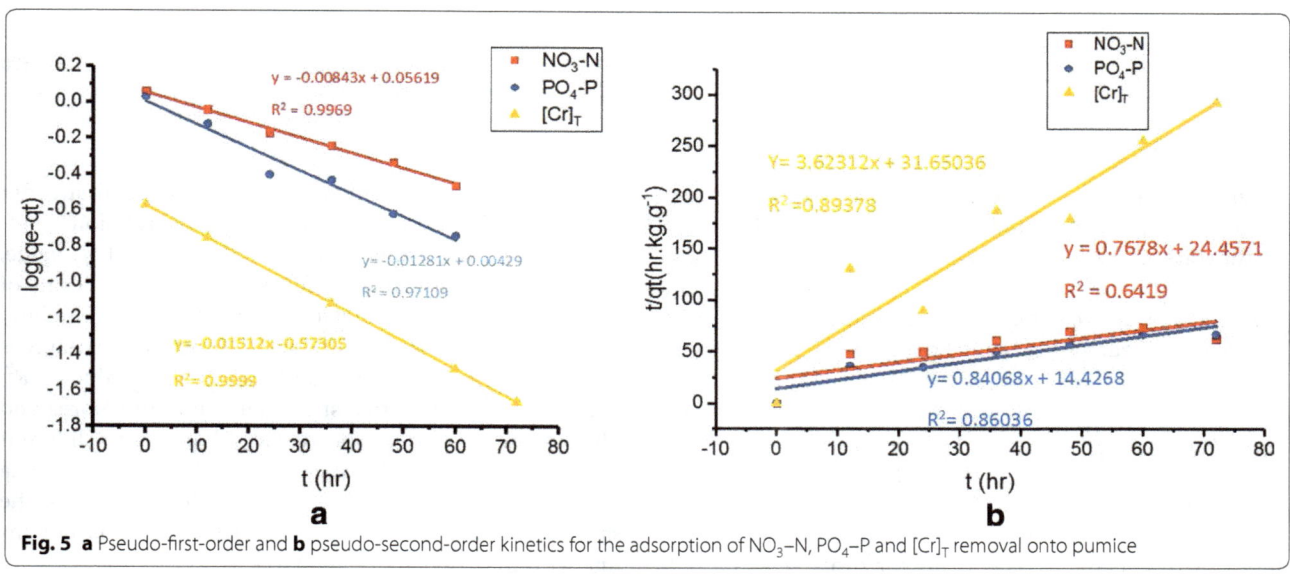

Fig. 5 **a** Pseudo-first-order and **b** pseudo-second-order kinetics for the adsorption of NO_3–N, PO_4–P and $[Cr]_T$ removal onto pumice

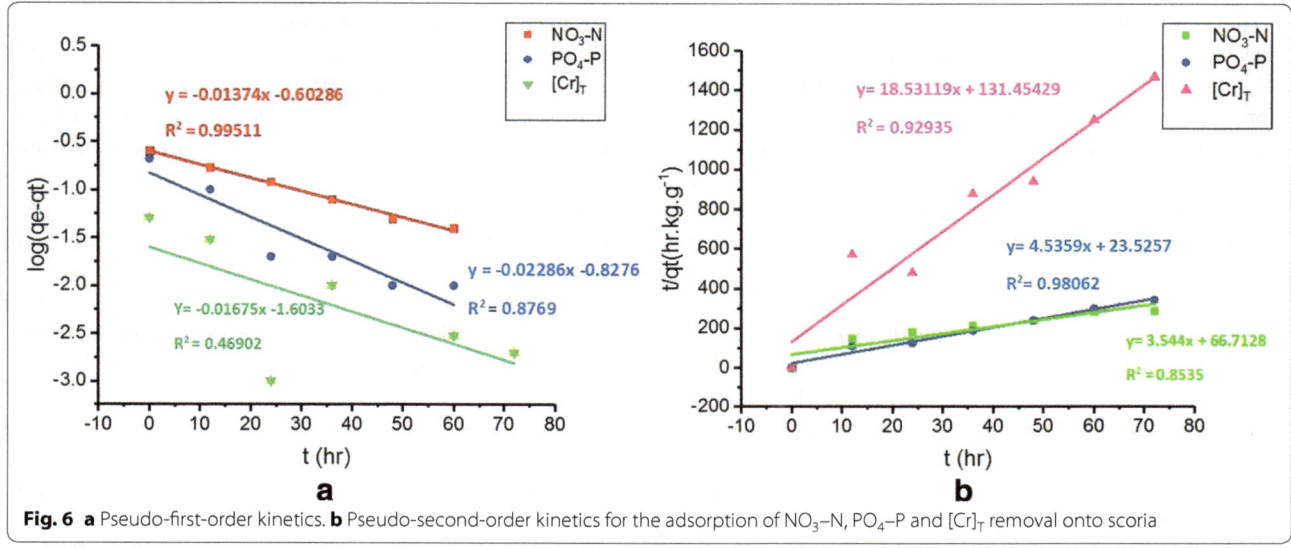

Fig. 6 **a** Pseudo-first-order kinetics. **b** Pseudo-second-order kinetics for the adsorption of NO_3–N, PO_4–P and $[Cr]_T$ removal onto scoria

Table 3 Kinetic model parameters for the adsorption of NO_3–N, PO_4–P and $[Cr]_T$ onto the pumice (pH = 9.1, $T° = 21$ °C)

Pollutants	Initial conc. (g/L)	Pseudo-first-order kinetic				Pseudo-second-order kinetic		
		q_e, exp. (g/kg)	q_e, cal. (g/kg)	K_1 (h^{-1})	R^2	q_e, cal. (g/kg)	K_2 (kg g^{-1} h^{-1})	R^2
NO_3–N	0.124	1.160	1.140	0.020	0.9969	1.302	0.024	0.6419
PO_4–P	0.168	1.080	1.010	0.030	0.9711	1.189	0.049	0.8604
$[Cr]_T$	0.0357	0.268	0.267	0.035	0.9999	0.276	0.415	0.8938

Table 4 Kinetic parameters for the adsorption of NO_3–N, PO_4–P and $[Cr]_T$ onto the scoria (pH = 9.1, $T° = 21$ °C)

Pollutants	Initial conc. (g/L)	Pseudo-first-order kinetic				Pseudo-second-order kinetic		
		q_e, exp. (g/kg)	q_e, cal. (g/kg)	K_1 (h^{-1})	R^2	q_e, cal. (g/kg)	K_2 (kg g^{-1} h^{-1})	R^2
NO_3–N	0.124	0.250	0.249	0.032	0.9951	0.282	0.19	0.8535
PO_4–P	0.168	0.210	0.149	0.053	0.8769	0.220	0.87	0.9806
$[Cr]_T$	0.0357	0.051	0.025	0.039	0.4690	0.539	2.61	0.9294

where q_e and q_t are the sorption capacity (g/kg) of pollutants at equilibrium and at time t respectively and k_2 is the rate constant for pseudo-second order sorption (kg g^{-1} h^{-1}). For the boundary conditions $t = 0$ to $t = t$ and $q_t = 0$ to $q_t = q_t$, the above equation has been integrated and linearized to make Eq. (6).

$$\frac{t}{q_t} = \frac{1}{(k_2 \times q_e^2)} + \left(\frac{1}{q_e}\right) \times t \qquad (6)$$

where k_2 (kg g^{-1} h^{-1}) is the adsorption rate constant of pseudo-second-order equation. The value of q_e and k_2 can be obtained from the slope and the intercept of the plot of (t/q_t) versus t respectively for those selected pollutants adsorption from the tannery wastewater onto pumice Fig. 5b and scoria Fig. 6b.

The results in Figs. 5b and 6b show linear plots with very high values of R^2 on the adsorption of PO_4–P and $[Cr]_T$ onto scoria. But the value R^2 for NO_3–N adsorption onto pumice was low. In addition the result showed that, good agreement between experimental and calculated values of q_e in the adsorption kinetics of PO_4–P and $[Cr]_T$ onto pumice adsorbent (Tables 3, 4). Therefore, the adsorption of PO_4–P and $[Cr]_T$ onto the scoria represents a good fit with pseudo-second-order kinetics. In this case the adsorption of PO_4–P and $[Cr]_T$ onto the scoria process suspected to be chemisorption Alemayehu et al. (2011).

In terms of R^2, Table 3 clearly shows, all the three pollutant removal process in the pumice containing filter tank strongly agree with pseudo-first-order kinetic model than second-order. A result of chromium adsorption kinetics and equilibrium study done by Pandey et al. (2010) showed the same trend with this kinetic study findings.

The NO_3–N is strongly agree with Pseudo first-order kinetic model by considering both correlation coefficients (R^2) and experimental and calculated q_e value, but PO_4–P agree with second-order kinetics in terms of both kinetic parameters using scoria adsorbent. Similarly $[Cr]_T$ removal process onto scoria substrate also agrees with the second-order model based on the value of R^2, which is comparable to a study done in South Korea and Nigeria on the adsorption of Cr(VI) ion onto different natural adsorbents (Ali et al. 2016; Owalude and Tella 2016). However, in this kinetic model for $[Cr]_T$ onto the same adsorbent (scoria), the experimental q_e is not in-line with the calculated one (Table 4).

Adsorption equilibrium isotherm modeling of chromium removal from tannery wastewater

In this study, two different adsorption isotherm models, namely Langmuir and Freundlich were used to evaluate the affinity of the two studied adsorbents (filter medium) for the removal of $[Cr]_T$ from the real tannery wastewater.

Langmuir isotherm

The Langmuir isotherm model assumes that a monolayer adsorption at specific homogenous sites such as $[Cr]_T$ in real tannery wastewater is adsorbed on the adsorbent surface in this specific case pumice and scoria. Equation (7) is the Langmuir expression.

$$\frac{C_e}{q_e} = \frac{1}{K_L q_m} + \frac{C_e}{q_m} \qquad (7)$$

where q_m (the maximum capacity of adsorption, g/kg) and K_L (a constant related to the affinity of the binding sites, L/kg) are the Langmuir isotherm constants. Both q_m and K_L can be determined from the slope and intercept of the plot C_e/q_e versus C_e which is the linear form of Langmuir equation that gives a straight line showed in Figs. 7a and 8a.

The Langmuir isotherm constants of $[Cr]_T$ onto pumice and scoria substrates were displayed in Tables 5 and 6 respectively. The high value of correlation coefficients $(R^2) = 0.9464$ in pumice substrate indicated minimal deviation from the fitted equation showing that

the adsorption of $[Cr]_T$ onto pumice follows Langmuir equation and also it shows that pumice has a maximum adsorption capacity of $[Cr]_T$ than scoria (Tables 5, 6). Even though the previous researchers used laboratory scale test with synthetic solution unlike this study, Alemayehu et al. (2011) obtained similar result with this finding.

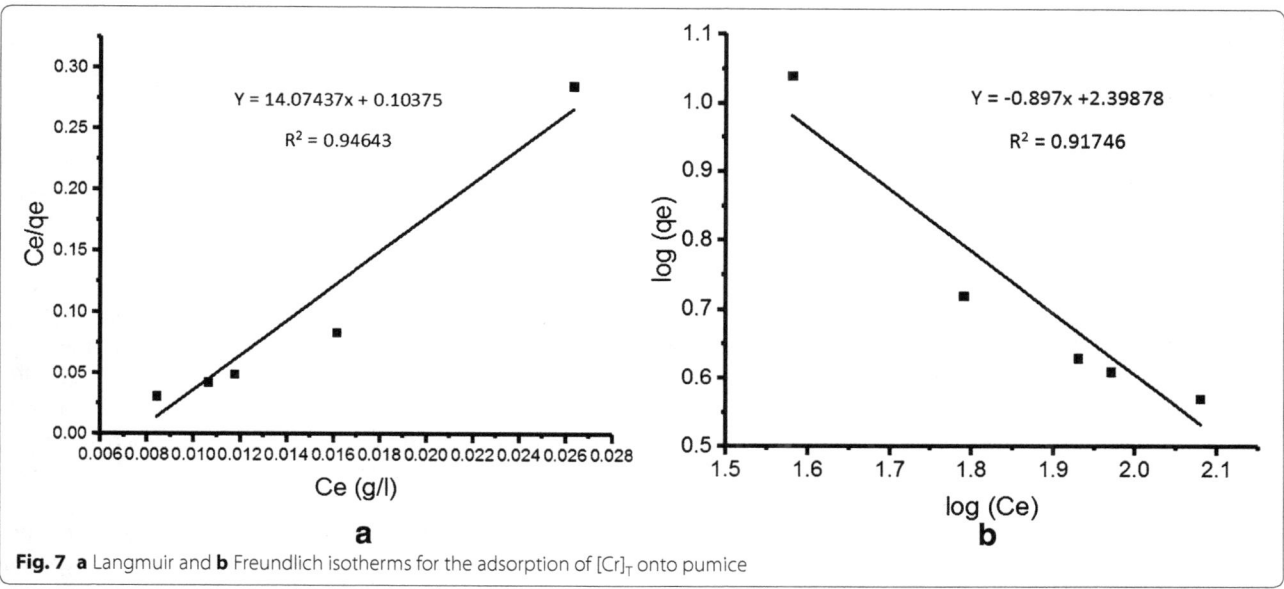

Fig. 7 a Langmuir and b Freundlich isotherms for the adsorption of $[Cr]_T$ onto pumice

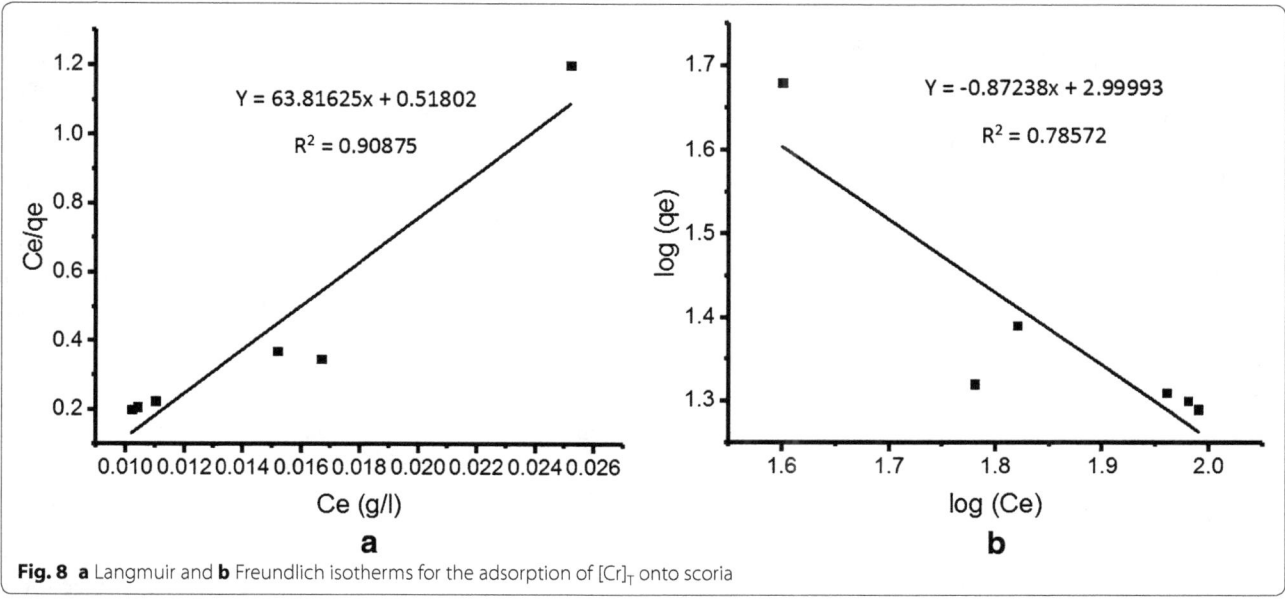

Fig. 8 a Langmuir and b Freundlich isotherms for the adsorption of $[Cr]_T$ onto scoria

Table 5 Langmuir and Freundlich isotherm constants for the adsorption of $[Cr]_T$ onto the pumice (pH = 9.1, $T° = 21 °C$)

Langmuir constants			Freundlich constants		
q_m (g/kg)	K_L (L/kg)	R^2	K_F (L/kg)	$1/n$	R^2
0.071	135.75	0.9464	250.5	− 0.897	0.9174

Table 6 Langmuir and Freundlich isotherm constants for the adsorption of $[Cr]_T$ onto the scoria (pH = 9.1, $T° = 21 °C$)

Langmuir constants			Freundlich constants		
q_m (g/kg)	K_L (L/kg)	R^2	K_F (L/kg)	$1/n$	R^2
0.016	120.652	0.9088	997.7	− 0.872	0.7857

Freundlich isotherm

The Freundlich exponential equation presumes that the adsorption process takes place on a heterogeneous surface. The linear form of Freundlich equation is given in Eq. (8).

$$\log q_e = \log K_F + \frac{1}{n} \log C_e \qquad (8)$$

where K_F (L/kg) is an indicator of the multilayer adsorption capacity and $1/n$ is the adsorption intensity and indicates both the relative distribution of energy and the heterogeneity of the adsorbent sites. Figures 7b and 8b representing the linear plot of log q_e versus log C_e at constant temperature. The value of K_F and $1/n$ (Tables 5, 6) were determined from the intercept and the slope of the plot of log q_e versus log C_e respectively.

Linear regression is the widely used approach in evaluating the fit between experimental data and the various isotherm models, this can be determined from the value of (R^2) (Ncibi 2008). In this study, R^2 (0.9174) of Freundlich isotherm on the adsorption of $[Cr]_T$ onto pumice is greater than the value of R^2 (0.7857) onto scoria substrate. On the other hand the value of R^2 is greater in the Langmuir isotherm model than Freundlich in both substrates. Therefore adsorption of $[Cr]_T$ onto both substrates fit with the Langmuir and Freundlich isotherm model with $R^{2>}$ 0.75.

Comparison of the efficiency of pumice and scoria substrate for tannery wastewater treatment

Wastewater from the leather industry is known to be heavily contaminated with inorganic and organic pollutants. Treatment processes for such type of high strength wastewaters is challenging. In this case adsorption and

filtration method are more effective than chemical and biological processes, but at the same time more expensive. So alternative adsorbent must be investigated. In this study two volcanic rocks pumice and scoria were identified and tested for their ability in removal of hazardous pollutants and compared the efficiency for the treatment of tannery wastewater by adsorption and filtration techniques.

Based on this finding the pollutant removal efficiency of pumice is better than scoria to remove BOD_5 at each retention time, but in terms of COD removal reveres result was obtained. In general the two substrates were poor to remove both BOD_5 and COD this is obvious that the organic matters cannot be removed by adsorption some may be removed biologically at the top of the filter medium. Better removal efficiency has been seen by Scoria to remove TSS, the minimum removal efficiency (65%) was achieved by pumice at RT = 24 h and the maximum removal efficiency (84%) was obtained by scoria at RT = 72 h. This better removal efficiency was seen because most of the TSS may be removed by the straining mechanism of filtration in both filter materials (Table 7).

In nutrient removal potential, scoria shows better efficiency than pumice for example nitrate removal efficiency of scoria and pumice were 99 and 95% respectively at RT = 72 h. Similarly the phosphate removal also better by scoria on the first 24 and 48 h, but at RT = 72 h pumice removed 66% and scoria removed 63% only. This study result is in line with study done in Jimma University Ethiopia using an aqueous solution running over 24 h contact time with similar filter media (Birhane et al. 2014) with a slight difference. This difference may be due to the phosphate concentration, experimental setup and the type of wastewater used for the test between the two studies.

In another study that was tested using four filter materials. The removal of phosphate ranged from 35 to 41% for calcite, 59 to 100% for zeolite, 49 to 100% for sand, and 73 to 100% for iron filings (Reddy et al. 2013). From the indicated four filter materials the result of this paper can be compared in the range of the result obtained from iron filings. That means pumice is similar adsorption characteristics with iron filings. The removal of nitrate and phosphate may be mainly attributed to adsorption, ion exchange and precipitation. The reverse trend was shown to remove sulfate and chromium that indicate pumice had a better efficiency than scoria in all given retention time (Table 7; Fig. 9).

Since industrial effluents like tannery wastewater containing sulfide and sulfate are toxic to aquatic environment, it is essential to reduce them and bring the discharge levels of these species to below the toxic limit. In this investigation the sulfate removal potential of pumice was greater than scoria the efficiency of pumice

Table 7 Comparison of scoria and pumice filter media efficiency in tannery wastewater treatment

Parameters	% removal at RT = 24 h		% removal at RT = 48 h		% removal at RT = 72 h	
	Pumice	Scoria	Pumice	Scoria	Pumice	Scoria
BOD$_5$	56	52	57	55	62	59
COD	27	48	30	50	45	54
TSS	65	75	68	83	70	84
NH$_4$–N	41	32	28	27	23	6
NO$_2$–N	88	98	95	98	97	97
NO$_3$–N	40	51	56	80	95	99
PO$_4$–P	41	57	51	60	66	63
Sulfide	70	73	71	75	72	77
Sulfate	84	77	83	76	83	75
Total chromium	76	71	76	71	70	69
Average	58.8	63.4	61.5	67.5	68.3	68.3

to remove sulfate was from 83 to 84% whereas scoria shows from 75 to 77%. In general both filter mediums were effective to remove sulfide and sulfate from tannery wastewater.

The better removal performance of pumice may be achieved based on the good sulfate adsorption nature of the pumice and its chemical composition. On the other hand metal ions from the filter material may react with dissolved sulfide ions to form metal sulfides as colloidal suspension, which were coagulated and precipitated and finally filtered out, this mechanism may be contributed

mainly to sulfide removal in the filtration tank. In a study carried out to investigate the effect of chemical modification method on sulfate removal efficiency of adsorbents, the removal of sulfate ion using Fe-modified carbon residue was notably higher compare with unmodified carbon residue and commercially available activated carbon (Runtti et al. 2016).

Batch experimental laboratory scale studies were undertaken in a different world on the potential of pumice to reduce the concentration of heavy metals including chromium from aqueous solution in the laboratory. But

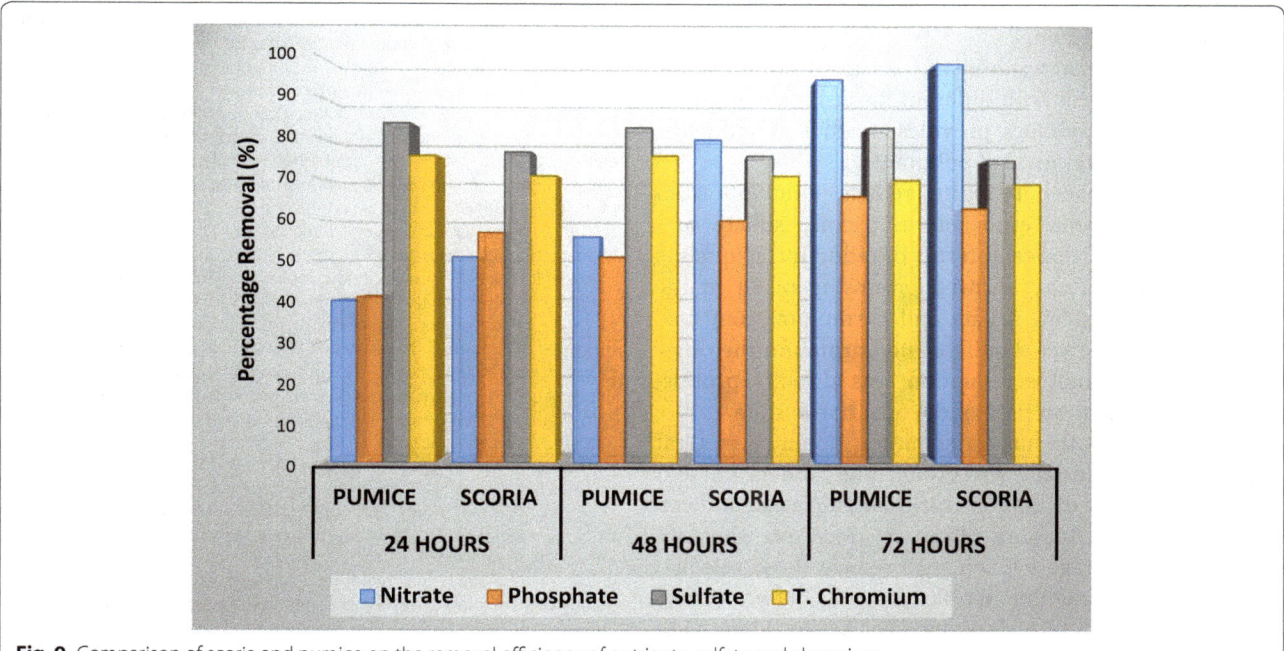

Fig. 9 Comparison of scoria and pumice on the removal efficiency of nutrients, sulfate and chromium

in this study the chromium removal efficiency of pumice and scoria was done from real tannery wastewater.

In this study pumice had better in chromium reduction potential than scoria. In the first 24 and 48 h retention time pumice and scoria achieved 76 and 71% in chromium reduction respectively (Table 7). A study done by (Alemayehu et al. 2011), showed similar result that was pumice had a better result to remove chromium than scoria. The maximum adsorption yield, 77% for scoria and 80% for pumice, was obtained at low pH in that study. This chromium removal potential difference between the two substrates may be due to the difference in chemical composition (Table 7; Fig. 9).

Considering all the selected tannery wastewater parameters for this study, the average treatment efficiency of pumice and scoria were 58.8 and 63.4% in RT = 24 h, 61.5 and 67.5% at RT = 48 h respectively, and equivalent efficiency was obtained at the third retention time that is over 72 h both filter materials showed 68.3%. Even though there were differences in different types of parameter reduction, the average tannery wastewater treatment efficiency of scoria was greater than pumice at the first 24 and 48 h, but it is not statistically significant at 95% confidence interval, p = 0.3 and 0.2 respectively. Equivalent percentage removal (68.3%) was shown in both substrates at RT = 72 h, but in the overall result, Scoria had shown greater efficiency than pumice but it is not statistically significant at 95% confidence interval, p = 0.2.

Conclusion

Based on the investigation the following major conclusions have been drawn:

Dire tannery wastewater characteristics were very high strength wastewater with different hazardous pollutants. The applied methodology provides an efficient and feasible approach for removal of pollutants.

Pumice had good potential to reduce pollutants from tannery wastewater, especially nutrients, sulfate and chromium but it is poor in terms of reduction of organic pollutants. Scoria also showed promising result in the reduction of nutrient, sulfate and chromium from tannery wastewater. Except sulfate and chromium the other wastewater parameter reduction was directly proportional to the retention time of the filtration system.

More or less both substrates have a potential to treat high strength industrial wastewater like tannery effluent by using different techniques including filtration and can be an alternative to sand filtration. However, when we compare the average efficiency to reduce those selected wastewater parameters scoria showed better results than pumice, even though pumice has better potential to reduce chromium and sulfate than scoria. Therefore the use of locally available low cost adsorbents may contribute to the low technology solution for sustainable wastewater management. Any interested company can use these substrates as a filter medium in the filtration bed or substitute by sand for available filtration bed for wastewater treatment.

Abbreviations
APHA: American Public Health Association; BOD: biological oxygen demand; COD: chemical oxygen demand; DO: dissolved oxygen; ECDSWCRL: Ethiopian Construction Design and Supervision Works Corporation Research Laboratory; EPA: environmental protection authority; ES: effective size; RT: retention time; TSS: total suspended solids; TWW: tannery wastewater; UC: uniformity coefficient; UNIDO: United Nations Industrial Development Organization; XRF: X-ray fluorescence.

Authors' contributions
All authors have made an essential intellectual contribution to this study. MBA designed the study, conducted the experiments, Collected, analyzed and interpreted the data and wrote the manuscript. SLA involved on the study design, supervised the experiment, provided comments and suggestion for the whole work. MMK supervised the work, drafting and revising the primary manuscript, edited the manuscript, provided pertinent comments and suggestion on the manuscript. All authors read and approved the final manuscript.

Authors' information
Mekonnen Birhanie Aregu is a Ph.D. Scholar in environmental pollution and sanitation, Center for Environmental Sciences, College of Natural Science, Addis Ababa University and lecturer at School of Public Health, Dilla University. He has given Public health, Environmental Health Science and Technology courses and also contribute in community services and problem solving applied research activities and published several articles in the internationally peer reviewed journals. He is also a senior consultant in environmental health, waste management and emission control.

Seyoum Leta Asfaw (Ph.D.) is an Associate Professor of environmental pollution and sanitation stream, Center for Environmental Sciences, Addis Ababa University. He has given various courses in the stream. He is also supervising and monitoring several Ph.D.s and M.Sc. students in the areas of environmental biotechnology, wastewater treatment and water quality studies, waste to energy (resource recovery and climate change mitigation and adaptation studies), bioremediation, phytoremediation, microbial ecological studies, environmental sanitation. He has published several peer reviewed papers in different international reputable journals. Currently, he is also executive Director for Horn of Africa Regional Environmental Centre and Network (HOARECN).

He has more than 20 years of experience in environmental science and technology studies and management. He had also been a regional program manager for the Bio-resources innovation network for Eastern Africa development. He has also been the principal investigator for a number of research projects such as "Development of innovative technologies for the sustainable treatment of high strength wastewater in East Africa", a regional research program involving Ethiopia, Kenya, Tanzania and Uganda funded by Sida. He developed an innovative, integrated pilot technology for the treatment of agro-process wastewater, generating biogas, biofertilizer, and clean water at Modjo Tannery, Addis Ababa, Ethiopia.

Professor Mohammed Mazharuddin Khan (Ph.D.) is a professor of Environmental Science at Center for Environmental Sciences, Addis Ababa University. He is the CHARLES DARWIN GOLD MEDAL was conferred on the 27th day of June, 2009 towards the contribution made in the field of Life Science by International Society for Ecological Communication at Vinoba Bhave university, Hazaribagh, Jharkhand, India. He has given several courses like Botany, Biotechnology, Environmental sciences, Ecology, Microbiology, and Environmental Microbiology at postgraduate and undergraduate level. He is also a founder of different organizations, moreover he is a member of more than eight international professional associations and societies and served in different positions. Professor Khan has published several internationally peer reviewed papers.

Acknowledgements
The principal investigator would like to thank Addis Ababa University for financial support.

Competing interests
The authors declare that they have no competing interests.

Funding
The first author is grateful to Addis Ababa University in supporting for expenditures during laboratory analysis.

References
Akbal F, Akdemir N, Onar AN (2000) FT-IR spectroscopic detection of pesticide after sorption onto modified pumice. Talanta 53:131–135
Alemayehu E, Lennartz B (2009) Virgin volcanic rocks: kinetics and equilibrium studies for the adsorption of cadmium from water. J Hazard Mater 169:395–401
Alemayehu E, Thiele-Bruhn S, Lennartz B (2011) Adsorption behavior of Cr(VI) onto macro and micro vesicular volcanic rocks from water. Sep Purif Technol 78:55–61
Ali A, Saeed K, Mabood F (2016) Removal of chromium (VI) from aqueous medium using chemically modified banana peels as efficient low-cost adsorbent. Alex Eng J 55:2933–2942
APHA (2005) Standard methods for the examination of water and wastewater, 21st edn. American Public health association, Washington
Arias CA, Del Bubba M, Brix H (2001) Phosphorus removal by sands for use as media in subsurface flow constructed reed beds. Water Res 35(5):1159–1168
Asaye K (2009) Evaluation of selected plant species for the treatment of tannery effluent in a constructed wetland system. Dissertation, Addis Ababa University
Asfaw A (2014) Heavy metals concentration in tannery effluents, associated surface water and soil at Ejersa area of East Shoa, Ethiopia. Herald J Geogr Reg Plan 3(3):124–130
Babel S, Kurniawan TA (2003) Low-cost adsorbents for heavy metal uptake from contaminated water. J Hazard Mater 97:219–243
Banuraman S, Meikandaan TP (2013) Treatability study of tannery effluent by enhanced primary treatment. Int J Mod Eng Res 3(1):119–122
Birhane M, Abebe A, Alemayehu E, Mengistie E (2014) Efficiency of locally available filter media on fluoride and phosphate removal for household water treatment system. Chin J Popul Res Environ 12(2):110–115
Boller MA, Kavanaugh MC (1995) Particle characteristics and head loss increase in granular media filtration. Water Res 29(4):139
Dabrowski A (2001) Adsorption—from theory to practice. Adv Coll Interface Sci 93:135–224
Durai G, Rajasimman M, Rajamohan N (2011) Kinetic studies on biodegradation of tannery wastewater in a sequential batch bioreactor. J Biotechnol Res 3:19–26
Elmorsi TM (2011) Equilibrium isotherms and kinetic studies of removal of methylene blue dye by adsorption onto miswak leaves as a natural adsorbent. J Environ Prot 2:817–827
Farizoglu B, Nuhoglu A, Yildiz E, Keskinler B (2003) The performance of pumice as a filter bed material under rapid filtration conditions. Filtr Sep 40(3):41–47
Ho YS, McKay G (1999) Pseudo-second order model for sorption processes. Process Biochem 34:451–465
Islam B, Musa AE, Ibrahim EH, Salma AA, Babiker M (2014) Evaluation and characterization of tannery wastewater. J Forest Prod Ind 3(3):141–150
Kwon JS, Yun ST, Kim SO, Mayer B, Hutcheon I (2005) Sorption of Zn(II) in aqueous solutions by scoria. Chemosphere 60:1416–1426
Kwon JS, Yun ST, Lee JH, Kim SO, Jo HY (2010) Removal of divalent heavy metals (Cd, Cu, Pb, and Zn) and arsenic (III) from aqueous solutions using scoria: kinetics and equilibria of sorption. J Hazard Mater 174(13):307–313
Lagergren SK (1898) About the theory of so-called adsorption of soluble substances. Kungliga Svenska Vetenskapsakademiens Handlingar 24(4):1–39
Leta S, Assefa F, Dalhammar G (2003) Characterization of tannery wastewater and assessment of downstream pollution profiles along Modjo River in Ethiopia. Ethiop J Biol Sci 2(2):157–168
Lofrano G, Aydn E, Russo F, Guida M, Belgiorno V, Meric S (2008) Characterization fluxes and toxicity of leather tanning bath chemicals in a large tanning district area. Water Air Soil Pollut 8:529–542
Mandal T, Dasgupta D, Mandal S, Datta S (2010) Treatment of leather industry wastewater by aerobic biological and Fenton oxidation process. J Hazard Mater 180:204–211
Martins AE, Pereira MS, Jorgetto AO, Martines MA, Silva RI, Saeki MJ, Castro GR (2013) The reactive surface of castor leaf (*Ricinus communis* L.) powder as a green adsorbent for the removal of heavy metals from natural river water. Appl Surf Sci 276:24–30
Moufti MR, Sabtan AA, El-Mahdy OR, Shehata WM (2000) Assessment of the industrial utilization of scoria materials in central Harrat Rahat, Saudi Arabia. Eng Geol 57:155–162
Ncibi MC (2008) Applicability of some statistical tools to predict optimum adsorption isotherm after linear and non-linear regression analysis. J Hazard Mater 153:207–212
Owalude SO, Tella AC (2016) Removal of hexavalent chromium from aqueous solutions by adsorption on modified groundnut hull. Beni-suef Univ J Basic Appl Sci 5:377–388
Pandey PK, Sharma SK, Sambi SS (2010) Kinetics and equilibrium study of chromium adsorption on zeoliteNaX. Int J Environ Sci Technol 7(2):395–404
Rao BH, Dalinaidu A, Singh DN (2007) Accelerated diffusion test on the intact rock mass. J Test Eval 35(2):111–117
Reddy KR, Xie T, Dastgheibi S (2013) Nutrients removal from urban storm water by different filter materials. Water Air Soil Pollut 225:1778
Rezic I, Zeiner M (2008) Determination of extractable chromium from leather. Monatshefte fur Chemie-Chemical 140(3):325–328
Runtti H, Tuomikoski S, Kangas T, Kuokkanen T, Rämö J, Lassi U (2016) Sulfate removal from water by carbon residue from biomass gasification: effect of chemical modification methods on sulfate removal efficiency. BioResources 11(2):3136–3152
Shen TT (1999) Industrial pollution prevention, 2nd edn. Springer, Berlin, p 40
Sivakumar P, Kanagappan M, Sam Manohar Das S (2015) Physicochemical characteristics of untreated effluent from tannery Industries in Tamil Nadu: a comparative study. Int J Pharm Bio Sci 6(1):446–451
Sugasini A, Rajagopal K (2015) Characterization of physicochemical parameters and heavy metal analysis of tannery effluent. Int J Curr Microbiol Appl Sci 4(9):349–359
Tadesse AT, Seyoum LA (2015) Evaluation of selected wetland plants for removal of chromium from tannery wastewater in constructed wetland, Ethiopia. Afr J Environ Sci Technol 9(5):420–427
UNIDO (2012) United Nations Industrial Development Organization Vienna, Technical assistance project for the upgrading of the Ethiopian leather and leather products industry, independent evaluation report Ethiopia. UNIDO project number: TE/ETH/08/008. https://open.unido.org/api/documents/4763121/download/Independent%20Evaluation%20Report%20-%20ETHIOPIA%20-%20Technical%20assistance%20project%20for%20the%20upgrading%20of%20the%20Ethiopian%20leather%20and%20leather%20products%20industry. Accessed Sept 2016
Wosnie A, Wondie A (2014) Bahir Dar tannery effluent characterization and its impact on the head of Blue Nile River. Afr J Environ Sci Technol 8(6):312–318

11

Effects of land use change on soil physicochemical properties in selected areas in the North West region of Cameroon

Valentine Asong Tellen[1]*[iD] and Bernard P. K. Yerima[2]

Abstract

Background: Land use changes from natural ecosystems into managed ecosystems may have deleterious effects on soil structure and quality. This study characterise the soils under, and assesses the effects of different land use systems on selected soil physicochemical properties in the North West region of Cameroon. Six land use systems including: natural forest, natural savanna, grazing land, afforested land, farmland and Eucalyptus plantation were identified. Ninety soil samples were collected from each at the 0–15 cm depth. Fifteen soil physicochemical properties were measured.

Results: The conversion of natural forest or savanna to farmland reduces the silt contents, moisture content, organic matter, organic carbon, total nitrogen, available phosphorus, pH, cation exchange capacity and exchangeable bases, but increases bulk density, electrical conductivity, exchangeable acidity and sand content significantly ($P < 0.05$). The results revealed that deforestation and subsequent cultivation of soil had negative effects on the measured soil properties.

Conclusions: Land use change has ruined soil quality in the North West region. To reverse soil degradation and promote restoration, emphases should be placed on promoting the use of sustainable land management practices within the savanna, grazing, agricultural and forest management systems.

Keywords: Soil quality indicators, Land use change, Soil degradation, Africa, Cameroon

Background

Land use/land cover (LULC) changes influence the biogeochemistry, hydrology, and climate of the earth. Elucidating the impact of LULC at the local to regional scales on soil quality status is not direct but rather complex to guarantee any generalizations (Hoogsteen et al. 2015). Across sub-Saharan Africa, natural resources remain central to rural people's livelihoods (Roe et al. 2009). Nonetheless, natural (rainfall and temperature) as well as anthropogenic (farming, grazing, burning) forces can exert pressure on these resources, thereby influencing spatial and temporal scale changes on a landscape. LULC changes are indicators of forest resource dynamics within a landscape. The dynamics of LULC change associated with the anthropogenic activities are occurring rapidly in tropical landscapes. Recent international concerns place high attention on monitoring changes in tropical resources and reporting on those factors (such as agriculture) influencing these changes (such as deforestation), for consideration of novel scientific and policy interventions (goal #15 of the 2030 Agenda for Sustainable Development). To understand the dynamics of ecological processes and the impacts related to these changes in LULC, an assessment of the effects of these changes on soil quality is important.

According to the United Nations Convention to Combat Desertification (UNCCD), 24 billion tons of fertile soils are lost due to erosion every year, while 12 million hectares of land are degraded through drought and

---

*Correspondence: tvasong@yahoo.com
[1] Department of Development Studies, Environment and Agricultural Development Program, Pan African Institute for Development-West Africa (PAID-WA), P.O. Box 133, Buea, South West Region, Cameroon

the encroachment of the desert (this is 23 hectares per minute) where 20 million tons of grain could have been grown. Epule et al. (2011) stated that Cameroon's forests are part of the Congo Basin and it is ranked the second largest tropical rainforest hot spot in the world after the Amazon Basin in Latin America largest. FAO (2010a) remarked that Cameroon's forest contains about 2696 million metric tons of carbon in living biomass. This indicates that deforestation is even more intimidating for the environment. Even so, FAO (2010a, b) reported that Cameroon forests occupies about 28 million hectares (ha) of land and of this number, about 220 thousand (ha) are lost each year; this is equivalent to about − 1.0% of annual forest cover loss. Evidence, though anecdotal, reveals that the population growth in Cameroon and scarcity of arable land has exacerbated food insecurity and water scarcity. FAO (2009) forecasted that the high population density with continued demand for arable land in Africa would amplify deforestation pressure. In fact, land degradation is a very serious challenge as it leads to hunger, poverty and is at the root of many conflicts (FAO 2017). Progress towards meeting the sustainable development goal # 15 of the 2030 agenda requires an understanding of the drivers of soil degradation.

It is reported that an ample surface area of African forests has been lost, with a significant influence resulting from small scale agriculture (FAO 2009; Harvey et al. 2004). In fact, Cheek et al. (2000) and Harvey et al. (2004) projected a 96.5% future loss of the original forest cover within the Bamenda Highlands, with its climate change implications. The need for sustainable land use-ecosystems conjures the protection and enhancement of soil quality through designing efficient site specific actions to control erosion and restore soil quality, thereby improving the conditions and productivity of the agro-silvipastoral landscapes in the western highlands region of Cameroon. The objectives of this study include: (i) to characterize the soils under the different land use systems; and (ii) to assess the influence of land use change on selected soil physicochemical properties in the North West region of Cameroon.

Methodology
Description of the study area
This study was conducted in the North West Region of Cameroon which lies between latitudes 5°45″ and 9°9″ N and longitudes 9°13″ and 11°13″ E. It covers an area of about 17,400 km^2 and is bordered to the North and West by the Republic of Nigeria, to the South by the West and South West Regions and to the East by the Adamawa Region (Manu et al. 2014). The topography of the Region varies greatly from depressions lower than 400 m above sea level to high mountains, 3000 m above

sea level. Three study sites were selected following a stratified random sampling technique. Each stratum represents a particular topographic zone [the lower altitude (< 900 m); the mid-altitude (900–1500 m) and the high altitude (> 1500 m)] within the North West Region. The representative study sites selected include Ndop (lower altitude), Nkwen (mid-altitude) and Awing (high altitude) (Fig. 1).

The topography greatly influences the climate with a tropical transition from the rainy humid and continuously warm climate in the South to an extremely unpredictable (regarding temperature and precipitation) but somewhat dry and hot climate of the North. Absolute annual average precipitation ranges from 1700 to 2824 mm. The high altitudinal areas are cold (< 15 °C) such as Awing and Santa whereas the low altitude zones are hot (average 27 °C) such as Ndop plain and Ako Sub Division. There are two distinct seasons: the rainy season (mid-March to mid-October) and the dry season (mid-October to mid-March). The vegetation here results from the prevailing soil conditions, altitude, human activities on the environment and climate. The region lies within the savannah zone where grasses and shrubs predominates. The dominant soil type is Oxisol (rich in oxides of Fe and Al and has a characteristic reddish color) which encourages erosion (pseudo sand and pseudo silt) and results to gullies on bare surfaces while the valleys are covered with alluvial deposits (Yerima and Van Ranst 2005a, b; Yerima 2011).

Land use/land cover systems identified in the North West Region
Six LULC systems were identified and presented as follows:

i.   *Farmland*: It is characterized by the cultivation of crops such as cabbages, onion, carrot, pumpkins, and green pepper. Annual crops such as maize, potatoes, beans, and pea, are most commonly cultivated (Fig. 2). Subsistence farming characterizes agriculture in the study area, and the main cropping system is mixed, although rotation, inter-cropping, mono-cropping and fallow systems are also common.

ii. *Natural forest land*: It is composed of various indigenous trees, shrubs, and bushes like *Podocarpus falcatus* (Zigba) (Fig. 3). The forest is usually found in protected areas where it is restricted from farming or livestock grazing. However, the culture has not allowed replanting (reforestation) and the newly germinating seedlings are being destroyed by farm encroachment and animals browsing and trampling. Due to high deforestation rates, these forests are found in patches, often located in valleys and small

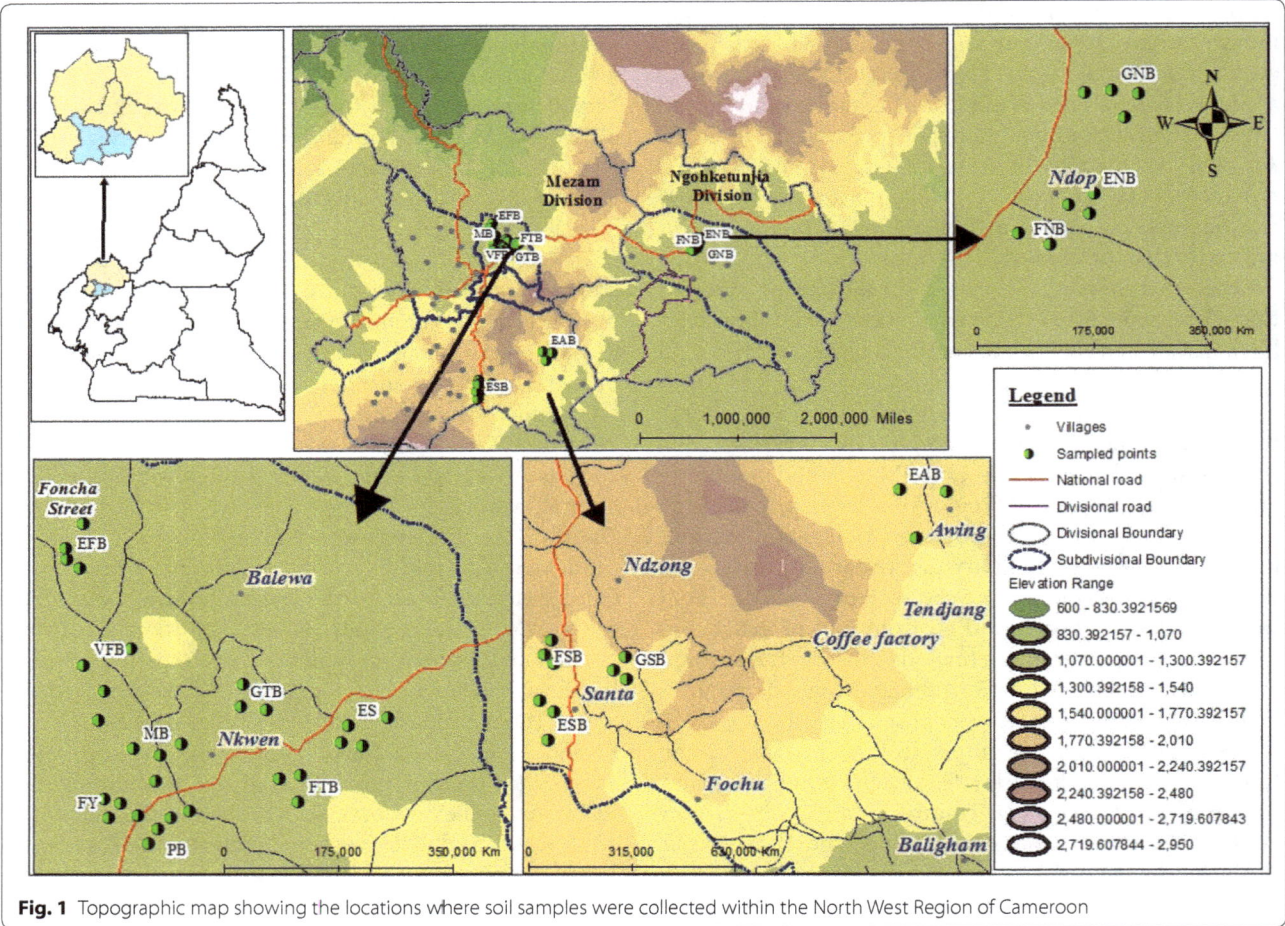

Fig. 1 Topographic map showing the locations where soil samples were collected within the North West Region of Cameroon

depressions which often harbor streams and other water bodies.

iii. *Natural savanna*: It is composed of short grasses and usually located within protected areas (Fig. 4). This land use system is used for grazing in areas where no property right exists. It is believed to have been created due to the shrinkage of the forest cover as a result of deforestation due to human and animal disturbance.

iv. *Eucalyptus plantation*: It predominantly consists of two commonly known exotic eucalyptus species in the region (*Eucalyptus salinga* and *Eucalyptus grandis*). These tree stands are indiscriminately planted on water catchments and are gradually replacing native tree species of the NW region (Fig. 5).

v. *Afforestation area*: It is represented by an afforested research unit created in 2010 (6–7 years ago), characterized by fast-growing environmentally friendly tree species for fuel and timber (Fig. 6) to reduce pressure on the endangered native tree species which are at risk of extinction. Some of these plant spe-

cies include: pine, Zigba *(Podocarpus falcatus)*, large diameter rattans (*Laccosperma secundiflorum* and *L. robustum* species), mahogany (*Swietenia macrophylla King*), iroko (*Milicia excelsa*), Pygeum (*Prunus Africana*), mango (*Mangifera indica*) and other plant species that have food, fuelwood, medicinal, timber, etc., attributes (Yerima 2011).

vi. *Grazing land*: It consists of short grasses (pasture) and used for cattle grazing and is considered a communal land (Fig. 7). Under such interference, it has become very difficult to find natural settings in the area (Yerima 2011).

Soil sampling and analysis

The soils were characterized following procedures proposed by (Yerima and Van Ranst 2005a). Soil samples were collected from the six main land use systems described above (Natural forest, natural savanna, farmland, afforested land, grazing land and Eucalyptus plantation). Under each land use, soil samples were obtained from a plot, with dimensions of 20 × 20 m (400 m²), at a

Fig. 2 Collection of soil samples from farmlands on varying geomorphic surfaces (**a**); a common tillage practice on farmlands in the North West Region (**b**)

Fig. 3 Protected man-made forest at the Yongka Western Highlands Research Garden Park (**a**); the collection of soil samples from a naturally protected forest adjacent to Lake Awing (**b**)

constant depth of 0–15 cm, following a "Z-layout" design. Soil samples were taken from the four corners and center of each layout. Approximately 1 kg of composite sample was collected from each location and placed into plastic bags. They were then transported, air-dried at room temperature, crushed, homogenized, and passed through a 2 mm sieve before laboratory analysis. A total of 90 soil samples (six land use types x five replicates per sample plots x one soil depth class: 0–15 cm × three altitudinal zones) were collected within the study area, from June to July 2015, for analysis. Undisturbed soil samples were taken with a core sampler that was 7.5 cm long and 6.4 cm in diameter for bulk density determination. Soil quality indicators comprising of the three standard groupings including: physical (moisture content, bulk density); chemical (pH, total nitrogen, available phosphorus, exchangeable bases, cation exchange capacity, C/N ratio, electrical conductivity) and biological (organic matter content) proposed by Yerima and Van Ranst (2005a), for soils in the tropics, were selected for analysis. The

soils were analysed at the Soil Science Laboratory at the University of Dschang, following standard procedures and methods as described below:

Physical properties

Moisture content (MC) was calculated using the gravimetric method where soil samples were placed into ceramic crucibles and weighed to get the fresh weight and then oven-dried at 105 °C to constant weight for about 24 h and the dry weight recorded. These values were then used to calculate the moisture contents of the soils using the formula:

$$MC\ (\%) = \frac{100\ (fw - dw)}{dw}$$

where MC soil moisture content (%), fw fresh weight (g) of soil sample, dw dry weight (g) of soil sample.

Bulk density was measured following the core method described by Yerima and Van Ranst (2005a, b), where samples contained in the core rings of known weight,

Fig. 4 Natural savanna grassland located near a palm plantation at Nkwen, Bamenda (**a**); the collection of soil samples from natural savanna land at the Yongka Park (**b**)

Fig. 5 *Eucalyptus salinga* plantation at Awing

height and diameter were weighed and the fresh weights recorded, then oven dried at 105 °C for 24 h, after which the dry weights were also recorded. The volume of the core was determined using the following formula:

$$V = \pi\, r^2 h \left(cm^3\right),$$

where V = volume of core (cm^3); π = 3.14; r = radius = diameter/2 (cm); and h = height (cm).

Soil bulk density was then determined using the following formula:

$$BD = M/V,$$

where *BD* bulk density, *M* mass of oven dry soil (g) and *V* volume of core (cm^3).

Soil textural fractions (sand, silt, and clay) were analyzed following the Bouyoucos hydrometer method, where 15 g of 2 mm air-dried soil was weighed into 500 ml beakers and subjected to treatments for removing organic matter using H_2O_2, followed by dispersing the soils with sodium hexametaphosphate (Pauwels et al. 1992). The resulting compositions were placed on a mechanical shaker and allowed to shake for 3 h. Suspensions were then transferred into sedimentation measuring cylinders and brought to the 1000 ml mark using distilled deionized water. The mixtures were well stirred using a mechanical rotator to bring the particles into suspension. A hydrometer was then used to obtain readings after 40 s (first reading, R1) and 2 h, (second reading, R2) respectively. Calculations were done using the following equations:

$$\% \, (Silt + Clay) = \frac{R1}{15} \times 100 \tag{A}$$

The first reading (R1) gave the silt + Clay content

$$But \, \% \, Clay = \frac{R2 - R1}{15} \times 100 \tag{B}$$

Therefore, % Silt = A − B while % Sand = 100 − % (Silt + Clay).

After getting the percentage sand, silt and clay, the soil textural triangle was used to classify the soil texture.

Chemical analysis

Moisture correction factor was calculated using the following formula:

$$mcf = \frac{100 + MC \, (\%)}{100}$$

where *mcf* moisture correction factor and *MC* soil moisture content (%).

Soil pH was measured both in water and KCl (a 1:2.5 soil: H_2O/KCl ratio) using a glass electrode Thermo-Russel pH meter, calibrated using buffer solutions of pH 7 and 4 for H_2O and KCl, respectively. Soil total nitrogen (TN) was determined using the Kjeldahl distillation method (Pauwels et al. 1992) where 1 g each of air-dried soil samples were placed into 500 ml Kjeldahl flasks, followed by the addition of 5 ml of distilled deionized water. A scoop of digestion accelerator mixture (sulphuric–salicylic acid mixture) was then added to each flask. Five milliliter of concentrated H_2SO_4 was added and the mixture

Fig. 6 Nine year old Eucalyptus saligna tree stand (**a**); 8 years old Podocarpus (*Podocarpus falcatus*) tree stand (**b**); collection of soil samples under a Jackfruit tree stand (**c**); collection of soil samples under a mixed fruit tree stand (*Prunus domestica, Mangifera indica*) at the Yongka Highland Research Garden Park (**d**)

was allowed to digest for 1 h in a fume cupboard by gentle heating until the vigorous effervescence subsided to ensure that the digest was free of charred organic matter. The digest was then allowed to cool followed by addition of 20 ml of distilled water. After ensuring settlement, the supernatant solutions were decanted into 100 ml volumetric flasks. The process was repeated and 5 ml of 40% NaOH and 100 ml of distilled water were added. The distillate was collected and mixed with 5 ml of the boric acid (H_3BO_3) solution-indicator mixture. The distillate was titrated with 0.01 M H_2SO_4 from green to pinkish endpoint and the titer value recorded. The soil TN was calculated using the formula:

$$Kjeldahl\ N\ (\%) = (T - B) \times M \times \frac{2.8}{S}$$

where T ml of standard acid with sample titration, B ml of standard acid with blank titration, M molarity of sulphuric acid, S weight of soil sample (g), and *2.8* a constant.

Available phosphorous (Av.P) was determined following Bray-II method where solutions of 0.1 N HCl and 0.03 N NH_4F were used to extract available phosphorus from the soil samples. Phosphorus was determined colorimetrically using the ammonium molybdate blue method. In this process, 2 g of < 2 mm air-dried soil samples were weighed into clean dried test tubes. Fourteen ml of the extracting solution was added and vigorously shaken for 30 s, using an electrical shaker and immediately filtered into other previously prepared test tubes carrying funnels and Whatman # 42 filter papers. For colour development, 5 ml of the extract and standards were pipetted into a set of test tubes. Then, 5 ml of colour development reagents (Ascorbic acid), and mixed reagent (Ammonium molybdate, Potassium Antimony tartrate and sulphuric acid) were added to each test tube. The samples were allowed to stand for 15 min for complete color development. Absorbance was then measured using a colorimeter, set at a wavelength of 882 nm.

Exchangeable bases (Ca^{2+}, Mg^{2+}, K^+, and Na) were extracted using 50 ml of ammonium acetate (1.0 M NH_4OAc) solution buffered at pH 7. Potassium (K^+) and sodium (Na^+) in the extract were determined using the

Fig. 7 Collection of soil samples from grazing and at Ntambang (**a**) and Santa (**b**)

flame photometer while magnesium (Mg^{2+}) and calcium (Ca^{2+}) in the extract were determined by complexometric titration.

Cation exchange capacity (CEC) was determined following the extraction method were soil samples were also saturated with ammonium acetate buffered at pH 7 to displace the exchangeable bases as explained in the case of the determination of exchangeable bases. However, for CEC determination, the column of each sample was then thoroughly washed with 95% alcohol to discard excess ammonium acetate that saturated the complex. This was verified using Nessler's reagent. Sixty milliliter of KCl was then added to each tube to allow potassium to replace ammonium ions on the exchange complexes. The filtrate containing the NH4$^+$ ions was collected into 100 ml volumetric flasks and brought up to 100 ml mark by the addition of KCl. Twenty-five milliliter of each sample were transferred into distiller's tubes and NaOH added followed by 2–3 drops of the end-point indicator, phenolphthalein. Forty ml of boric acid were placed into conical flasks and distilled water added to the 100 ml level. Each

sample was distilled and the distillate in the conical flasks titrated with 0.01 M H$_2$SO$_4$ from a burette. The CEC was then calculated using the following formula:

$$EC\,(100g)\,soil = (V - Vo) \times 1.6$$

where V the volume of sulfuric acid added to the sample, Vo the volume of sulfuric acid added to the blank, and 16 a constant.

Electrical conductivity (EC) was determined following standard procedures proposed by Pauwels et al. (1992) where a 1:5 soil-solution ratio (10 g of < 2 mm soil and 50 ml distilled ionized water) was agitated for an hour and the readings from an EC meter, calibrated with 0.01 N KCl, recorded.

Exchangeable acidity (EA) (H$^+$ + Al^{3+}) was determined for samples with pH < 5.5. In the procedure, 1 N KCl was added to the flasks containing 1 g of < 2 mm soil sample for displacement of the Al^{3+} + H$^+$ ions. EA was then determined by titration of extracts with 0.02 N NaOH for the neutralization of the acidic ions in the extract using three drops of phenolphthalein as an indicator. The exchangeable acidity was then calculated using the following formula:

$$EA\,meq\,100g^{-1} = 40 \times t \times (Vx - Vo)$$

where EA the exchangeable acidity (H$^+$ + Al^{3+}) (meq 100 g^{-1}), t the exact molarity of NaOH used, Vx the volume of NaOH added to the sample, Vo the volume of NaOH added to the blank.

C/N ratio was obtained by taking the ratio of S percent carbon to nitrogen in each sample as follows:

$$C/N = \frac{SOC\,(\%)}{TN\,(\%)}$$

where C/N the ratio of carbon to nitrogen, SOC the concentration to carbon (%) in the soil sample, TN the concentration of total nitrogen (%) in the soil sample.

Soil organic carbon (SOC) was determined following the Walkley and Black wet oxidation method (Walkley and Black 1934) where 5 g each, of the soil samples, were first placed into a wide-mouth Erlenmeyer flask, followed by the addition of 10 ml of 1 N Potassium dichromate (K$_2$Cr$_2$O$_7$) into each flask, in a fume cupboard. Twenty milliliter of concentrated sulphuric acid (H$_2$SO$_4$) was then added to each flask and the solution mixture was allowed to stand for at least 30 min. One hundred and fifty milliliter of distilled water was then added to each flask followed by one drop of the indicator Barium diphenylamine sulfate. The solution was then titrated with ferrous sulfate solution while stirring the mixture to the end-point (when the brown colour changes sharply to green). The amount of ferrous sulfate required for each sample for complete combustion was read and recorded.

The difference between the amounts of $FeSO_4$ added for the samples compared to that added to the blank titration determines the amount of combusted carbon. A correction factor of 0.39 was used to account for the incomplete combustion of organic carbon. The percent carbon content of the soil samples was then calculated using the following formula proposed by Van Reewijk (2002):

$$\% \ OC = M \times \frac{V1 - V2}{s} \times 0.39 \times mcf$$

where M the molarity of ferrous sulfate solution (from blank titration), $V1$ the ml of ferrous sulfate solution required for blank, $V2$ the ml of ferrous sulfate solution required for sample, S the weight of the air-dry sample in gram, mcf the moisture correction factor, while 0.39 a constant.

Statistical analysis

The data were analyzed using descriptive and multivariate statistics using SPSS version 21.0 for windows. Data distributions were checked for normality and then $\log_{10}$ transformed when they were skewed. Pearson's correlation was used to analyse the relationship between selected soil physical and chemical properties. The means and standard deviations of the selected parameters were compared to show their distribution across different land use/land cover systems and altitudes in the region. Analysis of variance ratio (ANOVA) was used to test for significant differences between the means, with treatments (land use) and group (elevation) set as the independent variables, to determine which parameters varied significantly with each treatment (Brejda et al. 2000a, b).

Results and discussion

Descriptions and characteristics of soils in the study area

In Awing and Santa (high altitude), the soils were presumably Inceptisols because they were young soils with some colour changes and have rocks at very shallow depths. In Bamenda (mid-altitude) and at the Yongka Park in particular, the soil type varied and was found to include: Oxisols, those that possessed loamy and clayey texture, slightly acidic, contain little or no weatherable minerals, traces of water dispersible clay and extreme weathering of most minerals other than quartz to kaolinite and free iron oxides, and have low CEC (< 16 cmol (+)/kg); Inceptisols, those that have rocks at vey shallow depth, and were young soils; and Entisols, those that lack pedogenic horizons and occur on slopes. The Oxisols presented their characteristic reddish-brown colour, indicating the presence of oxides of Fe, while Entisols were observed at the foot of the slope, in areas with frequent water saturation and on shoulder slopes where the rate of erosion was presumably higher than the rate of formation of pedogenic horizons. Hence, a characteristic lack of genetic horizons (Fig. 8).

Yerima (2011) stated that these soils differ in such characteristics as texture, effective depth, gravel content, compactness and water infiltration rates. At Ndop, (low altitude), the soils were observed to be rich in alluvial deposits and this is corroborated by the fact that Ndop plain is an intermontane basin in the Bamenda Highlands. The soil types observed here include; Inceptisols, Entisols, and Oxisols. It was suggested that due to these variations in soil types across elevation, important differences in physical, chemical and biological characteristics exist at regional scale.

Effect of LULC change on soil physical properties

Table 1 presents the means ($\pm$ SD) while Table 2 presents a summary of ANOVA for soil MC, BD and particle size distribution in the soil surface layer (0–15 cm) across different LULC systems and altitudes. The results for the individual soil quality attributes include:

Soil moisture content (MC) The results also show that soil moisture content significantly varied with land use types ($P < 0.01$) and with elevation ($P < 0.01$) (i.e. MC increases with increase in altitude (Table 1). Generally, at all the elevations, the soils under *Eucalyptus saligna* plantation had the highest moisture contents compared to the other land uses/land cover systems except for those under natural forest land cover systems at the high and low altitudes (Fig. 9). These results corroborate with the findings of Getachew et al. (2012).

Soil moisture content (%) at the surface (0–15 cm) showed significant differences ($P < 0.01$) between the soils of the different land uses/land cover for all the elevation (Table 2). At high altitude, the mean soil MC (%) differed significantly ($P < 0.01$) between farm and natural forest only. The study show empty patches with predominantly fern plants under Eucalyptus plantations compared to a dense, continuous layer of undergrowth including *Cynodon dactylon*, *Podocarpus* sp. found under native forest stands (Fig. 10). At mid-altitude, the mean soil MC (%) under natural savanna forest cover differed significantly ($P < 0.01$) from those under *E. saligna* plantation. Also, the mean soil MC (%) under the *E. saligna* plantation differed significantly ($P < 0.01$) from those under the afforestation plantations in the Yongka Western Highlands Research Garden Park and grazing land use systems. At low altitude, the mean soil MC (%) under Natural Forest cover differed significantly ($P < 0.01$) from those under all the other land use systems except *E. saligna* plantation. Also, the mean soil MC (%) under natural savanna forest cover and grazing land use differed significantly ($P < 0.01$) from those under *natural forest*. Generally, the lower MC

Fig. 8 Soil profiles showing; Entisol with thin A horizon (**a**); inceptisol (**b**); inceptisol with thick A horizon (**c**) and Entisol exposing stony horizons that constraint plant growth (**d**) in the North West Region (Source: Yerima (2011))

(%) in soils under *E. saligna* plantation compared to those under natural forest may be due to the observed sparse or absence of undergrowth (Fig. 11).

In agreement with Aweto and Moleele (2005), the sparse or absence of undergrowth and the light canopy of Eucalyptus trees in plantations can lead to higher rate of soil water evaporation, whereas the dense undergrowth in the native forest could lower soil temperature and reduce evaporation to enhance soil water infiltration. These results also agree with the findings of Cao et al. (2010) who reported low soil moisture contents, ranging from 20.2 to 30.5% in the topsoil (0–10 cm) under *Eucalyptus* spp. plantations, aged from 3 to 13 years in China. The sparse or absence of undergrowth is an indication of reduced biodiversity under Eucalyptus plantations.

Although House (1992) stated that the presence or absence of understorey is a factor of the density of the stand and of the rainfall regime, Zerga (2015) reported a similar finding in Ethiopia and indicated that the dominant leaf litterfall under Eucalyptus stands prohibits the growth of other plants due to its allelopathic effect. The latter explained that chemicals from the leaves of Eucalyptus trees reduce the soil nutrients that are necessary for undergrowths, hence, enhancing soil degradation through erosion, nutrient and water depletion. It was also observed in this study that the establishment of eucalyptus plantation not only suppresses undergrowths but also affects the performance of cultivated crops adjacent to the plantation stands, probably due to this allelopathic effects.

Table 1 Mean (± SD) of soil MC, BD and particle size distribution in the soil layer of 0–15 cm across different land use/land cover systems and altitudes

Soil property	Altitude	Land use types						P value
		Virgin forest	Virgin savana	Farming	Park afforestation	Grazing land	Eucalyptus forest	
MC (%)	High (> 1500 m)	8.69[a] ± 0.34	6.96[ab] ± 0.84	5.06[b] ± 1.33	–	6.96[ab] ± 0.84	6.97[ab] ± 3.31	*
	Mid (900–1500 m)	5.55[ab] ± 1.16	2.78[ac] ± 0.12	3.90[ac] ± 0.85	3.49[ad] ± 0.686	2.42[ce] ± 1.42	5.83[b] ± 2.02	*
	Low (< 900 m)	6.98[a] ± .1.984	2.31[b] ± 0.83	3.03[b] ± 1.56	–	2.30[b] ± 0.83	4.77[ab] ± 0.68	*
	Total	7.07 ± .1.81	4.01 ± 2.26	3.99 ± 1.41	3.49 ± 0.69	3.90 ± .2.438	5.86 ± 2.51	*
BD (g/cm³)	High (> 1500 m)	0.51[a] ± 0.03	0.84[b] ± 0.09	0.89[b] ± 0.09	–	0.84[b] ± .009	0.756[b] ± .151	*
	Mid (900–1500 m)	0.57[a] ± 0.11	0.87[b] ± 0.10	1.11[c] ± 0.24	0.94[bd] ± 0.09	1.19[c] ± 0.17	0.80[bd] ± .072	*
	Low (< 900 m)	0.79[a] ± 0.11	1.45[b] ± 0.00	1.08[c] ± 0.20	–	1.39[b] ± 0.08	0.80[d] ± 0.07	*
	Total	0.63 ± 0.15	1.05 ± 0.30	1.04 ± 0.21	0.94 ± 0.09	1.14 ± 0.26	0.78 ± 0.11	
Sand (%)	High (> 1500 m)	28.50[ab] ± 0.58	37.50[ab] ± 1.95	28.38[b] ± 7.15	–	37.50[ab] ± 1.91	37.0[ac] ± 5.35	*
	Mid (900–1500 m)	36.50[a] ± 5.80	34.00[a] ± 6.93	34.50[a] ± 4.98	36.92[a] ± 4.46	32.00[a] ± 4.97	35.4[a] ± 9.99	NS
	Low (< 900 m)	38.00[ab] ± 0.00	37.00[ab] ± 0.00	41.88[b] ± 5.69	–	31.50[ab] ± 8.43	35.2[ac] ± 6.702	*
	Total	34.33 ± 5.31	36.17 ± 4.09	34.86 ± 7.67	36.97 ± 4.46	33.67 ± 5.93	36.0 ± 7.427	
Silt (%)	High (> 1500 m)	57.00[ab] ± 0.00	48.00[ab] ± 2.00	56.13[b] ± 6.49	–	48.00[ab] ± 2.00	48.0[ac] ± 5.657	*
	Mid (900–1500 m)	47.00[a] ± 5.89	52.00[a] ± 5.77	48.00[a] ± 5.94	45.50[a] ± 4.10	53.00[a] ± 6.93	49.0[a] ± 10.7	NS
	Low (< 900 m)	43.00[ab] ± 0.00	47.00[ab] ± 0.00	42.00[b] ± 5.76	–	51.50[ab] ± 9.15	52.00[ac] ± 5.29	*
	Total	49.00 ± 6.88	49.00 ± 3.91	48.61 ± 7.98	45.50 ± 4.10	50.83 ± 6.46	49.20 ± 7.81	
Clay (%)	High (> 1500 m)	14.50[a] ± 0.58	14.50[a] ± 1.00	15.12[a] ± 0.99	–	14.50[a] ± 1.00	14.63[a] ± 2.77	NS
	Mid (900–1500 m)	16.25[a] ± 2.87	13.50[a] ± 1.73	17.17[a] ± 2.69	17.25[a] ± 3.77	14.75[a] ± 2.75	15.25[a] ± 2.96	NS
	Low (< 900 m)	18.50[a] ± 0.58	16.00[ab] ± 0.00	16.00[ab] ± 1.69	–	16.75[ab] ± 2.22	12.7[b] ± 3.50	*
	Total	16.42 ± 2.31	14.67 ± 0.50	16.25 ± 2.17	17.25 ± 3.77	15.33 ± 2.18	14.50 ± 2.98	

Means in the same row followed by the same letters (a, b or c) are not significantly different at 1% significance

NS non significance, *MC* moisture content, *BD* bulk density

* Significant at P < 0.01

Table 2 Summary of ANOVA for BD, MC, and particle size distribution in relation to land use and elevation

Source of variations	df	BD		MC		Clay		Silt		Sand	
		MS	P	MS	P	MS	P	MS	P	MS	P
Land use (LU)	5	0.485	0.000	23.725	0.000	17.158	0.023	37.487	0.391	20.638	0.689
Elevation (E)	2	0.653	0.000	71.025	0.000	10.801	0.181	111.44	0.049	55.171	0.197
LU * E	8	0.113	0.000	4.987	0.034	9.176	0.16	124.66	0.002	102.48	0.005
Error	80	0.019		2.246		6.177		35.486		33.329	

MS is the mean square, P is the p value, df is degree of freedom

Soil bulk density The results also showed that soil bulk density significantly varied with land use types (P < 0.01) and across elevation (P < 0.01) with significant interactions between subject effects (land use and elevation) (P < 0.01) (Table 2). Generally, at all altitudes, the soil under natural forest had the lowest bulk density (g/cm³) compared to the other land use s/land cover systems, followed by the soil under *E. saligna* plantation, while soil under farmland and grazing land had a higher bulk density (Fig. 10). These results corroborate the findings of Getachew et al. (2012)

and further depicts the altitudinal variations. In terms of absolute values however, the results for this research showed that soils under farmland had the highest bulk density (0.894 and 1.450 g/cm³), while those under the natural forest had the lowest bulk density, (0.517 and 0.790 g/cm³) in the top 0–15 cm soil layer at high and low altitudes respectively. At mid-altitude, soils under grazing land use system had the highest bulk density (1.185 g/cm³) while those under the natural forest land cover system had the lowest bulk density (0.570 g/cm³) in the top

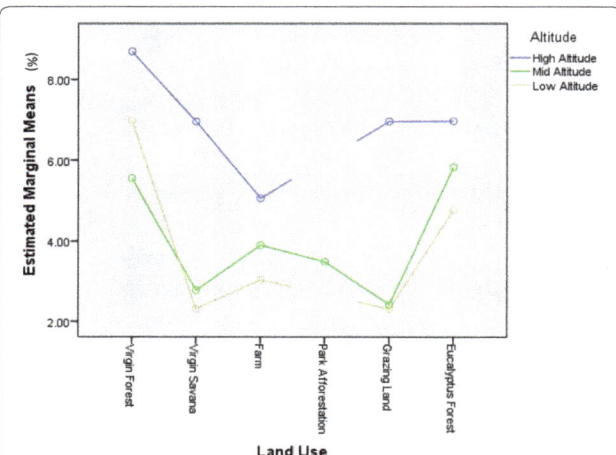

Fig. 9 Estimated marginal means of soil moisture content (%) in the soil layer of 0–15 cm across different land use/land cover systems and altitudes

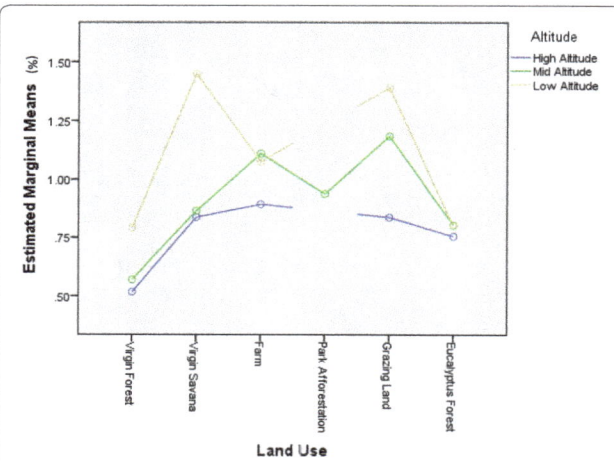

Fig. 10 Estimated marginal means of bulk density (g/cm³) in the of 0–15 cm soil layer across different land use/land cover systems and altitudes

0–15 cm soil layer. It can be suggested that deforestation and subsequent tillage practices resulted in soil compaction, low infiltration and hence increased in bulk density for surface soil in North West Region of Cameroon. Similar finding has been reported in other areas around the world (Getachew et al. 2012; Javad et al. 2014).

Nonetheless, bulk density (g/cm³) showed a significant difference ($P < 0.01$) at all altitudes, between the soils of the different land uses/land cover system for the surface 0–15 cm soil layer (Table 1). At high altitude, bulk density (g/cm³) differed significantly ($P < 0.01$) between the soils under natural forest and all the other land use/land cover systems except *E. saligna* plantations. However, at

this altitude, bulk density (g/cm³) showed no significant difference between soils under natural savanna and those under farmland, grazing land, and *E. saligna* plantation, respectively. Also, bulk density (g/cm³) showed no significant difference ($P > 0.01$) between the soils under farmland and *E. saligna* plantations. At mid-altitude, bulk density (g/cm³) also showed significant differences ($P < 0.01$) between the soils under natural forest and all the other land use/land cover systems except natural savanna and *E. saligna* plantations. At this altitude, bulk density (g/cm³) showed no significant difference ($P > 0.01$) between soils under natural savanna and all other LULC systems except grazing land. At low altitude, bulk density (g/cm³) differed significantly ($P < 0.01$) between the soils under natural forest and all the other land use/land cover systems except those under the *E. saligna* plantation. However, at this altitude, bulk density (g/cm³) showed no significant difference between soils under natural savanna and those under grazing land.

Soil bulk density represents a measure of soil compaction and health. Kakaire et al. (2015) stated that a higher soil bulk density means that less amount of water is held in the soil at field capacity, while a lower soil bulk density means soils are less compacted and are able to retain more water. These results corroborate the findings of Ravina (2012) who reported a higher soil bulk density of 1.24 g/cm³ under *Eucalyptus* spp. plantation compared to 0.66 g/cm³ under a native forest in a Brazilian soil (0–15 cm). In addition, Kolay (2000) indicated that bulk density of productive natural soils generally ranges from 1.1 to 1.5 g/cm³. Since the soil bulk densities found in all the land uses were lower and within this range, it can be concluded that the soil productivity in the area is good. Furthermore, since the soil bulk densities found in the Eucalyptus plantation, grazing land, and farmland were higher than those under native forest, it can be concluded that the conversion of forest to Eucalyptus plantations, farmland and grazing land increases soil bulk densities probably due to increased soil compaction. The findings from the study confirm those of Aweto and Moleele (2005), who concluded that *Eucalyptus* spp. plantations increased soil bulk density more than the native forest in Botswana.

Particle size distribution   Although not significant, the results show that particle size distribution varied with LULC systems across the different altitudes. However, the interactions between subject effects were only significant ($P < 0.01$) for sand and silt contents (Table 2).

Sand content

At high altitude, the soils under both natural savanna and grazing land use systems had the highest percentage of

Fig. 11 Eucalyptus plantation showing sparse undergrowth (**a**); absence of undergrowth with light canopy (**b**); native forest showing dense undergrowth (**c**); dark forest canopy with mushroom growing on plant remains (**d**)

sand content (37.5%), while those under natural forest and cropland had the lowest (28.5 and 28.4%, respectively) (Fig. 12). This i s p robably d ue t o t he f act t hat there was no existing land uses where savanna vegetation was protected against disturbance either by burning or grazing in the area. The lower sand content under farmland may be attributed to tillage practices and differential segregation by erosion on inceptisols. However, at this altitude, mean percentage sand content at the surface (0–15 cm) layer showed no significant difference ($P > 0.01$) between the soils under all the land use/land cover systems.

At mid-altitude, the soils under afforesta-tion land use systems in the Yongka Park had the high-est percentage of sand content (36.9%) while those under grazing land had the lowest (32%). Although the graz-ing land had low sand proportions at this altitude, mean percentage sand content at the surface (0–15 cm) layer showed no significant difference ($P > 0.01$) between the

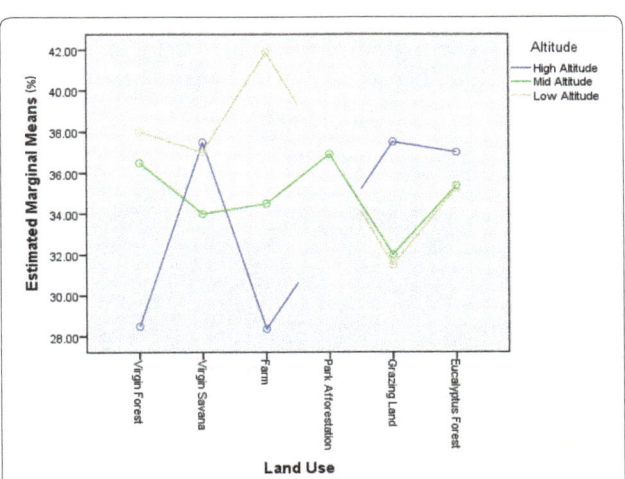

Fig. 12 Estimated marginal means of sand (%) in the 0–15 cm soil layer across different land use/land cover systems and altitudes

soils under all the land use/land cover systems. This can be due to the fact that the soils are oxisols, originating from a granitic parent material, characterized by colloidal fractions and dominated by low activity clays and sesquioxides (Yerima and Van Ranst 2005b).

The soils under afforestation land use system in the park are varied, which is probably a representation of the different stages of soil development. The park area was formerly a grazing land, with an unstable geomorphic surface, which resulted in soil erosion and exposure. The gullies observed under the afforestation stand with *Artocarpus heterophyllus* (Jackfruit) is associated with the concentration of runoff water from the road (Fig. 13).

At low altitude, the soils under farmland had the highest percentage of sand content (41.9%) while those under grazing land use systems had the lowest (31.5%). Again, at this altitude, mean percentage sand content at the surface (0–15 cm) layer showed no significant difference ($P > 0.01$) between the soils under all the land use/land cover systems. There was no existing land uses where natural forest cover and savanna vegetation was protected against disturbance either by burning or grazing in this area. Therefore there were high similarities with nonsignificant differences in soil properties under the different land uses in this area. Generally, the results showed that sand content increased when converting natural forest to cropland, and this is most likely resulting from the preferential removal of clay and silt and residual accumulation of sand in soil surface resulting from preferential segregation and evacuation of the smaller silt and clay particles, by accelerated water erosion. These results are in agreement with the findings of Javad et al. (2014) who attesting to the results of Ayele et al. (2013) reported that sand content is a physical parameter affected by soil erosion and, hence, can be measured and used as an indicator for evaluating soil degradation under different land use systems.

Silt content

At high altitude, the soils under the natural forest land cover system had the highest percentage of silt content (57%), while those under farmland was intermediate (56%) (Fig. 14). Those under the natural savanna, grazing land and *E. saligna* plantation had the lowest percentages and of equal values (48%). However, at this altitude, mean percentage silt content at the surface (0–15 cm) layer showed no significant difference ($P > 0.01$) between the soils under all the LULC systems.

It was noticed that the area under natural savanna forest cover at high altitude was also used for grazing. The frequent burning of grass during the dry season by

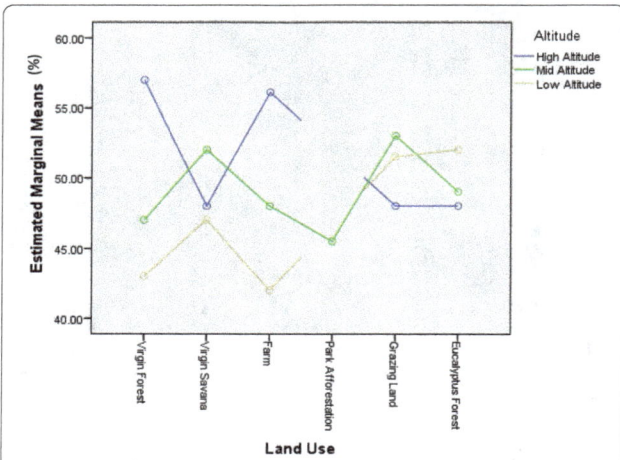

Fig. 14 Estimated marginal means of silt (%) in the 0–15 cm soil layer across different land use/land cover systems and altitudes

Fig. 13 *Artocarpus heterophyllus* afforested stand (**a**); gully erosion under the *Artocarpus heterophyllus* stand (**b**)

cattle herdsmen and trampling effects due to overgrazing is suggested to have influence soil structure as burning destroys and removes soil organic matter, thereby loosening of soil particles and encouraging water erosion on gentle slopes. In addition, the *E. saligna* plantation here was located on steep slopes, with abundant leaf litter and no vegetation understory. It can be suggested that the silt content under natural savanna, grazing land and *E. saligna* plantation were lower due to accelerated water erosion. The lack of ground cover and understory may have contributed to initiating erosion which selectively washes away clay and silt. At mid-altitude, the soils under grazing land use systems had the highest percentage of silt content (53%) while those under the afforestation land use projects in the Yongka Park had the lowest percentage (47%). At this altitude, mean percentage silt content at the surface (0–15 cm) layer showed no significant difference ($P > 0.01$) between the soils under all the land use/land cover systems. The soils under the afforestation stand were located on the shoulder of the slope and signs of severe erosion (rills and gullies) were observed, compared to those of grazing land. Geologically, on a midslope, the rate of soil erosion is increased and the topsoil layer is greatly reduced. At low altitude, the soils under the *E. saligna* plantation had the highest percentage of silt content (52%) while those under farmland use systems had the lowest (42%). At this altitude, mean percentage silt content at the surface (0–15 cm) layer showed no significant difference ($P > 0.01$) between the soils under all the land use/land cover systems. Here the soils under the *E. saligna* plantation were located on the toe slope (flat surface) and signs of severe deposition (siltation, floods and stagnant water bodies) were generally observed in the area after heavy rainfall. Geologically, on the toe slope, the rate of soil erosion is minimal and the topsoil layer is mostly comprised of mineral deposits transported principally by water (erosive agent) from top and shoulder slopes. However, soils under farmland use systems are prone to erosion comparatively.

Clay content

The result shows that at high altitude, the soils under farmland use system had the highest percentage of clay content (15%), while those under *E. saligna* plantation was intermediate (14.6%) (Fig. 15). Those under the natural savanna, natural forest, and grazing land had the lowest percentages and of equal values (14%). At this altitude, mean percentage clay content at the surface (0–15 cm) layer showed no significant difference ($P > 0.01$) between the soils under all the land use/land cover systems.

At mid-altitude, the soils under the afforestation land use projects in the Yongcak Park had the highest percentage of clay content (17.3%) while those under the

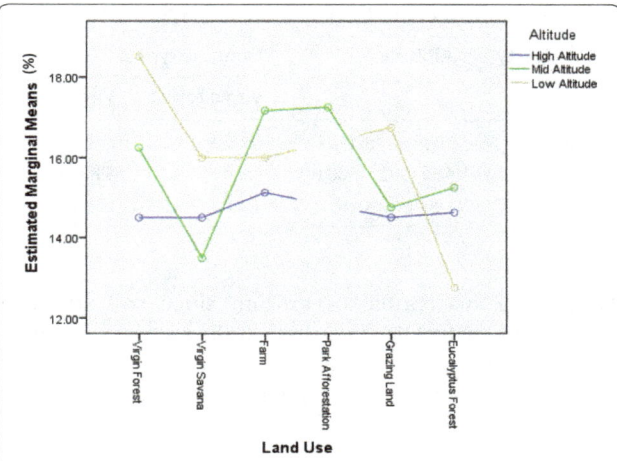

Fig. 15 Estimated marginal means of clay (%) in the 0–15 cm soil layer across different land use/land cover systems and altitudes

natural savanna land cover had the lowest percentage (13.5%). At this altitude, mean percentage clay content at the surface (0–15 cm) layer showed no significant difference ($P > 0.01$) between the soils under all the land use/land cover systems. At low altitude, the soils under the natural forest land cover system had the highest percentage of clay content (18.5%) while those under *E. saligna* plantation had the lowest (12.8%). At this altitude, mean percentage clay content at the surface (0–15 cm) layer showed no significant difference ($P > 0.01$) between the soils under all the land use/land cover systems. However, the percentage clay content was relatively lower than those of sand and silt in all the land use/land cover systems across the different altitudes. When fine particles of soils are high, EC may increase. However, increased EC in soils is predominantly due to the presence of soluble salts (Yerima and Van Ranst 2005a), but which is not found in the study area, and this may cause instability of soil structure.

Soil texture　Generally, the results show that the soil texture in the study area ranged from loam to silt loam, which is very good for agriculture. Specifically, the result shows that on one hand, the soil texture under natural vegetation cover and farmland use systems in the NW region changed from silt loam to loam, as elevation decreases (Table 3). On the other hand, the soil texture under grazing land use system and *E. saligna* plantation changed from loam to silty loam as elevation decreases. This is suggested to be due to pedogenic processes including degradation (surficial erosion) and aggradation (cumulization) (Yerima and Van Ranst 2005a).

Ideally, the conversion of forest into cropland is known to deteriorate soil physical properties and making the

Table 3 Soil texture in the 0–15 cm soil layer across different land use/land cover systems and altitudes

Soil property	Altitude	Land use types					
		Virgin forest	Virgin savana	Farming	Park afforestation	Grazing land	Eucalyptus forest
Texture	High (> 1500 m)	Silt loam	Loam	Silt loam	–	Loam	Loam
	Mid (900–1500 m)	Loam	Silt loam	Loam	Loam	Silt loam	Loam
	Low (< 900 m)	Loam	Loam	Loam	–	Silt loam	Silt loam

land more susceptible to erosion since soil structure (macroaggregates) is disturbed. Soil erosion can modify soil properties by reducing soil depth, changing soil texture, and by the loss of nutrients and organic matter (Lobe et al. 2001).

Effects of land use change on soil chemical properties

Generally, the chemical properties of soils show variations under the different land uses across the different altitudinal zones of the study area. Table 4 presents the mean (± SD) while Table 5 present the summary of

Table 4 Mean (± SD) of soil pH, SOC, TN, and Av.P in the 0–15 cm soil layer across different land use/land cover systems and altitudes

Soil property	Altitude	Land use types						ANOVA
		Virgin forest	Virgin savana	Farming	Park afforestation	Grazing land	Eucalyptus forest	
pH H$_2$O	High	5.35 ± .288	5.90 ± 0.20	5.21 ± 0.538		5.90 ± 0.200	5.57 ± 0.183	*
	Mid	5.63 ± 0.125	5.65 ± .057	5.48 ± 0.540	5.70 ± 0.159	6.10 ± .316	5.56 ± 0.130	*
	Low	6.50 ± 0.230	6.50 ± .000	5.87 ± 0.205		6.00 ± .496	5.77 ± 0.221	*
	Total	5.82 ± 0.551	6.02 = .380	5.51 ± 0.520	5.70 ± .159	6.00 ± 0.335	5.61 ± 0.183	
pH KCl	High	4.45 ± c.288	4.65 ± 0.06	4.34 ± 0.27		4.65 ± 0.06	4.36 ± 0.23	*
	Mid	4.65 ± 0.10	4.50 ± 0.00	4.58 ± 0.06	4.46 ± 0.12	4.60 ± 0.22	4.56 ± 0.11	*
	Low	5.55 ± 0.17	5.30 ± 0.00	5.01 ± 0.32		4.88 ± 0.30	4.58 ± 0.15	*
	Total	4.88 ± 0.53	4.81 ± 0.36	4.63 ± 0.34	4.46 ± 0.12	4.71 ± 0.23	4.49 ± 0.19	
Δ pH	High	− 0.90 ± 0 00	− 1.25 ± 0.17	− 0.88 ± 0.37		− 1.25 ± 0.17	− 1.21 ± 0.25	
	Mid	− 0.98 ± 0 15	− 1.15 ± 0.06	− 0.91 ± 0.56	− 1.24 ± 0.14	− 1.50 ± 0.34	− 1.00 ± 0.16	
	Low	− 0.95 ± 0.06	− 1.20 ± 0.00	− 0.86 ± 0.18		− 1.25 ± 0.38	− 1.20 ± 0.16	
	Total	− 0 94 ± 0.09	− 1.20 ± 0.10	− 0.89 ± 0.41	− 1.24 ± 0.14	− 1.29 ± 0.32	− 1.13 ± 0.22	
SOC (%)	High	6.50 ± 1.52	4.83 ± 0.36	3.31 ± 0.57		4.83 ± 0.36	5.79 ± 2.11	*
	Mid	4.58 ± 1.51	2.10 ± 1.15	3.04 ± 0.87	2.83 ± 0.88	2.50 ± 0.83	3.46 ± 0.95	*
	Low	3.10 ± 0.31	3.10 ± 0.00	1.83 ± 0.90		1.90 ± 1.18	3.05 ± 0.53	*
	Total	4.73 ± 1.93	3.34 ± 1.33	2.77 ± 0.99	2.83 ± 0.88	3.08 ± 1.52	4.31 ± 1.89	
C/N ratio	High	21.91 ± 0.76	14.58 ± 2.39	13.06 ± 4.50		14.58 ± 2.39	24.74 ± 15.25	
	Mid	16.95 ± 8.63	9.10 ± 3.82	17.68 ± 6.53	12.05 ± 4.76	13.37 ± 6.31	17.36 ± 7.18	
	Low	11.18 ± 3.34	23.85 ± 0.00	8.65 ± 3.94		11.13 ± 9.31	14.00 ± 4.18	
	Total	16.68 ± 6.67	15.84 ± 6.78	13.78 ± 6.43	12.05 ± 4.76	13.03 ± 6.18	19.64 ± 11.28	
TN (%)	High	0.30 ± 0.06	0.34 ± 0.07	0.27 ± 0.06		0.34 ± 0.06	0.27 ± 0.11	*
	Mid	0.29 ± 0.08	0.22 ± 0.03	0.18 ± 0.03	0.25 ± 0.07	0.20 ± 0.03	0.21 ± 0.05	Ns
	Low	0.28 ± 0.01	0.13 ± 0.00	0.21 ± 0.05		0.19 ± 0.06	0.23 ± 0.04	Ns
	Total	0.29 ± 0.05	0.23 ± 0.09	0.21 ± 0.06	0.25 ± 0.07	0.24 ± 0.08	0.24 ± 0.08	
Av.P (ppm)	High	16.1 ± 3.00	19.1 ± 13.3	20.1 ± 10.3		19.1 ± 13.3	13.4 ± 7.19	Ns
	Mid	14.6 ± 3.74	15.6 ± 4.91	15.5 ± 7.65	9.83 ± 3.53	9.33 ± 4.16	11.1 ± 4.31	Ns
	Low	13.3 ± 2.25	5.30 ± 0.00	13.2 ± 5.61		8.90 ± 2.40	11.9 ± 2.39	Ns
	Total	14.6 ± 3.02	13.3 ± 9.59	16.1 ± 8.17	9.83 ± 3.53	12.4 ± 8.87	12.2 ± 5.28	

Means in the same row followed by the same letters are not significantly different at 1% significance

Ns non significance, *MC* moisture content, *BD* Bulk density

* Significant at P < 0.01

Table 5 Summary of ANOVA for pH, SOC, TN, and available P in relation to land use and elevation

Source of variations	df	Av.P		SOC		TN		C:N ratio		pH H$_2$O	
		MS	P	MS	P	MS	P	MS	P	MS	P
Land use (LU)	5	71.178	0.167	8.659	0.000	0.010	0.018	96.951	0.076	0.669	0.000
Elevation (E)	2	302.101	0.002	40.292	0.000	0.065	0.000	103.16	0.116	1.943	0.000
LU * E	8	39.388	0.892	2.539	0.035	0.008	0.027	149.80	0.003	0.286	0.008
Error	80	44.179		1.14		0.004		46.552		0.101	

MS in the mean square, P is the p value, df is degree of freedom

ANOVA for soil pH, SOC, TN, and Av.P in the 0–15 cm soil layer across different LULC systems and altitudes in the study area.

Effects of land use change on soil pH H$_2$O, pH KCl and Δ pH The results also showed that soil pH significantly varied with land use types ($P < 0.01$) and across elevation ($P < 0.01$) with significant interaction between subjects effects ($P < 0.01$) (Table 4). According to Landon (1991) ratings, the soil pH for all the land uses in this study were low to medium (slightly acidic), probably due to the parent material (granitic) which is acidic in nature and are characteristic of oxisols. The results show that the soil under farmland use had relatively lower pH H$_2$O and pH KCl values and lower net charges compared to those of the other land uses/land cover systems in the area. In general, soil pH decreased with increase in altitude (Figs. 16, 17 and 18). This may be largely due to the use of chemical fertilizers including urea, potash, and N, K, P (20:10:10), as well as the high use of weedicides such as roundup, by farmers in the area which contain high amounts of cations that helps to neutralise the negative charges. The results show that there is a net negative charge for all the soils in

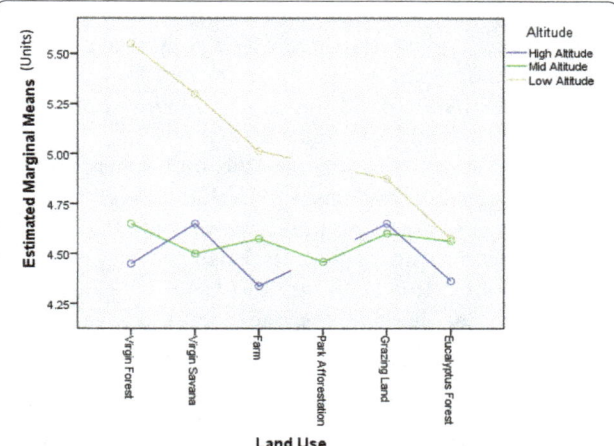

Fig. 17 Estimated marginal means of pH KCl values in the 0–15 cm soil layer across different land use/land cover systems and altitudes

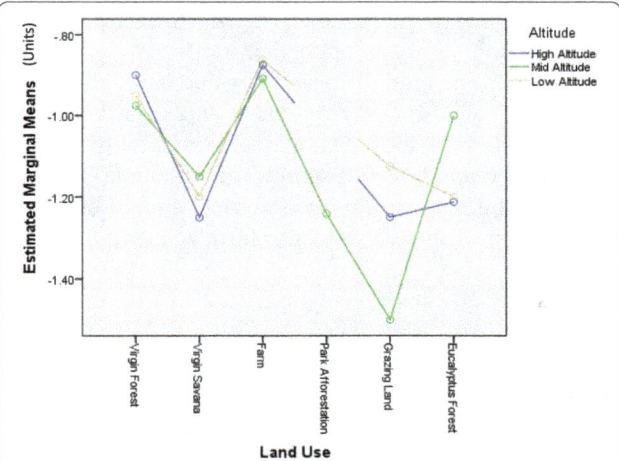

Fig. 18 Estimated marginal means of net charge (Δ pH) values in the 0–15 cm soil layer across different land use/land cover systems and altitudes

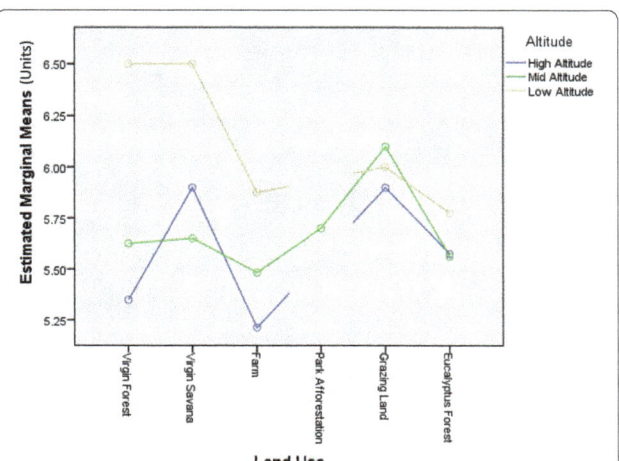

Fig. 16 Estimated marginal means of pH H$_2$0 values in the 0–15 cm soil layer across different land use/land cover systems and altitudes

the area under study. The weathering of the granitic parent material, which results in iron and aluminum oxides, as well as leaching of more soluble soil minerals and basic cations, may have caused the slight acidity of the soils in the study area. The net charge was less negative under native forest systems compared to grazing, savanna, Eucalyptus plantation and afforestation systems.

At high altitude, the soils under farmland use system had the lowest mean pH value (5.2) while those under natural savanna and grazing land had the highest mean pH value (5.9). At this altitude, mean pH values at the surface (0–15 cm) layer showed a significant difference ($P < 0.01$) between the soils under natural forest and natural savanna, as well as between those under farmland, natural savanna and grazing land use systems. The results also show that the soils under natural forest cover and *E. saligna* plantation had a relatively lower pH values (pH < 5.5) in this area. This is because Low pH slows down the breakdown of litter due to low microbiological activity (Yerima and Van Ranst 2005a).

Similarly, at mid-altitude, the soils under farmland use systems had the lowest mean pH value (5.4) followed by those under *E. saligna* plantation (5.5), while those under natural savanna and grazing land had the highest mean pH value (6.1). This similarity in pH values can be due to the fact that the land use under natural savanna was also used for cattle grazing. At this altitude, mean pH values at the surface (0–15 cm) layer showed a significant difference ($P < 0.01$) between the soils under farmland and grazing land use systems, as well as between soils under *E. saligna* plantation and grazing land use system. The relatively high pH values under grazing land may be due to the high cow dung wastes deposited on the fields. Interestingly, at low altitude, the soils under *E. saligna* plantation had the lowest mean pH value (5.7) while those under natural forest and natural savanna had the highest mean pH value (6.5). Humphrey and Amawa (2014) also reported a similar finding in the study area. At this altitude, mean pH values at the surface (0–15 cm) layer showed a significant difference ($P < 0.01$) between the soils under natural vegetation (forest and savanna) cover and those under farmland and *E. saligna* plantation.

Generally, the slightly acidic nature of the soils under all the land use/land cover systems may be due to the weathering of granitic parent materials and the intense leaching of basic cations. Also, the low pH values in farmland could be due to high tillage frequency, high rates of inorganic fertilizer applications (especially ammonium fertilizers), low amount of organic matter as a result of erosion or due to aluminum toxicity. In fact, the significant differences of mean pH values at the surface (0–15 cm) layer between soils under cultivated land and those under natural vegetation across different altitudes indicates that the conversion of natural vegetation cover (forest and savanna) to farmland decreases pH of the

soil, with increased efficiency as altitude increases, in the North West Region of Cameroon. This result contradicts the findings of Kizilkaya and Dengiz (2010) who reported a significant increase of pH from 6.03 in soils under natural forest to 7.71 in soils under cultivated land in Turkey. Although these are two different environments, this difference may be due to the fact that sustainable agricultural land management practices such as the application of organic manure, mulching, rotation and limited tillage were adopted in the cultivated land in Turkey while those in our study area did not adopt sustainable land management practices. It could also be due to the high basic fertilizer applications. Low soil pH impeds the CEC of the soil by altering the surface charge of colloids (finest clay particles and soil organic matter) (McCauley et al. 2005). Low CEC implies that soil will have less exchangeable cations required as crop nutrients, nutrients are weekly adsorbed and hence may be leached out.

Effects of land use change on soil organic carbon (SOC) and organic matter (SOM) The results also showed that SOC significantly varied with land use types ($P < 0.01$) and with elevation ($P < 0.01$) but the interaction between subjects effects was not significant ($P > 0.01$) (Table 5). Following the ratings by Yerima and Van Ranst (2005a), the results show that all the soils under the different land use systems across the different elevations were medium to high in SOC content. However, the percentage of OC concentration for soils under natural forest as well as those under *E. saligna* plantations were higher compared to all the other land use systems at all elevations (Fig. 19). In addition, the results indicate that OC in soils increased with increase in elevation in the N.W region. These results corroborate the findings of Yerima and Van Ranst (2005a).

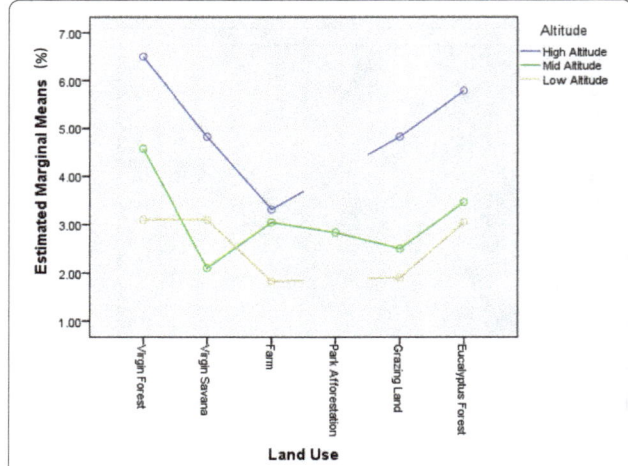

Fig. 19 Estimated marginal means of soil organic carbon (%) in the 0–15 cm soil layer across different land use/land cover systems and altitudes

At high altitude, the soils under natural forest land cover system (protected forest) had the highest mean soil organic carbon content (6.50%) followed by those under *E. saligna* plantations (5.79%) (Table 4). Those under farmland use systems had the lowest mean percentage of soil organic carbon content (3.31%). At this altitude, the mean percentage of SOC content at the surface (0–15 cm) layer showed a significant difference ($P < 0.01$) between the soils under natural forest and farmland. Also, the results showed a significant difference ($P < 0.01$) between the soils under *E. saligna* plantations and those under farmland. However, there were no significant differences ($P > 0.01$) between soils under natural savanna and grazing land. This may be due to the fact that the natural savanna land cover systems were observed to be subjected to conditions similar to grazing land use (burning and cattle grazing) in the area (Fig. 20). More so, the relatively higher soil organic carbon contents at higher elevations may be due to a slow down in the rate of organic

matter mineralization associated with low temperatures and decrease microbial activity.

At mid-altitude, the soils under natural forest land cover systems still had the highest mean SOC content (4.58) followed by those under *E. saligna* plantation (3.46%) (Table 4). Although the mean percentage of SOC contents for soils under farmland use systems were low (3.04%), those under the natural savanna land use systems had the lowest mean percentage (2.10%) compared to all the other land use/land cover systems. However, at this altitude, the mean percentage of SOC content at the surface (0–15 cm) layer showed significant differences ($P < 0.01$) between the soils under natural forest cover and savanna, afforestation, and grazing land use systems.

Although the natural savanna land cover system was within the Yongka Garden Park, the results showed that the soil was low in soil organic carbon content compared to those under the natural forest cover. This can be suggested to be due to the fact that the land was previously

Fig. 20 Sheep and cattle grazing on grazing land (**a**); cattle grazing on Eucalyptus plantation (**b**); fire disturbance on grazing land (**c**); burnt tree trunks showing evidence of fire disturbance on Eucalyptus plantation (**d**)

used for cattle grazing, and soils in the area were generally low in organic matter before the establishment of the park (9 years old, after many years of cattle grazing), due to unsustainable land use practices and soil degradation. Results from interview with researchers from the park, on the land use history of the area reveals that after several years of intensive exploitation, prior to the establishment of the Park, mining of the soil organic carbon and nutrient stocks was an issue, which led to a decline in fertility, compaction of surface soils and slow and poor regeneration of vegetation on land left for fallow. They added that issues of water and fuelwood scarcity in the nearby community were prominent at the time. This result is in line with the findings of Humphrey and Amawa (2014), who stated that intensification of agriculture and the use of inappropriate cultural practices including the cultivation on fragile (steep) hill-slopes, setting of bushfires, overgrazing by cattle, building of settlements, and increased consumption of the regions fuelwood, has lead to environmental and soil degradation in the area. In fact, Yerima (2011) elucidated that soils in the park are acidic in nature, have low nutrient contents with compact and dense structures that are an impediment to plant growth. These are tangible reasons attracting afforestation initiatives and other sustainable land use practices that promote conservation of biodiversity in the area. Hence, it is unarguable that the SOC values reported in this study only presents a picture of the ongoing land regeneration efforts that may require a longer period of time to show its actual image.

At low altitude, the soils under natural forest and those under natural savanna land cover systems had the highest mean soil organic carbon content (3.10%) followed by soils under *E. saligna* plantations (3.05%) (Table 4). Those under farmland use systems had the lowest mean percentage of soil organic carbon content (1.83%). However, at this altitude, the mean percentage of soil organic carbon content at the surface (0–15 cm) layer showed no significant differences ($P > 0.01$) between the soils under all the land use/land cover systems. This can be attributed to the higher soil temperatures, which increases the rate of mineralization due to increase in microbial activity. In addition, the area is located at the toe slope where the difference in slope gradient influences erosion, flooding and subsequent deposition of inorganic materials downslope. Deposition of inorganic materials through erosion buries the topsoil which is normally rich in organic matter, hence the true picture of soil organic carbon is blurred at low altitude. As the soils under natural savanna were subjected to grazing practices, it is possible that the soil samples collected were influenced by animal waste deposition, hence the high SOC value recorded. Furthermore, cultivated soils generally have low organic matter

content compared to native ecosystems, since cultivation increases aeration of soil, which enhances decomposition of soil organic matter (Kizilkaya and Dengiz 2010).

These results conform to the assertion that SOC stocks are sensitive to land use and cover change (Guo and Gifford 2002; Wiesmeier et al. 2012) probably because of the alterations of both carbon inputs (amount and quality of litter mass) and losses (decomposition and mineralization). Soil carbon improves the physical properties of soil, increases the cation exchange capacity (CEC) and water-holding capacity of the soil, and contributes to the structural stability of soils by helping to bind particles into aggregates (Leeper and Uren 1993). It can be suggested that, anthropogenic activities that accentuate SOC loss in the soil including tillage (hoeing, plowing), biomass burning, residual removal, overgrazing, and drainage, are responsible for the distribution of SOC contents observed, at different elevations, under the different land use/land cover systems in the N.W region.

Since carbon is a fundamental constituent of soil organic matter (Ogle et al. 2005), the trends in effects of soil organic carbon under the different land use/land cover systems in the N.W region mirrors that of SOM and hence carbon storage of soil. It is well recognized that SOM increases structural stability, resistance to rainfall impact, the rate of infiltration and faunal activities (Roose and Barthes 2001). SOM, of which carbon is a major part, holds a great proportion of nutrients cations and trace elements that are of importance to plant growth. According to Leu (2007), it prevents nutrient leaching and is integral to the organic acids that make nutrients accessible to plants while acting as a soil buffer to resist strong changes in pH. It is widely accepted that the carbon content of soil is a key element in its overall health (Yerima and Van Ranst 2005a).

Effects of land use change on soil total nitrogen   Soil total nitrogen is typically used as an important index for soil quality evaluation and reflects the soil nitrogen status (Sui et al. 2005). Similar to soil organic carbon, soil total nitrogen content (%) also exhibited obvious differences at the surface (0–15 cm) layer under different land use/land cover systems across the three elevations (Table 4). The results also showed that soil total nitrogen did not vary significantly with land use types ($P < 0.01$), but varied significantly with elevation ($P < 0.01$). However, the interaction between subjects' effects was not significant ($P < 0.01$) (Table 4). Generally, the results show that soil total nitrogen contents increase with an increase in elevation in the N.W region (Fig. 21).

The study showed that at high altitude, both the soils under savanna and grazing land use systems had the highest mean soil total nitrogen content (0.36%) followed

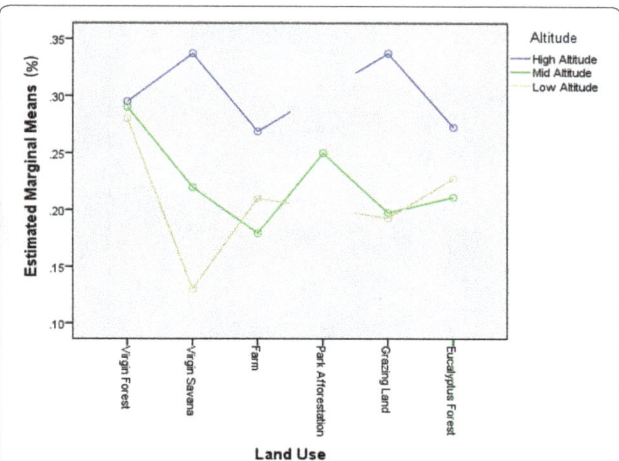

Fig. 21 Estimated marginal means of nitrogen (%) in the 0–15 cm soil layer across different land use/land cover systems and altitudes

by those under natural forest cover (0.30%), while those under farmland system had the lowest. At this altitude, mean soil total nitrogen at the surface (0–15 cm) layer showed no significant difference ($P > 0.01$) between the soils under all the land use/land cover systems. It was observed that pasture density under grazing land use systems at the higher altitude was higher compared to that at lower altitudes. This has a direct relationship with organic matter content. Again, the similarities in results between savanna and grazing land may probably be due to the similarities in observed conditions (burning and cattle grazing) under the two land use systems in the area. The high amounts of nitrogen in soils under grazing land may be due to the burning of grass that produce ash which is rich in nitrogen and other major nutrients. It is well known that fire simulates cycling of nitrogen and thus, relatively high amounts of nitrogen in the ash could be found under such disturbed land use sites. Grazing land is also subjected to deposition of cow dung waste, which enhances the soil organic matter content. A high soil organic matter content strongly correlates with high nitrogen content. However, strong fires on sandy soils may give long-lasting loss of soil surface humus and nitrogen, which may lead to site impoverishment. Also, studies have suggested that grazing can promote nutrient cycling because livestock feces and urine provide large amounts of soluble nitrogen that is readily available to plants for growth and livestock excretions can promote soil organic matter (SOM) mineralization rates (McNaughton et al. 1997).

At mid-altitude, the soils under natural forest land cover system had the highest mean soil total nitrogen content (0.29%) while those under farmland had the lowest (0.18%). At this altitude, mean soil total nitrogen at

the surface (0–15 cm) layer showed no significant difference ($P > 0.01$) between the soils under all the land use/land cover systems. Although the level of nitrogen fertilizer use as agric-input was high in the region, the total nitrogen contents are low in farms, probably because of the poor nitrogen retention ability of the soils under farmland uses and the loss of organic matter which is a source of nitrogen. At low altitude, the soils under natural forest land cover systems also had the highest mean soil total nitrogen content (0.28%), while those under savanna land use had the lowest values (0.19%). At this altitude, mean soil total nitrogen at the surface (0–15 cm) layer showed significant differences ($P < 0.01$) only between the soils under natural forest vegetation and those under natural savanna land cover systems. This can be due to the fact that soils under savanna cover were disturbed by human activities including; burning and grazing, which greatly influence the soil organic matter and hence soil nitrogen content.

Effects of land use change on soil C/N ratio　　The results also show that the C/N ratio had no significant differences with land use types ($P > 0.01$) and with elevation ($P > 0.01$) but the interaction between subjects effects was significant ($P < 0.01$) (Table 4). However, the results indicate an increase in C/N ratio with an increase in elevation in the N.W region except for soils under savanna and farmland use systems (Fig. 22). The quality of organic matter is expressed in form of the C/N ratio. According to ratings by Yerima and Van Ranst (2005a), the results show that the quality of organic matter in all the soils under the different land use systems across the different elevations ranged between good quality, medium, and low quality. At high altitude, the mean C/N ratio for soils under natural

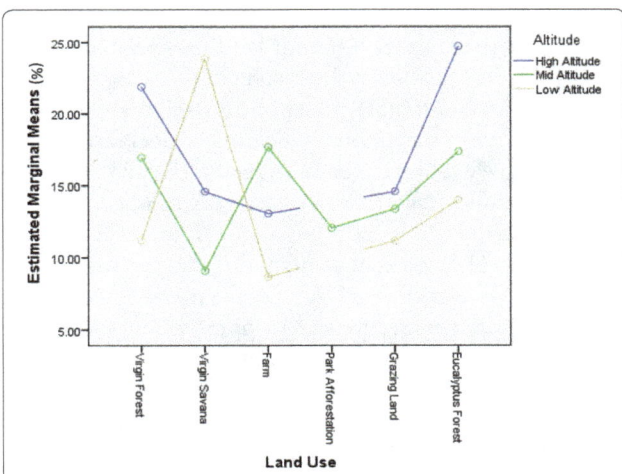

Fig. 22 Estimated marginal means of C/N ratio in the 0–15 cm soil layer across different land use/land cover systems and altitudes

forest, savanna, and grazing land, as well as those under *E. saligna* plantations, reveals that the quality of organic matter was low (C/N > 14). Only soils under farmland had the medium quality of organic matter (C/N = 10–14).

At mid-altitude, the mean C/N ratio for soils under savanna forest cover reveals that the organic matter was of good quality (C/N < 10), while those for soils under afforestation plantations and grazing systems were of medium quality (C/N = 10–14). However, those under natural forest, farmland and *E. saligna* plantations reveal that the quality of organic matter was low (C/N > 14). At low altitude, the mean C/N ratio for soils under farmland reveals that the quality of organic matter was good (C/N < 10), while those for soils under natural forest, *E. saligna* plantations and grazing land were of medium quality (C/N = 10–14). However, those under natural forest cover reveals that the quality of organic matter was low (C/N > 14).

Soil C/N ratio is a sensitive indicator of soil quality. The soil C/N ratio is usually considered as an indicator of soil nitrogen mineralization ability. High C/N ratios in soils can retard the rate of organic matter and organic nitrogen decomposition by limiting the ability of soil microbial actions, whereas low C/N ratios in soils could accelerate the process of microbial decomposition of organic matter and nitrogen. However, Wu et al. (2001) reported that low soil C/N ratio is not conducive to carbon sequestration. Therefore, it can be concluded that soils under natural forest, grazing land, and *E. saligna* plantations slow down the decomposition rate of organic matter and organic nitrogen by limiting the soil microbial activity ability and can best sequester carbon in the region as a means to combat climate change.

Effects of land use change on soil available phosphorus Though not significant, the soil available phosphorus also exhibited obvious differences at the surface (0–15 cm) layer under different land use/land cover systems (Table 4). However, the results indicate that concentrations of soil available phosphorus varied significantly ($P < 0.01$) with elevation, indicating a decrease with a decrease in elevation in the N.W region (Fig. 23) although the interaction effect between subjects was not significant ($P > 0.01$).

The study showed that at high altitude, the soils under farmland use systems had the highest mean soil available phosphorus concentrations (20.1 ppm) followed by those under savanna and grazing land use system (19.1 ppm), while those under *E. saligna* plantation land use had the lowest value (13.4 ppm) (Table 4). This can be due to the fact that farmers applied fertilizers such as Diammonium phosphate (DAP) on their farmlands. However, at this altitude, mean soil available phosphorus concentrations

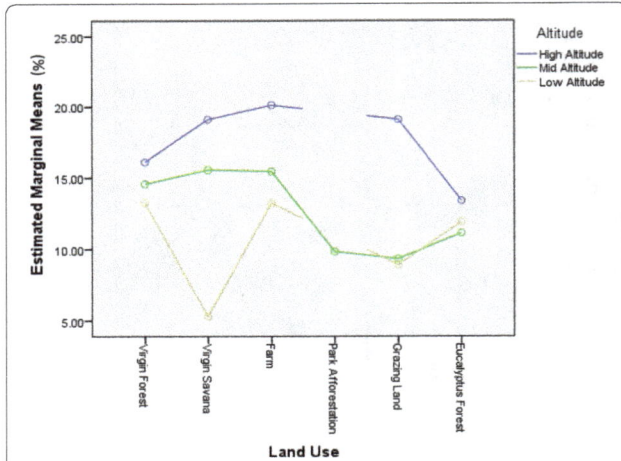

Fig. 23 Estimated marginal means of available phosphorus (ppm) in the 0–15 cm soil layer across different land use/land cover systems and altitudes

at the surface (0–15 cm) layer showed no significant difference ($P > 0.01$) between the soils under all the land use/land cover systems. These results are consistent with the findings of Awdenegest et al. (2013).

At mid-altitude, the soils under natural savanna, and those under farmland had the highest mean soil available phosphorus concentrations (15.6 and 15.5 ppm, respectively), while those under natural forest and *E. saligna* plantation land cover systems were intermediate (14.6 and 11.1 ppm, respectively). Those under afforestation and grazing land use had the lowest values (9.83 and 9.33 ppm, respectively). It can be concluded here that residue ash may have enhanced the P concentrations under natural savanna, and those under farmland because burning was a cultural phenomenon under these land use systems in the area. Also, the application of chemical P-fertilizers and organic manure (poultry manure and cow dung) may have enhanced the P concentrations in selected farmlands in this area. At this altitude, mean soil available phosphorus concentrations at the surface (0–15 cm) layer showed no significant difference ($P > 0.01$) between the soils under all the land use/ land cover systems.

At low altitude, the soils under natural savanna, and those under farmland had the highest mean soil available phosphorus concentrations (15.6 and 15.5 ppm, respectively), while those under natural forest land cover were intermediate (14.6 ppm), while those under savanna land use had the lowest values (5.30 ppm). At this altitude, mean soil available phosphorus concentrations at the surface (0–15 cm) layer showed no significant difference ($P > 0.01$) between the soils under all the LULC systems. It is suggested that the relatively lower available

Effects of land use change on soil physicochemical properties in selected areas in the North West...

173

phosphorus concentrations in the protected forest and *E. saligna* plantations at all elevations may be related to phosphorus fixation due to the relatively higher organic matter concentrations under these land use systems. This result agrees with the findings of Yimer et al. (2008) who reported higher concentrations of P in soils of the native forest than those of cropland and grazing in the Bale Mountains of Ethiopia.

According to ratings by Yerima and Van Ranst (2005a), available phosphorus across all land uses was very low, except those in the top 0–15 cm soil layer of farmlands, at high altitude which was low. In addition, the available phosphorus in soils under all the land uses in the study area falls below the medium sufficiency range of 26–54 mg/kg suggested by Carrow et al. (2004). The available phosphorus deficiency in soils of our study area may be due to the inherent low-P status of the parent material and erosion loss. This may also be due to the low soil pH which causes P-fixation. These results confirm the findings of Yerima and Van Ranst (2005b) who reported that the available phosphorus in most soils of the North West region is low due to P-fixation, crop harvest, and erosion by water.

Effects of land use change on cation exchange capacity (CEC, cmol (+)/kg soil) CEC also exhibited some differences at the surface (0–15 cm) layer under different land use/land cover systems, although not significant. Also, there were no significant differences ($P > 0.01$) with elevations and the interaction between subjects effects was not significant ($P < 0.01$) (Table 4). Generally, the results did not show a clear picture of the variation of CEC of soils under different land use/land cover systems with elevation, in the N.W region (Fig. 24).

The study shows that at high altitude, both the soils under savanna and grazing land use systems had the highest mean concentration of CEC (23.1 cmol (+)/kg soil), followed by those under farmland (19.7 cmol (+)/kg Soil), while those under *E. saligna* plantation and natural forest cover were relatively lower (17.6 and 16.3 cmol (+)/kg soil, respectively) (Table 4). However, at this altitude, mean CEC concentration of soils at the surface (0–15 cm) layer showed no significant difference ($P > 0.01$) between the soils under all the land use/land cover systems. These results contradict the findings of Awdenegest et al. (2013) who reported that the CEC concentration was low in oxisols under savanna and grazing land use systems compared to farmland use systems in Southern Ethiopia.

At mid-altitude, the soils under natural forest land cover systems had the highest mean concentration of CEC (25.8 cmol (+)/kg soil) followed by those under *E. saligna* plantations (24.7 cmol (+)/kg soil), while those

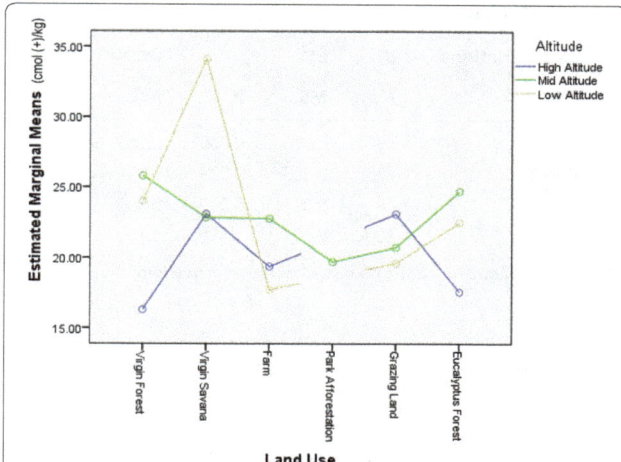

Fig. 24 Estimated marginal means of cation exchange capacity in the 0–15 cm soil layer across different land use/land cover systems and altitudes

under afforestation in the park had the lowest values (19.7 cmol (+)/kg soil). Also, at this altitude, mean concentration of CEC of soils at the surface (0–15 cm) layer showed no significant difference ($P > 0.01$) between the soils under all the land use/land cover systems.

At low altitude, the soils under savanna land cover systems had the highest mean CEC concentration (34.1 cmol (+)/kg soil) followed by those under natural forest (24.0 cmol (+)/kg soil) while those under farming land use systems had the lowest values (17.7 cmol (+)/kg soil). At this altitude, mean CEC concentration of soils at the surface (0–15 cm) layer showed a significant difference ($P < 0.01$) only between the soils under natural savanna vegetation cover and farmland use systems. Generally, according to ratings by Landon (1991), the CEC values in soils under all the land use/land cover systems, were medium except those under natural forest and savanna land use cover at mid and low elevations, respectively which were high.

Effects of land use change on electrical conductivity (mS/cm) EC values ranged from 0.05 mS/cm under grazing land use systems to 0.18 mS/cm under natural forest vegetation. Generally, the results show that EC varied significantly ($P > 0.01$) with land use, but showed no significant difference in elevation (Table 6). Also, the interaction effects between subjects were not significant. At all elevations, mean EC values at the surface (0–15 cm) layer showed significant differences ($P < 0.01$) between the soils under natural forest and those of the other land use/land cover systems (Fig. 10). More so, although the EC content in soils under farmland were not significantly different ($P > 0.01$) from those for soils under the other land use

Table 6 Summary of ANOVA for EC, exchangeable cations and CEC in relation to land use and elevation

Source of variations	df	EC		Ex. Na$^+$		Ex. K$^+$		Ex. Ca^{2+}		Ex. Mg^{2+}		CEC	
		MS	P	MS	P	MS	P	MS	P	MS	P	MS	P
Land use (LU)	5	0.016	0.000	0.063	0.681	3.842	0.65	6.734	0.248	2.958	0.091	92.832	0.332
Elevation (E)	2	0.003	0.029	0.077	0.472	13.786	0.001	51.454	0.000	11.39	0.001	105.58	0.271
LU * E	8	0.001	0.209	0.136	0.235	3.182	0.089	26.472	0.000	2.521	0.114	79.766	0.440
Error	80	0.001		0.102		1.768		4.949		1.493		79.446	

MS is the mean square, P is the p value, df is degree of freedom

systems (except those under natural forest vegetation) at all elevations (high, mid and low), the soils under natural forest had higher EC values (0.18, 0.17 and 0.12 mS/cm, respectively) than those of the other land use/land cover systems (Fig. 25). Therefore, the conversion of forest to cultivated land decreases EC in the study area. These results are in line with the findings of Kizilkaya and Dengiz (2010) who reported that changing forest to cultivated land increased EC values in their area of study due to high application rates of chemical fertilizers.

Although EC represents soil soluble salt components, it is believed that the use of basic chemical fertilizer such as ammonium phosphate and urea under farmlands in our study area did not lead to higher EC values above normal (EC > 0.15 mS/cm will affect plant growth and development) when compared to those under natural forest covers. Therefore, farmers must avoid complete reliance on chemical inputs but continue to rely more on organic fertigation to keep EC < 0.15 in soils. In this regards, the soils under natural forest at high and mid-altitudes in this study may affect the growth and development of only some EC sensitive plants species since their EC concentrations are slightly above normal.

Effects of land use change on exchangeable bases **Exchangeable sodium (Na$^+$) (cmol (+)/kg soil)**

Generally, the results show that mean soil exchangeable Na$^+$ had no significant differences with land use types (P > 0.01) and across all elevations (Table 6). The concentration of exchangeable Na$^+$ was the smallest component on the exchange complex. In addition, the interaction between subject effects was not significant (P > 0.01). Although there was no significant differences (P < 0.01) at the surface (0–15 cm) layer, at high altitude, soils under the protected forest had the highest mean soil exchangeable Na$^+$ concentrations (1.05 cmol (+)/kg soil) followed by the soils under farmlands (1.01 cmol (+)/kg soil), while those under savanna and grazing land use systems had lower values (0.65 and 0.67, respectively) (Fig. 26).

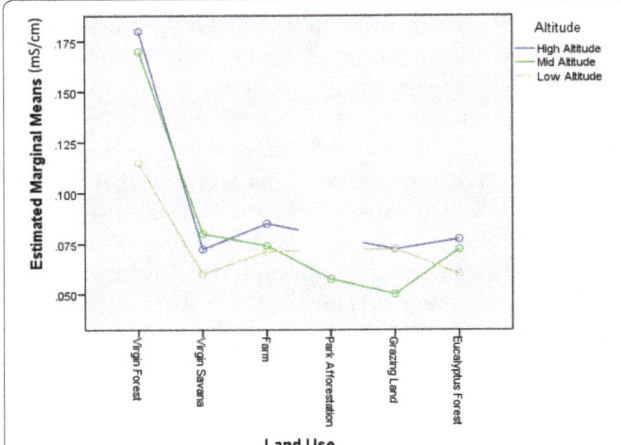

Fig. 25 Estimated marginal means of electrical conductivity in the 0–15 cm soil layer across different land use/land cover systems and altitudes

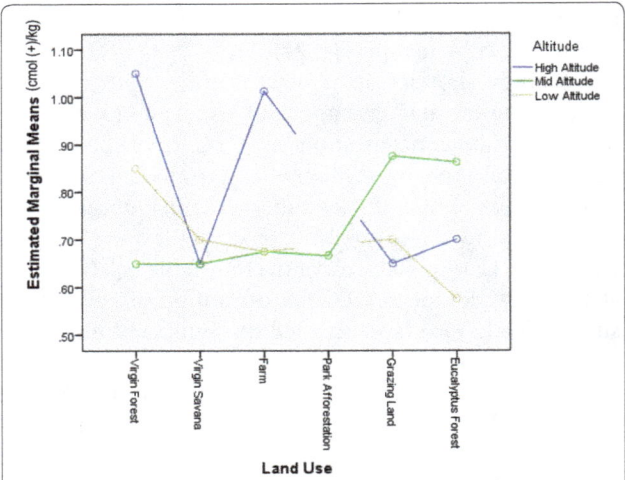

Fig. 26 Estimated marginal means of exchangeable Na$^+$ in the 0–15 cm soil layer across different land use/land cover systems and altitudes

This result corroborate the findings of Yimer et al. (2008) who reported that the concentration of soil exchangeable Na^+ was lower in cropland than in the grazing and native forest. Alem et al. (2010) also observed higher soil exchangeable Na^+ concentration in soils under *E. grandis* when compared to those of native forest in Ethiopia. Significantly high concentrations of exchangeable Na^+ in the soil in, particularly in proportion to the other cations present, can have an adverse effect on crops and physical conditions of the soil (Yerima and Van Ranst 2005a; Bashour and Sayegh 2007). Although Adetunji (1996) indicated that soils with exchangeable Na^+ of 1 cmol (+)/kg soil should be regarded as potentially sodic, those under native forest in the study cannot be regarded as sodic soils, since the soil pH was slightly acidic, and there were no existing evidence of soluble salts in the area. In fact, the concentration of exchangeable Na^+ in the other land use systems did not attain 1 cmol (+)/kg soil. The alternate wet and dry seasons and the topographic (drainage) conditions may be responsible for the potential sodicity value recorded under the protected forest systems in this study.

At mid-altitude, soils under the grazing land use had the highest mean soil exchangeable Na^+ concentrations (0.88 cmol (+)/kg soil) followed by the soils under *E. saligna* plantations (0.86 cmol (+)/kg soil), while those under protected forest and savanna land cover systems had the lowest concentrations (0.65 cmol (+)/kg soil). However, at this altitude, mean soil exchangeable Na^+ concentrations at the surface (0–15 cm) layer showed no significant difference ($P > 0.01$) between the soils under all the land use/land cover systems. At low altitude, soils under the natural forest land use system had the highest mean soil exchangeable Na^+ concentrations (0.85 cmol (+)/kg soil) while those under *E. saligna* plantation had the lowest concentration (0.58 cmol (+)/kg soil). This may be due to the fact that the low soil pH under the Eucalyptus plantation would lead to a decrease in soil base saturation, through immobilization of the exchangeable bases, and may result in soil exchangeable bases depletion over time (Aweto and Moleele 2005).

Exchangeable potassium (K^+, cmol (+)/kg soil)

The results showed that soil exchangeable K^+ did not significantly vary with land use types ($P > 0.01$) but varied significantly with elevation ($P < 0.01$) (Table 6). However, the interaction between subject effects was significant ($P > 0.01$). At high altitude, soils under the protected forest had the highest available potassium concentration (4.00 cmol (+)/kg soil) followed by the soils under farmlands (2.94 cmol (+)/kg soil), while those under savanna and grazing land use had the lowest concentrations (1.05 cmol (+)/kg soil) (Fig. 27). At this altitude, mean soil available potassium at the surface (0–15 cm)

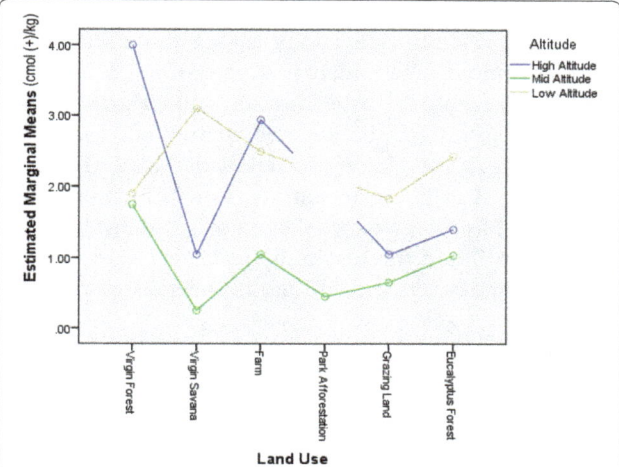

Fig. 27 Estimated marginal means of exchangeable K^+ in the 0–15 cm soil layer across different land use/land cover systems and altitudes

layer showed significant differences ($P < 0.01$) between the soils under natural forest and all the other land use systems except farmland. At mid-altitude, soils under the protected natural forest also had the highest mean soil exchangeable K^+ concentration (1.75 cmol (+)/kg soil) followed by the soils under farmlands (1.05 cmol (+)/kg soil), while those under savanna and grazing land use had the lowest concentration (0.25 cmol (+)/kg soil). However, at this altitude, mean soil exchangeable K^+ concentration at the surface (0–15 cm) layer showed no significant difference ($P > 0.01$) between the soils under all the land use/land cover systems.

At low altitude, soils under the savanna land use system also had the highest mean exchangeable K^+ concentration (3.10 cmol (+)/kg soil) followed by the soils under farmlands (2.49 cmol (+)/kg soil) while those under natural forest cover and grazing land use had the lowest concentrations (1.90 and 1.83 cmol (+)/kg soil, respectively). At this altitude, mean soil exchangeable K^+ concentration at the surface (0–15 cm) layer showed no significant difference ($P > 0.01$) between the soils under all the land use/land cover systems. The observed high concentrations of soil exchangeable K^+ under the natural forest land use system can be attributed to a relative pumping of potassium from the subsoil to topsoil by vegetation (Bohn et al. 2001). Also, the observed high concentration of soil exchangeable K^+ under the cultivation land use system can be attributed to the observed frequent application of household wastes, particularly wood ash, as well as burning of farm residues. These results are consistent with the findings of Bohn et al. (2001). According to ratings by Landon (1991), soil exchangeable K^+ concentration under the natural forest land use system and those under all the other land use systems was high across the different elevations except those in grazing land at mid-altitude which was medium. The medium soil exchangeable K^+ concentrations under grazing land could be associated with soil degradation and losses due to leaching as the grazing land was denuded of vegetation cover. A critical concentration of 0.12 cmol/kg soil is required for plant growth on oxisols (Yerima and Van Ranst 2005b) and the results indicate that exchangeable K^+ concentration is not limiting in the soils of the study area.

Exchangeable calcium (Ca^{2+}, cmol (+)/kg soil)

The results showed that soil exchangeable Ca^{2+} concentrations did not significantly vary with land use type ($P > 0.01$) but varied significantly with elevation ($P < 0.01$) (Table 6). However, the interaction between subject effects was significant ($P < 0.01$). At high altitude, soils under the protected forest had the highest mean exchangeable Ca^{2+} concentration (3.25 cmol (+)/kg soil) followed by the soils under farmlands (6.00 cmol (+)/kg soil), while those under savanna and grazing land use had the lowest (2.00 cmol (+)/kg soil) (Fig. 28). At this altitude, mean soil exchangeable Ca^{2+} at the surface (0–15 cm) layer showed significant differences ($P < 0.01$) between the soils under natural forest and all the other land use systems except those under farmland.

At mid-altitude, soils under the protected natural forest also had the highest mean soil exchangeable Ca^{2+} concentrations (3.68 cmol (+)/kg soil) followed by the soils under *E. saligna* plantation (2.20 cmol (+)/kg soil), while those under savanna land cover had the lowest concentration (0.80 cmol (+)/kg soil). However, at this altitude, mean soil exchangeable Ca^{2+} concentrations at

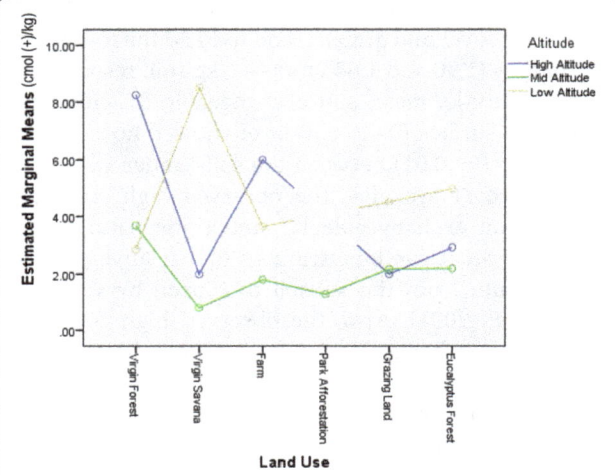

Fig. 28 Estimated marginal means of exchangeable Ca^{2+} in the 0–15 cm soil layer across different land use/land cover systems and altitudes

the surface (0–15 cm) layer showed no significant difference ($P > 0.01$) between the soils under all the land use/land cover systems. At low altitude, soils under the natural savanna land use systems had the highest mean exchangeable Ca^{2+} concentration (8.50 cmol (+)/kg soil), followed by the soils under *E. saligna* plantation (4.98 cmol (+)/kg soil), while those under natural forest cover had the lowest concentration (2.85 cmol (+)/kg soil). At this altitude, mean soil exchangeable Ca^{2+} at the surface (0–15 cm) layer showed significant differences ($P < 0.01$) between the soils under natural forest, savanna, and farmland.

According to ratings by Landon (1991), soil exchangeable Ca^{2+} concentrations in the protected forest and farmland as well as those under savanna and grazing land and *E. saligna* plantation in the high and low altitudes respectively, was medium, while the soil exchangeable Ca^{2+} concentrations under all the other land use systems across the different altitudes was low. The medium soil exchangeable Ca^{2+} in the protected forest, farmland, grazing land and *E. saligna* plantation was probably due to the application of household wastes (wood ash in particular) in the fields as well as the burning of floral and crop residues since ash is a good source of Ca^{2+}, K^+, P, and Mg^{2+} (Voundi et al. 1998) and pumping of bases from the subsoil by the vegetation and returning them into the topsoil (Yimer et al. 2008). On the other hand, the low soil exchangeable Ca^{2+} could be as a result of soil erosion and nutrient losses through leaching as the grazing land was denuded of vegetation cover. A critical concentration of 0.2 cmol/kg soil is required for plant growth in tropical soils (Landon 1991) and the results indicate that exchangeable Ca^{2+} is not limiting in the soil of study area.

Exchangeable magnesium (Mg^{2+}, cmol (+)/kg soil)

The results also showed that the concentrations of soil exchangeable Mg^{2+} did not vary significantly with land use type ($P > 0.01$) but significantly differed with altitude ($P < 0.01$) (Table 6) with the lowest concentrations at mid-altitude (Table 7). More so, the interaction between subject effects was not significant ($P > 0.01$). At high altitude, soils under protected forest had the highest mean exchangeable Mg^{2+} concentrations (3.35 cmol (+)/kg soil) followed by the soils under farmlands (2.69 cmol (+)/kg soil), while those under savanna and grazing land use had the lowest concentrations (0.93 cmol (+)/kg soil) (Fig. 29). At this altitude, mean soil exchangeable Mg^{2+} concentration at the surface (0–15 cm) layer showed significant differences ($P < 0.01$) between the soils under natural forest and all the other land use systems, except those under farmland.

At mid-altitude, soils under the protected natural forest also had the highest soil mean exchangeable Mg^{2+}

Table 7 Mean (± SD) of soil EC, exchangeable cations and CEC in the 0–15 cm soil layer across different land use/land-cover systems and altitudes

Soil property	Altitude	Land use types						ANOVA
		Virgin forest	Virgin savana	Farming	Park afforestation	Grazing land	Eucalyptus forest	
Ex. Na$^+$ (cmol (+)/kg Soil)	High (> 1500 m)	1.05 ± 0.17	0.65 ± 0.06	1.01 ± 0.51		0.67 ± 0.29	0.70 ± 0.21	Ns
	Mid (900–1500 m)	0.65 ± 0.33	0.65 ± 0.29	0.68 ± 0.23	0.67 ± 0.29	0.88 ± 0.06	0.86 ± 0.44	Ns
	Low (< 900 m)	0.85 ± 0.52	0.70 ± 0.00	0.68 ± 0.35		0.88 ± 0.21	0.58 ± 0.35	Ns
	Total	0.85 ± 0.38	0.67 ± 0.16	0.77 ± 0.38	0.67 ± 0.29	0.70 ± 0.29	0.74 ± 0.35	
Ex. K$^+$ (cmol (+)/kg Soil)	High (> 1500 m)	4.00 ± 0.23	1.05 ± 0.64	2.94 ± 1.82		0.74 ± 0.22	1.40 ± 1.40	*
	Mid (900–1500 m)	1.75 ± 1.57	0.25 ± 0.17	1.05 ± 0.46	0.45 ± 0.24	0.65 ± 0.30	1.04 ± 0.79	Ns
	Low (< 900 m)	1.90 ± 0.35	3.10 ± 0.00	2.49 ± 3.08		1.83 ± 1.59	2.43 ± 2.02	Ns
	Total	2.55 ± 1.37	1.47 ± 1.30	2.00 ± 2.04	0.45 ± 0.24	1.18 ± 1.04	1.46 ± 1.37	
Ex. Ca^{2+} (cmol (+)/kg Soil)	High (> 1500 m)	8.25 ± 0.17	2.00 ± 0.60	6.00 ± 3.24		2.00 ± 0.60	2.93 ± 2.87	*
	Mid (900–1500 m)	3.68 ± 2.79	0.80 ± 0.12	1.80 ± 0.79	1.28 ± 0.63	2.18 ± 1.69	2.20 ± 1.59	Ns
	Low (< 900 m)	2.85 ± 0.29	8.50 ± 0.00	3.63 ± 3.64		4.50 ± 3.54	4.98 ± 4.84	Ns
	Total	4.93 ± 2.88	3.77 ± 3.55	3.52 ± 3.09	1.28 ± 0.63	2.89 ± 2.39	3.05 ± 2.96	
Ex. Mg^{2+} (cmol (+)/kg Soil)	High (> 1500 m)	3.35 ± 0.17	0.93 ± 0.55	2.69 ± 1.63		0.93 ± 0.55	1.23 ± 1.20	
	Mid (900–1500 m)	1.60 ± 1.44	0.25 ± 0.17	0.98 ± 0.42	0.43 ± 0.21	0.60 ± 0.24	0.94 ± 0.20	
	Low (< 900 m)	1.70 ± 0.35	2.90 ± 0.00	2.36 ± 2.91		1.73 ± 1.46	2.28 ± 1.86	
	Total	2.22 ± 1.15	1.36 ± 1.21	1.86 ± 1.89	0.43 ± 0.21	1.08 ± 0.97	1.32 ± 1.23	
EC	High (> 1500 m)	0.18 ± 0.01	0.07 ± 0.01	0.08 ± 0.02		0.07 ± 0.02	0.08 ± 0.03	*
	Mid (900–1500 m)	0.17 ± 0.03	0.08 ± 0.01	0.07 ± 0.02	0.06 ± 0.01	0.05 ± 0.00	0.07 ± 0.03	*
	Low (< 900 m)	0.12 ± 0.06	0.06 ± 0.00	0.07 ± 0.04		0.07 ± 0.05	0.06 ± 0.01	*
	Total	0.16 ± 0.05	0.07 ± 0.01	0.08 ± 0.03	0.06 ± 0.01	0.06 ± 0.03	0.07 ± 0.02	
CEC (cmol (+)/kg Soil)	High (> 1500 m)	16.3 ± 5.32	23.1 ± 10.9	19.4 ± 12.6		23.1 ± 10.9	17.6 ± 6.78	Ns
	Mid (900–1500 m)	25.8 ± 14.2	22.8 ± 9.06	22.8 ± 6.88	19.7 ± 7.38	20.7 ± 1.79	24.7 ± 9.6	Ns
	Low (< 900 m)	24.0 ± 12.2	34.1 ± 0.00	17.7 ± 8.30		19.6 ± 11.5	22.5 ± 7.44	*
	Total	22.0 ± 11.1	26.7 ± 9.21	20.4 ± 9.11	19.7 ± 7.38	21.1 ± 8.49	21.4 ± 8.43	

Means in the same row followed by the same letters are not significantly different at 1% significance

Ns non significance, *MC* moisture content, *BD* Bulk density

* Significant at P < 0.01

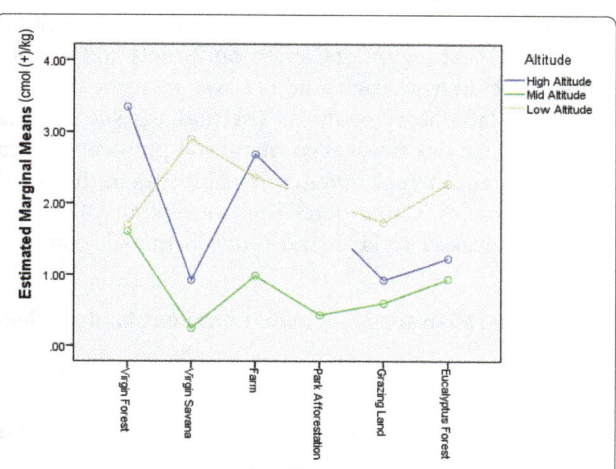

Fig. 29 Estimated marginal means of exchangeable Mg^{2+} in the 0–15 cm soil layer across different land use/land cover systems and altitudes

concentrations (1.60 cmol (+)/kg soil) followed by the soils under farmland (0.98 cmol (+)/kg soil), while those under savanna land cover had the lowest concentrations (0.25 cmol (+)/kg soil). More so, at this altitude, mean soil exchangeable Mg^{2+} concentrations at the surface (0–15 cm) layer showed no significant difference ($P > 0.01$) between the soils under all the land use/land cover systems. At low altitude, soils under the natural savanna land use system had the highest mean exchangeable Mg^{2+} concentration (2.90 cmol (+)/kg soil) followed by the soils under farmland (2.36 cmol (+)/kg soil), while those under natural forest cover had the lowest concentration (2.85 cmol (+)/kg soil). At this altitude, mean soil exchangeable Mg^{2+} at the surface (0–15 cm) layer showed no significant differences ($P < 0.01$) between the soils under all the land use/land cover systems.

Generally, the results for exchangeable bases follow a similar trend and hence the simultaneous explanation provided above is believed to provide justifications for these dynamics. According to ratings by Landon (1991), soil exchangeable Mg^{2+} concentration in all the land use/land cover systems was medium, except for those under savanna and afforestation in the mid-altitude, which was less than the critical level of 0.5 cmol (+)/kg soil. A concentration less than the critical level would require an application of magnesium limestone for management accordingly (Awdenegest et al. 2013).

Effects of LULC change on soil exchangeable acidity ($H^+ + Al^{3+}$) The laboratory results showed that some soils under natural forest cover, farmland and *E. saligna* plantations, particularly at high and mid-altitudes, had relatively lower pH values (pH < 5.5), though only mean pH concentrations of soils under farmland are indicated in Table 4. These soils were selected and analysed for exchangeable acidity (EA) and the results are presented in Table 8.

It has been well reported that high exchangeable acidity occurs in very acidic soils, with low pH values (Yerima and Van Ranst 2005a; Aweto and Moleele 2005). The hydrolysis of Al^{3+} ions that constitute part of the clay layers become exchangeable and contribute to the development of soil acidity (Yerima and Van Ranst 2005a; Oyedele et al. (2009). In fact, correlation analysis (Table 9) shows a significantly strong negative relationship (r = − 0.752, P < 0.01) between soil pH and soil exchangeable acidity. Frimpong et al. (2014) stated that at pH below 5.5, aluminum and manganese toxicities might occur.

At high altitude, soils under farmland and Eucalyptus plantations had high EA values (Ex. Acidity > 2), while at mid-altitude, only those under farmland had a high EA value (Table 8). These results indicate that at high altitudes, soils under farmland and Eucalyptus plantations as well as those under farmland at mid-altitude, present a high potential for aluminum toxicity to plants and may have immobilized soil essential nutrients. Aweto and Moleele (2005) reported that soils with higher exchangeable acidity cause immobilization of soil essential nutrients including; P, N, Ca, Mg, and K under *Eucalyptus* spp. The observed high values of EA indicates a difficulty to

Table 9 Correlation matrix table for selected soil pH and exchangeable acidity at surface 0–15 cm soil depth

	pH	EA
pH	1	− 0.752**
EA	− 0.752**	1

** Correlation is significant at the 0.01 level; EA is exchangeable acidity

manage the acidity problems, while the observed moderate and lower EA values indicates the easiness to also manage the acidity problem of the respective land uses and elevations, within the study area.

According to the Apal Agricultural Laboratory; soil test interpretation guide (2016), where extractable aluminum is > 2, sensitive plants will be affected. It also states that excess soluble/available aluminum (Al^{3+}) is toxic to plants and can cause a number of problems. The guide further explains that some issue caused by Al^{3+} toxicity can include: direct toxicity, primarily seen as stunted roots; reduction of the availability of phosphorus, through the formation of Al-P compounds; reduction of the availability of sulfur, through the formation of Al-S compounds; reduction of the availability of other cations (Ca^{2+} and Mg^{2+}) through competitive interactions; and reduced rhizobium levels on legumes. The high EA under eucalyptus plantations confirms the low yields observed for most crops planted around eucalyptus plantations and confirms the acidifying nature of eucalyptus leaves under decomposition. The high EA values in farmland are consistent with the acidification resulting from the application of ammonium fertilizers (Yerima and Van Ranst 2005a, b).

Conclusions

This study was aimed at assessing the effects of six land use systems on fifteen soil physicochemical properties in the North West region of Cameroon. Ninety soil samples were collected from each land use system at the 0–15 cm depth for laboratory analysis. The findings suggest that LULC change has influenced many soil physicochemical properties at different topographic altitudes in the North West region of Cameroon. The conversion of natural forest or savanna to farmland reduced the silt contents,

Table 8 Means of selected soil exchangeable acidity (H^++Al^{3+}) in the 0–15 cm soil layer across different land use/land cover systems and altitudes

Factor	Altitude	Land use types					
		Virgin forest	Virgin savana	Farming	Park afforestation	Grazing land	Eucalyptus forest
EA	High (> 1500 m)	1.5	–	3.07	–	–	3
	Mid (900–1500 m)	1.6	–	2.6	–	–	1.82
	Low (< 900 m)	–	–	–	–	0.4	–

moisture content, organic matter, soil organic carbon, total nitrogen, available phosphorus, pH, cation exchange capacity, and exchangeable bases, but increased the soil bulk density, electrical conductivity, exchangeable acidity and sand content significantly ($P < 0.05$). The results revealed that deforestation and subsequent cultivation of soil had negative effects on the measured soil properties. Therefore, it can be concluded that the conversion of natural forest or pasture land to cultivation land subjected soil physicochemical properties to degradation thereby sullying soil quality. To reverse soil degradation and promote restoration in the region, emphases should be placed on promoting site-specific, sustainable land management practices within the savanna, grazing, agricultural and forest management systems.

The scope of this research was limited to only three subdivisions (Santa, Bamenda, and Ndop) under just two divisions (Mezam and Ngoketungia) but gives a representation of the geomorphic surfaces in the NW region of Cameroon. The research also used only selected soil physical and chemical properties as indicators of soil quality under the influence of land use change. Generally, soil quality varies greatly with soil type, depth and over a long timescale, but this research was based on assessing the state of soil quality under different land uses at soil surface (0–15 cm) level.

Abbreviations
Av.P: available phosphorous; BD: bulk density; C/N: the ratio of carbon to nitrogen; CEC: cation exchange capacity; EA: exchangeable acidity; EB: exchangeable bases; EC: electrical conductivity; KCl: potassium chloride; ANOVA: analysis of variance ratio; LULC: land use/land cover systems; MC: moisture content; NW: North West; OC: organic carbon; SOC: soil organic carbon; SOM: soil organic matter; SPSS: Statistical package for social science; TN: total nitrogen.

Authors' contributions
VAT literature review, field survey, data collection and analysis and manuscript preparation. YBPK overall management of the field survey, technical support and reviewing the manuscript. Both authors read and approved the final manuscript.

Author details
[1] Department of Development Studies, Environment and Agricultural Development Program, Pan African Institute for Development-West Africa (PAID-WA), P.O. Box 133, Buea, South West Region, Cameroon. [2] Department of Soil Science, Faculty of Agronomy and Agricultural Science, University of Dschang, P. O. Box 222, Dschang, West Region, Cameroon.

Acknowledgements
Authors are thankful to the team of laboratory technicians in the Soil science laboratory at the University of Dschang, Cameroon for the technical support. Authors are also thankful to the anonymous reviewers for their valuable suggestions to improve the manuscript.

Competing interests
The authors declare that they have no competing interests.

Declaration
I, Valentine Asong Tellen, holder of ORCID number 0000-0001-8513-788X hereby declare that this research article is written by the authors whose names have been appropriately indicated.

Funding
Self-funded.

References
Adetunji MT (1996) Field soil tests for NO3, NH4, PO4, K, Ca and Na. Department of Soil Science and Mechanization, University of Abeokuta, Nigeria. In: Simple Soil, Water and Plant Testing Techniques for Soil Resource Management. Proceedings of a Training Course Held in Ibadan, Nigeria, 16–27 September 1996. International Institute of Tropical Agriculture. FAO, Rome

Alem S, Woldemariam T, Pavlis J (2010) Evaluation of soil nutrients under *Eucalyptus grandis* plantation and adjacent sub-montane rain forest. J For Res 21(4):457–460

Apal Agricultural Laboratory (2016) Soil test interpretation guide. http://www.apal.com.au/. Accessed 22 June 2016

Awdenegest M, Melku D, Fantaw Y (2013) Land use effects on soil quality indicators: a case study of Abo-Wonsho Southern Ethiopia. Appl Environ Soil Sci 2013:9 **(Article ID 784989)**

Aweto AO, Moleele NM (2005) Impact of *Eucalyptus camaldulensis* plantation on an alluvial soil in south eastern Botswana. Int J Environ Stud 62(2):163–170

Ayele T, Beyene S, Esayas A (2013) Changes in land use on soil physicochemical properties: the case of smallholders fruit-based land use systems in Arba Minch, southern Ethiopia. Int J Curr Res 5(10):3203–3210

Bashour II, Sayegh AH (2007) Methods of analysis for soils of arid and semi-arid regions. FAO, Rome, pp 1–128

Bohn HL, McNeal BL, O'Connor GA (2001) Soil chemistry, 3rd edn. Wiley, New York, p 322

Brejda JJ, Karlen DL, Smith JL, Allan DL (2000a) Identification of regional soil quality factors and indicators. II. Northern Mississippi loess hills and palouse prairie. Soil Sci Soc Am J 64:2125–2135

Brejda JJ, Moorman TB, Karlen DL, Dao TH (2000b) Identification of regional soil quality factors and indicators: I. Central and southern high plains. Soil Sci Soc Am J 64:2115–2124

Cao Y, Fu S, Zou X, Cao H, Shao Y, Zhou L (2010) Soil microbial community composition under *Eucalyptus* plantations of different age in subtropical China. Eur J Soil Biol 46(2):128–135

Carrow RN, Stowell L, Gelernter W, Davis S, Duncan RR, Skorulski J (2004) Clarifying soil testing: III. SLAN sufficiency ranges and recommendations. Golf Course Manag 72:194–198

Cheek M, Onana JM, Pollard BJ (2000) The plants of Mount Oku and the Ijim Ridge, Cameroon: a conservation checklist. RBG, Kew, p 211

Epule T, Peng C, Lepage L, Chen Z (2011) Forest loss triggers in Cameroon: a quantitative assessment using multiple linear regression approach. J Geograp Geol 3(1):30–40. https://doi.org/10.5539/jgg.v3n1p30

FAO (2009) State of the World's Forests. Rome, Italy. pp 3–5. http://www.fao.org/docrep/011/i0350e/i0350e00.htm

FAO (2017) Action against desertification. Background information available at: http://www.fao.org/in-action/action-against-desertification/background/en/. Retrieved May 2017.

Food and Agricultural Organization of the United Nations (2010a) FAOSTAT. http://faostat.org. Accessed Apr 2013

Food and Agricultural Organization of the United Nations (2010b) Forest Resource Assessment. FAO Forestry paper no. 163. FAO, Rome, http://www.fao.org. Accessed Feb 2013

Frimpong KA, Afrifa EKA, Ampofo EA, Kwakye PK (2014) Plant litter turnover, soil chemical and physical properties in a Ghanaian gold-mined soil revegetated with Acacia species. Int J Environ Sci 4(5):987

Getachew F, Abdulkadir A, Lemenih M, Fetene A (2012) Effects of different land uses on soil physical and chemical properties in Wondo Genet area, Ethiopia. N Y Sci J 5:110–118

Getachew F, Abdulkadir A, Lemenih M, Fetene A (2012) Effects of different land uses on soil physical and chemical properties in Wondo Genet area, Ethiopia. N Y Sci J 5:110–118

Guo LB, Gifford RM (2002) Soil carbon stocks and land use change. Glob Change Biol 8:345–360

Harvey Y, Pollard BJ, Darbyshire I, Onana JM, Cheek M (2004) The plants of Bali Ngemba Forest Reserve, Cameroon. Royal Botanic Gardens, Kew, p 154

Hoogsteen MJJ, Lantinga EA, Bakker EJ, Groot JCJ, Tittonell PA (2015) Estimating soil organic carbon through loss on ignition: effects of ignition conditions and structural water loss. Eur J Soil Sci 66(2):320–328

House APN (1992) Eucalyptus: curse or cure?. Australian Center for International Agricultural Research, Canberra, p 197

Humphrey NN, Amawa SG (2014) Environmental degradation and the emergence of agricultural frontiers in the North West of Cameroon. J Sustain Dev 7(5):111–127. https://doi.org/10.5539/jsd.v7n5p111

Javad RSA, Hassan K, Esmaeil HA (2014) Assessment the effects of land use changes on soil physicochemical properties in Jafarabad of Golestan province. Iran. Bull Environ Pharmaco Life Sci 3(III):296–300

Kakaire J, Makokha GL, Mwanjalolo M, Mensah AK, Menya E (2015) Effects of mulching on soil hydro-physical properties in Kibaale Sub-catchment, South Central Uganda. Appl Ecol Environ Sci 3(5):127–135

Kizilkaya R, Dengiz O (2010) Variation of land use and land cover effects on some soil physico-chemical characteristics and soil enzyme activity. ZemdirbAgric 97(2):15–24

Kolay AK (2000) Basic concepts of soil science, 2nd edn. New Age International Publishers, New Delhi. pp 78–80, 90–91, 102–105, 138

Landon JR (1991) Booker tropical soil manual: a handbook for soil survey and agricultural land evaluation in the tropics and subtropics. Longman Scientific and Technical, New York

Leeper GW, Uren NC (1993) Soil science, an introduction, 5th edn. Melbourne University Press, Melbourne. ISBN 0-522-84464-2

Leu A (2007) Organics and soil carbon: increasing soil carbon, crop productivity and farm profitability. In: 'Managing the Carbon Cycle' Katanning Workshop, 21–22 March, 2007. http://www.amazingcarbon.com

Lobe I, Amelung W, Du Preez CC (2001) Losses of soil carbon and nitrogen with prolonged arable cropping from sandy soils of the South African Highveld. Eur J Soil Sci 52:939101

Manu IN, Andu WN, Tarla DN, Agharih WN (2014) Causes of cattle theft in the North West Region of Cameroon. Sch J Agric Sci 4(4):181–187

McCauley A, Jones C, Jacobsen J (2005) Basic soil properties: soil and water management module I. Montana State University Extension Service. http://landresources.montana.edu/SWM/PDF/Final_proof_SW1.pdf. Retrieved 4 Mar 2014

McNaughton SJ, Banyikwa FF, McNaughton MM (1997) Promotion of the cycling of diet-enhancing nutrients by African grazers. Science 278:1798–1800

Ogle SM, Breidt FJ, Paustiani K (2005) Agricultural management impacts on soil organic carbon storage under moist and dry climatic conditions of temperate and tropical regions. Biogeochemistry 72:87–121

Oyedele DJ, Awotoye OO, Popoola SE (2009) Soil physical and chemical properties under continuous maize cultivation as influenced by hedgerow trees species on an alfisol in South Western Nigeria. Afr J Agric Res 4(7):736–739

Pauwels JM, Van Ranst E, Verloo M, Mvonco Ze A (1992) Méthodes d'analyses de sols et de plantes, équipements, gestion de stocks de verrerie et de produits chimiques. A. G. Bulding, Bruxelles

Ravina M (2012) Impact of Eucalyptus plantations on pasture land on soil properties and carbon sequestration in Brazil. Swedish University of Agricultural Sciences, Uppsala

Roe D, Nelson F, Sandbrook C (eds) (2009) Community management of natural resources in Africa: impacts, experiences and future directions, Natural Resource Issues No. 18, International Institute for Environment and Development, London

Roose E, Barthes B (2001) Organic matter management for soil conservation and productivity restoration in Africa: a contribution from francophone research. Nutr Cycl Agroecosyst 61:159–170

Sui YY, Zhang XY, Jiao XG, Wang QC, Zhao J (2005) Effect of long-term different fertilizer applications on organic matter and nitrogen of black farm-land. J Soil Water Conserv 19:190–192

Van Reewijk LP (2002) Procedures for soil analysis. Technical paper, 5th edn. International Soil Reference and Information Center (ISRIC), Wageningen

Voundi NJC, Demeyer A, Verbo MG (1998) Chemical effects of wood ash on plant growth in tropical acid soils. Bioresour Technol 63:251–260

Walkley A, Black IA (1934) An examination of Degtjareff method for determining soil organic matter and a proposed modification of the chromic acid titration method. Soil Sci 37:29–37

Wiesmeier M, Sporlein P, Geu U, Hangen E, Haug S, Reischl A, Schilling B, von Lutzow M, Kogel-Knabner I (2012) Soil organic carbon stocks in southeast Germany Bavaria, as affected by land use, soil type and sampling depth. Glob Change Biol 18:2233–2245. https://doi.org/10.1111/j.1365-2486.2012.02699.x

Wu HB, Guo ZT, Peng CH (2001) Changes in terrestrial carbon storage with global climate changes since the last interglacial. Q Sci 21:366–376

Yerima PKB (2011) Protective measures taken for preserving species endangered of extinction due to the climate change in sub-Sahara Africa: Case of the Yongka Western Highlands Research Garden-Park, Nkwen-Bamenda, NW Region, Cameroon. In: Research and Development in sub-Saharan Africa, Support Africa International

Yerima BPK, Van Ranst E (2005a) Introduction to soil science: soils of the tropics. Trafford Publishing, Victoria, p 440

Yerima BPK, Van Ranst E (2005b) Major soil classification systems used in the tropics: soils of Cameroon. Trafford Publishing, Victoria, p 312

Yimer F, Ledin S, Abdelkadir A (2008) Concentrations of exchangeable bases and cation exchange capacity in soils of cropland, grazing and forest in the Bale Mountains Ethiopia. Forest Ecol Manag 256(6):1298–1302. https://doi.org/10.1016/j.foreco.2008.06.047

Zerga B (2015) Ecological impacts of Eucalyptus plantation in Eza Wereda, Ethiopia. Int Invent J Agric Soil Sci. 3(4):47–51. ISSN: 2408-7254. http://internationalinventjournals.org/journals/IIJAS

12

Impact of land use land cover change on ecosystem services: a comparative analysis on observed data and people's perception in Inle Lake, Myanmar

Seema Karki[1,4], Aye Myat Thandar[2], Kabir Uddin[1], Sein Tun[3], Win Maung Aye[3], Kamal Aryal[1], Pratikshya Kandel[1] and Nakul Chettri[1*]

Abstract

Background: A healthy wetland provides a range of goods and services contributing to human wellbeing. Inle Lake, the first Biosphere Reserve in Myanmar, has been supporting the local inhabitants with ecosystem services (ES) including habitat for a wide range of biodiversity. In the recent years, influenced by land use land cover change (LULCC), the lake has witnessed changes with altered flow of ES, affecting human well-being. Communities' perceptions are often undermined, when it comes to research LULCC. We analyzed LULCC change data from 1989–2000 to 2000–2014 using Landsat imageries. This was then linked to ES considering dependency through qualitative data collated from participatory rural appraisal tools and structured questionnaires focusing on people's perception to understand the LULCC dynamics and its implication.

Results: During 25 years (1989–2014), there has been a sharp reduction of 164 km^2 perennial wetland area in the Inle Lake, which is 4.2-fold higher in 2014 to that of 1989. Similarly, forest area has been declined by 92 km^2 (8.56%) in last 25 years. Contrary to this, cropland area showed an increment of 60.67% in 2000 and 64.53% in the year 2014 alone giving a total increase by 268 km^2 over the last 25 years and an expansion of 40 km^2 seasonal freshwater area were observed showing periodic increment over the time. Communities from the three study areas, namely, Kyaung Taung, Zay Gon and Kyar Taw are found to have high dependence in their surrounding ecosystems. These villages utilizes 17 ES from forest ecosystem, 13 from agro-ecosystem, 10 from seasonal and 4 from perennial water body for their livelihood respectively. Around 93% of the respondents opined that forest ecosystem has decreased over the last 10 years. Around 40% of the respondents reflected an increase in area used for cropland; 43% conversely perceived a declination. About 63% of the respondents perceived such changes have brought huge reduction in availability of freshwater ES. A significant number of respondents (92%) perceived an enormous reduction in seasonal water body during the dry season.

Conclusion: Observed decreasing trends in forest and perennial wetland areas were consistent with people's perceived changes. Communities associate loss of forest and wetland area with reduced availability of ES as well as degraded health of the lake.

Keywords: Ecosystems, Land use land cover change, Ecosystem services, Communities' perception, Wetland

*Correspondence: Nakul.Chettri@icimod.org
[1] International Centre for Integrated Mountain Development (ICIMOD),
GPO Box 3226, Kathmandu, Nepal
Full list of author information is available at the end of the article

Introduction

A home to 40% of the world's species and 12% of all animal species (Mitsch and Gosselink 2000), wetlands cover around 6% of the world's land area (Zedler and Kercher 2005) of which the largest area (31.8%) is in Asia (Davidson et al. 2018). The wetland provides a wide array of provisioning, supporting, cultural and regulating services contributing to human wellbeing (Lamsal et al. 2015; Sharma et al. 2015; Chaudhary et al. 2016, 2017). Converting such benefits in economic terms, 12.8 million km^2 of the existing global wetland could yield 70 billion United States Dollar (USD, Schuijt and Brander 2004). The recent estimation for the total economic value of 63,000 km^2 of global wetland, a fraction of the total, revealed to be 3.4 billion USD per year (TEEB 2010). However, most of the wetlands across the globe are under stresses due to various drivers of change, including the land use land cover change (LULCC). Since 1900 AD, the wetland lost 64–71% of its original area and was faster for inland than coastal natural wetlands (Davidson 2014). As evident from the recent studies, the LULCC is one of the five major drivers of change for wetlands in Asia (Romshoo and Rashid 2014; Zorrilla-Miras et al. 2014; Chettri and Sharma 2016). As a result, wetland degradation and its conservation have been a subject of global concern (Gopal 2013; Reis et al. 2017; Davidson et al. 2018).

Inle Lake, the first Biosphere Reserve identified by the Man and the Biosphere Reserve Programme of the United Nations Organization for Education, Science and Culture (UNESCO) in 2015, is known among the global 200 ecoregions (Olson and Dinerstein 1998). With its 1.5 million years history of formation (Bertrand and Rangin 2003), the Inle Lake is lying at an average 884 m above mean sea level with high ecological significance (Su and Jassby 2000; Turner et al. 2000; Butkus and Myint 2001; Akaishi et al. 2006; Okamoto 2012). It provides numerous tangible and intangible ecosystem services (ES) to the local communities (Ma 1996). The lake regulates flow and supports natural water filtration, providing fresh water as one of the provisioning services to downstream (Thaw 1998) and is a major source of hydroelectric power for southern Myanmar (Su and Jassby 2000).

Designated as one of the freshwater biodiversity hotspot, Inle Lake is also habitat for numerous globally significant species (Annandale 1918; Roberts 1986; Ma 1996; Kottelat and Witte 1999; Groombridge and Jenkins 1998; Platt and Rainwater 2004; Lwin and Sharma 2012). It is the home for numerous threatened species like White-rumped vulture (*Gyps bengalensis*), Greater spotted eagle (*Clanga clanga*), Pallid harrier (*Circus macrourus*), Bare's pochard (*Aythya baeri*), Sarus crane (*Grus antigone*), and Ferruginous pochard (*Aythya nyroca*, Gyi et al. 2011). The lake is also an important nesting and breeding ground for amphibians and fishes (Ma 1967; Thant 1968; Kottelat 1986). More interestingly the lake is famous for floating garden or hydroponics cultivation (Myint and Maung 2000; Akaishi et al. 2006; Than 2007). The garden in the lake is a good source of vegetables and is an important tourist destinations in Myanmar (MoHT 2013). Considering the significance, the government supported tourism policy of 1996 has recognized Inle Lake as a major tourist hub (Butkus and Myint 2001; MoHT 2013). There is high number of tourists visiting lake, contributing to local economy (Ingelmo 2013; Munz and Molstad 2012; ICIMOD and MoNREC 2017).

Despite of being global significance, the lake and its catchment have undergone series of land use transformation over the years impacting its health (Lwin and Sharma 2012; Htwe et al. 2015). Deforestation in the mountains due to agricultural expansion and shifting cultivation, expansion of floating garden within the lake, sedimentation load and change in the water quality are some of the factors affecting the lake (Sidle et al. 2007). Those drivers have not only reduced the size of the lake, but have also affected ecosystem health and flow of ES, the major source of livelihood of the people.

In the recent global trend, understanding the linkages between ES with human wellbeing are emerging and also becoming a priority research area (Cardinale et al. 2011; Castro et al. 2014; Chaudhary et al. 2017; Ding et al. 2017; Omrani et al. 2017; Kandel et al. 2018). The concept of ES has been considered as products of coupled and nested social–ecological systems and emphasized to be measured in the complex context of those socio-ecological systems (Balvanera et al. 2006; Fisher et al. 2009; MA 2005; Mace et al. 2011; Bateman et al. 2013; Reyers et al. 2013; Scholes et al. 2013). However, the existing literature has limited integration with the broader social science literature about people's choices and behavior (Bryan et al. 2010; Milner-Gulland 2012). In response, the Intergovernmental Science-Policy Platform on Biodiversity and Ecosystem Services (IPBES) endorsed an ES approach that explicitly recognizes the benefits people gain from nature building support for sustainable development goals (de Groot et al. 2010; Diaz et al. 2015; Schmalzbauer and Visbeck 2017; Diaz et al. 2018). Therefore, assessments and sustainable management of ES require an understanding of both supply and demand considering the qualities, quantities, spatial scales and dynamics forming a bridge between ecological and social systems (Nahlik et al. 2012). So far, researchers in the Inle Lake have been generating knowledge in a sectorial approach, considering mainly biodiversity, LULCC and sedimentations to name a few. The understanding of drivers and its impacts on ES and the implication for human wellbeing has not been explored. This study is an attempt to bridge gaps between social and ecological understanding. To justify the above context, following three questions were developed and the research was oriented to answer following questions.

A. How LULCC (temporal and spatial) has changed over the period in the study area?
B. What are the states of major ecosystems in the given study area and how the local people are dependent on these ecosystems?
C. What are the people's perception in terms of the LULCC and its impact on the ES they are depended on?

Materials and methods

Study area

Inle Lake, situated on the Shan plateau of Myanmar, is part of the Shwenyaung rift valley, nourished by surrounded by catchment areas (Ma 1996; Su and Jassby 2000). Its immediate catchment is inhabited by about 200 villages (Butkus and Myint 2001) that serve as watershed for Nyaung Shwe Township with various ES (Akaishi et al. 2006; Lwin and Sharma 2012). The study was carried out in and the surrounding areas of the Inle Lake (Fig. 1). Three representative villages—namely, Kyaung Taung, Zay Gon and Kyar Taw were selected on the basis of origin of watershed and level of local community's livelihood dependence on Inle Lake. The

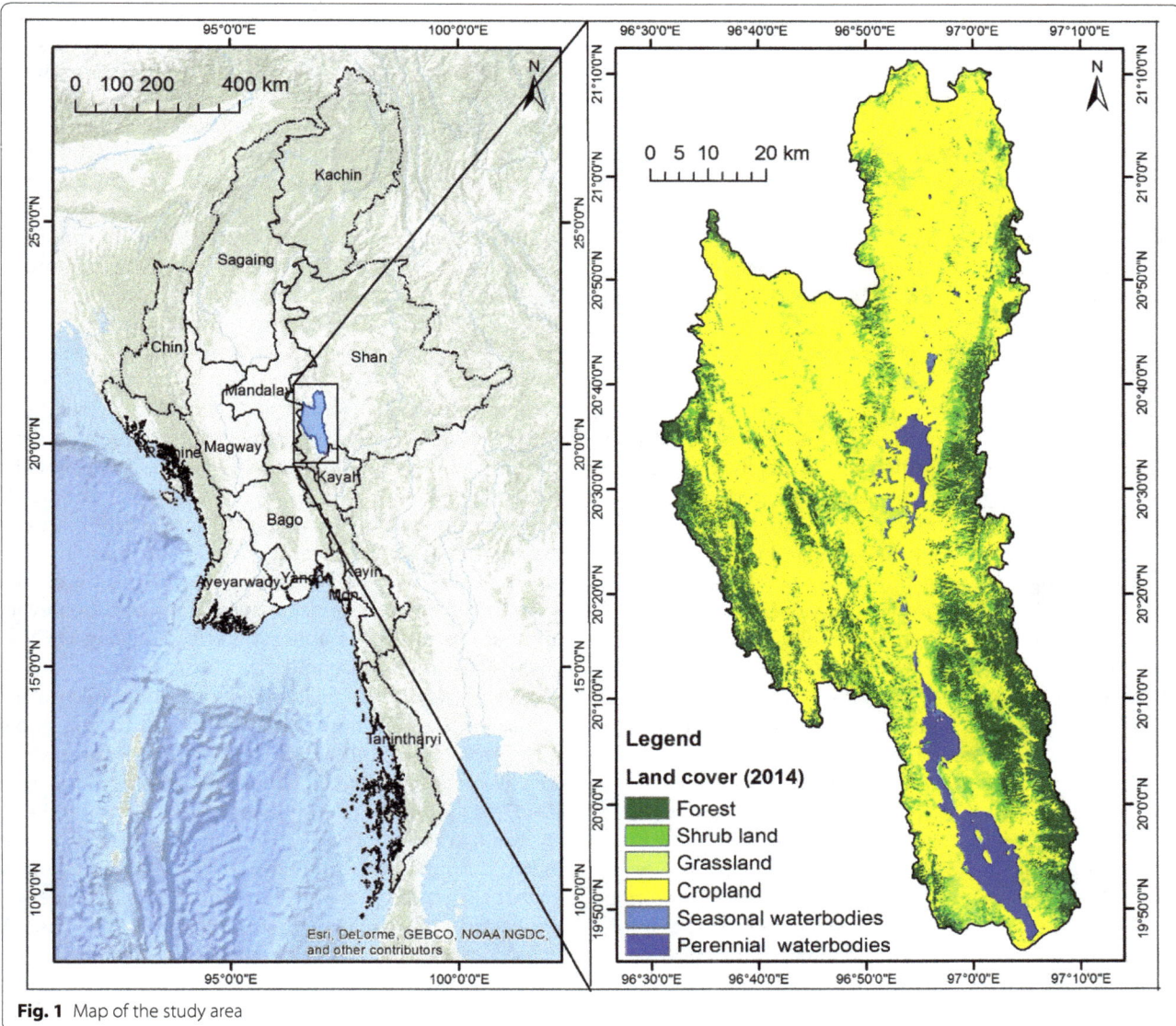

Fig. 1 Map of the study area

Kyaung Taung represents upstream catchment of the watershed and around 186 households inhibit in this area. Local communities in this village depend more on agricultural farming and livestock rearing. Rainfed farming is more prominent due to lack of irrigation facility. Zay Gon, also called as market area, is a middle stream comprising 168 households. It is a tradeoff zone where number of ES brought from Kyaung Taung village and Kyar Taw village are traded. Similarly, Kyar Taw, famous as floating garden represents downstream of the study area and consists of 173 households. These floating gardens have a unique feature called hydroponic cultivation which was introduced in the early 1960s (Sidle et al. 2007). The overall conceptual framework used in this study is presented in Fig. 2 along with the detail in the following section.

Land use land cover change analysis
To identify the spatio-temporal changes of Inle Lake over a period of 25 years, LULCC analysis was undertaken. For the analysis, we acquired medium spatial resolution landsat thematic mapper (TM) of 1989 and 2000; and Landsat 8 of 2014. A classification scheme was used with six major land classes such as forest, shrubland, grassland, cropland, seasonal and perennial water bodies. The Thematic Mapper (TM), and Landsat 8 images were rectified into Universal Transverse Mercator (UTM) Zone 47. After rectifying, eCognition developer software was used for OBIA (a methodological framework for machine-based interpretation of complex classes using both spectral and spatial information (Lang et al. 2011). The six land cover types were classified using a multiresolution segmentation algorithm which consecutively merged pixels by identifying image objects of one pixel and merging them with neighbours using relative homogeneity criteria (Blaschke and Hay 2001). A land water mask was created during class modelling using band ratio and texture information based on spectral values and vegetation indices like the Normalized Difference Vegetation Index (NDVI). An NDVI image was created in a pre-processing stage using customized features: $NDVI = (RED - IR)/(RED + IR)$. The land and water mask was created using the formula $IR/Green * 100$. The image objects were labeled according to attributes such as NDVI, land water mask, layer value, and color, and relative position to other objects, using user-defined rules. Objects with an area smaller than the defined minimum mapping unit were merged with other objects. The classified land cover map was then exported to a raster file format for further analysis. To validate the accuracy of the maps, both field sampling and references through high resolution map were used.

Participatory approach and tools
We used a few participatory rural appraisal (PRA) tools such as focus group discussion (FGD), resource mapping, transect walk along with a structured qualitative survey using pre-set questionnaire to understand the people's dependency on the major ecosystems and their ES. The major ES listed were further categorized into four groups following MA (2005). The collected qualitative data were then used to compare with LULCC maps.

Household survey
We adopted an 'Ecosystem Services Cascade' framework that enabled the study to rationalize importance and significance of ES to human wellbeing. As explained by MA (2005) and Costanza et al. (1997), we considered the tangible and intangible benefits provided by an ecosystem as provisioning, regulating, supporting and cultural services that people derived from four ecosystems mainly forest, agro-ecosystem, seasonal and perennial water bodies. Because of the seasonal variation affecting the water bodies, we classified rain fed water bodies into seasonal and perennial water bodies. Seasonal rain influencing fresh water bodies like inundation are considered as seasonal water bodies, excluding seasonal influences are considered as perennial water bodies. Survey questions mainly focused on (i) dependency on ES by communities for their livelihood, (ii) community's perception on state of LULCC and ES and (iii) long term changes over the flow of goods and services derived from these four ecosystems.

A questionnaire was designed following Chaudhary et al. (2017) with some adjustment for the local context. Systemic stratified sampling (SSS) approach was applied to conduct household survey. We divided the study sites into three strata as explained in the study area as upstream, middle stream and downstream sites. The SSS approach was used in such a way that selection of first household from sample list is at random and then every kth household in the sample list is selected using $k = N/n$, where N is total households in the study site and n = sample household. For example, if a 1st household on site is chosen, the next household would be 3rd household in the study area. Out of 527 households in three study sites, we selected 33% for household survey, where N = 178. Description of the sampling area for household survey is illustrated in Table 1. Household survey was conducted during morning and evening at home in the local language. The head of the household was interviewed irrespective of gender (above 18 years). The survey focused on the perceptions considering dependency on different ecosystems for ES, and the impact of LULCC on their supply. The average time per interview was 45-min.

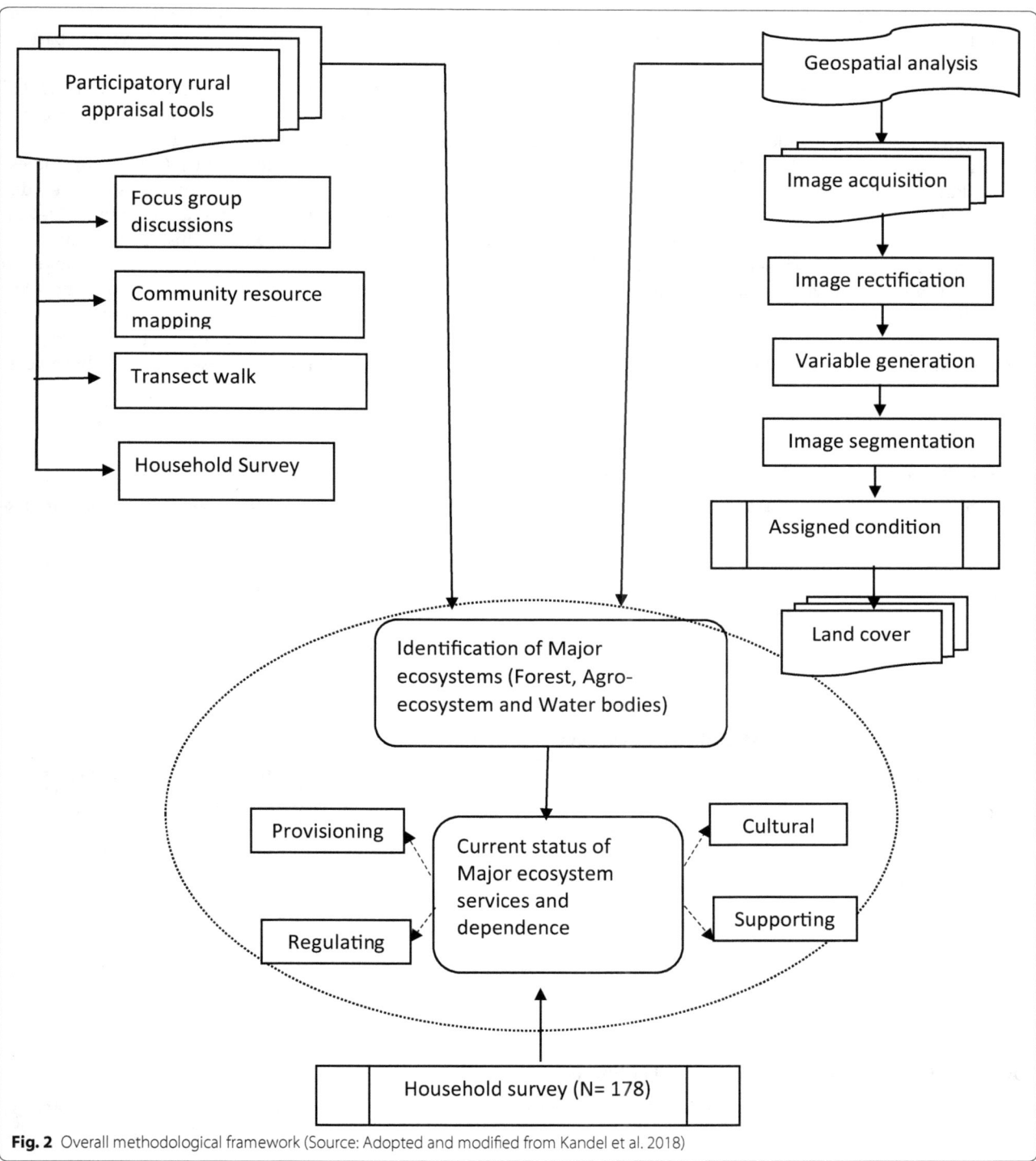

Fig. 2 Overall methodological framework (Source: Adopted and modified from Kandel et al. 2018)

The results, obtained from household survey on communities' dependency and their perceptions on changing LULCC and ES through qualitative analysis were then compared with the observed LULCC data for 1889–2000 and 2000–2014.

Results

Land use land cover change

Major land use land cover types in the study area consisted of forest, shrub-land, grassland, cropland, seasonal and perennial water bodies. In the year 2014, cropland was dominant land use types with 64.5% coverage followed by forest (18%) and the least was freshwater (4.2%). There have been a subsequent changes to these land use land cover over the period of 25 years (1989–2004, Table 2). We observed a sharp reduction of 164 km^2 seasonal water body area in Inle Lake in 2014 which is 4.2 smaller than in 1989. Similarly forest area has declined by 92 km^2, shrub land showed a negative change of 52 km^2 and 1 km^2 grass-land area has dropped down in last 25 years. Contrary to this, an increase of 268 km^2 cropland area and 40 km^2 perennial water body were observed (Table 2) showing periodic increment over the time. The periodic data of the year 1989 showed that the cropland was 59.5%. It further increased to 60.67% in 2000 and 64.53% in the year 2014 giving a total cropland

increment of 268 km^2 in 25 years. Similarly, the perennial water body has increased by 40 km^2 against the baseline year 1989.

To further segregate the periodic changes of LULCC, the breakdown of the observed results in the form of change matrix of land cover from 1989–2000 to 2000–2014 are presented in Tables 3 and 4. Comparing Tables 3 and 4, an overall forest area of 92 km^2 has reduced during 1989–2017 but in the later years during 2000–2014 the rate of forest loss is 115 km^2. This 115 km^2 forest loss is mainly because of the conversion of forest land into crop land. Declinations of 1 km^2 of shrub-land and 1 km^2 of grassland were observed. Cropland has increased by 268 km^2 that has invaded wetland, shrub-land and grassland in 25 years of timeline. However, over those years,

Table 1 Description of sampling areas for household survey

Study area	Upstream mountain area	Middle stream market area	Downstream floating garden	Total
Village track	Lat Maung Kwe	Nan Pan	Nan Pan	–
Village name	Kyaung Taung	Zay Gon	Kyar Taw	–
Location	N20°39′24″ E96°51′48″	N20°27′56″ E96°54′13″	N20°26′57″ E96°54′52″	
Total households	186	168	173	527
Sample size	58	60	60	178

Table 2 Summary of land cover statistics for 1989, 2000 and 2014

ID	Land cover	Year 1989		Year 2000		Year 2014		LC changes in km^2 (1989–2014)
		km^2	%	km^2	%	km^2	%	
1	Forest	1074	19.79	1097	20.23	982	18.10	− 92
2	Shrub land	321	5.92	268	4.94	269	4.96	− 52
3	Grassland	394	7.27	394	7.27	393	7.24	− 1
4	Cropland	3232	59.58	3291	60.67	3500	64.53	268
5	Seasonal water bodies	214	3.95	183	3.37	51	0.94	− 164
6	Perennial water bodies	189	3.49	191	3.52	229	4.22	40
7	Total	5424	100	5424	100	5424	100	

Table 3 2 Change matrix of land cover (km^2) in 1989 to 2000

Land cover (km^2)	Forest	Shrub land	Grassland	Cropland	Seasonal water bodies	Perennial water bodies	Total (1989)
Forest	1074	0	0	0	0	0	1074
Shrub land	24	268	0	25	3	1	321
Grassland	0	0	394	0	0	0	394
Cropland	0	0	0	3232	0	0	3232
Seasonal water bodies	0	0	0	34	177	3	214
Perennial water bodies	0	0	0	0	3	187	189
Total (2000)	1097	268	394	3291	183	191	5424

Table 4　Change matrix of land cover (km²) in 2000 to 2014

Land cover (km²)	Forest	Shrub land	Grassland	Cropland	Seasonal water bodies	Perennial water bodies	Total (2000)
Forest	982	0	0	115	0	0	1097
Shrub land	0	268	0	0	0	0	268
Grassland	0	0	393	0	0	1	394
Cropland	0	0	0	3279	3	8	3291
Seasonal water bodies	0	1	0	104	47	30	183
Perennial water bodies	0	0	1	2	0	189	192
Total (2014)	982	269	393	3500	51	229	5424

perennial water body has been altered and an increment of 38 km² was witnessed. Referring to Table 3, perennial water body has influenced grassland, seasonal water body, and cropland. Likewise, spatial-temporal changes of forest, shrubland, grassland, cropland, seasonal and perennial water bodies are presented in Fig. 3.

Utilization of ES for livelihood

Communities from the three study areas, namely-Kyaung Taung, Zay Gon and Kyar Taw showed varied dependency depending upon the proximity of the ecosystems (Fig. 4). It was observed that all the depended communities seem to use available ecosystems optimally. Our qualitative data revealed that the local inhabitants utilizes 17 types of ES from forest ecosystem, 13 from agro-ecosystem, 10 from seasonal and 4 from perennial water body for their livelihoods (Table 5). Almost all of the respondents in Kyaung Taung village mentioned that they consume mushroom (100%) and wild edible fruits/vegetables (97%) from forest ecosystem. About 83% of the same village collects fuelwood. Despite deforestation and degradation in the forest areas, forests still account for the supply of fuelwood in Kyaung Taung village. Only 7% of the respondents in Kyar Taw village and 8% in Zay Gon village consumed fuelwood from forests. Likewise, a wide range of wetland services are utilized by floating garden communities. About 91% of respondents use water for bathing, 66% for fishing, 28% as source for fodder, 24% as source for seaweed and 14% for irrigation. The agro-ecosystem seems very productive in mountain area. About 93% of the households cultivate vegetables, 87% cultivate paddy and mushroom, 65% collect fuelwood from agro-ecosystem in mountain area. Similarly, the agro-ecosystem in market area looms vegetable production (87%), ornamental plants (67%), fuelwood supply (38%) and wild and edible fruits (37%). In an average, fresh water (perennial and seasonal) attributed to drinking water supply (93%), water for bathing (61%) and water for irrigation (6%) in three study sites. Apart from the forest, study results elucidated that the fuelwood and

fodder requirements in the community are met from agro-ecosystems and wetlands.

Community perception on state of ES and LULCC

Figure 5 illustrates the communities' perception on the changes of flow of ES over the last decade. Around 93% of the respondents opined that forest ecosystem has decreased over the last 10 years. Fuelwood extraction, illegal logging, charcoal making, shifting cultivation, extension of agricultural land and population growth played an influential role to the exacerbated forest ecosystem. Also, the communities' claimed that almost no forest has remained in the village area. Around 40% of the respondents reflected an increase in area used for cropland; 43% conversely perceived a declination. Communities mentioned that maximum use of chemical fertilizer has affected the soil fertility and water. Interestingly, 17% mentioned there is no change in such practices.

About 63% of the respondents perceived such changes in four ecosystems have brought huge reduction in availability of freshwater. The reduction of freshwater has caused inland water transportation used for tourism and other use a challenge. Also, respondents reiterated that lake water is not potable since last 10 years and retrograding water quality has affected natural aquaculture. Major apprehensions are depletion of forests and increased soil erosion leading to sedimentation, erratic rainfall and drying out of rain water collection pond. About 30% also mentioned that reforestation had somewhat contributed to reduce those negative changes. A significant number of respondents (92%) perceived an enormous reduction in seasonal water body in dry season (Fig. 5).

Comparison of LULCC and perceived changes in ES

Observed loss in forest area and seasonal water body through LULCC are consistent with community's perceived changes. Around 93% of the households mentioned flow of ES from forest ecosystems has declined. Comparing this information with the LULCC (Table 2), 92% of the forest area has been lost in the last 25 years

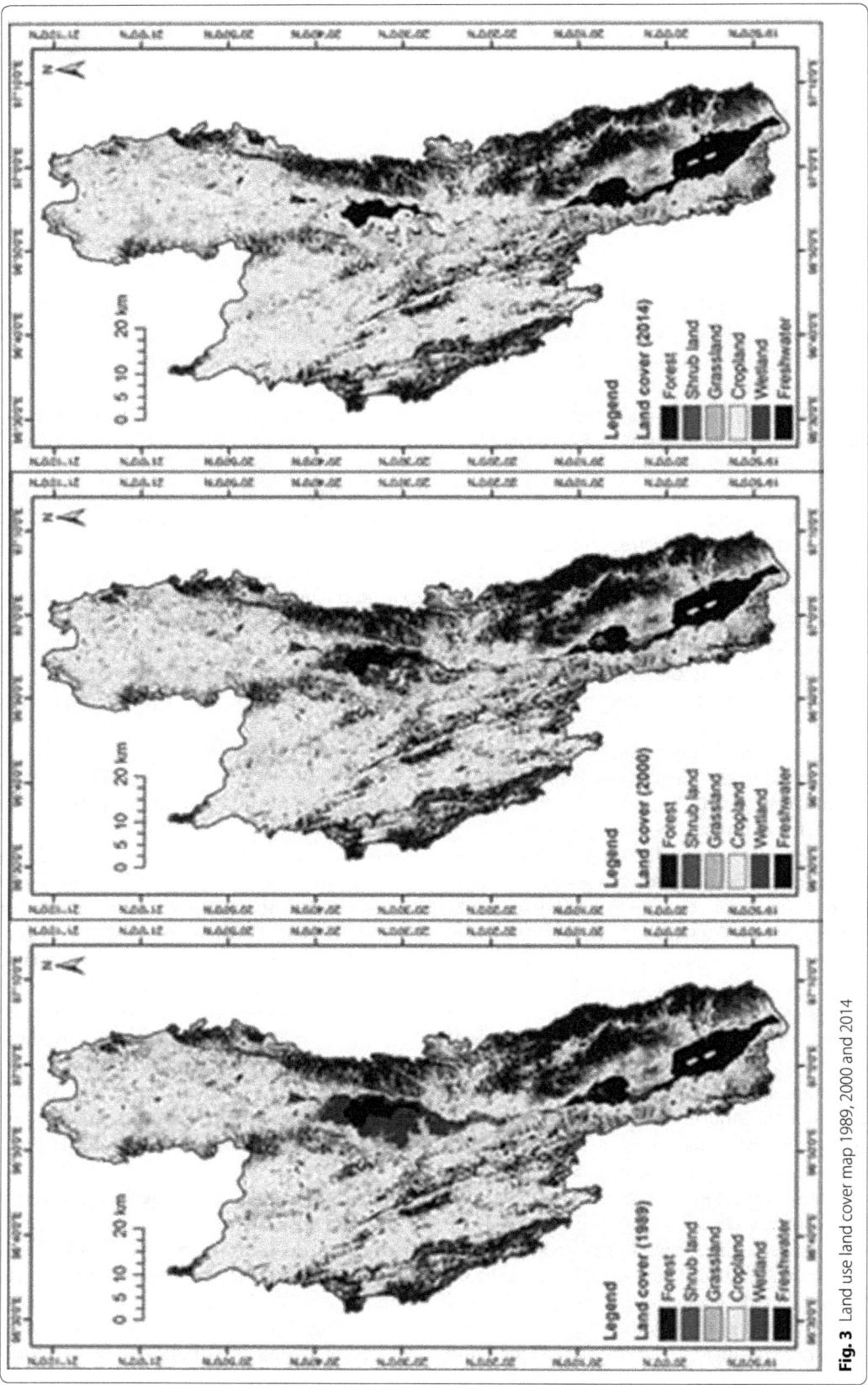

Fig. 3 Land use land cover map 1989, 2000 and 2014

Impact of land use land cover change on ecosystem services: a comparative analysis on observed...

189

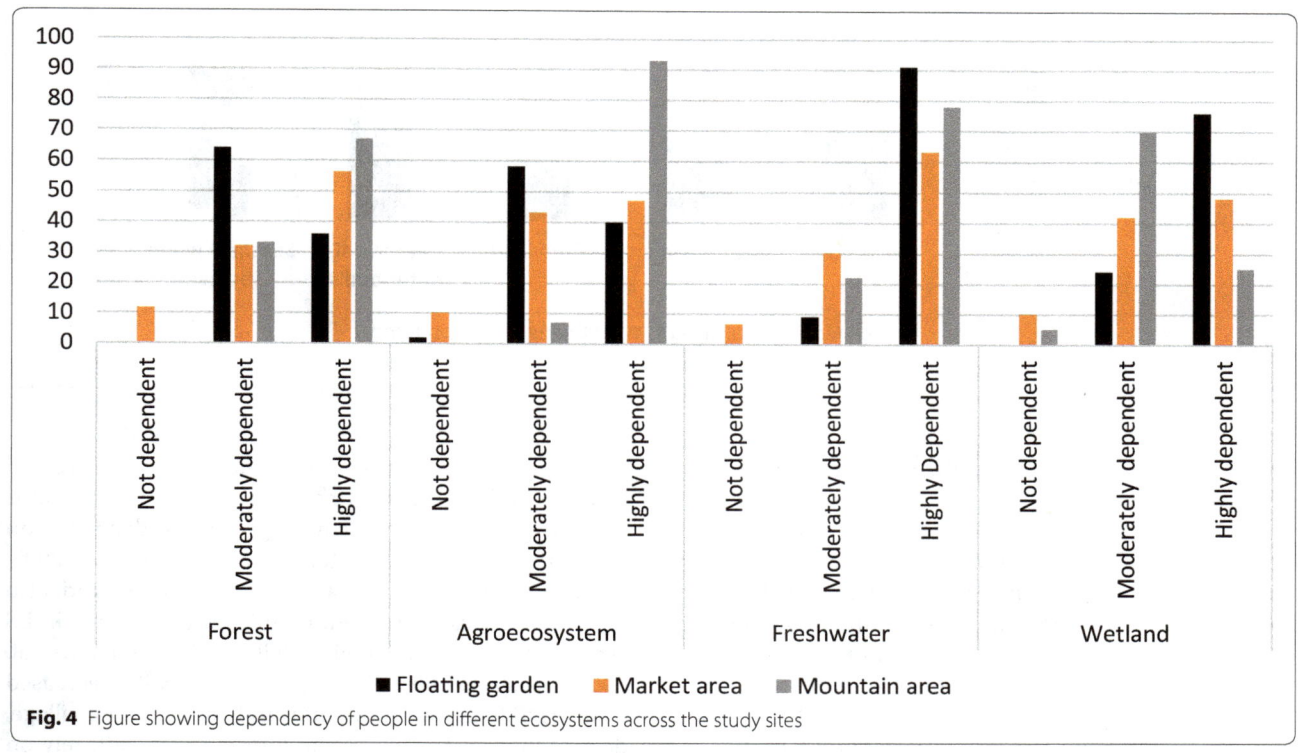

Fig. 4 Figure showing dependency of people in different ecosystems across the study sites

Table 5 Number of provisioning ES utilized by local communities for their livelihoods

Ecosystem	Forest (17)	Agro-ecosystem (13)	Perennial water bodies (4)	Seasonal water bodies (10)
Provisioning goods and services	Fuel wood	Fuel wood	Fish	Ornamental plants
	Fodder	Fodder	Drinking water	Wild edible vegetables
	Grazing	Grazing	Water for bathing	Fruits
	Timber	Timber/poles	Water for irrigation	Fish
	Poles	Medicinal plants		Drinking water
	Medicinal plants	Ornamental plants		Water for bathing
	Ornamental plants	Wild edible vegetables/		Water for irrigation
	Wild edible vegetables/	mushrooms		Silt soil
	mushrooms	Fruits		Source for seaweed
	Fruits	Fiber		Source for fodder
	Fiber	Thatch		
	Thatch	Dyes		
	Bush meat	Paddy		
	Paddy	Cereals		
	Cereals			
	Drinking water			
	Water for bathing			
	Water for irrigation			

which is evident to the community's belief on declining ES from forest ecosystems. Furthermore, this has been evident from the visible change observed on the ES during the last 25 years (see Fig. 6). Communities associated loss of forest and water body area with reduced ES that they are receiving. They mentioned that ES listed in Table 5, are nowadays in declining trend. An observed data of increased cropland area by 8.3% (Table 2)

reflected a mixed perception (Fig. 5). However, observed increased perennial water body area through LULCC analysis contradicts to 63% of the communities' belief. Communities believed that the availability of freshwater in all three study sites has been reducing (Fig. 5). But LULCC analysis (Tables 3, 4) showed that perennial water bodies have increased. In terms of changes in flow of goods and services from agro-ecosystem (Fig. 5), 43%

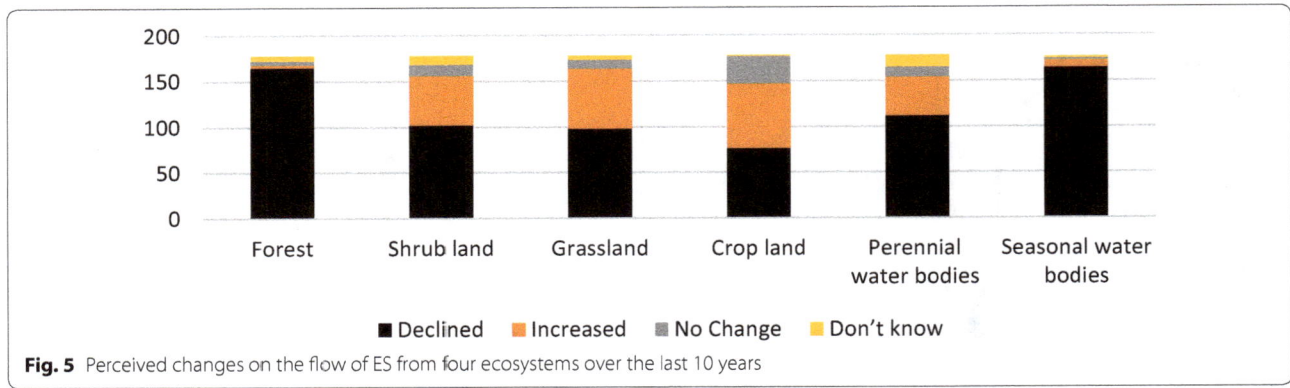

Fig. 5 Perceived changes on the flow of ES from four ecosystems over the last 10 years

of the households mentioned the agricultural productivity have increased but 40% expressed that the productivity has reduced and 17% mentioned there has been no changes in the agricultural productivity. Community's perception over such changes might be mainly due to the degraded land converted into agricultural land and economic return from land conversion over to the population growth. Due to limited outmigration from Inle Lake, the growing population demand more land and irrigation for farming, but less water availability for irrigation results into less productivity. However, there is a clear indication of increased in crop area by 8.29% from 1989 to 2014 suggesting agricultural intensification.

Discussion

How the LULCC (temporal and spatial) has changed over the period in the study area?

The LULCC has been identified as one of the main drivers of change worldwide (Pandit et al. 2007; Chettri and Sharma 2016). Such LULCC, as a continued socioecological disturbance, changes the flow of ES (Janssen and Anderies 2007). A widespread deforestation and unplanned LULCC threatens natural ecosystems (United Nations 2002; Sidle et al. 2007), decreases multi-functionality (Kandziora et al. 2014) and limits the habitat of globally important species (Chettri et al. 2013). Myanmar has been witnessing major LULCC in the recent years (Htwe et al. 2015) and our study also validate it. The Forest Department Statistics of Myanmar showed 37.4% of 343,587 km^2 natural forest area deforested in 1998 (United Nations 2002). A similar trend of 40.4% of forest cover loss from 2001 to 2012 has been reported by (Khaing 2014). The rate of deforestation and degradation were −1.17% from 1990 to 2000, −0.90% from 2000–2005 to −0.95% 2005–2010 (FRA 2010) showing increasing trend of deforestation lately.

Major LULCC in the study area depicted a reduction in forest, seasonal water body and increase in cropland

and perennial water body. Such changes have increased siltation in the lake, affecting fresh water hotspots that is home to worldwide threatened species and depended on the health of these ecosystems (Leimgruber et al. 2005; Htwe et al. 2015). Such changes, on the other hand, also bring challenges to the communities with changed in ES availability needed for their livelihood (Chaudhary et al. 2016). This also affects hydropower plant with decreased flow (ADB 2006). Communities in Kyaung Taung village, despite huge reduction in the forested area, still rely on forests to get forest products. In the study site some good initiations like stall feeding to reduce open grazing practices and government providing free seedling to motivate communities in conservation has started. However, due to the low survival rate of these planted seedlings, results from such good initiative seems insignificant.

Crop land expansion due to increase in intensity of agriculture in the forested catchment areas, sediment load from tributaries from the catchment areas, siltation inflow to the lake and marshland transformed into agricultural areas as hydroponic expansion are major determinants of reduced wetland area in the study site. Our study showed that Inle Lake is experiencing an expansion of cropland by 268 km^2 from 1989 to 2014. Similar significant changes have also been reported on the lake and its surrounding catchments by many earlier studies (Ma 1967; Thiha 2005; Htwe et al. 2015; Pradhan et al. 2015). Thus, it symbolizes a continuous transformation in size of the lake. Additionally, sediment load from tributaries amounting 2.63106 m^3/year (Su and Jassby 2000) and siltation from inflowing stream equivalent to 6,23,000 m^3/year clearing the natural vegetation for cultivation (Akaishi et al. 2006) are other factors in shrinking wetland area. Also, our result was supported by an estimated decline of open water surface in Inle Lake by 32.4% between 1935 and 2000 (Sidle et al. 2007). Furthermore, houses and restaurants built inside lake with poor sanitation and improper management of waste are also

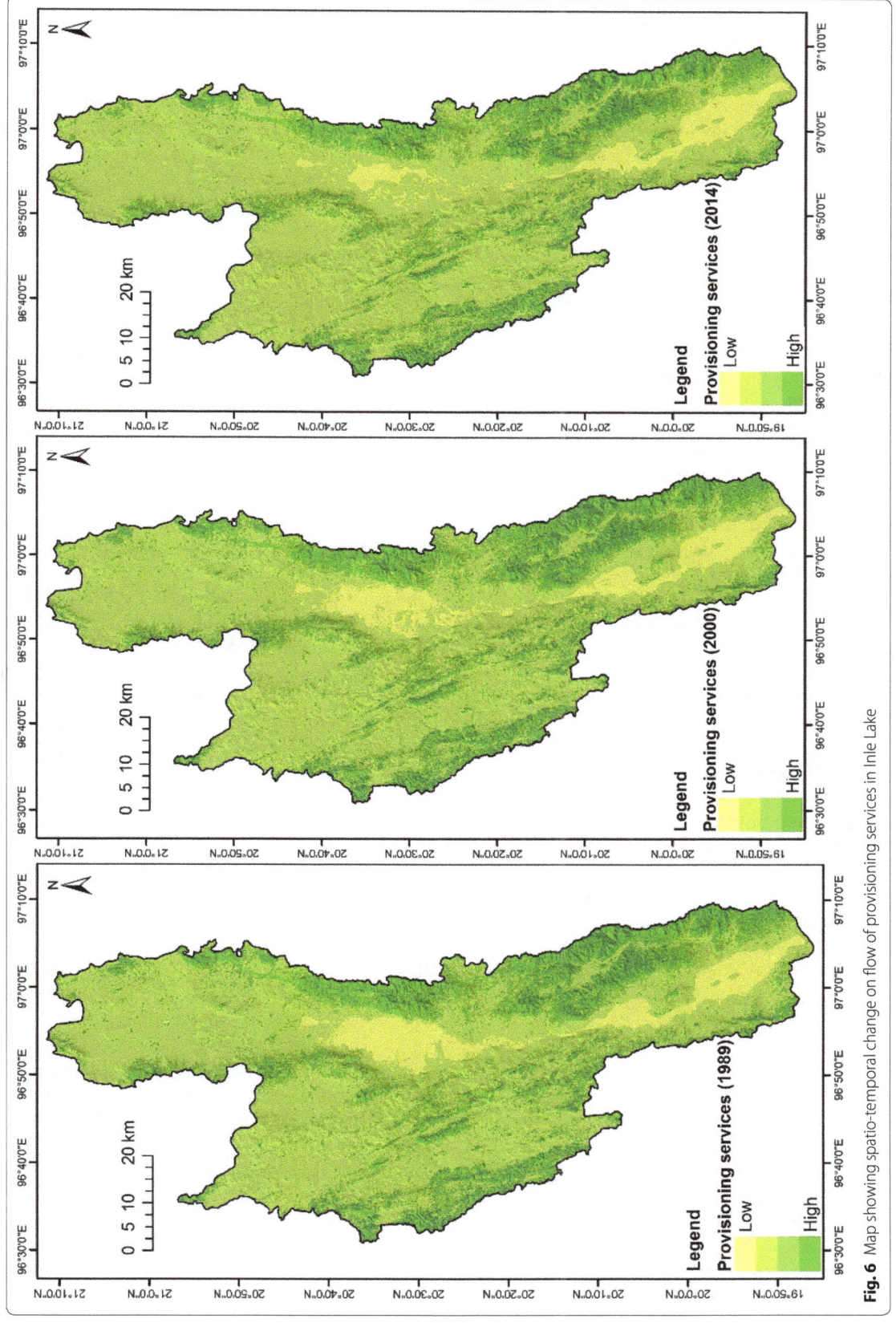

Fig. 6 Map showing spatio-temporal change on flow of provisioning services in Inle Lake

adding challenges on the health of the lake (ADB 2006; May 2007; San and Rapera 2010; Lwin and Sharma 2012). Simultaneously, the heavy rainfall trend (DMH 2016) has increased the rate of landslide impacting the lake further.

The reduction in wetland area not only threatens and limits the habitats of globally important species but also adds the leeching of agrochemicals from cropland and hydroponic cultivation into lake, further affecting water quality and promoting algal bloom in the lake (Ma 1996; Akaishi et al. 2006; Gyi et al. 2011). Increased population with double digit in Nyaung Shwe and Taunggyi Township from 1968 to 2010, limited out migration and local economic development opportunities could be other prime transforming agents converting water and marshland into agriculture (Lambin et al. 2001). An increased agricultural production rate degrades 40% of the land area posing great threat to biodiversity (Foley et al. 2011). A study conducted in China showed total food production and expanded arable land secured a negative effect on biodiversity (Hou et al. 2015). Intensive cultivation techniques and use of herbicides increasingly affect the landscapes' natural capacities in maintaining biodiversity and ecosystem functioning, including supply of ES abrading health of the perennial and seasonal water bodies.

What are the states of major ecosystems in the study area and how much the local people are dependent on these ecosystems?

Human life largely depend on forest and agriculture as important economic resources and means of development (SDG, Agenda 15). This is more relevant to Eastern Himalaya for local people with limited livelihood options (Chaudhary et al. 2015). A maintained resilient ecosystem for a continued flow of ES requires a harmonized relationship between human and nature (Gómez-Baggethun and Kelemen 2008). Communities in Inle Lake, largely depend on ES derived from forest, agro-ecosystem, and perennial and seasonal water bodies.

Comparatively higher dependency on ES in the mountain regions are well documented facts due to limited options (Chaudhary et al. 2016). Interestingly, our study found that dependency on the ES varied as per the proximity of ecosystems. Since ES have been shaped through human history by land allocation and management choices (Crouzat et al. 2015), our study illustrated that Kyaung Taung communities have more agro-ecosystem productivity and have higher access to forested area, while Kyar Taw village largely depends on ES generated from perennial and seasonal water bodies in Inle Lake. However, being a trading zone, Zay Gon sells some of the ES collected from Kyar Taw, in addition to the ES derived from their own agro-ecosystems. Livelihood of the communities in Zay Gon largely depend on trading, thus, a subtle change in supply of ES from Kyaung Taung and Kar Taw village could affect their livelihoods. These relationship clearly indicates the existing social and ecological linkages as well as the highland and lowland linkages. Both the communities living in forest ecosystems and wetland ecosystems were directly or indirectly dependent on the urban (market area) ecosystem for ES flow and trade-off and *vice versa*.

What are the people's perception in terms of the LULCC and its impact on the ES they are dependent on?

Wetland has been facing the major brunt due to LULCC in Asia (Romshoo and Rashid 2014; Zorrilla-Miras et al. 2014; Chettri and Sharma 2016). There has been significant reduction in wetland area globally making it a subject of global concern (Gopal 2013; Reis et al. 2017; Davidson et al. 2018). The perception and the observed data in Inle Lake showed consistency with the observed trend. The people's perception and LULCC analyses data revealed that there is significant change in the area as also reported by others (Htwe et al. 2015; Gyi et al. 2011). Communities in the study sites reiterated that quantity and quality of potable water has been worsen since the last decade as also reported by Akaishi et al (2006). The amount and quality of water could easily impact on possible crop yields as well as a direct impact on human health (Burkhard et al. 2015). Also communities in the study area mentioned that two perennial water bodies have dried up and people nowadays purchase drinking water. Enduring fish population loss as a poor water quality has forced fishermen to shift their occupation to farming in the study area. Additionally, in dry season reduced water level in the lake has affected the boat rowing and travelling.

Conclusion

The significance of the biodiversity and ES of Inle Lake to the local communities is important for livelihood and has been recognized by the UNESCO's man and Biosphere Reserve programme by notifying it as the first Biosphere Reserve in Myanmar. The study reiterated that LULCC is happening and it has implication to the sustained flow of ES for human wellbeing. The main drivers seem to be expansion of cropland manifested by increased siltation from the catchment area and chemical leeching that has affected the world's threatened floral, faunal and endemic species in the lake. Similarly, the rate of deforestation and forest degradation is increasing. As a result, the local communities are exploring adaptive measures to tackle the challenges. Interestingly, the people's perceptions are also supportive to the observed analysis of LULCC with some exceptions.

Impact of land use land cover change on ecosystem services: a comparative analysis on observed...

193

Our study showed that the local communities living in Inle Lake and its surrounding catchments have high dependence on the ES supplied by forest, agro-ecosystem, seasonal and perennial water bodies. The provisional ES use pattern vary as per the proximity of the ecosystems and availability of the alternative options. Moreover, the study also showed a strong upstream-downstream linkages in terms of trade-off among the communities living at different ecosystems. The study suggests following actions to address the changing effect of LULCC. First of all, looking into the tourism driven local economy, and people's high dependency on ES, demand and supply chain gap from need special attention with socio-ecological system approach. Second, restoration of the degraded areas through the inspection and regular monitoring of survival rate of planted seedlings. Third, alternative energy (improved cooking stoves, biogas) installation would add significant results to reduce further pressure on the resources. Fourth, an investment on establishing natural water ponds might be some viable options to collect rain water runoff to cope up with water scarcity to some extent. Lastly, an effort of establishing Payment for Ecosystem Services (PES) may further address the issue of siltation that is affecting hydropower plant and electricity generation. In order to draw detailed conclusions for decision-making and management of ecosystems in the study site, a socio-ecological linkage would give a better picture. A socio-ecological system approach would enable a clear policy reformulation that would support to keep ecosystems a healthy.

Abbreviations
ES: ecosystem services; FGD: focus group discussion; LULCC: land use land cover change; NDVI: Normalized Difference Vegetation Index; OBIA: object-based image analysis; PRA: participatory rural appraisal; SSS: systemic stratified sampling; TM: thematic mapper; UNESCO: United Nations Organization for Education, Science and Culture; USD: United States Dollar; UTM: Universal Transverse Mercator.

Authors' contributions
NC, KA and KU conceptualised the study. SK, AMT, ST, WMA and KA collected and analysed the data, SK, PKKU and NC wrote the manuscript. All authors read and approved the final manuscript.

Author details
[1] International Centre for Integrated Mountain Development (ICIMOD), GPO Box 3226, Kathmandu, Nepal. [2] Christian Albrechts University Kiel, Olshausenstr. 75, 24118 Kiel, Germany. [3] Inle Lake Biosphere Reserve, Nyaung Shwe, Myanmar. [4] Department of Infrastructure Engineering, Melbourne School of Engineering, The University of Melbourne, Parkville, Australia.

Acknowledgements
This study is jointly conducted by Ministry of Natural Resources and Environmental Conservation (MoNREC) and International Centre for Integrated Mountain Development (ICIMOD) under Himalica project funded by the European Union. We also would like to thank Mr. Madhav Dhakal, Associate Hydrologist for providing analyzed hydrometeorology data for Myanmar. The views and interpretations in this publication are those of the authors and there is no conflict of interest from any of the authors to publish this Manuscript.

Competing interests
The authors declare that they have no competing interests.

Funding
This paper is prepared under the Support to Rural Livelihoods and Climate Change Adaptation in the Himalaya (Himalica) Programme (ASIE/2012/292-464) funded by the European Union to ICIMOD.

References
Akaishi F, Satake M, Otaki M, Tominaga N (2006) Surface water quality and information about the environment surrounding Inle Lake in Myanmar. Limnology 7(1):57–62. https://doi.org/10.1007/s10201-006-0165-1

Annandale N (1918) Fish and fisheries of the Inle Lake. Records Indian Mus 14:33–64

Asian Development Bank (2006) Myanmar National Environmental Performance Assessment (EPA) report. National performance assessment and sub-regional strategic environment framework in the greater Mekong sub-region. ADB T. A. No. 6069–REG. National Commission for Environmental Affairs, Yangon

Balvanera P, Pfisterer AB, Buchmann N, He JS, Nakashizuka T, Raffaelli D, Schmid B (2006) Quantifying the evidence for biodiversity effects on ecosystem functioning and services. Ecol Lett 9(10):1146–1156

Bateman IJ, Harwood AR, Mace GM, Watson RT, Abson DJ, Andrews B, Binner A, Crowe A, Day BH, Dugdale S, Fezzi C, Foden J, Hadley D, Haines-Young R, Hulme M, Kontoleon A, Lovett AA, Munday P, Pascual U, Paterson J, Perino G, Sen A, Siriwardena G, van Soest D, Termansen M (2013) Bringing ecosystem services into economic decision–making: land use in the United Kingdom. Science 341:45–50. https://doi.org/10.1126/science.1234379

Bertrand G, Rangin C (2003) Tectonics of the western margin of the Shan plateau (central Myanmar): implication for the India-Indochina oblique convergence since the Oligocene. J Asian Earth Sci 21:1139–1157. https://doi.org/10.1016/S1367-9120(02)00183-9

Blaschke T, Hay GJ (2001) Object-oriented image analysis and scale-space: theory and methods for modeling and evaluating multiscale landscape structure. Int Archiv Photogramm Remote Sens 34(4):22–29

Bryan BA, Raymond CM, Crossman ND, Macdonald DH (2010) Targeting the management of ecosystem services based on social values: where, what, and how. Landscape Urban Plann 97(2):111–122. https://doi.org/10.1016/j.landurbplan.2010.05.002

Burkhard B, Müller A, Mueller F, Grescho V, Anh Q, Arida G, Bustamante JVJ, Van Chien H, Heong KL, Escalada M, Marquez L (2015) Land cover–based ecosystem service assessment of irrigated rice cropping systems in Southeast Asia: an explorative study. Ecosyst Serv 14:76–87. https://doi.org/10.1016/j.ecoser.2015.05.005

Butkus S, Myint S (2001) Pesticide use limits for protection of human health in the Inle Lake (Myanmar) watershed, Technical document. Living Earth Institute Olympia (NPO), Washington

Cardinale BJ, Matulich KL, Hooper DU, Byrnes JE, Duffy E, Gamfeldt L, Balvanera P, O'Cornor ML, Gonzalez A (2011) The functional role of producer diversity in ecosystems. Am J Bot 98(3):572–592. https://doi.org/10.3732/ajb.1000364

Castro AJ, Verburg PH, Martín-López B, Garcia-Llorente M, Cabello J, Vaughn CC, López E (2014) Ecosystem service trade–offs from supply to social demand: a landscape-scale spatial analysis. Landscape Urban Plann 132:102–110. https://doi.org/10.1016/j.landurbplan.2014.08.009

Chaudhary S, MacGregor K, Houston D, Chettri N (2015) The evolution of ecosystem services: a time series and discourse–centred analysis. Environ Sci Policy 54:25–34. https://doi.org/10.1016/j.envsci.2015.04.025

Chaudhary S, Chettri N, Uddin K, Khatri TB, Dhakal M, Bajracharya B, Ning W

(2016) Implications of land cover change on ecosystem services and people's dependency. A case study from the Koshi Tappu Wildlife Reserve, Nepal. Ecol Complex 1:1. https://doi.org/10.1016/j.ecocom.2016.04.002

Chaudhary S, Tshering D, Phuntsho T, Uddin K, Shakya B, Chettri N (2017) Impact of land cover change on a mountain ecosystem and its services: case study from the Phobjikha valley, Bhutan. Ecosyst Health Sustain 3:1–12. https://doi.org/10.1080/20964129.2017.1393314

Chettri N, Sharma E (2016) Reconciling the mountain b odiversity conservation and human wellbeing: drivers of biodiversity loss and new approaches in the Hindu-Kush Himalayas. Proc Ind Nat Sci Acad. 82:53–73

Chettri N, Uddin K, Chaudhary S, Sharma E (2013) Linking spatio-temporal land cover change to biodiversity conservation in Koshi Tappu Wildlife Reserve, Nepal. Diversity 5:335–351

Costanza R, Arge R, de Groot R, Farber S, Grasso M, Hannon B, Limburg K, Naeem S, Oneill RV, Paruelo J, Raskin RG, Sutton P, van den Belt M (1997) The value of the world's ecosystem services and natural capital. Nature 387:253–260. https://doi.org/10.1038/387253a0

Crouzat E, Mouchet M, Turkelboom F, Byczek C, Meersmans J, Berger F, Verkerk PJ, Lavorel S (2015) Assessing bundles of ecosystem services from regional to landscape scale: insights from the French Alps. J Appl Ecol 52(5):1145–1155. https://doi.org/10.1111/1365-2664.12502

Davidson NC (2014) How much wetland has the world lost? Long-term and recent trends in global wetland area. Marine Freshwater Res 65(10):934–941. https://doi.org/10.1071/MF14173

Davidson NC, Fluet-Chouinard E, Finlayson CM (2018) Global extent and distribution of wetlands: trends and issues. Marine Freshwater Res 69(4):620–627. https://doi.org/10.1071/MF17019

deGroot RS, Alkemade R, Braat L, Hein L, Willemen L (2010) Challenges in integrating the concept of ecosystem services and values in landscape planning, management and decision making. Ecol Complex 7:260–272. https://doi.org/10.1016/j.ecocom.2009.10.006

Díaz S, Demissew S, Carabias J, Joly C, Lonsdale M, Ash N, Larigauderie A, Adhikari JR, Arico S, Báldi A, Bartuska A (2015) The IPBES Conceptual Framework—connecting nature and people. Curr Opinion Environ Sust 14:1–16

Díaz S, Pascual U, Stenseke M, Martín-López B, Watson RT, Molnár Z, Hill R et al (2018) Assessing nature's contributions to people. Science 359(6373):270–272

Ding XW, Hou BD, Xue Y, Jiang GH (2017) Long-term effects of ecological factors on nonpoint source pollution in the upper reach of the Yangtze River. J Environ Informat 30(1):17–28

DMH (2016) Temperature and precipitation data recorded from 1989–2013 at Heho airport. Department of Meteorology and Hydrology, Myanmar

Fisher B, Turner RK, Morling P (2009) Defining and classifying ecosystem services for decision making. Ecol Econ 68:643–653. https://doi.org/10.1016/j.ecolecon.2008.09.014

Foley JA, Ramankutty N, Brauman KA, Cassidy ES, Gerber JS, Johnston M, Mueller ND, O'Connell C, Ray DK, West PC, Balzer C (2011) Solutions for a cultivated planet. Nature 478(7369):337–342

FRA (2010) Global Forest Resource Assessment. http://www.fao.org/forestry/fra/fra2010/en/

Gómez-Baggethun E, Kelemen E (2008) Linking institutional change and the flows of ecosystem services. Case studies from Spain and Hungary. In: Proceedings of the 2nd THEMES Summer School 118-145

Gopal B (2013) Future of wetlands in tropical and subtropical Asia, especially in the face of climate change. Aquatic Sci 75(1):39–61

Groombridge B, Jenkins M (1998) Freshwater biodiversity a preliminary global assessment, WCMC biodiversity series No. 8. World Conservation Press, Cambridge

Gyi MM, Lwin LL, Khin MT, Oo KS (2011) Spatial habitat degradation due to human inhibition in respective areas of Inle Lake. http://a-a-r-s.org/acrs/administrator/components/com_jresearch/files/publications/SC03-0260_Full_Paper_ACRS2013_Maung_Maung_Gyi.pdf (accessed 21 November 2015)

Hou Y, Muller F, Li B, Kroll F (2015) Urban-rural gradients of ecosystem services and the linkages with socioeconomics. Landscape Online. https://doi.con/10.3097/lo.201539

Htwe TN, Kywe M, Buerkert A, Brinkmann K (2015) Transformation processes in farming systems and surrounding areas of Inle Lake, Myanmar, during the last 40 years. J Land Use Sci 10(2):205–223. https://doi.org/10.1080/1747423X.2013.878764

ICIMOD and MoNREC (2017) A multi-dimensional assessment of ecosystems and ecosystem services at Inle Lake, Myanmar. ICIMOD Working Paper 2017/17. Kathmandu: ICIMOD

Ingelmo IA (2013) Design and development of a sustainable tourism indicator based on human activities analysis in Inle Lake, Myanmar. Proc Soc Behav Sci 103:262–272. https://doi.org/10.1016/j.sbspro.2013.10.334

Janssen MA, Anderies JM (2007) Robustness trade-offs in social–ecological systems. Int J Commons 1(1):43–66

Kandel P, Tshering D, Uddin K, Lhamtshok T, Aryal K, Karki S, Sharma B, Chettri N (2018) Understanding social–ecological interdependencies through ecosystem services value perspectives in Bhutan. Ecosphere, Eastern Himalaya. https://doi.org/10.1002/ecs2.2121

Kandziora M, Dörnhöfer K, Oppelt N, Müller F (2014) Detecting land use and land cover changes in northern German agricultural landscapes to assess ecosystem service dynamics. Landscape Online 1:35

Khaing DAA (2014) MIID himalica pilot myanmar land resource assessment. ICIMOD, Kathmandu

Kottelat VM (1986) The fish fauna of Inle Lake in Burma. Aquatic Terres Zool 39:403–406

Kottelat M, Witte KE (1999) Two new species of Microrasbora from Thailand and Myanmar, with two new generic names for small Southeast Asian cyprinid fishes (Teleostei: Cyprinidae). J South Asian Nat Hist 4(1):49–56

Lambin EF, Turner BL, Geist HJ, Agbola SB, Angelsen A, Bruce JW, Coomes OT, Dirzo R, Fischer G, Folke C, George P (2001) The causes of land-use and land-cover change: moving beyond the myths. Global Environ Change 11(4):261–269. https://doi.org/10.1016/S0959-3780(01)00007-3

Lamsal P, Pant KP, Kumar L, Atreya K (2015) Sustainable livelihoods through conservation of wetland resources: a case of economic benefits from Ghodaghodi Lake, western Nepal. Ecol Soc 20(1):10. https://doi.org/10.5751/ES-07172-200110

Lang S, Pernkopf L, Vanden JB, Förster M, Haest B, Buck O, Frick A (2011) Fostering Sustainability in European Nature Conservation—NATURA 2000 habitat monitoring based on earth observation services. In: Proceeding of 1st world sustainable forum 1–8

Leimgruber P, Kelly DS, Steininger MK, Brunner J, Müller T, Songer M (2005) Forest cover change patterns in Myanmar (Burma) 1990–2000. Environ Conserv 32(04):356–364. https://doi.org/10.1017/S0376892905002493

Lwin Z, Sharma MP (2012) Environmental management of the Inle Lake in Myanmar. Hydro Nepal J Water Energy Environ 11:57–60. https://doi.org/10.3126/hn.v11i0.7164

Ma KT (1967) Fishes and fishing gear of Inle Lake. University Press, Rangoon

Ma TDW (1996) Floating island agriculture (Ye–chan) of Inle Lake. M.A. thesis, University of Yangon, Yangon

Mace GM, Norris K, Fitter AH (2011) Biodiversity and ecosystem services: a multilayered relationship. Trends Ecol Evol 27:19–26. https://doi.org/10.1016/j.tree.2011.08.006

May SY (2007) Changes of water quality and water surface area in Inle Lake: facts and perception. Ph.D. thesis, University of Yangon, Myanmar

Millennium Ecosystem Assessment (MA (2005) Ecosystems and human well-being: synthesis. Island Press, Washington, DC, USA

Milner-Gulland EJ (2012) Interactions between human behaviour and ecological systems. Phil Trans Royal Soc B: Biol Sci 367:270–278. https://doi.org/10.1098/rstb.2011.0175

Mitsch WJ, Gosselink JG (2000) The value of wetlands: importance of scale and landscape setting. Ecol Econ 35(1):25–33. https://doi.org/10.1016/S0921-8009(00)00165-8

MoHT (2013) Myanmar tourism statistics 2012. Ministry of hotel and tourism. Nay Pyi Daw, Maynmar

Munz, A, Molstad A (2012) Working paper for tourism development for Inlay Lake. Consultant report for the Institute of International Development Project 'Inlay Lake: a plan for the future'. Yangon, Myanmar: IID

Myint DKW, Maung UKW (2000) Floating islands of the Inle Lake. Myanmar Persp 16(7):20

Nahlik AM, Kentula MA, Fennessy MS, Landers DH (2012) Where is the consensus? A proposed foundation for moving ecosystem service concepts into practice. Ecol Econ 77:27–35. https://doi.org/10.1016/j.ecolecon.2012.01.001

Okamoto I (2012) Coping and adaptation against decreasing fish resources: Case study of fishermen in Lake Inle, Myanmar. Institute of Developing Economies, Japan External Trade Organization (JETRO). https://ideas.repec.org/p/jet/dpaper/dpaper329.html

Olson DM, Dinerstein E (1998) The Global 200: a representation approach to conserving the earth's most biologically valuable ecoregions. Conserv Biol 12(3):502–515. https://doi.org/10.1046/j.1523-1739.1998.012003502.x

Omrani H, Abdallah F, Tayyebi A, Pijanowski B (2017) Modelling land-use change with dependence among labels. J Environ Informat 30(2):107–118

Pandit MK, Sodhi NS, Koh LP, Bhaskar A, Brook BW (2007) Unreported yet massive deforestation driving loss of endemic biodiversity in Indian. Himalaya Biodivers Conserv 16:153–163

Platt SG, Rainwater TR (2004) Inle Lake turtles, Myanmar with notes on Intha and Pa–O ethnoherpetology. Hamadryad 29:5–14

Pradhan N, Habib H, Venkatappa M, Ebbers T, Duboz R, Shipin O (2015) Framework tool for a rapid cumulative effects assessment: case of a prominent wetland in Myanmar. Environ Monit Assessm 187(6):1–18. https://doi.org/10.1007/s10661-015-4508-4

Reis V, Hermoso V, Hamilton SK, Ward D, Fluet-Chouinard E, Lehner B, Linke S (2017) A global assessment of inland wetland conservation status. Bioscience 67(6):523–533

Reyers B, Biggs R, Cumming GS, Elmqvist T, Hejnowicz AP, Polasky S (2013) Getting the measure of ecosystem services: a social–ecological approach. Front Ecol Environ 11:268–273. https://doi.org/10.1890/120144

Roberts TR (1986) Danionella translucida, a new genus and species of cyprinid fish from Burma, one of the smallest living vertebrates. Environ Biol Fishes 16(4):231–241

Romshoo SA, Rashid I (2014) Assessing the impacts of changing land cover and climate on Hokersar wetland in Indian Himalayas. Arab J Geosci 7(1):143–160

San CC, Rapera CL (2010) The on-site cost of soil erosion by the replacement cost methods in Inle Lake watershed, Nyaung Shwe Township, Myanmar. J Environ Sci Manag 13(1):67–81

Schmalzbauer B. Visbeck, M (2017) The Sustainable Development Goals-conceptual approaches for science and research projects. In: 19th EGU general assembly, EGU2017, proceedings from the conference held 23–28 April, 2017 in Vienna, Austria, p. 5312. http://adsabs.harvard.edu/abs/2017EGUGA..19.5312S

Scholes RJ, Reyers B, Biggs R, Spierenburg MJ, Duriappah A (2013) Multiscale and cross-scale assessments of social–ecological systems and their ecosystem services. Curr Opinion Environ Sustain 5:16–25. https://doi.org/10.1016/j.cosust.2013.01.004

Schuijt K, Brander L (2004) The economic value of the world's wetlands. WWF Living Waters: Conserving the Source of Life, Gland, p 31

Sharma B, Rasul G, Chettri N (2015) The economic value of wetland ecosystem services: evidence from the Koshi Tappu Wildlife Reserve, Nepal. Ecosyst Serv 12:84–93. https://doi.org/10.1016/j.ecoser.2015.02.007

Sidle RC, Ziegler AD, Vogler JB (2007) Contemporary changes in open water surface area of Lake Inle, Myanmar. Sustain Sci 2(1):55–65

Su M, Jassby AD (2000) Inle: a large Myanmar lake in transition. Lakes Reserv Res Manag 5(1):49–54

TEEB (2010) The economics of ecosystems and biodiversity: Mainstreaming the economics of nature—a synthesis of the approach, conclusions and recommendations of TEEB. TEEB Consortium (c/o UNEP), Geneva

Than MM (2007) Community activities contribution to water environment conservation of Inle Lake. Irrigation Department, Ministry of Agriculture and Irrigation, Union of Myanmar, Yangon

Thant K (1968) Checklist of fishes in the Inle Lake. Tekatho Pyinapade tha 2

Thaw K (1998) The industrial Inthas of Inle Lake. Myanmar Perspect 4:4

Thiha A (2005) Land–use adjustment based on watershed classification using remote sensing and GIS a study of Inle watershed, Myanmar. In: Zoebisch M, Cho KM, Hein S, Mowla R (eds) Mowla Integrated watershed management: studies and experiences from Asia. AIT, Bangkok

Turner RK, Van Den Bergh JC, Söderqvist T, Barendregt A, van der Straaten J, Maltby E, van Ierland EC (2000) Ecological–economic analysis of wetlands: scientific integration for management and policy. Ecol Econ 35(1):7–23. https://doi.org/10.1016/S0921-8009(00)00164-6

United Nations (2002) Myanmar country profile, Technical report to United Nations, Agenda 21, CP2002, Myanmar: http://www.un.org/esa/agenda21/natlinfo/wssd/myanmar.pdf. Accessed 11 Dec 2015

Zedler JB, Kercher S (2005) Wetland resources: status, trends, ecosystem services, and restorability. Ann Rev Environ Resour 30:39–74. https://doi.org/10.1146/annurev.energy.30.050504.144248

Zorrilla-Miras P, Palomo I, Gómez-Baggethun E, Martín-López B, Lomas PL, Montes C (2014) Effects of land-use change on wetland ecosystem services: a case study in the Doñana marshes (SW Spain). Landscape Urban Plann 122:160–174

13

Hydrochemical characteristics of surface water and ecological risk assessment of sediments from settlements within the Birim River basin in Ghana

Noah Kyame Asare-Donkor[*] [ID], Japhet Opoku Ofosu and Anthony Apeke Adimado

Abstract

Background: Geogenic and anthropogenic activities such as Artisanal and illegal gold mining continue to have negative impacts on the environment and river basins in China. This work studied the hydrogeochemical characteristics of surface water from the Birim River basin and assess the quality of water for human consumption and agricultural activities. In addition, the ecological risk assessment for Cd, Zn, Pb and As in sediment was evaluated using pollution indices.

Results: The results show that the turbidity, temperature, colour and iron concentration in the water samples were above the World Health Organization guidelines. Multivariate analysis explained five components that accounted for 98.15% of the overall hydrogeochemistry and affected by anthropogenic and geogenic impacts. The surface water was observed to range from neutral to mildly acidic, with the dominance of HCO_3^-, Cl^-, Ca^{2+}, Mg^{2+}, and Na^+ in ionic strength. The Piper diagram reveals five major surface water types: $Na-HCO_3-Cl$, $Na-Cl-HCO_3$, $Na-Ca-Mg-HCO_3$, $Na-Ca-Mg-HCO_3$ and $Ca-Na-Mg-HCO_3$. The Gibbs plot showed that the major ion chemistry of surface water was mostly influenced by atmospheric precipitation and the water quality index showed that the majority of the surface water from settlements within the Birim River basin were of poor quality for drinking and other domestic purposes. However, irrigation suitability calculations with reference to sodium adsorption ratio, residual sodium carbonate, and magnesium ratio values, together with Wilcox and USSL models indicated that the surface water within the area under study was suitable for agriculture. The potential ecological risk for single heavy metals pollution and potential toxicity response indices gave low to considerable ecological risks for the sediments, with greater contributions from Cd, Pb and As. Whilst geo-accumulation indices indicated that the sediments ranged from unpolluted to moderately polluted Modified degree of pollution and Nemerow pollution index calculations which incorporate multi-element effects, however, indicated no pollution.

Conclusion: There are some levels of both potential ecological risks and health hazards in the study area. Hence continuous monitoring should be undertaken by the relevant agencies and authorities so that various interventions could be put in place to prevent the situation from deteriorating further in order to protect the inhabitants of the settlements within the Birim River basin.

Keywords: Ecological risk assessment, Hydrogeochemical, Birim River basin, Water quality index, Multivariate analysis

*Correspondence: asaredonkor@yahoo.co.uk
Department of Chemistry Kwame, Nkrumah University of Science and Technology, Kumasi, Ghana

Background

Surface water which is the most significant inland water resource for human consumption, agricultural activities, recreational and industrial purposes (Razmkhah et al. 2010) has always been the end point of wastewater disposal from the adjacent areas. Lithology of river basins, anthropogenic inputs, climatic and atmospheric conditions affects the quality of surface water at any point. A series of organic, inorganic and biological pollutants, such as highly toxic heavy metals (Demirak et al. 2006; Moore and Ramamoorthy 2012) or non-toxic, biodegradable materials, such as faeces, food waste and wastewater can affect the quality of surface water (Bain et al. 2014).

The hydrogeochemical characteristics of surface water rely on the chemical composition of rock-forming minerals, such as sulphide, carbonate and silicate, as well as the physical process of erosion, which generates favourable conditions for mineral dissolution. Consequently, water resources are being enriched with metals, metalloids or ions, which often tend to be toxic to mankind as well as the natural environment (Lang et al. 2006; Négrel et al. 2006; Robinson and Ayotte 2006). Other geochemical activities, such as sorption, redox reactions, ion exchange, and complexation may alter it's hydrogeochemistry and subsequently affect the water quality. The hydrogeochemical characterization may be accomplished via several techniques, and among them are the typical hydrogeochemical ratios, which can evaluate the dominant and origin processes of water resources (Zhu et al. 2007), as well as multivariate statistical analysis (Saleem et al. 2015; Purushothaman et al. 2014). These two techniques may be valuable in identifying the factors that influence surface water chemistry, particularly hydrogeological and complex geological systems.

Information on the physicochemical parameters of water for the endurance of organisms, such as fauna and flora is essential for evaluating the quality and type of water (Liu et al. 2010). Temperature and pH are among essential features of the environment since they affect nutrition, growth, metabolic activity and human reproduction. Heavy metals have been revealed to pose a serious threat to human health due to their toxicity, persistence, bioaccumulation in the food chain and non-destructible nature in the environment (Asare-Donkor et al. 2015; Asefi and Zamani-Ahmadmahmoodi 2015; Boateng et al. 2015; Bortey-Sam et al. 2015; Singh et al. 2013; Zhang et al. 2014). Essential metals, such as Fe, Mn, Zn and Cu play a vital role in the biological processes but turn to be toxic above certain concentrations. Several studies have revealed that chlorine contents in water are related to senility, heart disease as well as cancer of the urinary tract, pancreas, liver, colon, and osteosarcomas (Kim et al. 2011; Comber et al. 2011).

There are diverse reports in literature which assess the geochemical characteristics of surface water systems, as well as anthropogenic contamination influenced by factors such as agricultural fertilizers, sewage effluents, evapotranspiration, water–rock interactions and ion-exchange in several parts of the world (Abdesselam et al. 2013; Alaya et al. 2014; Iranmanesh et al. 2014; Kim et al. 2015; Khashogji and El Maghraby 2013; Nandimandalam 2012; Singh et al. 2016). However, in Ghana, there are few such reports (Boateng et al. 2016; Helstrup et al. 2007; Fianko et al. 2010; Yidana et al. 2012). The determination of surface water composition is of extreme significance for the assessment of its suitability for drinking, irrigation and domestic purposes. The primary objective the study has been to evaluate the hydrochemical characteristics of surface water from the Birim River basin assess the ecological risk and the suitability for domestic and irrigation uses.

Methods

Study area

The Birim River basin is located between latitudes 5°45′N and 6°35′N; and longitudes 0°20′W and 1°15′W. The Birim River takes its source from the Atewa Hills, Eastern Region of Ghana and follows a course of 175 km^2 southwards to join the Pra River. The river drains an area of approximately 3895 km^2 with the major tributaries being Adim, Amaw, Kade and Si. It has an estimated area of 3875 km^2 (Ansah-Asare and Asante 2000). The rainfall pattern in the drainage area varies seasonally with major peaks from June to September and dry spells from December to January. It has a temperature range of 25–30 °C and relative humidity of 70–80% throughout the year. The area is endowed with mineral deposits, such as gold, bauxite, diamond, and others. The main occupations the settlers are farming and small-scale mining activities known as "Galamsey", mostly along the banks and inside of the Birim River.

Sampling

Forty-three samples of water and sediment were collected from 10 settlements (Fig. 1) within the Birim River basin. The sampling stations were Apapam (The source of the Birim River), Kibi, Adukrom, Bunso, Nsutam, Nsuapemso, Osino, Ankaase, Mampong, and Anyinam. Prior to sampling, water sample containers of 500 mL polyethylene bottles were rinsed with detergents and then washed thoroughly several times with distilled water. They were further soaked in 10% (v/v) HNO_3 and left overnight. The surface water samples were collected at the mid-stream at 30 cm depths. Two sets of surface water samples (one for heavy metal and the other for physicochemical parameters) were taken from each sampling point. Electrical

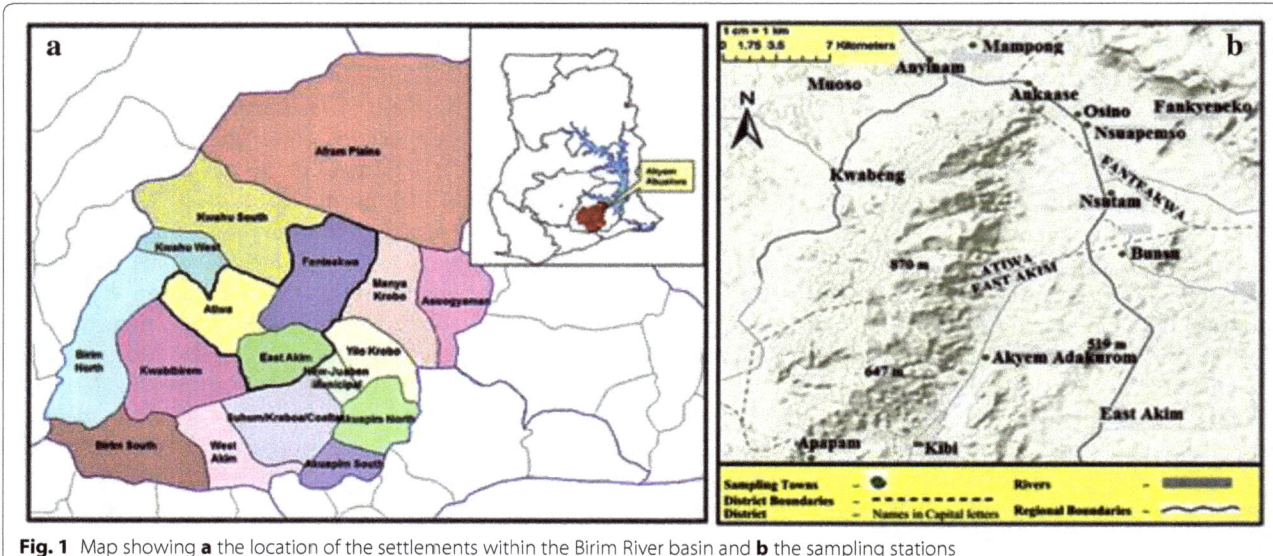

Fig. 1 Map showing **a** the location of the settlements within the Birim River basin and **b** the sampling stations

conductivity (EC), total dissolved solids (TDS) and pH were measured in situ. The samples for heavy metal analysis were filtered through pre-washed 0.45 m Millipore nitrocellulose filters to remove any suspensions treated with 2 mL of Analar HNO_3 at pH < 2 to preserve them. The water samples were appropriately labelled and immediately transported in an ice chest to the laboratory and kept in the refrigerator at -4 °C prior to analysis.

Sediment samples were taken from the top surface layer of sediment at a depth of 0–6 cm with a plastic trowel into open-mouth plastic containers. Sediment samples were obtained from the same location and at the same point where the water samples were taken. A sediment weight of about 30 g each was collected from each sampling station.

Sample analysis
Physicochemical parameters
Physical parameters, such as EC, pH and TDS were determined using cyberscan PC 650 multimeter series. Turbidity was measured using HANNA HI 93414 turbidity meter while Colour was determined using a spectrophotometer set at 465 nm. The temperature was determined using Mercury in Glass Thermometer and Total alkalinity was measured by the titrimetric method. The carbon dioxide content in the sample was determined by computations from the sample pH and total alkalinity. Total hardness was measured using EDTA titration and thus calculated as $CaCO_3$ content (mg/L). Chloride concentration was determined using potentiometric titration with an $AgNO_3$ solution with a glass and Ag–AgCl electrode system. The UV technique was used to determine the

nitrogen concentration in the water samples as nitrates at a wavelength of 220 and 275 nm. Magnesium concentration was calculated as the difference between total hardness and calcium hardness multiplied by 0.243. The flame photometric method was used to analyse the potassium and sodium ion concentrations at a wavelength of 766.5 and 589 nm respectively.

Heavy metals
Water samples were digested by a method described by Asare-Donkor et al. (2015) and Arnold and Lenores (1989). Briefly, 100 mL of each water sample was transferred into a 250 mL beaker and 5 mL of conc. HNO_3 was added. The mixture was gently heated on a hot plate after a few boiling chips had been added and evaporated to about 20 mL. Another 5 mL of conc. HNO_3 was added and then heated for 10 min and allowed to cool. About 5 mL conc. HNO_3 was used to rinse the sides of the beaker and the solution was quantitatively transferred into a 100 mL volumetric flask and made up to the mark with deionized distilled water.

The sediment samples were initially air-dried and further dried in an oven at 70 °C for 48 h to obtain a constant weight. The sediment was then crushed and ground into fine particles and further passed through a 2 mm sieve. The sediments were digested by a method described by Kouadia and Trefry (1987). About 1.0 g of the finely ground sediment sample was accurately weighed into a beaker and equal amounts of concentrated nitric acid, hydrofluoric acid and perchloric acid were added. The beaker was covered and set aside for several hours and evaporated to a few drops on a hot plate. Then 5 mL of

$HClO_4$ was added and evaporated to dryness. Conc HCl (10 mL) was added and the mixture was heated until the solution was clear and the fumes had ceased. Deionized distilled water was added and the digested material was filtered into a 100 mL volumetric flask, then the residue was washed several times with deionized distilled water and made to the mark. The heavy metals, such as Fe, Pb, Zn Cd and As were analysed with the VARIAN AA240FS atomic absorption spectrophotometer with an air-acetylene flame. Requisite lamps with appropriate operating absorption wavelength and other operating parameters for each element was employed for the determination. Each of the analysis was done in triplicate, in which the mean values were calculated.

Quality assurance

Replicate blanks and Standard Reference Material (SRM) of Fluka Analytical (Sigma-Aldrich Chemie GmbH, Product of Switzerland) were used for the quality control and method validation. Analytical results indicated a good agreement between those reported in this study and the certified value.

Statistical analysis

IBM SPSS-20 program was also used to analyse the Person correlation, principal component analysis (PCA) and cluster analysis (CA). XLSTAT '2016 statistical software and Origin 2016 Data Analysis and Graphing Software were used for the geochemical analysis.

Risk assessment methods
Enrichment factor (EF)

The enrichment factor (EF) is a convenient measure of geochemical trends and is used to characterize the degree of anthropogenic pollution through the establishment of enrichment ratios (Zakir et al. 2008). To evaluate the extent of contamination in the environment, the factors EF were computed relative to the abundance of species in the source material and to that found in the Earth's crust (Sinex and Helz 1981).

$$EF = (C_M/C_x\ sample)/(C_M/C_x\ Earthcrust) \qquad (1)$$

where C_M is the content of metal studied and C_X is the content of the immobile element, immobile elements may be Fe (Zhang et al. 2007). In this study, the background geochemical compositions by Taylor and McLennan (1995) were used as the background values for the calculation. Five classes of enrichment factors have been identified EF < 2, depletion to mineral, $2 \leq EF < 5$: moderate, $5 \leq EF < 20$: significant, $20 \leq EF < 40$: very high and EF > 40: extremely high (Sutherland 2000).

Index of geo-accumulation (I_{geo})

The I_{geo}, which is the geochemical benchmark to assess heavy metals pollution in sediments was in was calculated based on Eq. (2):

$$I_{geo} = \log_2\left(\frac{C_n}{1.5 \times B_n}\right) \qquad (2)$$

where, B_n and C_n represent the geochemical background concentration and the measured concentration of the studied heavy metal n in the sediment, respectively. Because of the possible dissimilarities in background concentrations of any given heavy metal and anthropogenic influences, a factor value of 1.5 was used.

Modified degree of contamination

The modified degree of contamination (mC_d) which has an advantage over single element indices since it takes account the synergistic effect of the contaminants at a study site (Brady et al. 2015) was calculated as follows:

$$mCd = \frac{1}{n}\sum_{i=1}^{n} C_f^i \qquad (3)$$

where $C_f^i = C^i / C_{ref}^i$, C^i is the heavy metal concentration in sediment samples; C_{ref}^i the reference value of the element (Turekian and Wedepohl 1961; Hankason 1980) and C_f^i the contamination factor of each element.

Nemerow pollution index

The Nemerow pollution index takes into account the comprehensive effects of heavy metals (Yan et al. 2016) and can be used to interpret heavy metal pollution at particular sites (Duodu et al. 2017). The equation for the calculation is given by Eq. (4).

$$P_N = \sqrt{\frac{\overline{C_f}^2 + C_{fmax}^2}{2}} \qquad (4)$$

where P_N is the Nemerow pollution index, $\overline{C_f}$, the arithmetic mean of contamination factors of all heavy metals, and C_{fmax} the maximum contamination factor among the heavy metals.

Potential ecological risk index

The total ecological risk index (RI) for heavy metals in sediments was calculated using Eq. (5)

$$RI = \sum_{i=1}^{n} E_f^i, \quad \text{where } E_f^i = C_f^i \times T_f^i \qquad (5)$$

where C_f^i is the contamination factor, T_f^i is the toxicity coefficient of metal i.

Table 1 shows the classification of heavy metal modified degree of contamination (mCd), Nemerow pollution index (P_N), Index of geo-accumulation (I_{geo}) and potential ecological risk used for the sediment.

Pollution load index

The pollution load index (PLI) was evaluated as the nth root of n multiplied contamination factor (C_f), as shown in Eq. (6):

$$PLI = (C_{f1} \times C_{f2} \times C_{f3} \times \cdots \times C_{fn})^{\frac{1}{n}} \tag{6}$$

PLI mainly allows a qualitative comparison between sites. *PLI* values < 1 signifies no pollution, *PLI* = 1 signifies baseline levels of pollutants and *PLI* > 1 signifies progressive site pollution. The pollution load index value of 0.08 indicated no deterioration of the site quality (PLI < 1).

Water quality index

Water quality index (*WQI*) was used in the evaluation of the status of water sources from the communities. In this process, each water quality parameter was assigned a specific weight (W) based on their relative significance on the water quality and the relative weight (W_i) was evaluated based on Eq. (7).

$$w_i = \frac{w_i}{\sum_{i=1}^{n} w_i} \tag{7}$$

where w_i represent the assigned weight of each parameter, W_i is the sum of the assigned weights of all the parameters and the number of parameters (Table 2). A maximum weight of five was assigned to NO_3^- and Cl^- owing to their significance to water quality, as well as human health.

Water quality index (WQI) allows easier illustrations of complex data to determine the status of water systems. The WQI calculations are depended on significant water quality parameters by providing a specific number to represent the overall water quality (Pius et al. 2012). According to WHO (2011), the quality rating scale (q_i) for each of the water parameter was evaluated following Eq. (8) and finally *WQI* was calculated using Eqs. (9) and (10):

$$q_i = \frac{C_i}{S_i} \times 100 \tag{8}$$

$$SI_i = W_i \times q_i \tag{9}$$

$$WQI = \sum SI_i \tag{10}$$

where C_i, q_i and S_i, respectively, represent the concentration of each chemical parameter in the water sample, the quality rating and the concentration of each parameter as per the WHO (2011) standard Table 2; SI_i and W_i represent the sub-index of the ith parameter and the sum of the assigned weights of all the parameters (Table 2), respectively.

Suitability for irrigation purposes

The suitability of surface water for irrigation purpose was evaluated by calculating the soluble sodium percentage (SSP), sodium adsorption ratio (SAR), Permeability index, Kelly's ratio (KR), residual sodium carbonate (RSC) and magnesium ratio (MR). The SSP, SAR, KR, RSC and MR were calculated using Eqs. (11) to (15) and the values together with the PIs are given in Additional file 1: Table S3.

$$SAR = \frac{Na^+}{\sqrt{Ca^{2+} + Mg^{2+}}} \tag{11}$$

$$KR = \frac{Na^+}{Ca^{2+} + Mg^{2+}} \tag{12}$$

$$RSC = \left(HCO_3^- + CO_3^-\right) - \left(Ca^{2+} + Mg^{2+}\right) \tag{13}$$

Table 1 Classification of heavy metal modified degree of contamination (mCd), nemerow pollution index (P_N), index of geo-accumulation (I_{geo}) and potential ecological risk

Class	mCd	Cont deg	P_N	Cont deg	I_{geo}	Cont deg.	RI	Ecological risk
0	< 1.5	Unpolluted	< 1	Unpolluted	< 0	Unpolluted	< 110	Low risk
1	$1.5 \leq Cd < 2$	Slightly polluted	$1 \leq P_N < 2.5$	Slightly polluted	0–1	Slightly to moderately polluted	$110 \leq RI < 200$	Moderate risk
2	$2 \leq Cd < 4$	Moderately polluted	$2.5 \leq P_N < 7$	Moderately polluted	1–2	Moderately polluted	$200 \leq RI < 400$	Considerable risk
3	$4 \leq Cd < 8$	Moderately-heavily polluted	≥ 7	Heavily polluted	2–3	Moderately to highly polluted	≥ 400	Severe risk
4	$8 \leq Cd < 16$	Heavily polluted			3–4	Highly polluted		
5	$16 \leq Cd < 32$	Severely polluted			4–5	Highly to extremely polluted		
6	≥ 32	Extremely polluted			5–6	Extremely polluted		

Table 2 Weights and relative weights of physiochemical parameters

Parameters	Units	WHO standard (2011)	Weight (wi)	Relative weight $W_i = \frac{w_i}{\sum_{i=1}^n w_i}$
pH		6.5–8.5	4	0.093
TDS	mg/L	500	4	0.093
SO_4^{2-}	mg/L	250	4	0.093
Ca	mg/L	75	2	0.047
Mg	mg/L	50	2	0.047
Total harness	mg/L	500	3	0.070
NO_3^-	mg/L	45	5	0.116
Fe	mg/L	0.3	3	0.070
Zn	mg/L	3.0	3	0.070
Na^+	mg/L	200	1	0.023
K^+	mg/L	12	1	0.023
Cl^-	mg/L	250	5	0.116
HCO_3^-	mg/L	300	3	0.070
EC	μS/cm	500	3	0.070
			$\sum w_i = 43$	$\sum w_i = 1$

$$SSP = \frac{Na^+ + K^+}{Ca^{2+} + Mg^{2+} + Na^+ + K^+} \times 100 \qquad (14)$$

$$MR = \left[\frac{Mg^{2+}}{Ca^{2+} + Mg^{2+}} \right] \times 100 \qquad (15)$$

Results and discussion

Physicochemical parameters

Physicochemical parameter values of the surface water samples analysed in the Birim River basin are presented in Table 3. The water temperature ranged between 25.00 and 28.00 °C with a mean value of 26.26 °C. The colour of the water samples ranged from 364.62 to 34615.35 PtCo, with an average 14,583.67 PtCo. The levels of colour in all the surface water samples exceeded the WHO permissible limit of 15 PtCo. Highly collared water may be owned by the decaying vegetation in the water resources (Karikari and Ansa-Asare 2006). The pH values of all the surface water samples ranged between 6.36 and 7.02, with a mean pH of 6.68, indicating acidic to neutral to nature of Birim River basin water. The acidic pH was attributable to the naturally occurring silicate minerals. The pH values of all the surface water samples were below the WHO acceptable limits of 6.5–8.5 for drinking and other domestic purposes (WHO 2011). The TDS and EC concentrations in the surface water samples ranged from 29.03 to 64.33 mg/L and 58.03 to 128.97 μS/cm, with a mean of 42.60 mg/L and 25.21 μS/cm, respectively. The

low TDS concentration in the water samples is due to the short residence time of the underground rocks as well as the slow weathering of granitic (Patel et al. 2016). The EC and TDS concentrations of all the river water samples are very much below the WHO acceptable limit of 500 μS/cm and < 600 mg/L, respectively (WHO 2011). Total alkalinity (TA) is a measure of HCO_3^- and CO_3^{2-} ions. TA refers to the ability of water to neutralise strong acids. The alkalinity of the surface water samples ranged between 20.00 and 161.00 mg/L, with a mean of 79.20 mg/L. The alkalinity of all the surface water samples fell below the WHO permissible limits of 200.00 mg/L. Total hardness (TH) represent alkaline earth elements, such as magnesium and calcium within the water resources. The TH ranged from 51.76 to 168.57 mg/L, with a mean value of 92.31 mg/L. TH concentrations of all the river water samples were below the WHO guideline of 500 mg/L. Turbidity has an average value of 1652.86 NTU that ranges from 31.60 to 3000.00 NTU. All the surface water samples had turbidity concentration above the WHO acceptable limit of 5 NTU. The high turbidity can be ascribed to larger particles, such as dissolved solids and organic matter in the surface water samples (Schafer et al. 2010). The calcium hardness of the sampled surface water varied from 30.74 to 100.26 mg/L, with a mean of 55.15 mg/L, while magnesium hardness ranged between 21.02 and 68.31 mg/L, with a mean of 37.16 mg/L All the analysed surface water samples were below the WHO permissible limit of 200 mg/L for magnesium hardness. The free and total CO_2 content in the surface water samples ranged from 17.57 to 44.63 and 35.17 to 172.11 mg/L with a mean of 30.02 and 99.24 mg/L, respectively.

Major ions and heavy metals

The concentrations of major ions, such as K^+, Na^+, Ca^{2+}, Mg^{2+}, Cl^-, SO_4^{2-}, HCO_3^-, and CO_3^{2-}) are summarised in Table 3. The sources of sodium ion (Na^+) are from the weathering products of silicate rocks and its movement from the absorbed complex of soil and rocks by magnesium and calcium. The concentrations of Na^+ varied between 2.90 and 5.10 mg/L, with a mean of 3.81 mg/L. The Na^+ concentrations were below the WHO recommended limits of 200 mg/L. The levels of K^+ in the surface water samples ranged between 0.40 and 3.90 mg/L, with a mean value of 1.00 mg/L and the levels were below the permissible limit of 12 mg/L by WHO (2011). The low concentration of K^+ in the surface water samples was due to its affinity to be immobilised by clay minerals to partake in forming secondary minerals. The levels of Ca^{2+} ion in the Birim River basin ranged from 12.80 to 41.60 mg/L, with a mean 22.40 mg/L. The levels of Mg^{2+} varied between 5.13 and 16.66 mg/L, with a mean level of 8.90 mg/L. The concentration of Ca^{2+} and Mg^{2+}

Table 3 Physicochemical parameter of surface water samples collected from settlements within the Birim River basin

	Minimum	Maximum	Mean	Std. deviation	Variance	Skewness	Kurtosis	WHO standard
pH	6.36	7.02	6.68	0.18	0.03	0.27	0.84	6.5–8.5
Conductivity (µS/cm)	58.03	128.97	85.37	25.21	635.70	0.65	−0.79	500
Turbidity (NTU)	31.60	3000.00	1652.86	1263.92	1597.38	−0.12	−1.89	5
Alkalinity (mg/L)	20.00	161.00	79.20	42.63	1817.29	0.44	0.14	200
TDS (mg/L)	29.03	64.33	42.60	12.60	158.74	0.64	−0.79	< 600
Temperature (°C)	25.00	28.00	26.26	0.88	0.78	0.95	0.80	< 25 °C
TH (mg/L)	51.76	168.57	92.31	41.05	1685.14	1.12	0.15	500
Colour (PtCo)	364.62	34,615.35	19,071.33	14,583.67	2126.65	−0.12	−1.89	15
CaH (mg/L)	30.74	100.26	55.15	24.40	595.43	1.09	0.08	–
MgH (mg/L)	21.02	68.31	37.16	16.66	277.41	1.17	0.26	200
NO_3^- (mg/L)	bd	1.00	0.38	0.31	0.10	0.55	−1.22	45
K^+ (mg/L)	0.40	3.90	1.00	1.07	1.15	2.66	7.50	12
Na^+ (mg/L)	2.90	5.10	3.81	0.79	0.63	0.80	−0.64	200
Cl^- (mg/L)	11.98	69.96	27.76	19.35	374.46	1.66	1.80	250
Mg^{2+} (mg/L)	5.13	16.66	8.98	4.04	16.33	1.19	0.30	50
Ca^{2+} (mg/L)	12.80	41.60	22.40	10.12	102.39	1.19	0.28	75
SO_4^{2-} (mg/L)	bd	1.00	0.41	0.33	0.11	−0.06	−2.08	250
CO_3^{2-} (mg/L)	0.01	0.16	0.05	0.05	bd	1.95	4.34	–
HCO_3^- (mg/L)	19.99	160.84	78.63	43.20	1866.47	0.41	0.02	300
Free CO_2 (mg/L)	17.57	44.63	30.02	9.20	84.63	−0.34	−0.82	–
Total CO_2 (mg/L)	35.17	172.11	99.24	44.21	1954.48	−0.14	−0.72	–
Fe (mg/L)	0.63	6.44	3.41	1.88	3.55	−0.10	−0.95	0.3
Zn (mg/L)	bd	0.02	0.01	0.00	0.00	−0.24	−0.31	3.0

ions, respectively, in all the river water samples, fall below the WHO maximum acceptable limit of 75 and 150 mg/L. Calcium (Ca) and magnesium (Mg) are serious hydrochemical elements in water (Razowska-Jaworek 2014). Ca is a significant ion that affects the hardness of water and has been the most abundant elements found in water resources. The anion chemistry of Birim River basin was dominated by HCO_3^- followed by Cl^-, SO_4^{2-}, NO_3^- and CO_3^{2-}. The Cl^- content in the Birim water samples varied from 11.98 to 69.96 mg/L, with a mean of 27.76 mg/L and all the surface water samples were below the acceptable limits of 250 mg/L recommended by the WHO (2011). Sulphate (SO_4^{2-}) ions are limited by Ca^{2+} ions in surface water. The major source of sulphate in water resources is sedimentary rock, such as anhydride and gypsum. The sulphate concentration in the Birim River basin ranged between bd and 1.00 mg/L, with a mean value of 11.98 mg/L. The SO_4^{2-} contents in the Birim River samples were below the acceptable limits of 250 mg/L recommended by WHO. The HCO_3^- and CO_3^{2-} concentrations ranged from 19.99 to 160.84 mg/L and 0.01 to 0.16 mg/L, with mean values of 78.63 and 0.05 mg/L, respectively. The levels of nitrate (NO_3^-) in surface water can be attributed to the oxidation of nitrogenous waste products in animal and human excreta,

wastewater disposal and agricultural activity. The concentrations of NO_3^- in surface water can vary depending on surface runoff of fertiliser, denitrification by bacteria and uptake by phytoplankton. The level of NO_3^- in surface water was moderately low, ranged from bd to 1.00 mg/L, with a mean value of 0.38 mg/L. The levels of NO_3^- were below the acceptable limits of 45 mg/L. The level of iron in the surface water ranged between 0.63 and 6.44 mg/L, with a mean value of 3.41 mg/L. The level of zinc in the surface water ranged between bd and 0.02 mg/L, with a mean level of 0.01 mg/L. The levels of zinc were below the acceptable limit of 3 mg/L as per the WHO (2011) standard. The levels of As, Pb and Cd were however observed to be below detection in all the water samples analysed.

Correlation of water quality parameters

There was a strong correlation between Ca^{2+} and Mg^{2+} with TH indicates that Ca^{2+} and Mg^{2+} are the major contributors to water hardness. Ca^{2+} and Mg^{2+} has a significant positive correlation ($r = 1.000$, $p < 0.01$), indicating the contribution of dolomite to the levels of Ca^{2+} and Mg^{2+} ions in the surface water. The positive correlation of Cl^- with Na^+ and Ca^{2+} indicates the dominance of soluble salts. High positive correlation of EC with CO_3^{2-} and

Hydrochemical characteristics of surface water and ecological risk assessment of sediments...

203

HCO_3^- ions signify high mobility of CO_3^{2-} and HCO_3^- ions in the surface water. In addition, the strong significant correlation between TDS and EC ($r = 1.000$, $p < 0.01$) might be ascribed to ions in TDS conducting electricity. The strong positive correlation between Mg^{2+} and SO_4^{2-} ions ($r = 0.703$) indicates agricultural activities, such as chemical and organic fertilisers in the study area. The strong correlation between NO_3^- and SO_4^{2-} signifies the influences of agricultural activities, evaporation, marine sources and poor drainage conditions on the surface water system. Surface water HCO_3^- was strongly correlated with pH, EC, alkalinity CO_3^{2-} and TDS. Total CO_2 was strongly correlated with HCO_3^-, CO_3^{2-} and free CO_2.

Hierarchical cluster analysis

The water quality differences between the sampling stations are given by dendrogram as shown in Fig. 2. Cluster 1, representing Adukrom, Nsutam, Osino and Anyinam are areas with high water pollution. Cluster 2 representing Kibi, Nsuapemso and Apapam are areas with moderate water pollution. Cluster 3 representing Ankaase, Mampong and Bunso are areas with lower water pollution.

Principal component analysis

The physicochemical parameters in the surface water samples were statistically evaluated with R-mode PCA to understand the correlation of the analysed parameters and identify the significant factors influencing the study area. Suitability of components for PCA was tested by performing Kaiser–Meyer–Olkin (KMO) tests. KMO test

is an index used in relating the magnitude of the observed correlation and partial coefficients with a value of > 0.5. In this study, KMO was 0.83, which indicates the suitability of the physicochemical parameters results for component analysis. The component loadings are classified as strong, moderate and weak with absolute loading values of > 0.75, 0.75–0.50 and 0.50–0.30, respectively. The PCA on the composite data sets extracted five components with eigenvalue > 1 and these components explained 98.15% of the total variance as shown in Additional file 1: Table S1.

The first component (PC1) accounted for 49.14% of total variance and includes strong positive loadings of pH, turbidity, alkalinity, total hardness, colour, calcium hardness, magnesium hardness, Mg^{2+}, Ca^{2+}, SO_4^{2-}, CO_3^{2-}, HCO_3^{2-} and total CO_2; and moderate positive load of TDS, EC, free CO_2 and Fe. This component signifies dissolution of carbonate minerals as well as geogenic attributes, such as surface runoff during the rainy season, soil erosion, and weathering of mineral bearing rocks. The second component (PC2) explaining 20.68% of the total variance, have a strong positive loading of temperature; moderate positive loadings of TDS and Fe and weak loadings of EC. The moderate positive load of Fe is influenced by geological activities (Adamu et al. 2015). The third component (PC3) explaining 13.71% of total variance, have a moderate loading of SO_4^{2-}, colour, and turbidity, while NO_3^- and K^+ has a weak positive load. The PC3 indicates agriculture practices, such as the use of chemical fertilisers, as well as anthropogenic contaminants induced by domestic wastes (Jalali 2007). The fourth component (PC4) which explained 9.05% of total variance has a moderate positive loading of Zn. The fifth component (PC5) explaining 5.56% of total variance, have a moderate and weak positive loading of Cl^- and Na^+, respectively.

Surface water evolution mechanisms

The geochemical evolution of surface water was analysed by plotting the levels of major cations and anions in a Piper trilinear diagram. The relative abundance of the anions and cations are shown in Fig. 3. The plot reveals five types of facies Ca–Na–Mg HCO_3 (39%), Na–Cl–HCO_3 (15%), Na–Ca–Mg–HCO_3 (23%), Na–Ca–Mg–HCO_3 (4%) and Na–HCO_3–Cl (8%) of which Ca–Na–Mg HCO_3 is the predominant facies type.

Gibbs plots were used to access the hydrochemical processes, such as evaporation, rock-water interaction and precipitation on surface water chemistry. Gibbs plots have been extensively utilised to evaluate the underlying mechanisms behind water evolution. All the studied water samples are in the lower part of the diagram (Fig. 4), signifying precipitation interactions as the leading factor controlling the surface water chemistry. Thus,

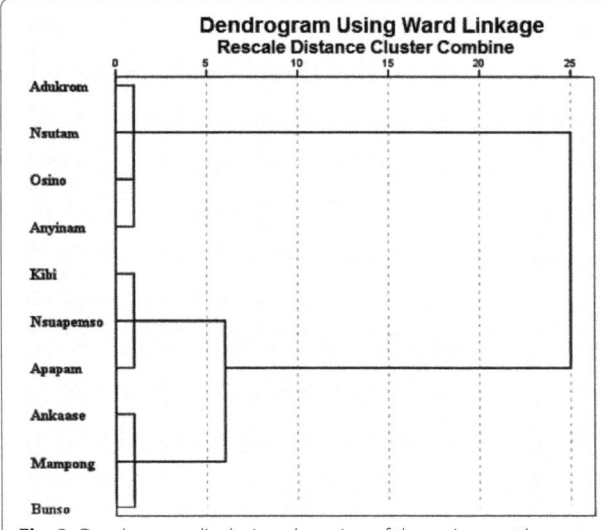

Fig. 2 Dendrogram displaying clustering of the various settlements within the Birim River basin

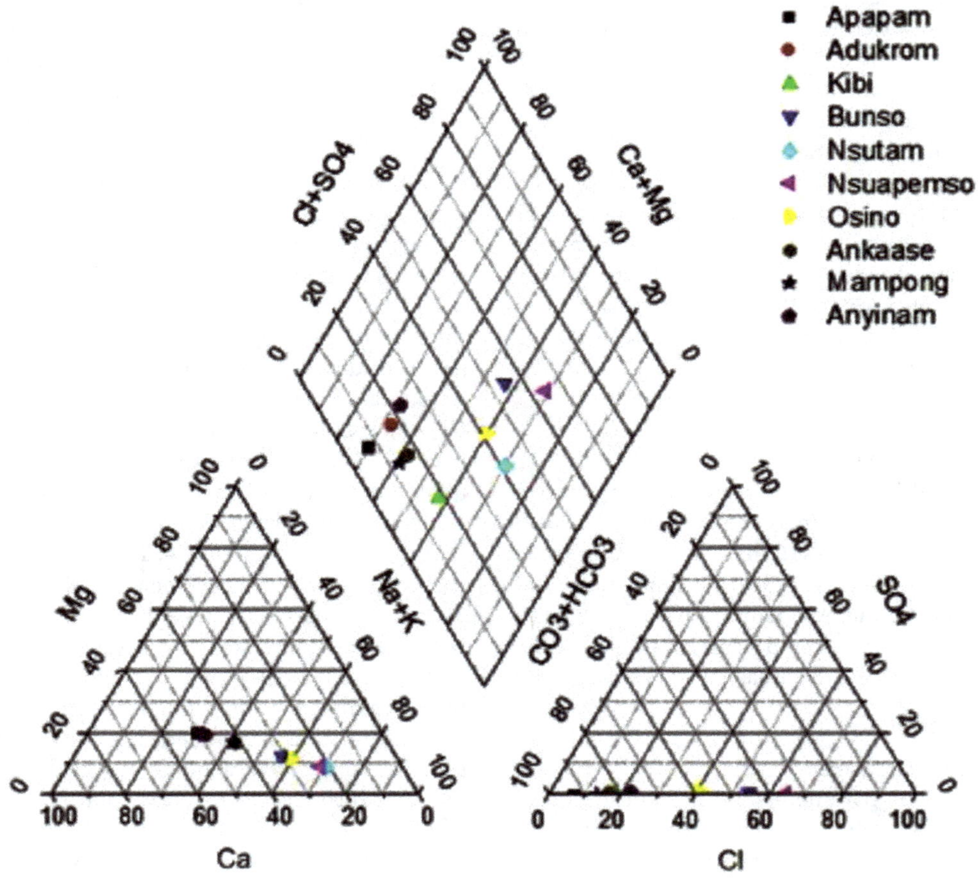

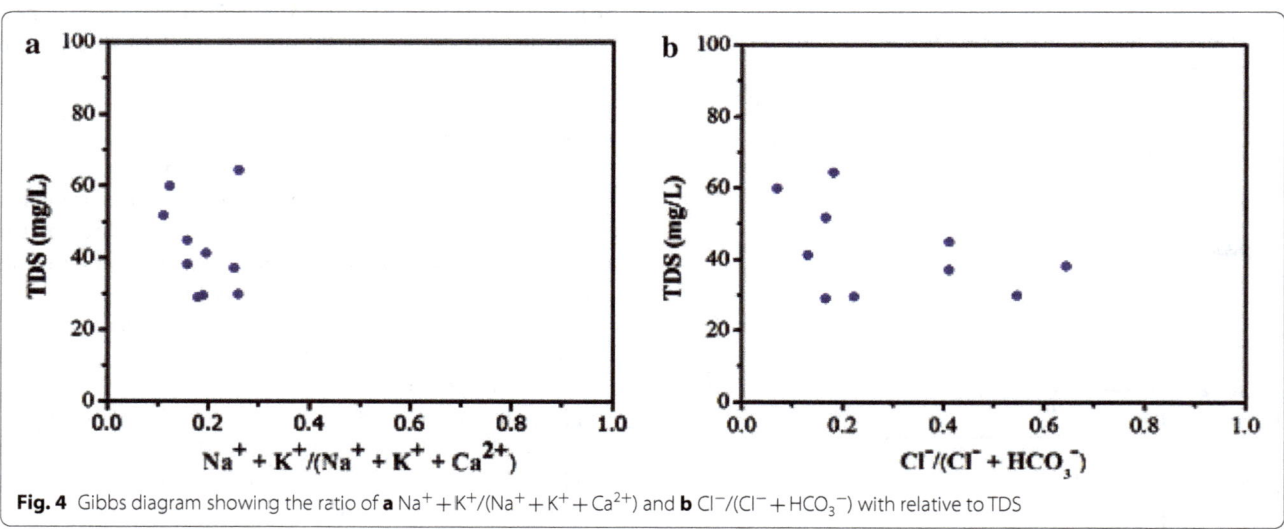

Fig. 3 Piper diagram for the surface water samples of the Birim River

Fig. 4 Gibbs diagram showing the ratio of **a** $Na^+ + K^+/(Na^+ + K^+ + Ca^{2+})$ and **b** $Cl^-/(Cl^- + HCO_3^-)$ with relative to TDS

Hydrochemical characteristics of surface water and ecological risk assessment of sediments...

205

there is a close connection between the surface water chemistry and atmospheric precipitation in the Birim River basin.

Hydrogeochemical facies

Molar ratios of major ions have been extensively utilised to ascertain the hydrogeochemical formation and process mechanisms of water resources (Murkute 2014; Singh et al. 2013; Marghade et al. 2011). Ca^{2+}/Mg^{2+} ratio is normally used to evaluate the source of Ca^{2+} and Mg^{2+} in the water systems. The ratio of 1, signifies dissolution of dolomite, a ratio greater than 1 (>1), indicates calcite

contribution and a ratio greater than 2 (>2), indicates dissolution of silicate minerals (Singh et al. 2013). All the water samples have $Ca^{2+}+Mg^{2+}$ ratio greater than 2 (>2) (Fig. 5a), indicating silicate minerals responsible for Ca^{2+}/Mg^{2+} contribution. The bimodal process of Ca enhancement and weathering process (carbonate vs silicate) are shown in Fig. 5b, displaying the ratio of $(Ca^{2+}+Mg^{2+})/(HCO_3^-+SO_4^{2-})$. If Ca^{2+}, Mg^{2+}, HCO_3^- and SO_4^{2-} ions are from the dissolution of dolomite, gypsum, and calcite, a 1:1 stoichiometry of $(Ca^{2+}+Mg^{2+})$ to $(HCO_3^-+SO_4^{2-})$ might occur (Singh et al. 2014). Most of the water samples, however, deviate from 1:1, which

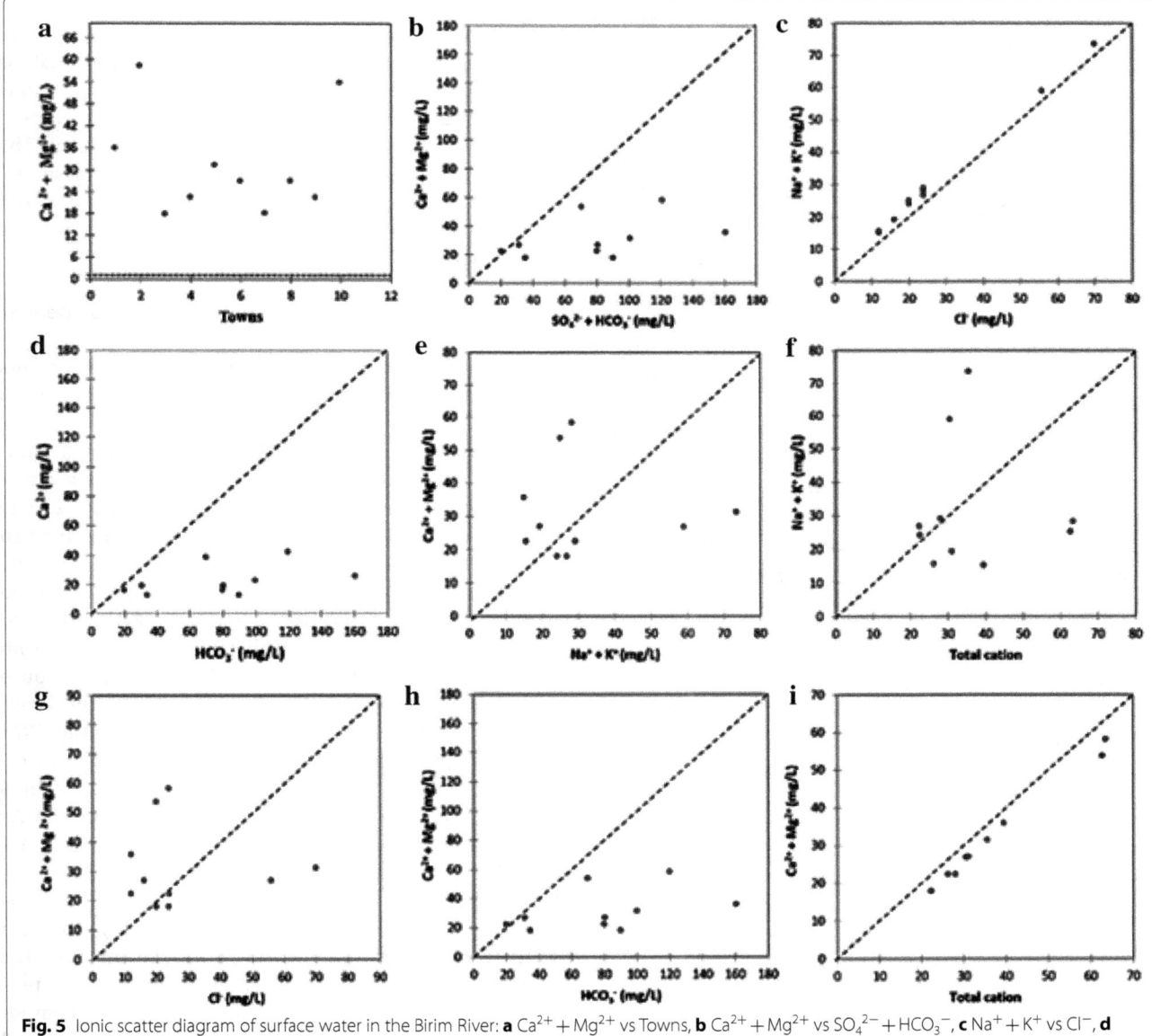

Fig. 5 Ionic scatter diagram of surface water in the Birim River: **a** $Ca^{2+}+Mg^{2+}$ vs Towns, **b** $Ca^{2+}+Mg^{2+}$ vs $SO_4^{2-}+HCO_3^-$, **c** Na^++K^+ vs Cl^-, **d** Ca^{2+} vs HCO_3^-, **e** $Ca^{2+}+Mg^{2+}$ vs Na^++K^+, **f** Na^++K^+ vs total cation, **g** $Ca^{2+}+Mg^{2+}$ vs Cl^-, **h** $Ca^{2+}+Mg^{2+}$ vs HCO_3^-, **i** $Ca^{2+}+Mg^{2+}$ vs total cation

indicates that Ca^{2+} and Mg^{2+} primarily occur from the dissolution of gypsum, dolomite, and calcite. The molar ratio also signifies that the sources of Ca^{2+} and Mg^{2+} were not only from carbonate since the ratio was not 1:2. The $(Na^+ + K^+)/Cl^-$ molar ratios in the analysed surface water from the Birim River basin were greater than 1 (>1) (Fig. 5c), indicating that halite, as well as silicate weathering, such as potash plagioclase and sodium plagioclase, was the source Na^+ and K^+ ions (Lin et al. 2016a). If Ca^{2+} and HCO_3^- in the surface water originate mainly from dolomite and calcite, the molar ratio of Ca^{2+} and HCO_3^- ions within the surface water will be 1:2 and 1:4, respectively.

In the Ca^{2+}/HCO_3^- plot (Fig. 5d), the Ca^{2+} and HCO_3^- molar ratio for some of the surface water was between 1:1 and 1:2, signifying that calcite was the only source of Ca^{2+} and HCO_3^- ions in the surface water. Nonetheless, most of the surface water samples had a molar ratio less than 1:2 ($<1:2$), suggesting dolomite as the dominant sources of Ca^{2+} and HCO_3^- ions. A high ratio of $(Ca^{2+} + Mg^{2+})/(Na^+ + K^+)$ and $(Na^+ + K^+)/\text{total cation}$ demonstrate that the chemical composition of surface water in the Birim River basin was mainly influenced by carbonate weathering with a small contribution of silicate weathering as shown in Fig. 5e, f. The plot of $Ca^{2+} + Mg^{2+}/Cl^-$ obviously showed that salinity decrease with increase in $Ca^{2+} + Mg^{2+}$, and this phenomenon can be ascribed to ion exchange (Fig. 5g). $Ca^{2+} + Mg^{2+}/HCO_3^-$ plot (Fig. 5h) revealed a horizontal trend line, signifying that $Ca^{2+} + Mg^{2+}/HCO_3^-$ ratio does not alter during the increase of HCO_3^-. Thus, the contribution of $Ca^{2+} + Mg^{2+}$ and HCO_3^- are from different sources. Enrichment of HCO_3^- and depletion of $Ca^{2+} + Mg^{2+}$ may be ascribed to cation exchange. The scatter diagram of $Ca^{2+} + Mg^{2+}$ versus total cations revealed that the data falls below the 1:1 trend line, which demonstrates an increased contribution of Na^+ and K^+ as TDS increases (Fig. 5i).

Water quality index

According to Sahu and Sikdar (2008), WQI values are grouped into five classifications: excellent (<50), Good (50.1–100), poor (100.1–200), very poor (200.1–300) and unfit for drinking (<300.1).

The observed WQI in analysed surface water samples ranged between 32.80 and 167.71 (Table 4). Based on the surface water quality index, 50% fall under poor type, 20% are the moderate type and 30% shows excellent water type. The WQI results indicate that the majority of the surface water from the Birim River are of poor quality for drinking and other domestic uses and may pose health problems to communities in the area. Generally high values of water quality indices indicate that most

Table 4 Water quality index values and classification for the individual settlements

Settlements	WQI values	Classification type
Apapam	32.80	Excellent water
Adukrom	118.77	Poor water
Kibi	38.34	Excellent water
Bunso	69.25	Good water
Nsutam	167.71	Poor water
Nsuapemso	59.59	Excellent water
Osino	126.90	Poor water
Ankaase	92.85	Good water
Mampong	126.04	Poor water
Anyinam	123.21	Poor water

of the study area have excessive levels of one or more water quality parameters. From the sensitivity analysis it was observed that the WQI did not vary much due the removal of an individual parameter with the exception of iron where the variation was very significant. The high WQI values were particularly observed in Nsutam, Osino, Mampong, Anyinam and Adukrom where iron concentrations were observed to be very high and some mining activities have been taking place.

Heavy metals concentrations in sediments and component analysis

Results of heavy metals concentration in sediments samples at the different sampling sites are presented in Table 6.

From Table 5, heavy metals levels in the sediment samples follow the order: Fe > Zn > Pb > Cd > As. The concentrations of As and Zn in all the sediment samples were below the corresponding values of effect range median (ERM), effect range low (ERL), Interim sediment quality guideline (ISQG) and probable effect level (PEL). Lead contents at all sites, was higher than ERL, ERM, PEL and ISQG guidelines, suggesting that Pb in sediments from the Birim River might be periodically expected to induce adverse biological effects on the biota. Cadmium levels in almost all the communities were higher than the ERM and ISQG guidelines except at Anyinam. Pearson correlation coefficients among the heavy metals in sediments showed significant positive correlations among Zn/As ($r = 0.734$, $p < 0.05$), Zn/Fe ($r = 0.812$, $p < 0.01$) and As/Fe ($r = 0.774$, $p < 0.01$), signifying that these heavy metals were related to each other and could have a common natural or anthropogenic source (Hu et al. 2013). The data were normalised to minimise the grain size effect on the heavy metals. The rotated component matrices are presented in Additional file 1: Table S2. The first two components with eigenvalues > 2 account for 77.77% of the total

Table 5 Mean Heavy metals concentration (mean ± SD) in sediments from settlements within the Birim River basin

Settlements	Fe	Pb	Cd	Zn	As
Apapam	549.842 ± 0.003	7.348 ± 0.001	1.532 ± 0.001	5.422 ± 0.002	0.801 ± 0.001
Adukrom	573.950 ± 0.014	6.000 ± 0.004	2.032 ± 0.001	6.371 ± 0.001	0.951 ± 0.001
Kibi	564.975 ± 0.021	6.857 ± 0.014	1.582 ± 0.001	7.471 ± 0.001	0.851 ± 0.001
Bunso	582.691 ± 0.001	7.052 ± 0.007	1.982 ± 0.001	9.671 ± 0.001	1.400 ± 0.000
Nsutam	563.142 ± 0.002	8.947 ± 0.000	1.881 ± 0.000	4.721 ± 0.001	0.550 ± 0.000
Nsuapemso	578.995 ± 0.007	5.399 ± 0.002	1.782 ± 0.001	11.121 ± 0.001	1.251 ± 0.001
Osino	548.945 ± 0.007	5.998 ± 0.001	1.782 ± 0.001	5.171 ± 0.001	0.553 ± 0.004
Ankaase	578.843 ± 0.004	7.047 ± 0.071	1.631 ± 0.000	12.021 ± 0.001	1.101 ± 0.001
Mampong	574.741 ± 0.001	6.348 ± 0.001	1.782 ± 0.001	9.870 ± 0.000	0.750 ± 0.000
Anyinam	572.641 ± 0.001	7.647 ± 0.000	1.232 ± 0.001	9.572 ± 0.002	0.951 ± 0.001
Mean ± SD	568.880 ± 11.900	6.860 ± 1.010	1.720 ± 0.240	8.140 ± 2.640	0.920 ± 0.280
ERL	–	46.7	1.2	150	8.2
ERM	–	218	9.6	410	70
ISQC	–	30.2	0.68	124	7.24
PEL	–	112	4.21	271	41.6

All results are in mg/kg

ERL effect range low (NOAA), *ERM* effect range medium (NOAA), *ISQC* Interim sediment quality guidelines (Environment Canada), *PEL* probable effect level (Environment Canada)

Table 6 The computed EF values for heavy metals in sediments samples collected from the Birim River

Community	Pd	Cd	Zn	As
Apapam	2.34	0.10	0.49	3.40
Adukrom	1.83	0.13	0.55	3.87
Kibi	2.12	0.10	0.65	3.51
Bunso	2.12	0.12	0.82	5.61
Nsutam	2.78	0.12	0.41	2.28
Nsuapemso	1.63	0.11	0.95	5.04
Osino	1.91	0.12	0.46	2.35
Ankaase	2.13	0.10	1.02	4.44
Mampong	1.93	0.11	0.85	3.04
Anyinam	2.34	0.08	0.82	3.88
Mean	2.11	0.11	0.71	3.77

variability. PC1 has very high positive loadings of Zn, Fe, and As. Therefore, this component is supposed to reflect the contribution of natural geological and anthropogenic sources, such as industrial, agricultural and transportation and can originate from similar pollution sources. PC1 results concur with the correlation analysis results. PC2 has very high positive loadings of Cd, which suggested that persistent application of phosphate fertilisers in the study area may have led to an increase in Cd accumulation in the sediments (Guo and He 2013). PCA has been employed to ascertain the hypothetical source of heavy metals (Sun et al. 2010; Yi et al. 2011).

Heavy metal contamination and its ecological risks in sediments

Enrichment factor

On the basis of the magnitude of the enrichment factor, five contamination categories have been recognized generally (Sutherland 2000). Additionally, if the EF value of an element is greater than unity it indicates that the metal is more abundant in the sample relative to that found in the Earth's crust. EF values less than 5, though of less significance, are indicative of metal accumulation since such small enrichments may arise from differences in the composition of sample material with respect to the reference Earth's crust ratio values used in the EF calculations. If the EF values are greater than 5, samples are considered contaminated (Atgin et al. 2000). In accordance with the above suppositions, sediments from the Birim River basin (Table 6) are not contaminated with the metals studied (EF < 5), but show moderate enrichment ($2 \leq EF < 5$) for only Pb and As. The mean values of EF in the sediments were 2.11, 0.11, 0.71 and 3.77 for Pb, Cd, Zn and As, respectively. These signify the anthropogenic source of Pb and As, which is in agreement with supposition by Zhang and Liu (2002) who stated that EF values greater than 1.5 suggest that the sources are more likely to be anthropogenic. It is generally presumed that high EF values are indicative of the anthropogenic source of heavy metals.

Table 7 Geo-accumulation Index, contamination degree, pollution load index and contamination factor of the surface sediment in the Birim River

	I_{geo}				PLI	C_f				Sum C_f	mC_d	P_N
	Pb	Cd	Zn	As		Pb	Cd	Zn	As			
Apapam	− 2.07	1.77	− 4.51	− 4.94	0.07	0.21	0.04	0.15	0.02	0.42	0.105	0.17
Adukrom	− 2.36	2.17	− 4.28	− 4.69	0.08	0.17	0.06	0.18	0.03	0.44	0.110	0.15
Kibi	− 2.16	1.81	− 4.05	− 4.85	0.08	0.19	0.04	0.21	0.02	0.46	0.124	0.17
Bunso	− 2.12	2.14	− 3.67	− 4.13	0.10	0.20	0.06	0.27	0.04	0.52	0.130	0.21
Nsutam	− 1.78	2.06	− 4.71	− 5.48	0.07	0.25	0.05	0.13	0.02	0.45	0.113	0.19
Nsuapemso	− 2.51	1.99	− 3.47	− 4.29	0.10	0.15	0.05	0.31	0.03	0.54	0.135	0.24
Osino	− 2.36	1.99	− 4.58	− 5.47	0.07	0.17	0.05	0.14	0.02	0.38	0.095	0.14
Ankaase	− 2.13	1.86	− 3.36	− 4.48	0.10	0.20	0.05	0.34	0.03	0.62	0.155	0.26
Mampong	− 2.28	1.99	− 3.64	− 5.03	0.08	0.18	0.05	0.28	0.02	0.53	0.133	0.22
Anyinam	− 2.01	1.45	− 3.69	− 4.69	0.09	0.21	0.03	0.27	0.03	0.54	0.235	0.21

Geo-accumulation and pollution indices

The I_{geo}, which is the geochemical benchmark to assess heavy metals pollution in sediments Because of the possible dissimilarities in background concentrations of any given heavy metal and anthropogenic influences, a factor value of 1.5 was used. The I_{geo} values of the surface sediments in the Birim river varied from − 2.51 to − 1.78 (average − 2.18) for Pb, 1.45 to 2.17 (average 1.92) for Cd, − 4.71 to − 3.36 (average − 3.99) for Zn, and − 5.48 to − 4.13 (average − 4.81) for As (Table 7). The I_{geo} index values for Pb, Zn and As were negative and could be included in the unpolluted status (class 0), but pollution from Cd was moderately contaminated in the study area (I_{geo}: 1–2). PLI mainly allows a qualitative comparison between sites. PLI values < 1 signifies no pollution, $PLI = 1$ signifies baseline levels of pollutants and $PLI > 1$ signifies progressive site pollution. The pollution load index value of 0.08 indicated no deterioration of the site quality (PLI < 1).

Modified degree of contamination and Nemerow pollution index

In the environment, heavy metals co-exist together with other organic compounds and their negative effect results from their combined effects. Since the single element indices do not adequately take into account this synergistic effect this study employed the multi-element indices such as the combined degree of contamination and Nemerow pollution indices. The results of both mC_d and P_N as shown in Table 7 indicate that they fall within the zero class < 1.5 and < 1 respectively (Table 1). Hence the sediments in the settlements under study are unpolluted.

Table 8 Potential ecological risk assessment of heavy metals in sediments samples collected from the Birim River

Community	E_f^i				RI
	Pb	Cd	Zn	As	
Apapam	2.62	120.95	0.08	0.80	124.45
Adukrom	2.14	160.42	0.10	0.95	163.61
Kibi	2.45	124.89	0.11	0.85	128.31
Bunso	2.52	156.47	0.14	1.40	160.54
Nsutam	3.20	148.50	0.07	0.55	152.32
Nsuapemso	1.93	140.68	0.17	1.25	144.03
Osino	2.14	140.68	0.08	0.55	143.06
Ankaase	2.52	128.76	0.18	1.10	132.56
Mampong	2.27	140.68	0.15	0.75	143.58
Anyinam	2.73	97.26	0.14	0.95	101.09
Mean	2.45	135.93	0.12	0.92	139.36

Ecological risk assessment

The potential ecological risk assessment properly combines the ecological effects and toxicology. This approach was employed to analyse the level of pollution of heavy metals in the surface sediments of Birim River basin. The overall potential ecological index and potential ecological risk from the single metals (E_f^i) results are shown in Table 8.

The E_f^i values range from 1.93 to 3.20 for Pb, 97.26–160.42 for Cd, 0.07–0.18 for Zn, and 0.55–1.40 for As, with mean values of 2.45, 135.93, 0.12 and 0.92, respectively. Based on the Er^i results, the ecological risks associated with As, Zn and Pb are generally low whiles that associated with Cd are a considerable risk. The RI values are 101.09–163.61, with an average of 139.36, indicative of a moderate risk. Cd has high (average

97%) contributions to the RI values in the sediments of Birim River basin, and these results are similar to those reported in China (Li et al. 2016; Lin et al. 2016b).

Suitability for irrigation purposes

The evaluation of the suitability of surface water for irrigation purpose relies on TDS, EC, and relative concentration of Ca^{2+}, Mg^{2+}, Na^+ and HCO_3^- ions. The high content of salts, particularly Na^+ ion in the irrigation water influences the soil structure, reduces aeration and permeability, as well as resulting in alkaline soil, which can affect plant growth. Sodium adsorption ratio and percent sodium (% Na) is the sodium hazard induced by an excessive Na^+ ion in the irrigation water (Alam 2014). High Na^+ ion concentrations in irrigation water can negatively affect soil physical properties, causing soil particle dispersion when large amounts of Na^+ ion are adsorbed onto the soil particles (Bob et al. 2016; Al-Omran et al. 2016; Arveti et al. 2011). SAR is a measure of the extent to which Na^+ ion in the water systems might be absorbed by the soil. Generally, the higher the SAR value, the larger the risk of sodium hazard on plant growth. SAR values > 2.0, signify unsuitability of water for irrigation purpose (Ayuba et al. 2013; Vasanthavigar et al. 2010). The SAR value in the study area ranged between 0.73 and 1.52, with a mean value of 1.01.

The % Na content is also another parameter for evaluating the suitability of water for agriculture purposes. Since the combination of sodium with CO_3^{2-} or Cl^- results in the formation of alkaline or saline soils. The % Na value ranged between 29.61 and 70.10, with a mean value of 48.91. Hence, virtually all the water samples are suitable for irrigation except at Nsutam and Nsuapemso with % Na values slightly higher than the permitted limit of 60% for irrigation purposes.

KR is the amount of Na^+ ions measured against Ca^{2+} and Mg^{2+}. A KR greater than unity (KR > 1) is indicative of an excess amount of Na^+ ions in the water and is considered as of alkali hazard to the soil thereby making the water unsuitable for irrigation. The KR value for all the water samples was above unity, indicating unsuitability of this water for irrigation purpose.

RSC occurs when the excess CO_3^{2-} combine with Na^+ ion to form $NaHCO_3$. RSC shows the potential to eliminate Ca^{2+} and Mg^{2+} ions from the soil solution. High RSC value in irrigation water can lead to solidification and salinization of agricultural soils (Zaidi et al. 2015). However, as the soil solution becomes more concentrated, there is a high affinity for Ca^{2+} and Mg^{2+} ions to precipitate out as CO_3^{2-} which increases the relative proportion of sodium (Ravikumar et al. 2013). According to Li et al. (2016), water for irrigation purposes are classified as suitable, not suitable and marginally suitable

for irrigation when RSC value are < 1.25, > 2.5 and 1.25–2.5 meq/L, respectively. In this study, RSC values ranged between —2.42 and 125.15 meq/L, signifying that not almost all the water samples are suitable for irrigation except at Bunso sampling sites.

Magnesium adsorption ratio expresses the relationship between calcium and magnesium concentration in surface water (Ayuba et al. 2013). Moreover, the excess amount of Mg^{2+} ion can affect the quality of soil, since high levels of Mg^{2+} ions in the soil can cause infiltration problems. MAR value > 50, represent unsuitable of water for irrigation purposes (Ayuba et al. 2013). The MR value in the surface water samples ranged from 28.53 to 28.77. All the MR in the surface water samples were below the acceptable limit of 50, indicating their suitability for irrigation purpose.

Permeability is greatly affected by Na^+, Ca^{2+}, Mg^{2+}, HCO_3^- and Cl^- contents in the soil. On the basis of PI, water for irrigation purposes can be classified as suitable, marginally suitable and unsuitable for irrigation when the PI values are 100% maximum permeability (class I), 75% maximum permeability (class II) and 25% maximum permeability (class III), respectively. The calculated PI value ranged between 22.88 and 61.65 (Additional file 1: Table S3). According to the classification of PI, almost all the water samples fall under the class II, indicating that they are marginally suitable for irrigation purposes with the exception of those from Adukrom and Anyinam.

This study utilised diagrams established by the United States Salinity Laboratory (USSL) as well as Wilcox to evaluate water quality for irrigation purposes (Fig. 6).

The USSL diagram as shown in Fig. 6a signifies that the alkalinity and salinity values of all the surface water samples were very low (C1–S1), and thus no alkali hazard to crop growth. The Wilcox diagram (Fig. 6b) signifies that all the surface water samples fell under the "Excellent to good" class. Thus both Wilcox and USSL diagrams demonstrated the suitability of all the surface water samples from the Birim River basin for irrigation.

Conclusion

The results of this study provided valuable information about the hydrochemistry and water quality of surface water as well as ecological risks of some heavy metal contents of sediment from different settlements within the Birim River basin. The hydrogeochemical analysis of the surface water samples revealed that the water was neutral to mildly acidic and the hydrochemical facies of the area was dominated by the Ca–Na–Mg–HCO_3 (39%) and Na–Ca–Mg–HCO_3 (23%) water type. According to the geo-accumulation and Pollution indexes for the studied metals, settlements in the Birim River basin have moderately contaminated sediment. However multi-element

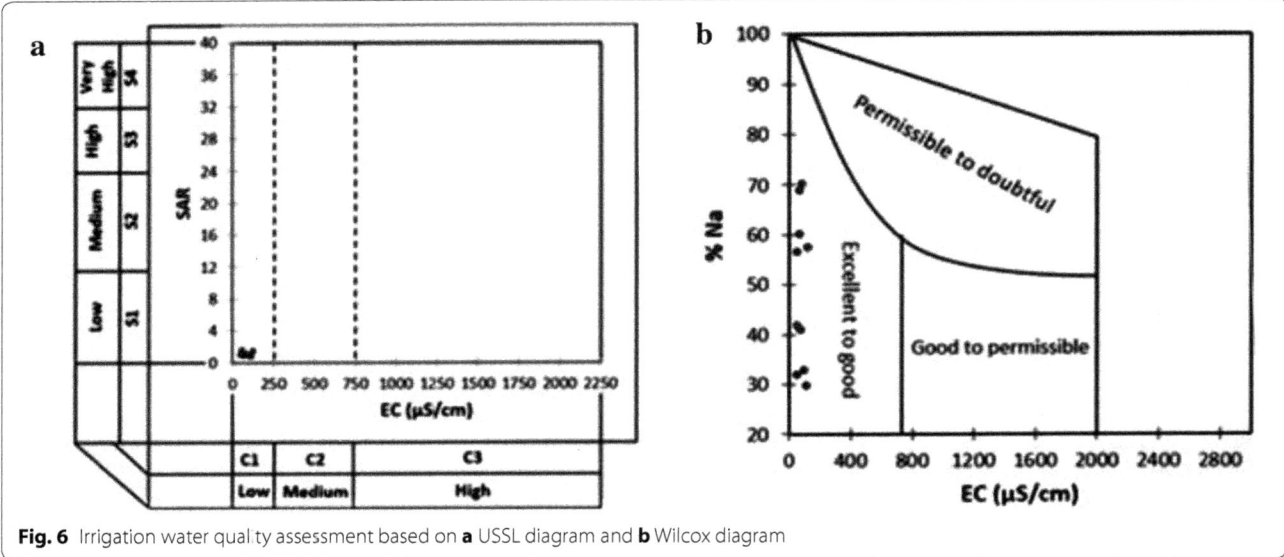

Fig. 6 Irrigation water quality assessment based on **a** USSL diagram and **b** Wilcox diagram

indices such as the modified degree of contamination and nemerow pollution indices indicate no pollution in the study area so far as the heavy metals studied are concerned. Based on the Er^i results, the ecological risks associated with As, Zn and Pb are generally low whiles that associated with Cd are a considerable whilst the RI values are indicative of a moderate risk. All these observations notwithstanding a systematic investigation are recommended to monitor the hydrochemistry, metal loading and change in the surface water, as well as sediment quality as both legal and illegal mining activities, is intensifying in the area.

Abbreviations
WQI: water quality index; SSP: soluble sodium percentage; SAR: sodium adsorption ratio; KR: Kelly's ratio; RSC: residual sodium carbonate; MR: magnesium ratio; ERM: effect range median; ERL: effect range low; ISQG: interim sediment quality guideline; PEL: probable effect level; PLI: pollution load index; CF: contamination factor; CD: contamination degree; I_{geo}: Geo-accumulation Index; PERI: potential ecological risk index; ER: ecological risk; T_i: toxicity coefficient; RI: total ecological risk index; USSL: United States Salinity Laboratory; KMO: Kaiser–Meyer–Olkin tests; PCA: principal component analysis.

Authors' contributions
JOO carried out the sampling, laboratory analyses and participated in the drafting of the manuscript. NKA and AAA conceived the study, participated in its design and coordination, performed the statistical analysis and helped to draft the manuscript. All authors read and approved the final manuscript.

Acknowledgements
The authors are very grateful to the National Council for Tertiary Education (NTCE), Ghana for a research Grant under the Teaching and Learning Innovation Fund (TALIFKNUSTR/3/005/2005).

Competing interests
The authors declare that they have no competing interests.

Funding
The National Council for Tertiary Education (NTCE) under the Teaching and Learning Innovation Fund funded the collection of samples, cost of analysis and interpretation of data with Grant Number TALIFKNUSTR/3/005/2005.

References
Abdesselam S, Halitim A, Jan A, Trolard F, Bourrié G (2013) Anthropogenic contamination of groundwater with nitrate in arid region: case study of southern Hodna (Algeria). Environ Earth Sci 70:2129–2141
Adamu CI, Nganje TN, Edet A (2015) Heavy metals contamination and health risk assessment associated with abandoned barite mines in Cross River State, southeastern Nigeria. Environ Nanotechnol Monit Manag 3:10–21
Alam F (2014) Evaluation of hydrogeochemical parameters of groundwater for suitability of domestic and irrigational purposes: a case study from central Ganga Plain, India. Arab J Geosci 7(10):4121–4131
Alaya MB, Saidi S, Zemni T, Zargouni F (2014) Suitability assessment of deep groundwater for drinking and irrigation use in the *Djeffara aquifers* (Northern Gabes, south-eastern Tunisia). Environ Earth Sci 71:3387–3421
Al-Omran AM, Aly AA, Al-Wabel MI, Sallam AS, Al-Shayaa MS (2016) Hydrochemical characterization of groundwater under agricultural land in arid environment: a case study of Al-Kharj, Saudi Arabia. Arab J Geosci 9:1–17
Ansah-Asare O, Asante K (2000) The water quality of Birim river in South-East Ghana. West Afr J Appl Ecol 1:23–34
Arnold G, Lenores C (1989) Standard methods for examination of water and waste water, 17th edn. American Public Health Association, New York, pp 3–8
Arveti N, Sarma MRS, Aitkenhead-Peterson JA, Sunil K (2011) Fluoride incidence in groundwater: a case study from Talupula, Andhra Pradesh, India. Environ Monit Assess 172:427–443
Asare-Donkor NK, Kwaansa-Ansah EE, Opoku F, Adimado AA (2015) Concentrations, hydrochemistry and risk evaluation of selected heavy metals along the Jimi River and its tributaries at Obuasi a mining enclave in Ghana. Environ Syst Res 4:1

Asefi M, Zamani-Ahmadmahmoodi R (2015) Mercury concentrations and health risk assessment for two fish species, *Barbus grypus* and *Barbus luteus*, from the Maroon River, Khuzestan Province, Iran. Environ Monit Assess 187:1–10

Atgin RS, El-Agha O, Zararsiz A, Kocatas A, Parlak H, Tuncel G (2000) Investigation of the sediment pollution in Izmir Bay: heavy elements, Spectrochim. Acta B 55(7):1151–1164

Ayuba R, Omonona O, Onwuka O (2013) Assessment of groundwater quality of Lokoja basement area, North-Central Nigeria. J Geol Soc India 82:413–420

Bain R et al (2014) Global assessment of exposure to faecal contamination through drinking water based on a systematic review. Trop Med Int Health 19:917–927

Boateng TK, Opoku F, Acquaah SO, Akoto O (2015) Pollution evaluation, sources and risk assessment of heavy metals in hand-dug wells from Ejisu–Juaben Municipality, Ghana. Environ Syst Res 4:1

Boateng TK, Opoku F, Acquaah SO, Akoto O (2016) Groundwater quality assessment using statistical approach and water quality index in Ejisu–Juaben Municipality, Ghana. Environ Earth Sci 75:1–14

Bob M, Rahman NA, Elamin A, Taher S (2016) Assessment of groundwater suitability for irrigation in Madinah City, Saudi Arabia. Arab J Geosci 9:1–11

Bortey-Sam N, Nakayama SMM, Ikenaka Y, Akoto O, Baidoo E, Mizukawa H, Ishizuka M (2015) Health risk assessment of heavy metals and metalloid in drinking water from communities near gold mines in Tarkwa, Ghana. Environ Monit Assess 187:1–12

Brady JP, Ayoko GA, Martens WN, Goonetilleke A (2015) Development of a hybrid pollution index for heavy metals in marine and estuarine sediments. Environ Monit Assess 187(5):306

Comber H, Deady S, Montgomery E, Gavin A (2011) Drinking water fluoridation and osteosarcoma incidence on the island of Ireland. Cancer Causes Control 22:919–924

Duodu GO, Ogogo KN, Mummullage S, Harden F, Goonetilleke A, Ayoko GA (2017) Source apportionment and risk assessment of PAH in Brisbane River sediment, Australia. Ecol Indicators 73:784–799

Demirak A, Yilmaz F, Tuna AL, Ozdemir N (2006) Heavy metals in water, sediment and tissues of *Leuciscus cephalus* from a stream in southwestern Turkey. Chemosphere 63(9):1451–1458

Fianko JR, Nartey VK, Donkor A (2010) The hydrochemistry of groundwater in rural communities within the Tema District, Ghana. Environ Monit Assess 168:441–449

Guo R, He X (2013) Spatial variations and ecological risk assessment of heavy metals in surface sediments on the upper reaches of Hun River, Northeast China. Environ Earth Sci 70:1083–1090

Hankason L (1980) An ecological risk index for aquatic pollution control: a secimentological approach. Water Res. 14(8):975–1001

Helstrup T, Jørgensen NO, Banoeng-Yakubo B (2007) Investigation of hydrochemical characteristics of groundwater from the Cretaceous–Eocene limestone aquifer in southern Ghana and southern Togo using hierarchical cluster analysis. Hydrogeol J 15:977–989

Hu B, Li G, Li J, Bi J, Zhao J, Bu R (2013) Spatial distribution and ecotoxicological risk assessment of heavy metals in surface sediments of the southern Bohai Bay, China. Environ Sci Pollut Res 20:4099–4110

Iranmanesh A, Ii RAL, Wimmer BT (2014) Multivariate statistical evaluation of groundwater compliance data from the Illinois Basin–Decatur Project. Energy Procedia 63:3182–3194

Jalali M (2007) Hydrochemical identification of groundwater resources and their changes under the impacts of human activity in the Chah basin in western Iran. Environ Monit Assess 130:347–364. https://doi.org/10.1007/s10661-006-9402-7

Karikari AY, Ansa-Asare OD (2006) Physico-chemical and microbial water quality assessment of Densu River of Ghana. West Afr J Appl Ecol 10:87–100

Khashogji MS, El Maghraby MMS (2013) Evaluation of groundwater resources for drinking and agricultural purposes, Abar Al Mashi area, south Al Madinah Al Munawarah City, Saudi Arabia. Arab J Geosci 6:3929–3942

Kim FM et al (2011) An assessment of bone fluoride and osteosarcoma. J Dent Res 90:1171–1176

Kim K-H, Yun S-T, Mayer B, Lee J-H, Kim T-S, Kim H-K (2015) Quantification of nitrate sources in groundwater using hydrochemical and dual isotopic data combined with a Bayesian mixing model. Agric Ecosyst Environ 199:369–381

Kouadia L, Trefry JH (1987) Sediment heavy metal contamination in the Ivory Coast West Africa. Water Air Soil Pollut 32:145–154

Lang Y-C, Liu C-Q, Zhao Z-Q, Li S-L, Han G-L (2006) Geochemistry of surface and ground water in Guiyang, China: water/rock interaction and pollution in a karst hydrological system. Appl Geochem 21:887–903

Li P, Wu J, Qian H (2016) Hydrochemical appraisal of groundwater quality for drinking and irrigation purposes and the major influencing factors: a case study in and around Hua County, China. Arab J Geosci 9:1–17

Lin M-L, Peng W-H, Gui H-R (2016a) Hydrochemical characteristics and quality assessment of deep groundwater from the coal-bearing aquifer of the Linhuan coal-mining district, Northern Anhui Province, China. Environ Monit Assess 188:1–13

Lin Q, Liu E, Zhang E, Li K, Shen J (2016b) Spatial distribution, contamination and ecological risk assessment of heavy metals in surface sediments of Erhai Lake, a large eutrophic plateau lake in southwest China. CATENA 145:193–203

Liu Y-Q, Moy B, Kong Y-H, Tay J-H (2010) Formation, physical characteristics and microbial community structure of aerobic granules in a pilot-scale sequencing batch reactor for real wastewater treatment. Enzyme Microb Technol 46:520–525

Marghade D, Malpe DB, Zade AB (2011) Geochemical characterization of groundwater from northeastern part of Nagpur urban, Central India. Environ Earth Sci 62:1419–1430. https://doi.org/10.1007/s12665-010-0627-y

Moore JW, Ramamoorthy S (2012) Heavy metals in natural waters: applied monitoring and impact assessment. Springer, Berlin

Murkute YA (2014) Hydrogeochemical characterization and quality assessment of groundwater around Umrer coal mine area Nagpur District, Maharashtra, India. Environ Earth Sci 72:4059–4073. https://doi.org/10.1007/s12665-014-3295-5

Nandimandalam JR (2012) Evaluation of hydrogeochemical processes in the Pleistocene aquifers of middle Ganga plain, Uttar Pradesh, India. Environ Earth Sci 65:1291–1308

Négrel P, Casanova J, Brulhet J (2006) REE and Nd isotope stratigraphy of a late jurassic carbonate platform, Eastern paris basin, France. J Sediment Res 76(3):605–617

Patel P, Raju NJ, Reddy BSR, Suresh U, Gossel W, Wycisk P (2016) Geochemical processes and multivariate statistical analysis for the assessment of groundwater quality in the Swarnamukhi River basin, Andhra Pradesh, India. Environ Earth Sci 75:1–24

Pius A, Jerome C, Sharma N (2012) Evaluation of groundwater quality in and around Peenya industrial area of Bangalore, South India using GIS techniques. Environ Monit Assess 184:4067–4077. https://doi.org/10.1007/s10661-011-2244-y

Purushothaman P, Rao MS, Rawat YS, Kumar CP, Krishan G, Parveen T (2014) Evaluation of hydrogeochemistry and water quality in Bist-Doab region, Punjab, India. Environ Earth Sci 72:693–706

Ravikumar P, Mehmood MA, Somashekar R (2013) Water quality index to determine the surface water quality of Sankey tank and Mallathahalli lake, Bangalore urban district, Karnataka, India. Appl Water Sci 3:247–261

Razmkhah H, Abrishamchi A, Torkian A (2010) Evaluation of spatial and temporal variation in water quality by pattern recognition techniques: a case study on Jajrood River (Tehran, Iran). J Environ Manage 91:852–860

Razowska-Jaworek L (ed) (2014) Calcium and magnesium in groundwater: occurrence and significance for human health. CRC Press, New York. https://doi.org/10.1201/b17085

Robinson GR, Ayotte JD (2006) The influence of geology and land use on arsenic in stream sediments and ground waters in New England, USA. Appl Geochem 21:1482–1497

Sahu P, Sikdar PK (2008) Hydrochemical framework of the aquifer in and around East Kolkata Wetlands, West Bengal, India. Environ Geol 55:823–835

Saleem M, Jeelani G, Shah RA (2015) Hydrogeochemistry of Dal Lake and the potential for present, future management by using facies, ionic ratios, and statistical analysis. Environ Earth Sci 74:3301–3313

Schafer A, Rossiter H, Owusu P, Richard B, Awuah E (2010) Developing country water supplies: physico-chemical water quality in Ghana. Desalination 251:193–203

Sinex SA, Helz GR (1981) Regional geochemistry of heavy elements in Chesapeake Bay sediment. Environ Geol 3:315–323

Singh NKS, Devi CB, Sudarshan M, Meetei NS, Singh TB, Singh NR (2013) Influence of Nambul River on the quality of fresh water in Loktak Lake. Int J Water Resourc Environ Eng 5:321–327

Singh CK, Rina K, Singh RP, Mukherjee S (2014) Geochemical characterization and heavy metals contamination of groundwater in Satluj River basin. Environ Earth Sci 71:201–216

Singh S, Raju NJ, Gossel W, Wycisk P (2016) Assessment of pollution potential of leachate from the municipal solid waste disposal site and its impact on groundwater quality, Varanasi environs, India. Arab J Geosci 9:1–12

Sun Y, Zhou Q, Xie X, Liu R (2010) Spatial, sources and risk assessment of heavy metals contamination of urban soils in typical regions of Shenyang, China. J Hazard Mater 174:455–462

Sutherland RA (2000) Bed sediment-associated heavy metals in an urban stream, Oahu, Hawaii. Environ Geol 39:611–627

Taylor SR, McLennan SM (1995) The geochemical evolution of the continental crust. Rev Geophys 33:241–265

Turekian KK, Wedepohl KH (1961) Distribution of the elements in some major units of the Earth's crust. Geol Soc Am Bull 72(2):175–192

Vasanthavigar M et al (2010) Application of water quality index for groundwater quality assessment: thirumanimuttar sub-basin, Tamilnadu, India. Environ Monit Assess 171:595–609

WHO (2011) Guidelines for drinking water quality, 4th edn. World Health Organization, Geneva

Yan N, Liu WB, Xie HT, Gao LR, Han Y, Wang MT, Li HF (2016) Distribution and assessment of heavy metals in the surface sediments of Yellow River, China. J Environ Sci China 39:45–51

Yi Y, Yang Z, Zhang S (2011) Ecological risk assessment of heavy metals in sediment and human health risk assessment of heavy metals in fishes in the middle and lower reaches of the Yangtze River basin. Environ Pollut 159:2575–2585

Yidana SM, Banoeng-Yakubo B, Sakyi PA (2012) Identifying key processes in the hydrochemistry of a basin through the combined use of factor and regression models. J Earth Syst Sci 121:491–507

Zaidi FK, Nazzal Y, Jafri MK, Naeem M, Ahmed I (2015) Reverse ion exchange as a major process controlling the groundwater chemistry in an arid environment: a case study from northwestern Saudi Arabia. Environ Monit Assess 187:1–18

Zakir HM, Shikazono N, Otomo K (2008) Geochemical distribution of heavy metals and assessment of anthropogenic pollution in sediments of Old Nakagawa River. Tokyo Jpn Am J Environ Sci 4(6):661–672

Zhang J, Liu CL (2002) Riverine composition and estuarine geochemistry of particulate metals in China-weathering features, anthropogenic impact and chemical fluxes. Estuar Coast Shelf Sci 54(6):1051–1070

Zhang LP, Ye X, Feng H (2007) Heavy metal contamination in Western Xiamen Bay sediments and its vicinity, China Mar. Pollut Bull 54:974–982

Zhang R, Jiang D, Zhang L, Cui Y, Li M, Xiao L (2014) Distribution of nutrients, heavy metals, and PAHs affected by sediment dredging in the Wujin'gang River basin flowing into Meiliang Bay of Lake Taihu. Environ Sci Pollut Res 21:2141–2153

Zhu GF, Li ZZ, Su YH, Ma JZ, Zhang YY (2007) Hydrogeochemical and isotope evidence of groundwater evolution and recharge in Minqin basin, Northwest China. J Hydrol 333:239–251

14

The impact of future climate and land use/cover change on water resources in the Ndembera watershed and their mitigation and adaptation strategies

Canute B. Hyandye[1,2]* , Abeyou Worqul[3], Lawrence W. Martz[4] and Alfred N. N. Muzuka[1]

Abstract

Background: Land use/cover and climate changes have a great influence on the hydrological processes in the watershed. The impacts of land use/cover and climate change are set to increase in the future due to the increased clearance of virgin forest lands for agriculture and the rise of global warming. The way in which the future climate will interact with the land use changes and affect the water balance in the watersheds requires more attention. This study was carried out in the Ndembera river watershed in Usangu basin, Tanzania, whereby the Soil and Water Assessment Tool was used to (i) assess the impact of near future (2010–2039) climate and 2013–2020 land use/cover change on the water balance and streamflow and (ii) evaluate the effectiveness of four land and water management practices as the mitigation and adaptation strategies for the impacts of climate and land use/cover changes. The 2020 land use/cover was predicted using Markov Chain and Cellular Automata models based on 2006 and 2013 land use/covers. The near-future climate scenario was generated from the Coupled Model Intercomparison Project 5 General Circulation Models.

Results: During the period from 2013 to 2020, the agricultural land and evergreen forests will increase by nearly 10 and 7%, respectively. Mixed forests will decrease by 12%. Such land use/cover changes will decrease the total water yield by nearly 13% while increasing evapotranspiration and surface runoff by approximately 8 and 18%, respectively. This moisture balance changes will be aggravated by warmer near-future mean annual temperatures (1.1 °C) and wetter conditions (3.4 mm/year) than in the baseline period (1980–2009). The warmer future climate will increase evapotranspiration and decrease water yield by approximately 35 and 8%, respectively. The management practices such as filter strips can reduce the annual evapotranspiration by 6%, and increase stream-flow by 38% in February.

Conclusion: The future land use/cover changes will interact with the near-future warmer temperatures and reduce water availability in the Ndembera watershed. Land and water management practices have great potential to mitigate the impacts of future climate and land use/cover changes on water resource, thus increasing its availability.

Keywords: Water balance, Soil and Water Assessment Tool, Climate change, Land and water management practices, Usangu catchment, Tanzania

Background

Land use/cover and climate changes have a great influence on the hydrological response of a watershed

(Kashaigili and Majaliwa 2013; Kirby et al. 2016). The hydrological processes which are affected by such changes include evapotranspiration, infiltration, surface runoff, groundwater flow and stream discharge regime (Natkhin et al. 2015). The effects of the land use/cover and climate change on hydrological processes are set to increase in the future due to the increased clearance of virgin forest lands for agriculture and the rise of global

*Correspondence: hyandyec@nm-aist.ac.tz; chyandye@irdp.ac.tz
[1] Department of Water and Environmental Science and Engineering (WESE), The Nelson Mandela African Institution of Science and Technology (NM-AIST), P.O. Box 447, Arusha, Tanzania

warming (Fischer 2013). Thus, the way in which the future climate will interact with the land use changes and affect the water balance in the watersheds requires more attention.

There is evidence to suggest that Tanzania is among the countries in Africa which is most at risk of being impacted by climate change (Hatibu et al. 1999). Further, it is anticipated that the farming sector will experience more impacts resulting in decreased production of different crops due to decreased water availability and the shift of growing seasons (Kangalawe and Lyimo 2013; Laderach and Eitzinger 2012). Although the climate change is predicted to cause adverse impacts on freshwater resources, especially in the dry sub-tropical regions (Pervez and Henebry 2015; Serdeczny et al. 2016), there is a little consideration given to the impacts in the process of planning of future water resource use and management (McCartney et al. 2012). Studies of the impacts of climate change on water resource are, therefore, encouraged to ensure its sustainability.

Usangu catchment, a part of the Rufiji Drainage Basin, is important for rice production in Tanzania. It produces more than 30% of the country's rice and supports over 30,000 rice-producing households on approximately 45,000 ha of irrigated land (SMUWC 2001). Rice and onions have captured good market price in the recent years which, in turn has brought a huge influx of Tanzanians into the catchment area to grow these crops. As a result, forest and wetlands along the rivers including the Ndembera river are converted into small irrigated farms. Kashaigili (2008) showed that there is a clear linkage between land use/cover changes and the changes in the hydrological regime for the Usangu wetlands and the Great Ruaha river in the Usangu catchment. It is therefore important to assess the changes in water balance and river discharge resulting from the land use/cover changes taking place in the Ndembera river watershed.

The historical land use/cover changes and their impacts on water resources in Usangu Catchment have been widely studied (Kikula et al. 1996; Kashaigili et al. 2006a; Kashaigili 2008). For example, Kashaigili et al. (2006a) observed a decline in water inflow into the Usangu Wetland due to the expansion of agricultural activities through forests clearing in the upstream. Indeed, some few land and water management practices have been applied in Usangu Catchment to improve the moisture holding capacity of soils. These practices include afforestation programs (Kashaigili et al. 2009), growing trees and shrubs (Malley et al. 2009) and contour bunds and terraces (Mwanukuzi 2011). Generally, the effectiveness of land and water management practices such as filter strips, contour and terracing and grassed waterways in improving the moisture holding capacity have been widely studied in many areas (Arnold et al. 2013; Fiener and Auerswald 2003; Taniguchi 2012; Wallace 2000). Nevertheless, there is limited information about their effectiveness in reducing the impact of climate and land use/cover change on water resources in Usangu Catchment.

This study was, therefore, carried out to assess the impact of the future (2020) land use/cover change and the near future (2010–2039) climate change effects on water balance and streamflow of the Ndembera river watershed in Usangu basin. Further, the effectiveness of land and water management practices in mitigating the impacts of future climate and land use changes on water balance and streamflow was evaluated.

Methods

Study area

The Ndembera river watershed is located in Usangu Basin, in the southern highlands of Tanzania (Fig. 1). It covers an area of about 1705 km^2. Ndembera is one of the five perennial rivers in the Usangu Catchment, in the Rufiji Basin (SMUWC 2001). The river originates from the springs of Udumka Village (Fig. 1), at an elevation of about 2060 meters above sea level (m.a.s.l. This river accounts for about 15% of the total flow of the Great Ruaha river (Elzein 2010). The section of the watershed which is considered in this study drains from Udumka Village to a river gauging station (1ka33). Water from Ndembera river supports about 16 large scale farms and irrigation schemes located at Udumka, Ihemi, Ifunda, Muwimbi, Igomaa, Mkunywa and Mahango villages as well as the Madibira irrigation scheme located near Madibira town (Fig. 1).

The climate of the watershed is seasonal, with one wet period (Early November to the end of April) and one dry period (Early May to the end of October). The precipitation is uni-modal, and its spatial distribution is strongly influenced by topography. The highlands receive precipitation of about 1600 mm/year, while the plains receive around 500–700 mm/year (Shu and Villholth 2012).

Preparation of the ArcSWAT model inputs

The input data for SWAT model (soil, land use, slope and weather data) were pre-processed in ArcMap 10.1 environment to obtain the data format required by ArcSWAT12 database. The SRTM 30 m Digital Elevation Model (DEM) of the study area (Fig. 2) was downloaded from the USGS database at https://earthexplorer.usgs.gov/. The DEM was used for delineating the study area's watershed and stream networks. The slope map (Fig. 3) was derived from the DEM using the Spatial Analyst tool.

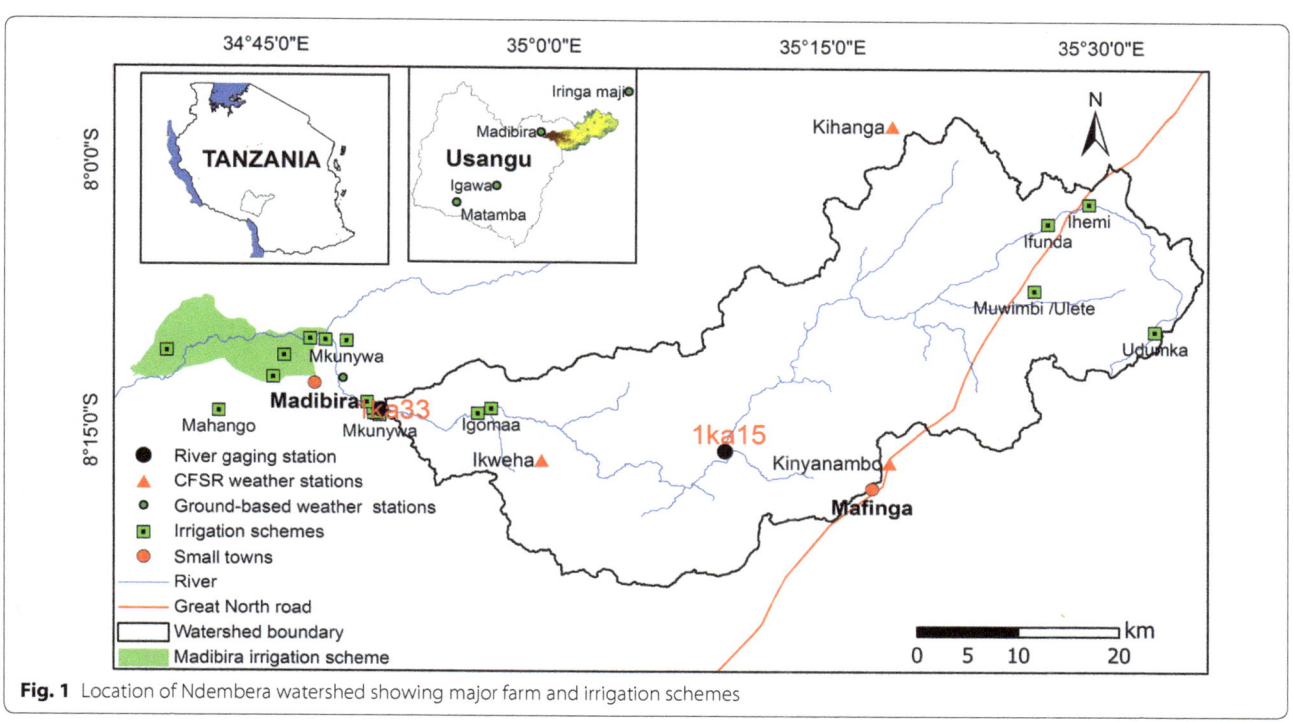

Fig. 1 Location of Ndembera watershed showing major farm and irrigation schemes

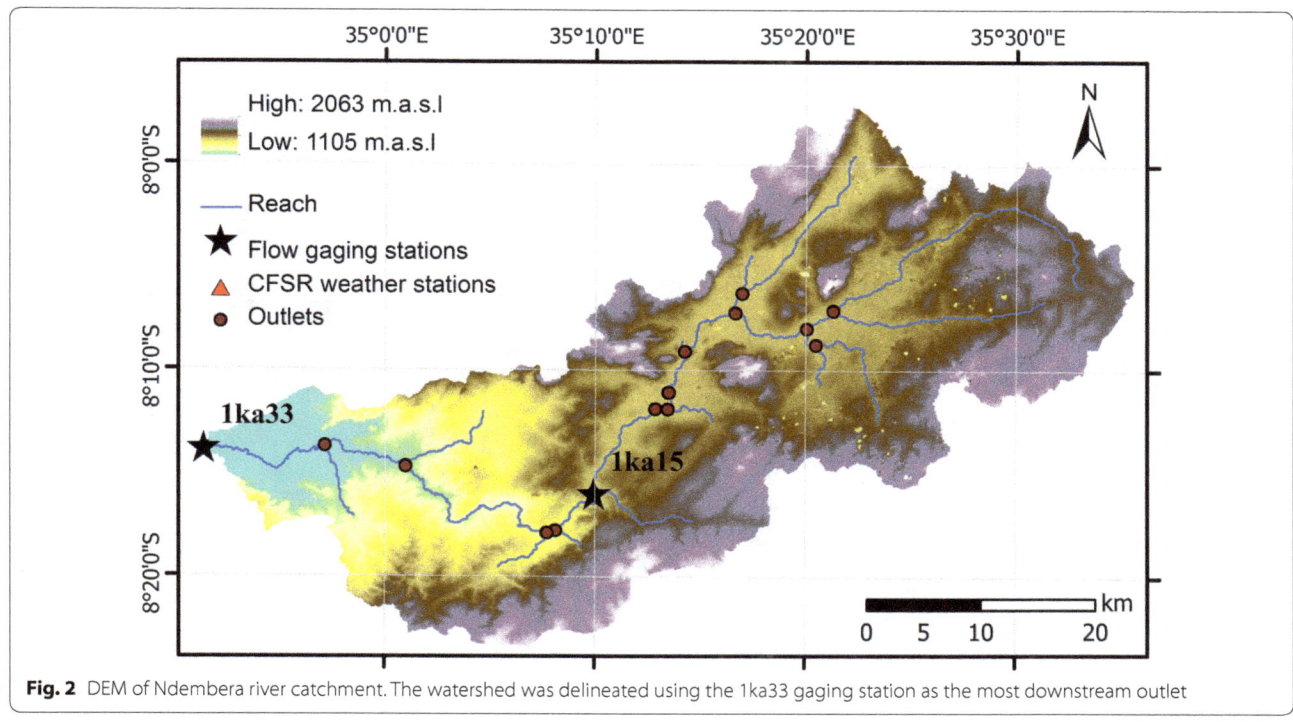

Fig. 2 DEM of Ndembera river catchment. The watershed was delineated using the 1ka33 gaging station as the most downstream outlet

Soil data in Fig. 4 was downloaded from the FAO Har-monized global soils database at http://www.waterbase.org/download_data.html. The watershed boundary was used to extract the soil data from the FAO soil database of the African soils slice. The attributes of these soils in Fig. 4 were updated using a "usersoil" table from the

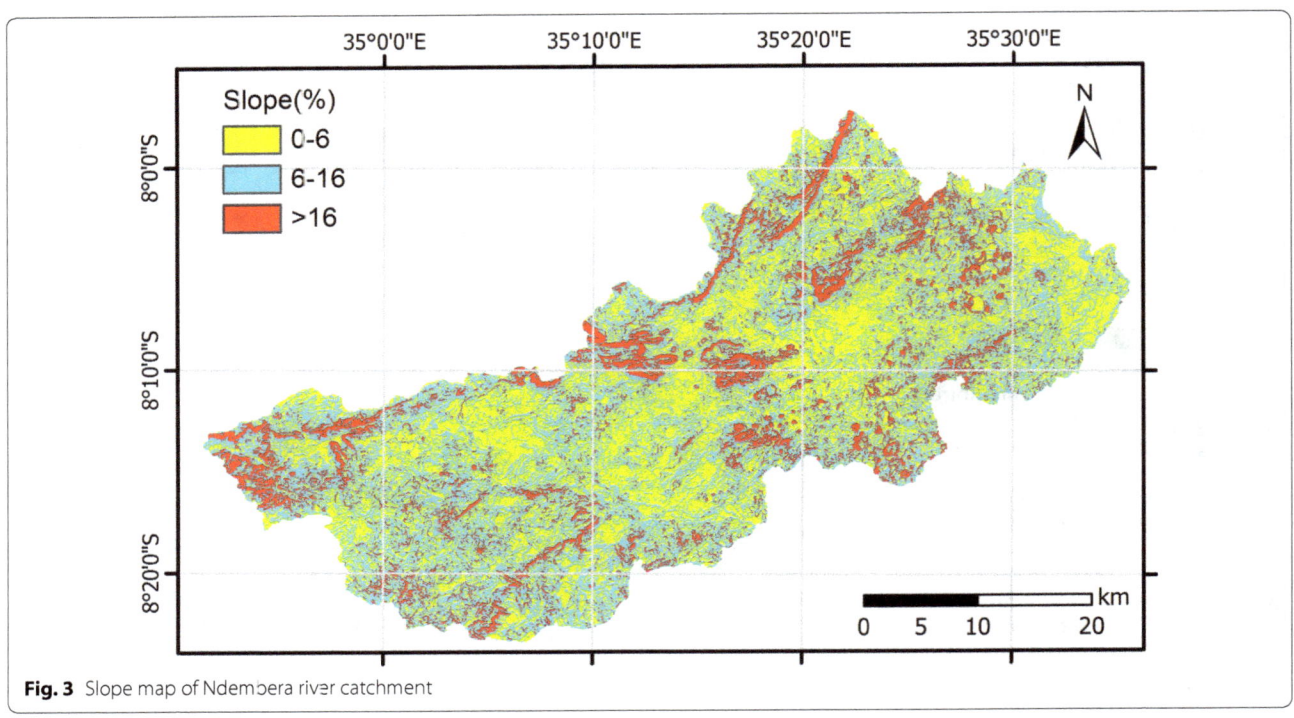

Fig. 3 Slope map of Ndembera river catchment

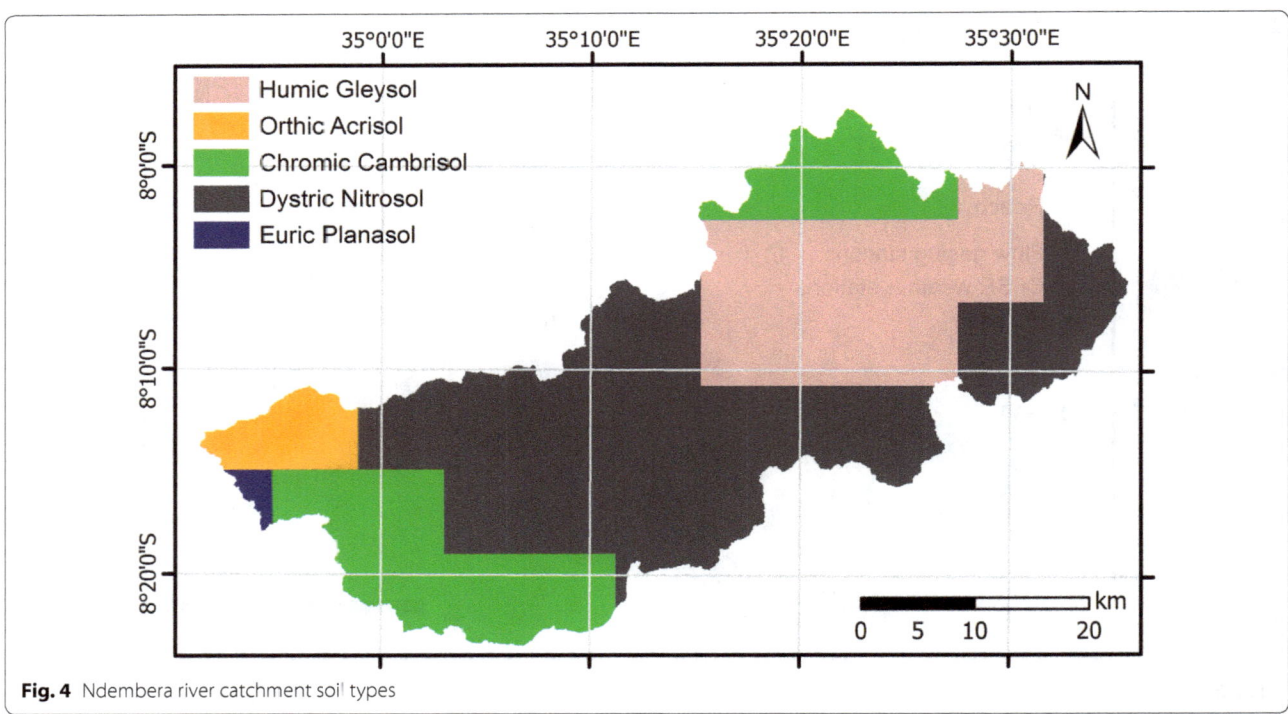

Fig. 4 Ndembera river catchment soil types

MapWindow SWAT12 database due to the fact that the "usersoil" table of ArcSWAT12 soil database contains USA soils only.

The baseline land use/cover map of 2013 (Fig. 5) was prepared by classifying Landsat images obtained from three path and rows (path168/row066, path169/row065

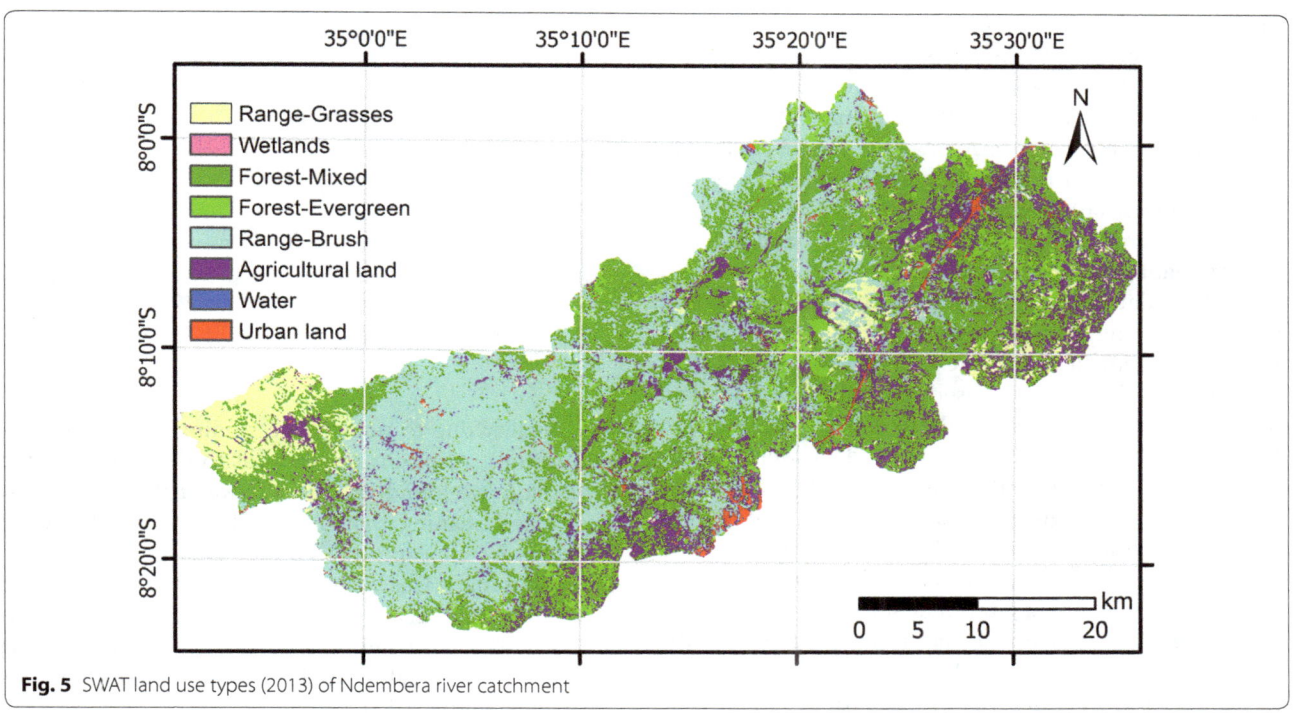

Fig. 5 SWAT land use types (2013) of Ndembera river catchment

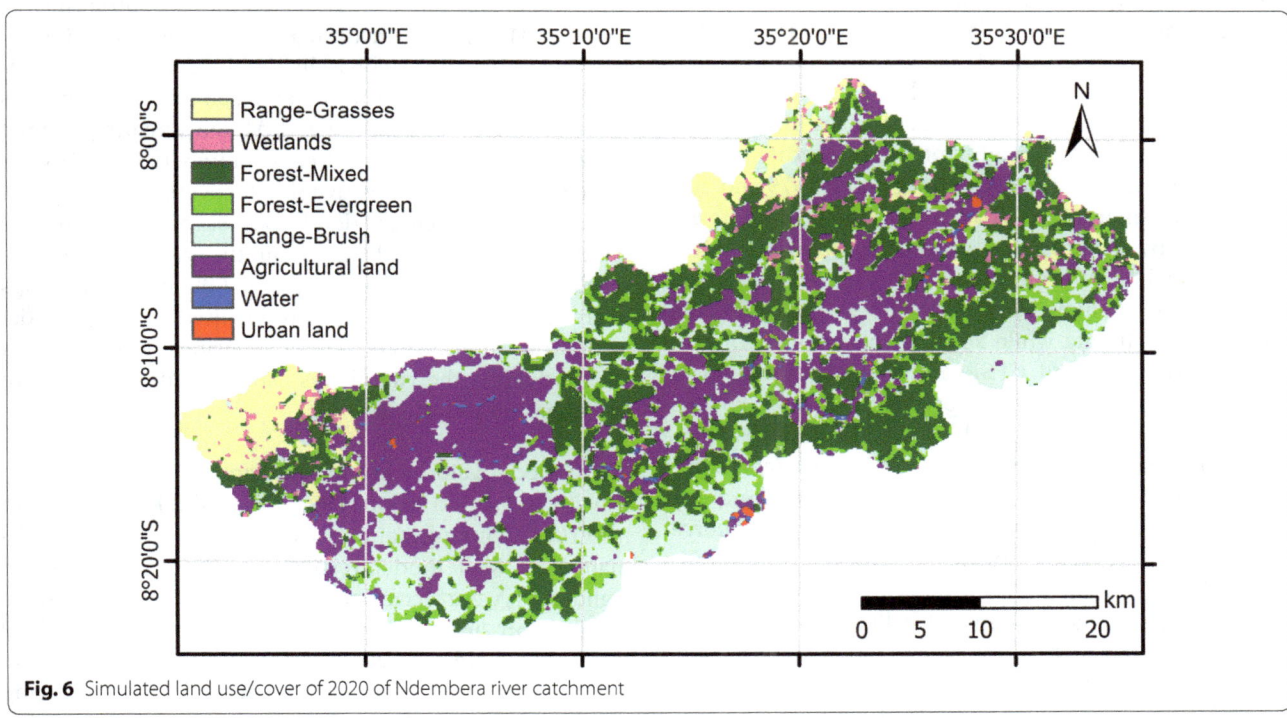

Fig. 6 Simulated land use/cover of 2020 of Ndembera river catchment

and path169/row066). The 2020 land use/cover of the Ndembera river watershed (Fig. 6) was extracted from the 2020 land use/cover of Usangu Catchment which was predicted using the Markov Chain and Cellular Automata models (CA-Markov) (Hyandye and Martz 2017). The CA–Markov model is one of the spatial transition-based models whereby the current trends of land use/cover evolution from "$t-1$" to "t" is used to project probabilities of land use and cover changes for the future date "$t+1$" (Behera et al. 2012; Eastman 2012; Houet and Hubert-Moy 2006). CA–Markov model is able to simulate changes in multiple land use types (Houet and Hubert-Moy 2006), hence giving possibilities of simulating the transition from one category of land use and cover to another (Behera et al. 2012).

Weather data was obtained from the ground-based weather gauging station (Iringa Maji) and the Climate Forecast System Reanalysis (CFSR) global weather data for SWAT. The ground-based gauging stations data were provided by the Rufiji Basin Water Organization (RBWO); while the CFSR weather data were downloaded from http://globalweather.tamu.edu/. Since the three operational ground-based weather stations in Usangu Catchment, namely, Matamba, Iringa Maji and Igawa station were located outside Ndembera watershed (Fig. 1), and Ndembera Auto met station located within the Ndembera watershed has not functioned since 2008, it was necessary to use CFSR global rainfall data. This CFSR data was obtained for Ikweha, Kihanga and Kinyanambo stations (Fig. 1). The CFSR data often captures the rainfall pattern very well, however, it often overestimates the gauged rainfall (Worqlul et al. 2014, 2017a); hence, Iringa Maji station data, a nearby weather station which had daily series data from 1962 to 2013, were used to perform bias correction of the CFSR precipitation data. The CFSR of the National Centres for Environmental Prediction (NCEP) readily provides weather data for any geographic location on earth between 1979 and 2014 (Roth and Lemann 2016). The bias of the CFSR data was corrected by a linear bias correction approach which is described by Worqlul et al. (2017b). This approach reduces the volume difference between CFSR and gauged rainfall data while keeping the pattern. The two datasets (uncorrected CFSR and gauged rainfall data) involved in the linear bias correction process covered the same time window (1999–2013). The annual volume difference between the observed and bias corrected data was minimized to zero. In addition, the mean monthly data (observed and corrected CFSR) were highly correlated due to their close mean monthly values as shown in Fig. 7.

The discharge data for model calibration and validation period (2000–2010) for Ndembera river at gauging station (1ka33) was obtained from the Rufiji Basin Water Organisation. Two major challenges were encountered. Firstly, the data series had gaps of up to 6 months, and sometimes for more than a year, as they were observed in 1998–1999. Secondly, the data series had ambiguous data units. It is acknowledged that data at 1ka33 were not of

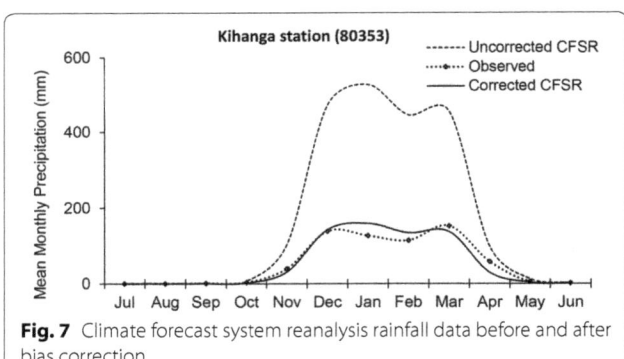

Fig. 7 Climate forecast system reanalysis rainfall data before and after bias correction

good quality due to poor gauge reading (personal communication). Also, during our visit to the station in the dry season, we found out that the water flow was below the gauge, meaning that, in the dry season, the water discharge is not captured. During data processing, the ambiguous data was deleted. Then, the data from the upstream station 1ka15 (Fig. 2) was used to fill data gaps at station 1ka33 for the period from 2000 to 2010 using the simple interpolation and linear regression methods (Koch and Cherie 2013).

SWAT model set-up and parameterization

The whole watershed was divided into 29 sub-basins. A total of 411 and 470 hydrological response units (HRU) were generated for the baseline and future land use scenarios, respectively. The general sub-basin parameters such as initial leaf area index (LAI_INI), plants potential heat units (PHU_PLT) and curve numbers (CN_2) were updated for the respective land covers. Management operations for specific land uses such as agricultural lands were scheduled based on the growing calendar used by farmers in the watershed (Table 1). The crops were auto-irrigated and auto-fertilised. The potential evapotranspiration was estimated using the Hargreaves method while the runoff was estimated using the curve number method.

SWAT model calibration and validation

River discharge data of 1999–2006 and 2007–2010 from the 1ka33 gauging station was used to calibrate and validate the SWAT model, respectively. The calibration and validation were accomplished using a semi-automatic Calibration and Uncertainty Programme; SWAT-CUP SUFI-2 (Arnold et al. 2012). A warm-up period of 3 years (1999–2001) was used for model initialization during model calibration. The calibration process was preceded by the sensitivity analysis of the parameters which control the observed river flows. The sensitivity analysis considered the range of parameters suggested by Holvoet et al. (2005). A total of twenty-three (23) parameters

Table 1 Crops growing calendar in Usangu Basin in a hydrological year

	Nov	Dec	Jan	Feb	Mar	Apr	May	Jun	Jul	Aug	Sep	Oct
Cereal and Grains												
Paddy(Rice)												
Maize												
Oil seeds (sunflower and groundnuts)												
Vegetables												
Onions												
Others (Tomatoes, vegetables, potatoes												

Legend

Sowing seeds		Farms Preparations	Trans- planting	1st weeding	2nd weeding
Mid-season		Harvest	Mixed activities		

Hydrological year starts in November and ends in October of the following year. Activities related to tomatoes, vegetables, and sweet potatoes vary from one location to another

pre-selected during manual calibration were used in the SWAT-CUP SUFI2 to run one thousand simulations. A total of nineteen (19) influential parameters of the stream flow were identified for model calibration (Table 2), whereby the sensitivity was evaluated using the t-statistic (t-stat) and p values.

SWAT model performance evaluation

The model performance in predicting the catchment conditions was evaluated using the statistical analysis parameters such as coefficient of determination (R^2), Nash–Sutcliffe efficient (NSE), r-factor and p-factor (Arnold et al. 2012) and graphical analysis. In addition, the percentage bias (PBIAS) was also considered. The r-factor refers to the thickness of the 95% prediction uncertainty envelope, while the p-factor is a percentage of observations covered by the 95% prediction uncertainty.

Calibration and validation of future climate data

The Near-term (2010–2039) climate scenario of precipitation and temperatures was generated from the 29 GCMs using the procedures described in the Guide for Running AgMIP Climate Scenario Generation Tools with R (Hudson and Ruane 2013). These GCMs were sourced from the Coupled Model Intercomparison Project 5 (CMIP5). The GCMs included ACCESS1-0, bcc-csm1-1, BNU-ESM, CanESM2, CCSM4, CESM1-BGC, CSIRO-Mk3-6-0, GFDL-ESM2G, GFDL-ESM2 M, HadGEM2-CC, HadGEM2-ES, inmcm4, IPSL-CM5A-LR, IPSL-CM5A-MR, MIROC5 and MIROC-ESM-CHEM. Others were MPI-ESM-LR,

MPI-ESM-MR, MRI-CGCM3, NorESM1-M, FGOALS-g2, CMCC-CM, CMCC-CMS, CNRM-CM5, HadGEM2-AO, IPSL-CM5B-LR, GFDL-CM3, GISS-E2-R and GISS-E2-H. The models were provided by Sokoine University of Agriculture, Tanzania and the National Aeronautics and Space Administration Goddard Institute for Space Studies (NASA-GISS), USA. The Simple Delta Method was used for statistical downscaling of the GCMs. This method preserves the observed patterns of temporal and spatial variability from the gridded observations (Hamlet et al. 2010).

The statistical downscaling of the 29 GCMs involved the calculation of the change factor (the ratio between a mean value, in the future, and historical run) using the delta change algorithm which was acquired together with the CMIP5-GCMs (Fig. 8). This change factor was then applied to the observed time series (1980–2009) to transform it into a time series representing the future climate. The 2010–2039 climate scenario was analysed under the Representative Concentration Pathway 8.5 Greenhouse Gas Emission scenario. This concentration pathway is characterized by increasing Greenhouse gases emissions over time, a representative of scenarios leading to high Greenhouse gases concentration levels (Chaturvedi et al. 2012).

The sub-selection of the five representative GCMs for hot/wet, hot/dry, cold/wet, cold/dry and middle (Ensemble mean) future climatic conditions was based on a scatter diagram approach (Fig. 9). The diagram represented the changes in mean monthly temperatures against the percentage in mean monthly change in future

Table 2 Parameters sensitive to streamflow, their default range, fitted value during calibration and final values used in SWAT model for streamflow simulations

Parameter	Description	SUFI2 fitted value	Default SWAT range	Final value in SWAT model
v__SURLAG.bsn	Surface runoff lag time (days)	20.76	0.05 to 24	20.76
v__ESCO.bsn	Soil evaporation compensation factor	0.83	0 to 1	0.83
a__GWQMN.gw	Threshold depth of water in the shallow aquifer for return flow to occur (mm H_2O)	217.08	0 to 5000	1217.08
a__GW_REVAP.gw	Groundwater "revap" coefficient	0.03	0.02 to 0.2	0.05
a__REVAPMN.gw	Threshold depth of water in the shallow aquifer for "revap" to occur (mm H_2O)	236.07	0 to 1000	986.07
v__ALPHA_BF.gw	Baseflow alpha factor (days)	0.83	0 to 1	0.83
v__SHALLST.gw	Initial depth of water in the shallow aquifer	1263.00	0 to 5000	1263.00
v__RCHRG_DP.gw	Deep aquifer percolation fraction	0.63	0 to 1	0.63
v__GW_DELAY.gw	Groundwater delay (days)	18.25	0 to 500	18.25
v__CH_N2.rte	Manning's "n" value for the main channel	0.11	− 0.01 to 0.3	0.11
v__EPCO.hru	Plant uptake compensation factor	0.50	0 to 1	0.50
a__OV_N.hru	Manning's "n" value for overland flow	28.86	0.01 to 30	28.86
a__CANMX.hru	Maximum canopy storage (mm H_2O)	10.78	0 to 100	10.78
a__SLSUBBSN.hru	Average slope length (m)	40.45	10 to 150	131.91
a__HRU_SLP.hru	Average slope steepness (m/m)	0.44	0.3 to 0.6	0.47
a__SOL_AWC().sol	Available water capacity of the soil layer (mm H_2O/ mm soil)	− 0.06	0 to 1	0.04
a__SOL_K().sol	Saturated soil hydraulic conductivity (mm/h)	21.37	0 to 2000	39.02
a__CH_K1.sub	Effective hydraulic conductivity in main channel alluvium (mm/h)	85.56	0 to 300	85.56
a__CN2.mgt	Initial SCS runoff curve number for moisture condition II	− 1.87	35 to 98	83.13

a__means absolute; a given value is added to the existing parameter value during the calibration

v__means replace; the existing parameter value is to be replaced by a given value during the calibration

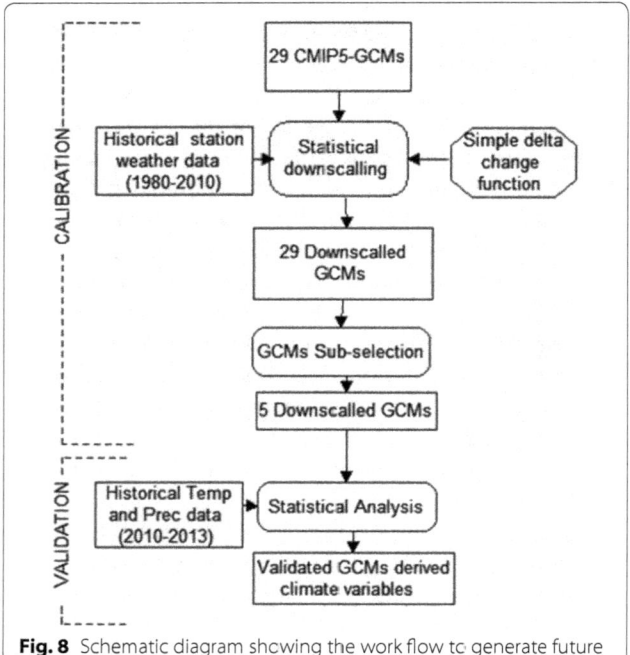

Fig. 8 Schematic diagram showing the work flow to generate future climate variables

precipitation from the baseline scenario. The GCM falling close to the median of each quadrant was selected according to Subash et al. (2016). The GCMs selected for each weather station in the watershed (Table 3) were averaged to obtain a site specific GCMs-derived climate data (Fig. 9).

The mean GCMs-derived precipitation and temperatures were validated by comparing them with the observed historical station data 1980–2009 and 2010–2013. The validation involved a graphical and statistical analysis. The statistical analysis included the Mean error (ME), correlation (R), Median, Mean and Standard deviation (SD).

Simulation of the impacts of future land use/cover, climate and land management practices on water balance

The 2020 land use/cover of the Ndembera river watershed was introduced in a calibrated SWAT model to replace the baseline land use/cover of 2013. The model was then run to simulate the water balance conditions using the 2020 land use/cover scenario without changing other SWAT input data (weather, soils, and slope). The following assumptions were made: (i) land use/cover will

Table 3 Selected CMIP5-GCMs for forcing the SWAT model

Station ID	Local name	Lat	Long	Elevation	GCM ID	GCM name	Future condition
80353	Kihanga	− 7.962	35.313	1589	D	CanESM2	Hot/wet
					B	bcc-csm1-1	Cool/wet
					S	MRI-CGCM3	Cool/dry
					2	GISS-E2-R	Hot/dry
					R	MPI-ESM-MR	Middle
83353	Kinyanambo	− 8.274	35.313	1803	M	IPSL-CM5A-LR	Hot/wet
					X	CNRM-CM5	Cool/wet
					S	MRI-CGCM3	Cool/dry
					A	ACCESS1-0	Hot/dry
					R	MPI-ESM-MR	Middle
83350	Ikweha	− 8.274	35.00	1427	M	IPSL-CM5A-LR	Hot/wet
					F	CESM1-BGC	Cool/wet
					S	MRI-CGCM3	Cool/dry
					R	MPI-ESM-MR	Hot/dry
					Q	MPI-ESM-LR	Middle

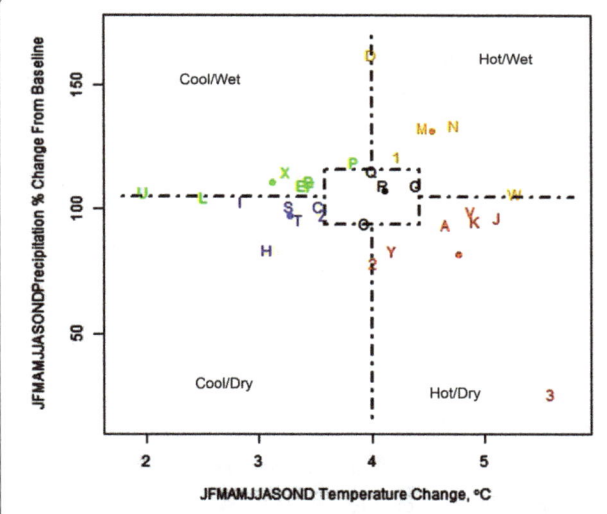

Fig. 9 Temperature-and-Precipitation change scatter diagram for Kinyanambo station (83353)

change as expected, and (ii) the hydrological and atmospheric conditions, soils and slope will remain unchanged over the next decade. The future GCMs-derived climate data was then introduced into the Ndembera SWAT model that had been updated with the 2020 land use/cover. This was done in order to simulate the combined effect of future climate and land use/cover changes on water balance. Lastly, the same SWAT model setup containing the 2020 land use/cover and the 2010–2039 GCMs-derived climate data was used to simulate the impacts of land and water management practices on water balance and streamflow. The practices included

terracing and contouring, filter strips, grassed waterways and deep ripper subsoiler tillage. The choice of these management practices was based on the nature of the landscape (slope) and the expected land use/cover and climate changes on water balance in the watershed such as reduced infiltration and increased evapotranspiration.

The effects of the four management practices on water balance were simulated in SWAT by activating the sub-models of the respective management practices. Table 4 shows the parameters and their final values used for water balance simulation under each management practice. These values were applied in the hydraulic response units with agricultural land use only.

Results

Sensitivity analysis results

The top six most sensitive parameters to the river discharge were the groundwater delay factor (GW_DELAY), average slope steepness (HRU_SLP), groundwater "revap" coefficient (GW-REVAP), available water content of the soil (SOL_AWC), average slope length (SLSUBBSN), and the curve number (CN_2) (Fig. 10). Compared to others, these six parameters showed higher t-Statistic values ($|\cong2$ to $16|$) and lower p values (0 to < 0.05). Four parameters (CN_2, SLSUBBSN, SOL_AWC and HRU_SLP) which are the surface flow response parameters showed high sensitivity while two channel response parameters (CH_K1 and CH_N2) showed very low sensitivity to the river discharge.

Performance of the SWAT model

The coefficients of determination (R^2), the Nash-Sutcliff efficient index (NSE) and PBIAS values showed that

Table 4 List of SWAT input parameters and values used for each management practice

Management practice	SWAT input table	Parameters	Parameter description	Value
Filter strips	Management (.Mgt)	FILTERW	Width of the edge of field filter strips (m)	30*
	Operations (.Ops)	FILTER_RATIO	Ratio of field area to filter strip area (unitless)	40
	Operations (.Ops)	FILTER_CON	Fraction of the HRU which drains to the most concentrated ten percent of the filters strip area	0.5
	Operations (.Ops)	FILTER_CH	Fraction of the flow within the most concentrated ten percent of the filter strip which is fully channelized (dimensionless)	0
Grassed waterways	Operations (.Ops)	GWATN	Manning's "n" value for overland flow	0.35
	Operations (.Ops)	GWATL	Grass waterway length (km)	1000
	Operations (.Ops)	GWATW	Average width of grassed waterway (m)	10*
	Operations (.Ops)	GWATD	Depth of grassed waterway channel from top of bank to bottom (m)	1
	Operations (.Ops)	GWATS	Average slope of grassed waterway channel	HRU slope × 0.75
Terraces and contour	Operations (.Ops)	TERR_CN	Curve number	62**
	Operations (.Ops)	TERR_SL	Average slope length (m)	61***
	Operations (.Ops)	CONT_CN	Initial SCS curve number II value	62**
Deep ripper subsoiler	Management (.Mgt)	CNOP	SCS runoff curve number for moisture conditions II	62**

* User defined value based on knowledge from the field. ** Adopted from Table 20-1 of runoff curve numbers for cultivated agricultural lands (Arnold et al. 2013). *** Values from the calibrated SWAT model. Other values without the asterisk (*) were default values from the SWAT database

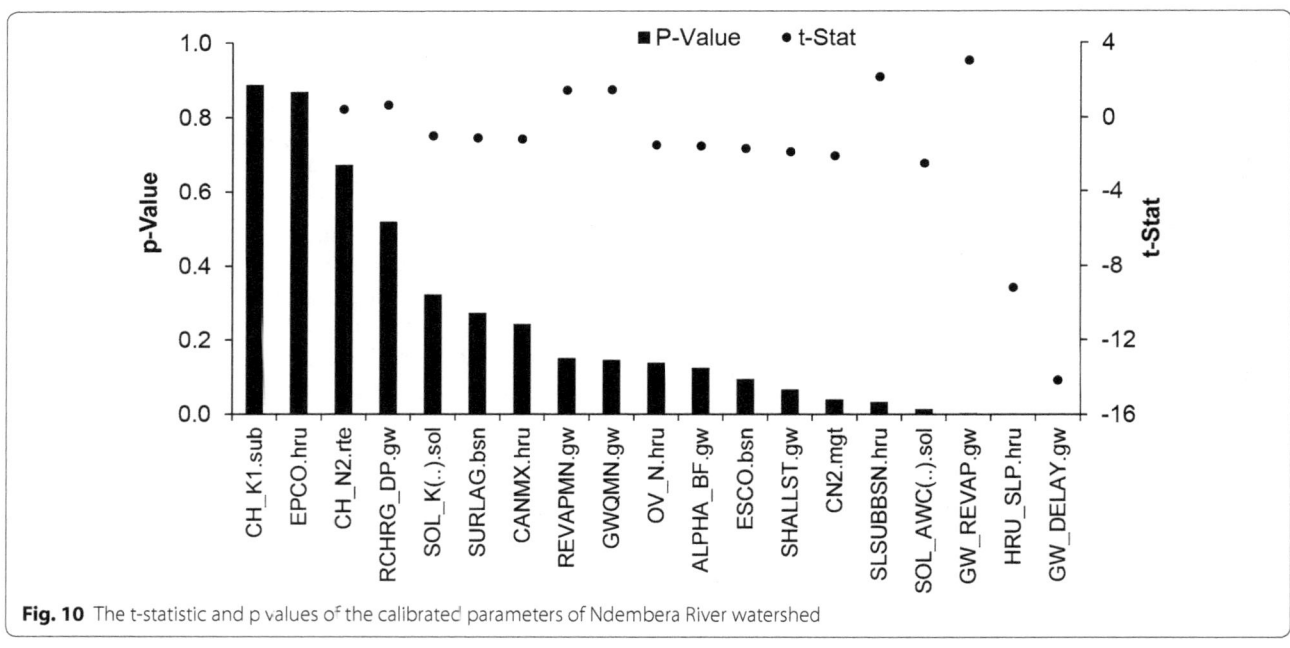

Fig. 10 The t-statistic and p values of the calibrated parameters of Ndembera River watershed

there was a good fit between the observed and the simulated flow both in the calibration and validation period (Table 5). The R^2 for calibration and validation periods were 0.79 and 0.80, respectively. The NSE ranged from 0.78 to 0.76. The respective p-factor and r-factor statistic were 0.63 and 0.61 for the calibration period and 0.48 and 0.65 for the validation period. The differences between the simulated and observed discharge at 1ka33, expressed as the percentage bias (PBIAS) was very small

for the calibration and validation period (3.8 and −4.6, respectively).

A comparison between SWAT simulated and observed discharge of the Ndembera river at 1ka33 on monthly time step in Fig. 11 showed a good capture of both ascent and recession of the river hydrograph. The hydrograph matched well with the precipitation rhythm of the catchment. The peaks of the simulated streamflow appeared to be underestimated both during the calibration and

Table 5 SWAT model's calibration and validation statistical analysis results

Parameter	Statistical values	
	Calibration	Validation
NSE	0.78	0.76
R²	0.79	0.80
p-Factor	0.63	0.48
r-Factor	0.61	0.65
PBIAS	3.8	− 4.6

validation periods except in 2010 where the peak flows of simulated flow were greater than the observed flow. Generally, the simulated streamflow from the calibrated model slightly underestimated the actual streamflow during the low flows. Given the high values for R² and NSE and good match hydrographs, the model was considered to be suitable for water balance simulation in this study.

Evaluation of the GCMs-derived climate data

The mean monthly precipitation and temperature from the downscaled GCMs for both the baseline and verification periods showed a good match with the observed data (Figs. 12, 13). In these figures, the curves of the GCMs-derived climate data captured very well the pattern (peaks and troughs) of the inter-annual variations in the observed data series from the three weather stations. Results in Table 6 depicted good linear relationship between observed and GCM-derived climate data (R = 0.92–0.99 for precipitation and 0.77–0.99 for temperature).

The comparison between observed and GCMs-derived temperature in Fig. 13a, b showed a slight underestimation of GCMs-derived temperature while Fig. 13c indicated an overestimation. The degree to which the GCMs-derived climate variables were underestimated or overestimated were denoted by positive and negative mean error values (ME) in Table 6. Kinyanambo station mean error values were 0.95 and 1.24 for the baseline and verification periods, respectively; while, Ikweha station had negative mean error values of − 1.19 and − 0.91, respectively. The mean error values of precipitation from all three weather stations ranged from − 0.19 to 0.05. Generally, the mean error values for both temperature and precipitation were relatively small. In addition, the distribution parameters, the mean, standard deviation (SD) and median of the GCMs-derived climate data were close to those from the observed station data.

Current and future trends of mean climate in the watershed

The annual mean precipitation and mean monthly temperature of the Ndembera watershed for the past three decades (1980–2009) showed an increasing trend (Figs. 14a, 15a). The same trend was observed in the near-future period (Figs. 14b, 15b). Whereas the annual mean precipitation increase from 1980–2009 was about 2.4 mm/year (Fig. 14a), that of the near-term period was about 3.2 mm/year (Fig. 15a). Based on the regression equation in Fig. 14b, the watershed annual mean precipitation was shown to be about 801.4 mm/year at the end of 2039. Regarding temperature, the watershed mean monthly temperature changed from 16.8 °C in 1980 to about 17.6 °C at the end of 2009 (Fig. 15a), an increase

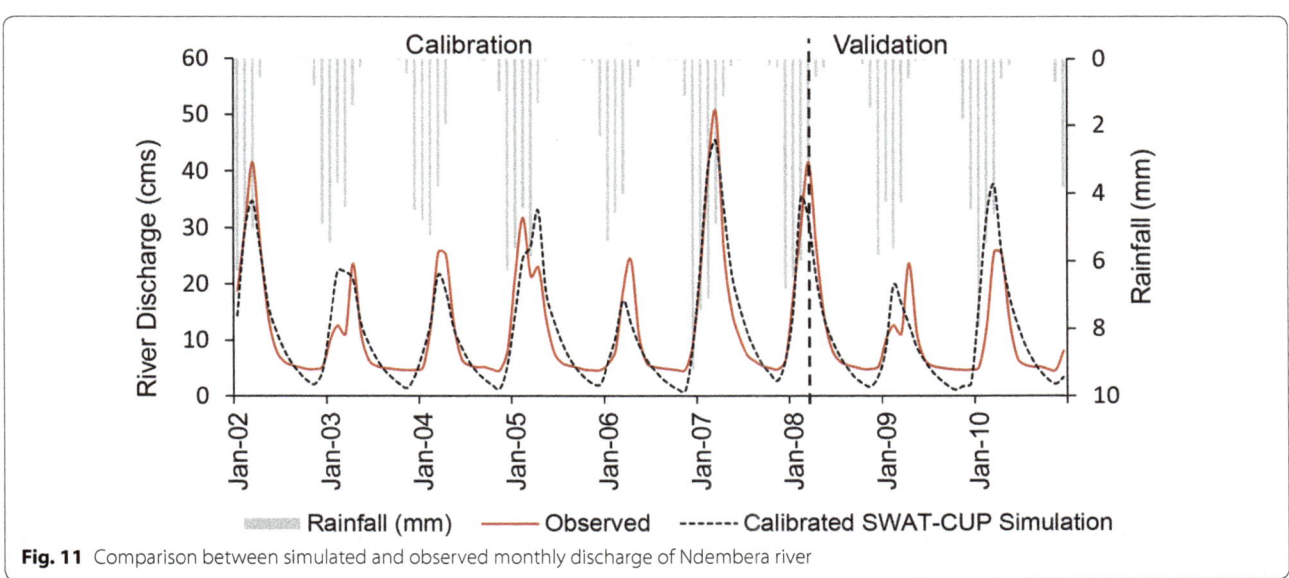

Fig. 11 Comparison between simulated and observed monthly discharge of Ndembera river

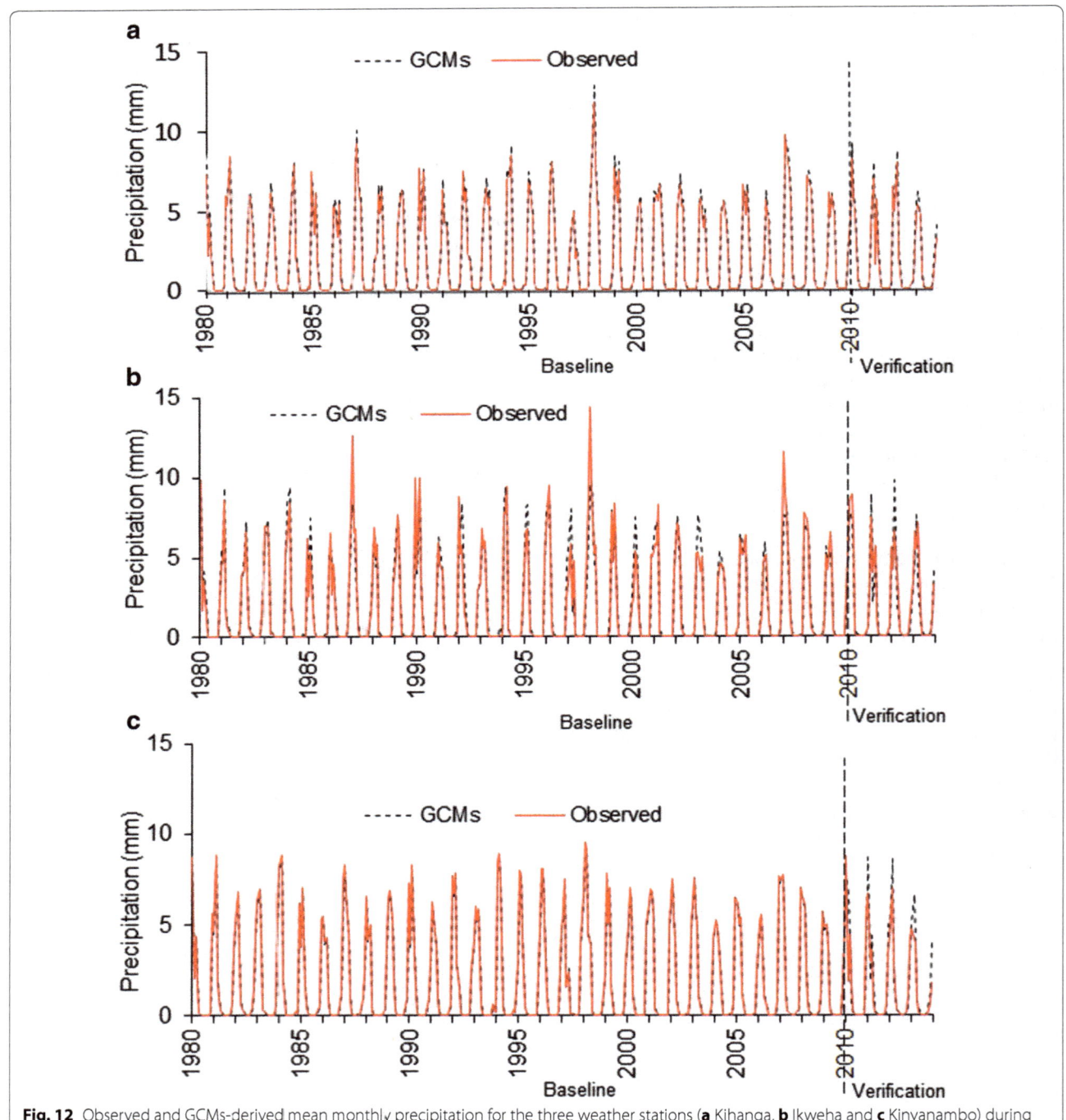

Fig. 12 Observed and GCMs-derived mean monthly precipitation for the three weather stations (**a** Kihanga, **b** Ikweha and **c** Kinyanambo) during baseline and verification periods

of about 0.8 °C over the past 30 years. The temperature of the period between 2010 and 2039 are expected to increase by 1.1 °C (from 18.1 °C in 2010 to 19.2 °C) (Fig. 15b).

A comparison of the amount of precipitation for each month averaged over a 30 years period, both in the baseline and near-term periods, showed higher amounts of precipitation in the wet months of January–April in the near-term than in the baseline period (Fig. 16). The change in the amount of precipitation in these months for the two periods ranged from 2 to 6 mm/month (~2–7%). In the early months of the wet season (November and

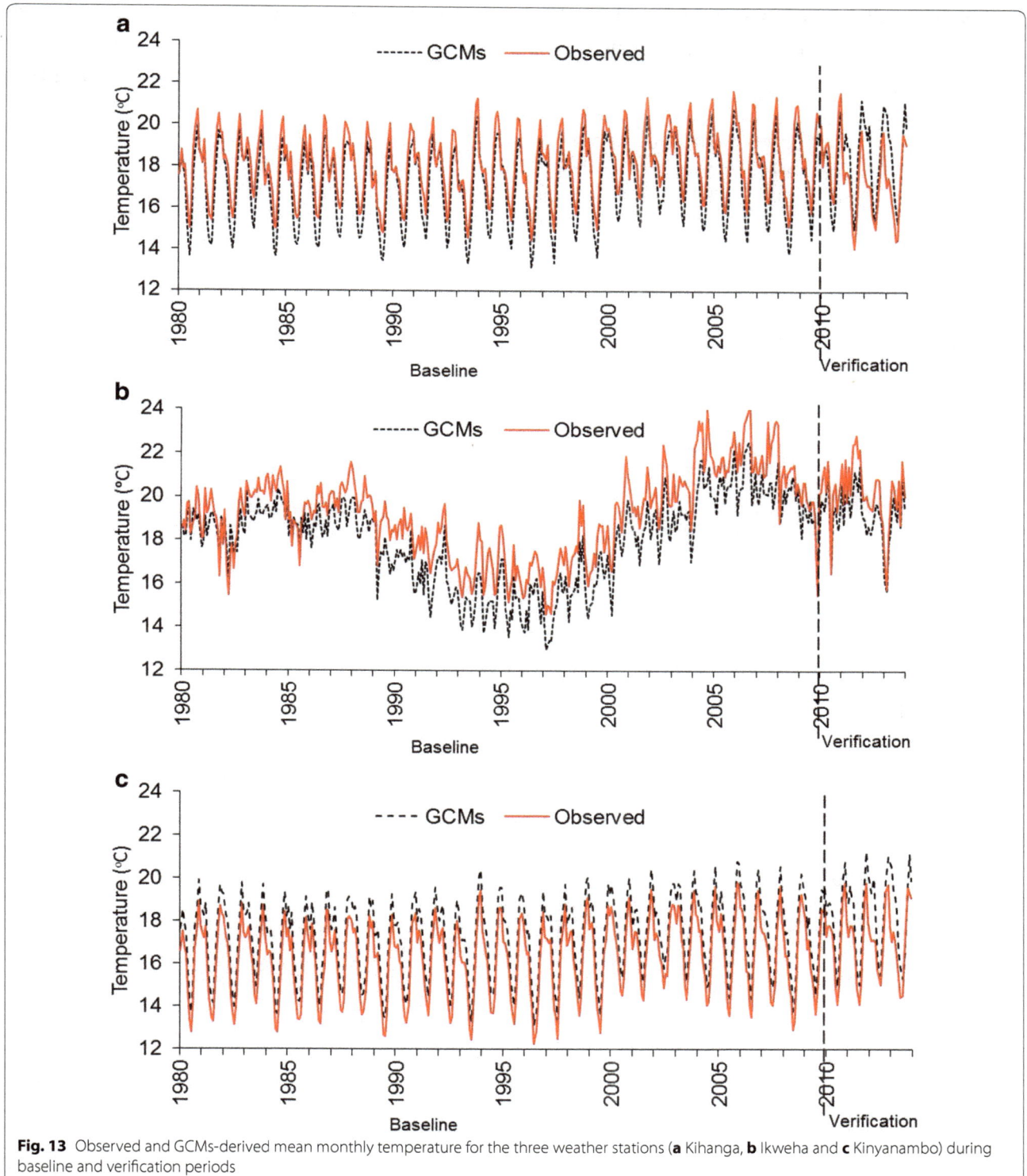

Fig. 13 Observed and GCMs-derived mean monthly temperature for the three weather stations (**a** Kihanga, **b** Ikweha and **c** Kinyanambo) during baseline and verification periods

December), the near-term showed less amounts of precipitation than the baseline period. The precipitation decrease was 6 mm/month (~20%) in November and 4 mm/month (~3%) in December (Fig. 16). The minimum and maximum monthly mean temperature for the watershed were higher in the near-term than in the

Table 6 Statistical analysis results of the monthly precipitation and temperature for the baseline and verification periods

Station	Parameter	Period	ME	R	Mean		SD		Median	
					GCM	Observed	GCMs	Observed	GCMs	Observed
Kihanga (80353)	Precipitation	Baseline	−00.09	0.99	2.12	2.03	2.80	2.69	0.29	0.28
		Verification	−0.19	0.98	2.00	1.81	2.66	2.48	0.30	0.37
Ikweha (83350)	Temperature	Baseline	−0.79	0.96	17.30	18.10	1.82	1.60	17.65	18.10
		Verification	0.81	0.77	18.32	17.52	1.83	1.71	18.73	17.45
	Precipitation	Baseline	0.05	0.96	2.08	2.02	2.80	2.93	0.27	0.11
		Verification	−0.15	0.92	1.95	2.10	2.72	2.75	0.35	0.28
	Temperature	Baseline	−1.19	0.95	17.95	19.14	2.03	1.99	18.43	19.23
		Verification	−0.91	0.92	19.31	20.22	1.17	1.40	19.36	20.16
Kinyanambo (83353)	Precipitation	Baseline	−0.05	0.99	2.00	2.05	2.67	2.74	0.29	0.30
		Verification	−0.19	0.92	1.86	1.66	2.55	2.22	0.39	0.31
	Temperature	Baseline	0.95	0.99	17.29	16.35	1.82	1.79	17.66	16.70
		Verification	1.24	0.91	18.33	17.09	1.83	1.57	18.72	17.21

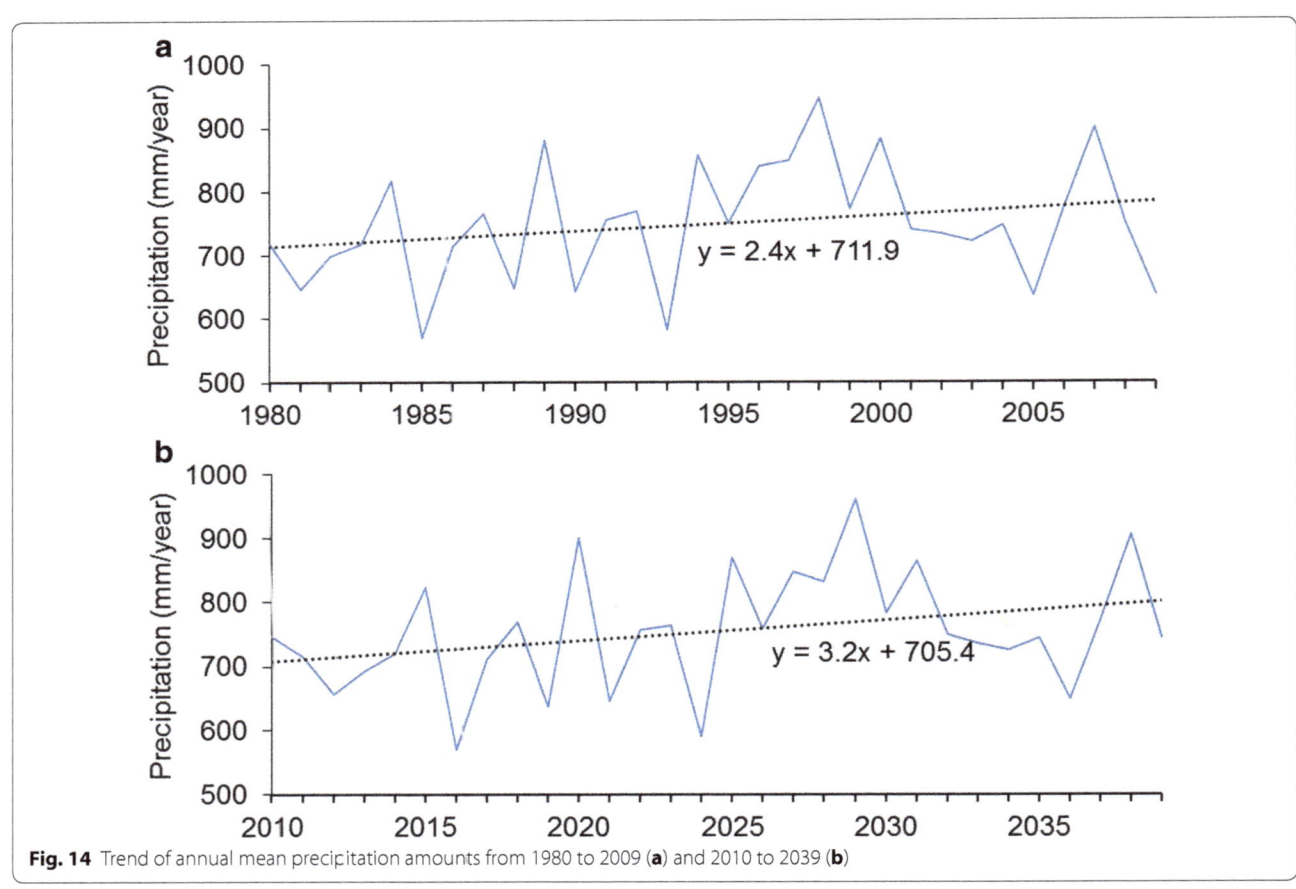

Fig. 14 Trend of annual mean precipitation amounts from 1980 to 2009 (**a**) and 2010 to 2039 (**b**)

baseline period throughout the year (Fig. 17). The highest temperature change between the two periods was 1.5 °C in November for maximum temperature and 1.4 °C in June and July for the minimum temperature.

Land use/cover change and its impact on the catchment water balance

The largest future land use/cover change between 2013 and 2020 were observed in the mixed forest lands (Fig. 18). The mixed forest land decreased by 12%.

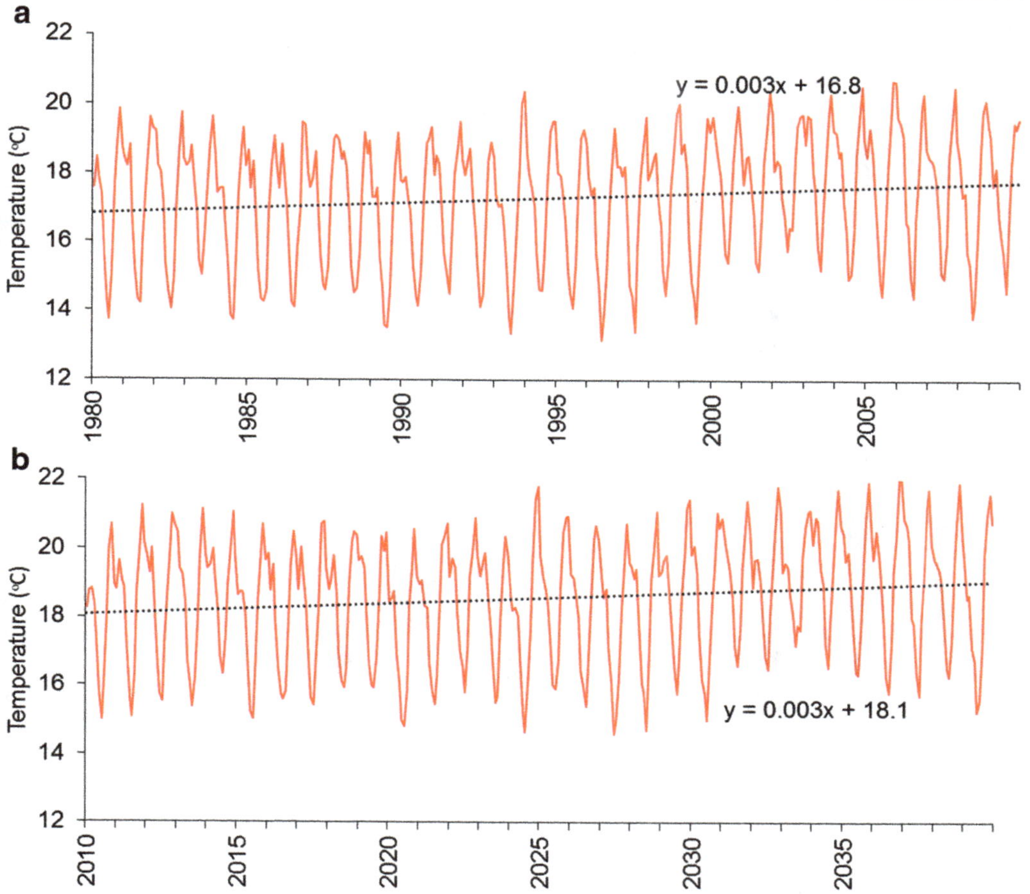

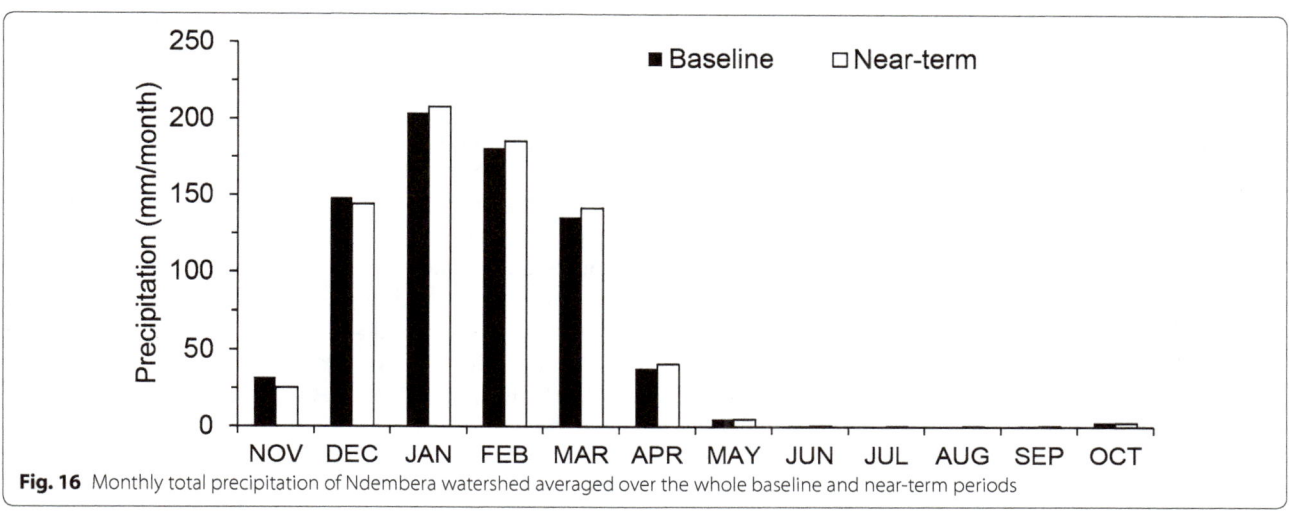

a

$y = 0.003x + 16.8$

b

$y = 0.003x + 18.1$

Fig. 15 Mean monthly temperature trend from 1980 to 2009 (**a**) and 2010 to 2039 (**b**)

■ Baseline　　□ Near-term

Fig. 16 Monthly total precipitation of Ndembera watershed averaged over the whole baseline and near-term periods

The areas under agricultural land were projected to increase by 10%, while evergreen forests increased by 7%. The shrub lands (Range-Brush) and grasslands (Range-Grasses) showed a decrease of 6 and 1%, respectively. Very small changes were observed for urban land and wetlands.

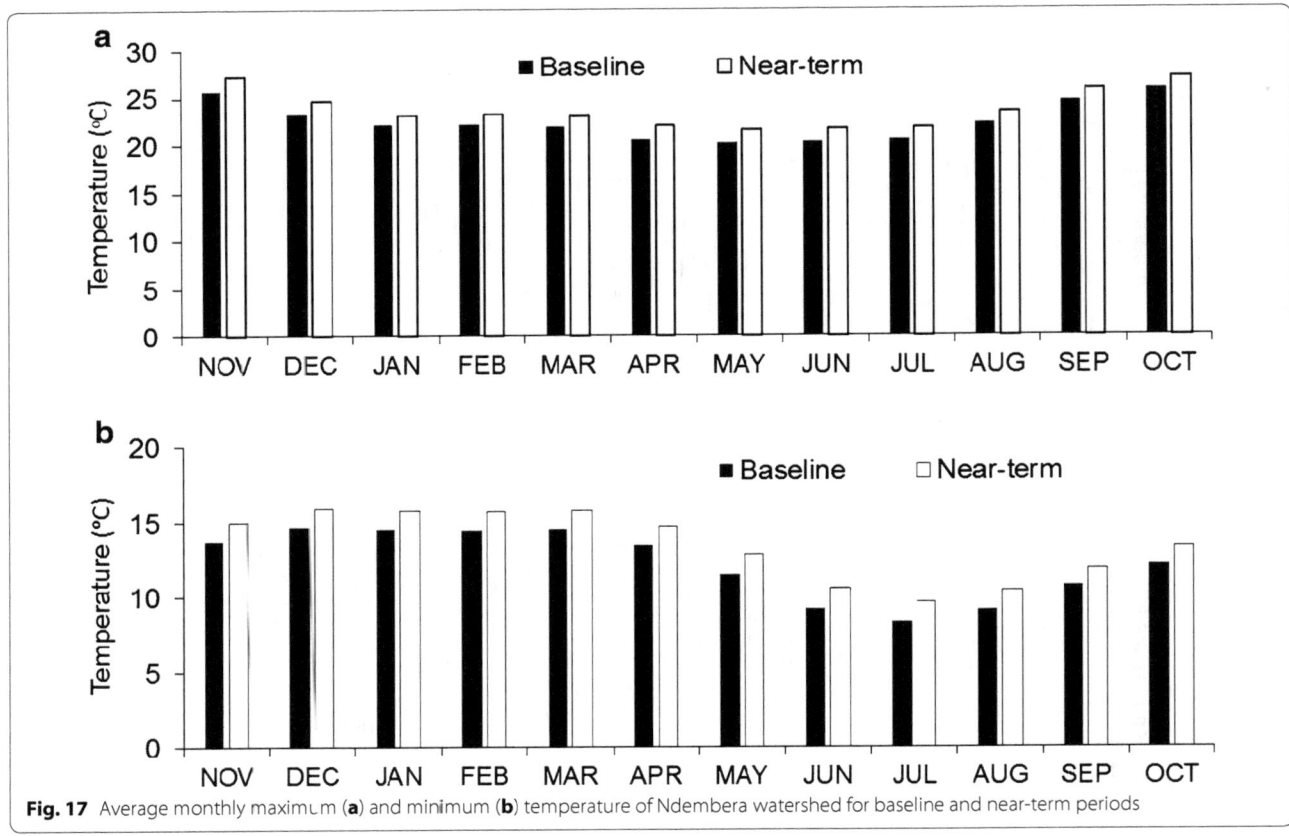

Fig. 17 Average monthly maximum (**a**) and minimum (**b**) temperature of Ndembera watershed for baseline and near-term periods

The observed land use/cover changes from 2013 to 2020 scenario affected the water balance of the watershed in a number of ways. These include a decrease of the total water yield and the lateral flow components by 32 mm/year (~13%) and 34 mm/year (~49%) (Fig. 19). On the contrary, the evapotranspiration and surface runoff increased by 30 mm/year (~8%) and 5 mm/year (16%), respectively. The amount of water loss from the channels during downstream flow (transmission losses) increased by 20 mm/year (88%). Some very minor changes were noted for the rest of the components such as total aquifer recharge, shallow and deep groundwater flow and water percolating out of the soil.

Water balance ratios, the Streamflow/Precipitation and baseflow/total flow indicated negative changes of −0.05 and −0.14, respectively (Table 7). On one hand, the baseflow contribution in the watershed total flow under 2013 and 2020 land use/cover scenarios were 0.78 and 0.64, respectively. On the other hand, the surface runoff contributed only 0.22 to 0.36, respectively. The ET/precipitation and Surface Runoff/Total flow ratios showed positive changes of 0.04 and 0.14, respectively. Furthermore, land use changes had no effect on the amount of precipitation that was partitioned into percolation and deep recharge.

A comparison of the average monthly river discharge at 1ka33 for the baseline and future land use/cover scenarios showed small differences, both in the mean discharge values and the timing of ascent and recession of the hydrograph (Fig. 20). During the ascent of the hydrographs in December and January, the simulated mean monthly discharge under the 2020 land use/cover was above the baseline discharge, that is, 5 and 14 m^3/s compared with 4.5 and 13 m^3/s, respectively. On the contrary, during the recession of the hydrograph, from March to April (wet season) through June and October (dry season), the simulated discharge under the 2020 land use/cover was lower than the baseline discharge. In addition, there was a horizontal shift of the discharge hydrograph to the left, showing some early ascending and recession of the hydrograph in the wet months (Fig. 20).

Impact of future climate change on the catchment water balance

Near-term climate change affected water balance components by decreasing the lateral water flow from 70 to 35 mm/year (~100% change) and the total water yield from 260 mm/year to 240 mm/year, a decrease of ~8% (Fig. 21). The evapotranspiration increased from

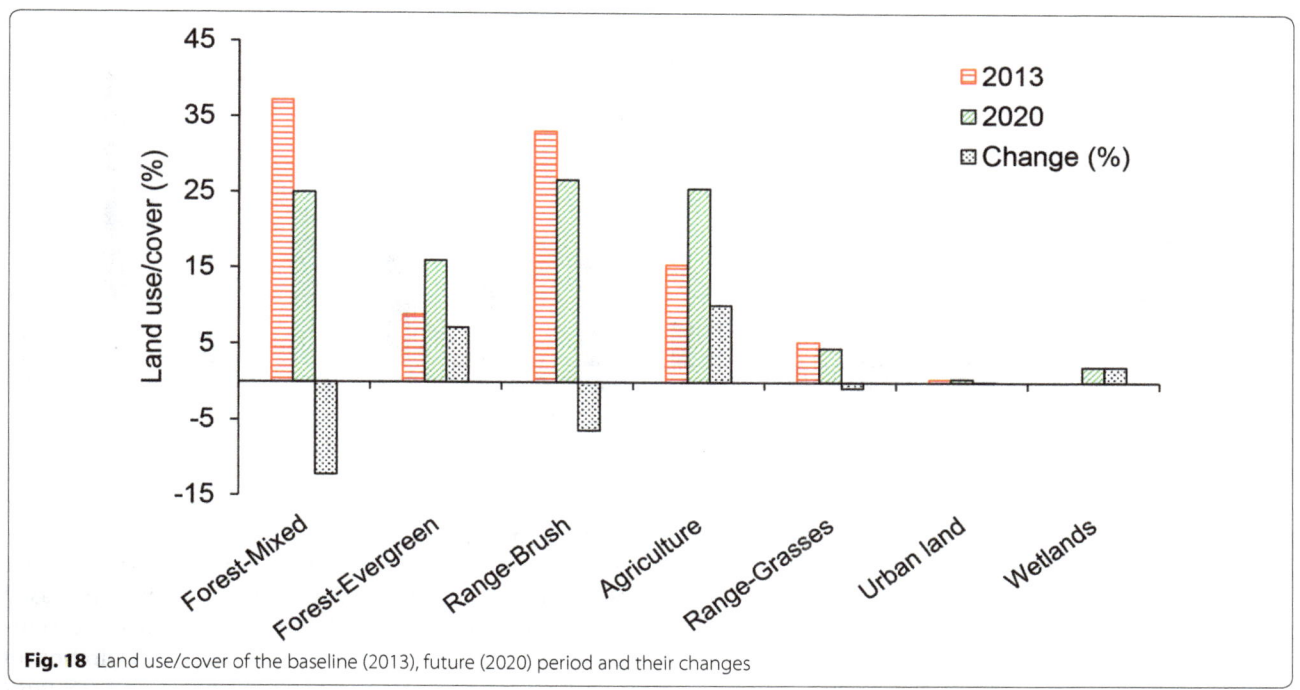

Fig. 18 Land use/cover of the baseline (2013), future (2020) period and their changes

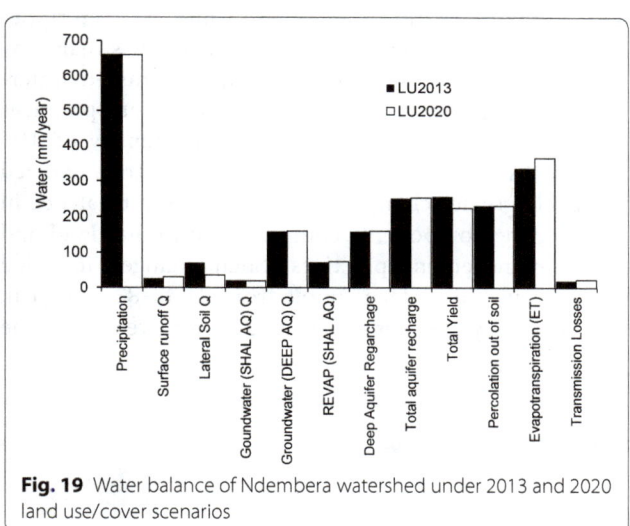

Fig. 19 Water balance of Ndembera watershed under 2013 and 2020 land use/cover scenarios

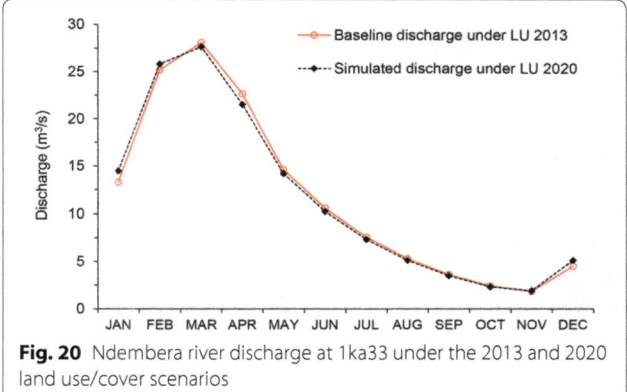

Fig. 20 Ndembera river discharge at 1ka33 under the 2013 and 2020 land use/cover scenarios

Table 7 Changes in the water balance ratios under 2013 and 2020 land use scenarios

Water balance ratios	LU2013	LU2020	Change
Streamflow/precipitation	0.18	0.13	− 0.05
Baseflow/total flow	0.78	0.64	− 0.14
Surface runoff/total flow	0.22	0.36	0.14
Percolation/precipitation	0.38	0.38	0
Deep recharge/precipitation	0.24	0.24	0
ET/precipitation	0.51	0.55	0.04

336 mm/year in the baseline to 453 mm/year (about 35% change). Compared with the baseline climate scenario, the surface runoff, total losses of water in channels during downstream flow and the Revap from shallow aquifer showed a relatively small decrease under the future climate change scenario. The shallow aquifer flow, deep aquifer recharge, groundwater recharge and percolation increased by a relatively small amount under the future climate scenario.

The near-term climate change scenario will increase the mean monthly river discharge for the near-term period relative to the baseline period (Fig. 22). Large differences between the baseline and future period flows were noted

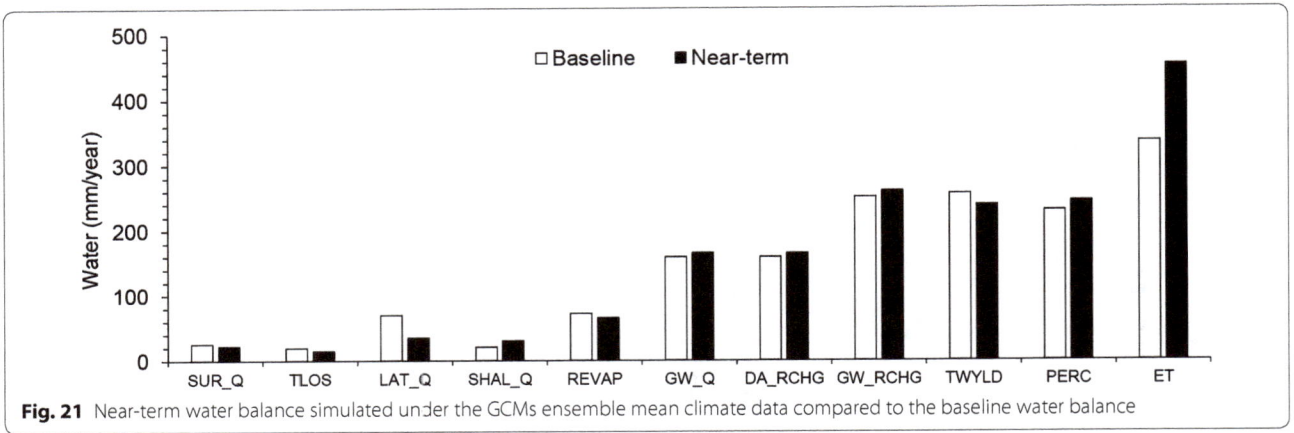

Fig. 21 Near-term water balance simulated under the GCMs ensemble mean climate data compared to the baseline water balance

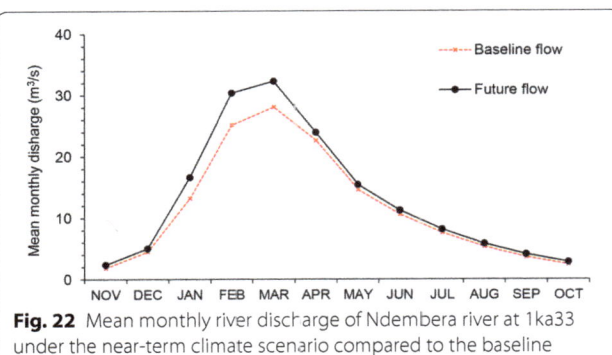

Fig. 22 Mean monthly river discharge of Ndembera river at 1ka33 under the near-term climate scenario compared to the baseline

during the wet season, especially in the months of January–March. The February river discharge will increase from 25 m³/s in the baseline period to 30 m³/s in the future period (~21% increase). Similarly, in the month of March, the discharge will increase from 28 to 32 m³/s (~15% change).

Impacts of land and water management practices on water balance and river discharge

The simulation of water balance of the watershed under the four land and water management practices increased the amount of water in almost all water balance components (Fig. 23). The amount of evapotranspiration of water under filter strips decreased by almost 26 mm/year (~6%). Furthermore, filter strips decreased the surface runoff by 12 mm/year (~54%) while the deep ripper tillage decreased the transmission losses in streams by 10 mm/year (~66%). Unexpectedly, the grassed waterways and terrace and contouring management practices increased the surface runoff by equal intensity of 31% (~21–28 mm/year). In general, the filter strips showed relatively greater changes in the annual water balance in most of the components compared with other land and water management practices. Such changes included increase in the surface runoff from 21 to 83 mm/year, total water yield from 234 to 315 mm/year and the

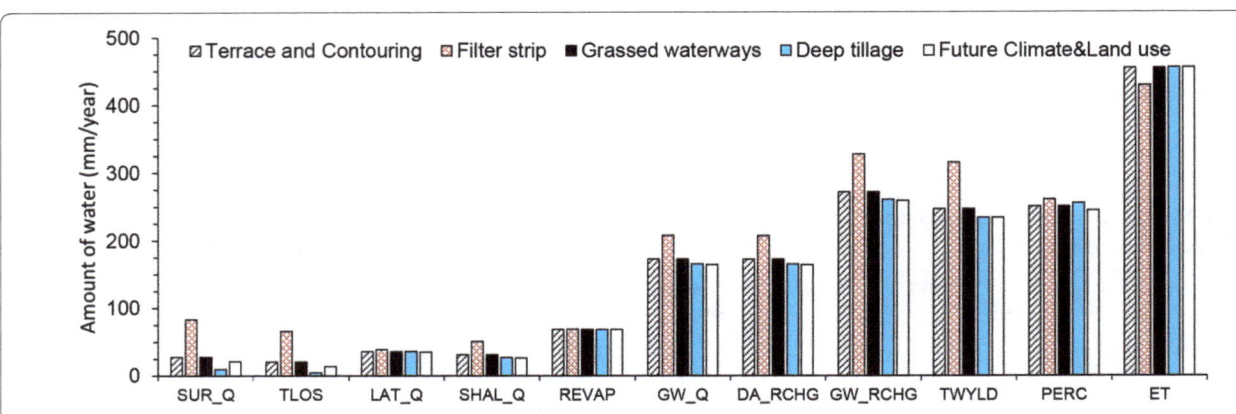

Fig. 23 Water balance under specific land and water management scenarios compared to the reference water balance scenario derived from near-term climate

percolation from 234 to 261 mm/year. The deep aquifer recharge and total groundwater recharge an increased by 43 and 68 mm/year, respectively.

The river discharge increased for almost all management practices compared with the discharge under the near-term climate and 2020 land use discharge scenario except for the deep ripper tillage (Fig. 24). The simulated flow under the filter strip practices scenario showed higher discharge than other management scenarios throughout the year. The river discharge under the filter strips were as higher as 42 m³/s in February and March compared with 26 and 30 m³/s of the reference discharge (discharge under the effect of near-term climate change and 2020 land use/cover) during the same months. These changes represent a change of about 62 and 40%, respectively. The hydrographs of the terracing and contouring as well as grassed waterways were slightly lower than the one under filter strips throughout the year but higher than the reference discharge during the wet season (January–April).

Discussion
Model parameterization and sensitivity analysis
The sensitivity analysis from other watersheds in the East African region had, among the top six sensitive parameters, few parameters similar to the one found in the Ndembera watershed. For example, in Murchison Bay Catchment, Uganda, the CN_2, GW_DELAY and GW_REVAP were among the top six sensitive parameters to river discharge (Anaba et al. 2017). The CN_2 and SOL_AWC were also among the most sensitive parameters reported in the Simiyu river Catchment in Tanzania (Mulungu and Munishi 2007).

The parameters presented in Table 2 and their sensitivity information in Fig. 10 are significant to the body of scientific knowledge in two ways. Firstly, allow a more stringent evaluation of the reality of the parameters used in the model parameterization and calibration as well as the models itself as suggested by van Griensven et al. (2012). Secondly, the results serve as a starting point of SWAT model parameterization in the subsequent studies within or in the nearby watershed which may need SWAT as a tool for hydrological processes related analysis. However, the parameter values may need some minor customizations because the catchments differ in their physical characteristics (Schmalz and Fohrer 2009).

Adequacy of the SWAT model and GCMs-derived data for hydrological processes simulation
The NSE values in Table 5 are far greater than the acceptable values; NSE > 0.5 (Moriasi et al. 2007). The PBIAS values −4.6 and 3.8 are within PBIAS < |25%|, can be said to describe a satisfactory model performance for monthly data of the stream flow (Dourte 2011; Moriasi et al. 2007). Small PBIAS values in this study indicate a good mass balance in terms of volume between the observed and simulated discharge. Although the p-factor and r-factor should be > 0.8 and < 1, respectively (Abbaspour 2007), the p-factor value of 0.48 (∼ 0.5) in Table 5 is also sufficient under less stringent model quality requirements (Schuol et al. 2008).

The satisfactory model calibration and validation results in Table 5 as well as a good match of the observed and simulated flows in Fig. 11 are interpreted as the outcomes of good model input data, notably rainfall and model parameterization. Rainfall is an important data input for hydrological models (Strauch et al. 2012). The distributed rainfall information increases the simulation accuracy and predictive capacity of the model (Dwarakish and Ganasri 2015).

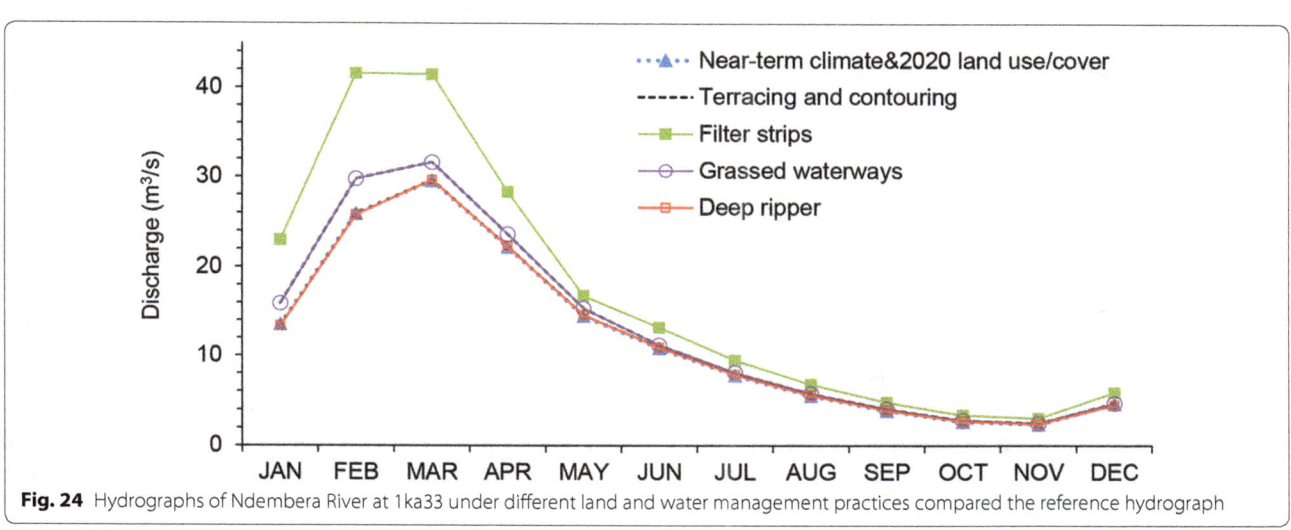

Fig. 24 Hydrographs of Ndembera River at 1ka33 under different land and water management practices compared the reference hydrograph

A good match between the GCMs-derived and the observed climate data (Figs. 12 and 13), may be associated with the strength of the quadrant method for CMIP5-GCMs sub-selection which was also previously used by Subash et al. (2016). Indeed, this observation is evidenced by high correlation values, small mean error values and closely related distribution statistical parameters (Table 6). The mean error values in Table 6 are better than those reported by Mutayoba and Kashaigili (2017) who evaluated the performance of the CORDEX Regional Climate Models in simulating rainfall characteristics over Mbarali river, a river catchment very close to Ndembera within the same Usangu Catchment. The mean error values ranged from −19.2 to 35.4. Nevertheless, the magnitude of correlation values from the same study ranged from 0.77 to 0.99, similar to this study. These good validation results in this study could also be attributed to ready-made functions used for the CMIP5-GCMs statistical downscaling and climate projection (Hudson and Ruane 2013). Underestimated GCM climate in Fig. 13a, b could be partly attributed to the poor accuracy of the observed temperature data.

The trend of climate change in the Ndembera watershed

The increasing trend of the warmer conditions (1.1 °C) in the near future (Fig. 15b), is mainly attributed to the increased CO_2 concentration and other greenhouse gases in the atmosphere at a global level (Edenhofer et al. 2011). The trend and magnitude of temperature change in the near future period observed in this study does not differ much from 1.3 °C by 2030 observed in a previous study in Tanzania (Laderach and Eitzinger 2012). The same study reported an increasing trend of the annual precipitation as observed in the current study. In addition, the recent climate change projections study by Serdeczny et al. (2016) has also reported a warming trend in sub-Saharan Africa and increased precipitation in East Africa.

The impact of land use/cover change on water balance and river discharge

The decrease in the total water yield and lateral water flow and the increase in evapotranspiration and surface runoff depicted in Fig. 19 could be explained by the observed 2013–2020 land use/cover changes (Fig. 18). For example, the increase in evapotranspiration by 30 mm/year in Fig. 19 could be attributed to the 10% increase in agricultural land (Fig. 18), mainly the irrigated onion and rice farms. It is worth noting that the irrigated rice in small river watersheds in Usangu basin has higher water demand than most crops because of the pre-saturation of the soil profile and the need for a standing water layer (Lankford and Franks 2000). The large surface standing water in agricultural fields creates high

possibility of water loss through evaporation. Another argument accounting for the increased evapotranspiration is the increase in evergreen forest (7%, Fig. 18) mainly the eucalyptus and pine commercial trees in Mufindi District. This implies the increase of more plant biomass and high leaf area index due to increased canopy. Usually, the canopy stores more water when the precipitation is intercepted and, therefore, making more amount of water available for evaporation (Wang and Kalin 2011). The increase in evaporation from canopy and irrigated farms can also explain for the observed decrease in total water yield (Fig. 19).

The increase of surface runoff/total flow ratio (Table 7) is translated as the result of the increased agricultural area, urban land and a decrease of tree and grass cover in forests and rangelands (Fig. 18). The removal of forest trees and grass cover tends to increase storm runoff and decrease infiltration to groundwater and baseflow of streams (Kiersch 2006). The removal of forests reduces the infiltration opportunities which, in turn increases the amounts of water leaving the area as storm runoff and reduces the gain in baseflow. This usually diminishes the dry season flow (Kashaigili 2008). According to Dagar et al. (2016), the use of machinery for various tillage practices causes the compaction of the soils. This scenario may account for the decrease in infiltration to groundwater, increased surface runoff and decline in baseflow observed in the current study. The increased surface runoff leads to an increased amount of water flowing to the streams in wet season and it can ultimately be lost through evaporation (Arnold et al. 2009).

The decrease in the baseflow component from 0.78 to 0.64 under the 2013 and 2020 land use/cover scenarios in Table 7 complements the findings of continuous declining trend of baseflow across the Usangu catchment in the period between 1960 and 2009 (Shu and Villholth 2012). The authors reported deforestation, irrigation and groundwater abstraction as the main factors causing the baseflow decline. Land use/cover change has been associated with the declining the baseflow in other studies conducted within the Usangu catchment (Kashaigili 2008) and in other countries such as Botswana and South Africa (Palamuleni et al. 2011).

The decrease in total water yield observed in Fig. 19 as a result of land use/cover change is mainly attributed to the decrease in forest cover. According to Palamuleni et al. (2011), the destructive land cover change may disrupt the hydrological cycle either through increasing or diminishing the water yield. In the current study, water yield decreased with decreasing forest cover contrary to the previous studies where water yield increased with forest reduction (DeFries and Eshleman 2004; Feng et al. 2012). The contradicting findings suggest that water yield

The impact of future climate and land use/cover change on water resources in the Ndembera...

233

is a function of factors other than the forest cover which in this case appear to have had greater influence on water yield. Such factors may include the increased evapotranspiration from increased agricultural areas (Fig. 18) or increased evapotranspiration due to the unaccounted recent decade-long temperature increase.

The horizontal shift of the river discharge pattern depicted in Fig. 20 is an indication of increased surface runoff due to the removal of vegetation cover as a result of expanding agricultural area and decrease of mixed forests (Fig. 18). The Great Ruaha river showed similar discharge pattern as a result of land use/cover change in the areas around the Ihefu Wetland (Kashaigili 2008). The observed low flows in the dry season under the 2020 land use/cover scenario (Fig. 20) signify reduced agricultural production especially irrigated rice at the Madibira and Mkunywa irrigation schemes. As reported earlier, the shortage of water downstream will have negative consequences on wildlife in the Ruaha National Park (Kashaigili et al. 2006b).

The impact of near-term climate change on water balance and river discharge

The continued increase in evapotranspiration from 336 to 453 mm/year under the influence of the near-future climate change scenario (Fig. 21) is mainly attributed to the observed change of mean watershed temperature of 1.1 °C (Fig. 15b) and the increase of both the mean monthly minimum and maximum temperatures (Fig. 17). The results imply that the evapotranspiration was underestimated when the impacts of land use/cover change were evaluated in isolation. The increased evapotranspiration is known to increase the water demand of plants and increase water stress which may reduce crop yields (Jensen 1968). The possibility of a decrease in crop yields in the watershed due to the effects of global warming and the increase in temperature is supported by the findings from a previous study in Tanzania (Laderach and Eitzinger 2012). In this study, it was found out that a change of temperature by +1.3 °C will decrease areas suitable for coffee cultivation by 20–50% in 2050. Globally, 1 °C increase in temperature in the developing countries will lower the growth in agricultural output by 2.66% (Dell et al. 2012). One of the reasons for this is the increased evaporative loss (Beck and Bernauer 2011). Nevertheless, increased evaporative loss may lead to increased yield of some crops. For example, a study by Jones et al. (2015) predicted that increased evapotranspiration by 6% coupled with increased CO_2 concentration fertilization will result in increased yield of irrigated sugarcane in South Africa during the 2070–2100 period. This implies that climate change brings with it potentials for crop production if appropriate adaptation measures are taken.

Despite the fact that the future climate scenario showed wetter and warmer conditions in the near-future than the baseline period (Figs. 16, 17), the impact of precipitation seemed not to counter the effect of temperature on evapotranspiration. This resulted in the decrease of total water yield shown in Fig. 21. The observed increase in percolation, groundwater recharge, groundwater flow, shallow aquifer flow and reduced revap from the shallow aquifer in Fig. 21 is most probably the result of higher precipitation amounts in the future compared to the baseline. The major reason being that the precipitation is a major component of water balance (Beeson et al. 2011). The change in the amount of precipitation has also some implications on other water balance components (Arnold et al. 2009).

The increase in the future river discharge in the wettest months of January to April (Fig. 22) corresponds well with the increase of the near-term mean monthly precipitations in the same months (Fig. 16). These results are in line with the observation by Taniguchi (2012) and Wambura (2014) that the increase in streamflow depends on the amount of precipitation. The future climate simulation also showed increased discharge of rivers during high flow in Bangladesh (Kirby et al. 2016) and the Sahelian regions (Amogu et al. 2010; Descroix et al. 2012). The increase in high flows in the Ndembera river observed in Fig. 22 may have resulted from the combined effect of future climate and the 2020 land use. This scenario has high and positive potential for boosting irrigated agriculture in the downstream. These irrigated crops could be the high-temperature tolerant type such as sugarcane (Jones et al. 2015).

The impact of land and water management practices as mitigation strategies

The observed changes in water balance such as decreased evapotranspiration and increased percolation, groundwater flow and recharge as well as total water yield (Fig. 23) are dependent on land and water management practices. Contours and terracing reduce the steep slope of the land and, therefore, reduce and delay the surface runoff and allows a long time for rainwater percolation (Dou et al. 2009). Deep ripper tillage increases soil depth and enhances percolation and ultimately reduces overland surface flow (Lacey 2008). Grassed waterways reduce runoff volumes due to their comparably high infiltration rates and the reduction in runoff velocity. The grassed waterways reduced runoff by 10 and 90%, respectively, in the two watersheds in Munich (Fiener and Auerswald 2003). This is contrary to the unexpected increased surface runoff in the current study of about 31% under both grassed waterways and terrace and contouring (Fig. 23). Nevertheless, in this study, the surface runoff under filter strips was higher than in other management interventions (Fig. 23). This is due to the fact that filter strips do not affect the surface runoff in SWAT (Arnold et al. 2013). Moreover, high increase in percolation, shallow aquifer and groundwater flow, groundwater recharge and ultimately total water yield under the filter strips compared to other management interventions is attributed to this increased infiltration rate (Arnold et al. 2013).

The increase in streamflow under most of the management interventions observed in Fig. 24 is linked to the increased groundwater recharge due to the decrease of evapotranspiration as shown in Fig. 23. The reduced evapotranspiration and increased streamflow brought about by management practices contributes greatly to reducing the dependency on the river for irrigation and reduce water competition between users (Lankford and Franks 2000). In addition, these management practices will potentially reduce the stress caused by the decline in available water resource due to climate change (Carpenter et al. 1992).

Conclusion

The land use/cover changes in Ndembera river watershed from 2013 to 2020 will interact with the projected near-term warmer temperatures (1.1 °C) and affect water balance by increasing evapotranspiration and surface runoff and decrease in water yield. Changes in these components decreased the baseflow and streamflow, which ultimately decreased the availability of water within the watershed. Major land use/cover of concern are the increase in the areas under agricultural lands, increase in evergreen forests and decrease in mixed forests.

The simulation of combined effects of land use/cover and climate change on water balance generated larger changes than when land use/cover effects are analyzed in isolation. The future warmer climate will exacerbate the water losses in Ndembera river watershed, in turn, will make the watershed unsuitable for producing high temperature sensitive crops. Nevertheless, planting of high temperature tolerant crops such as sugarcane in the watershed could be one of the adaptation strategies. The success of growing crops that are tolerant to elevated temperatures will be realized by adopting land and water management practices which reduce loss and make more water available for crops.

Land and water management practices evaluated in this study have proved to be effective mitigation and adaptation measures for the observed adverse hydrological impacts of future climate and land use/cover changes. Among the four management practices which were evaluated, three of them namely filter strips, terracing and contouring and grassed waterways were the most effective. These practices had great effect in increasing

groundwater recharge, groundwater flow, percolation and total water yield. Notably, filter strips were the most effective measures in reducing the evapotranspiration.

Ndembera watershed experience the loss of tree cover especially in the mixed forest areas. This reduces the potentials of the watershed to perform carbon dioxide gas sequestration function as well as loss of water through increased evapotranspiration and surface runoff. The replacement of trees should be encouraged, especially those ones which are adapted to the soil and climate of the planting area. In addition, the trees should be those with the moderate to aggressive development to occupy the site quickly. These trees should be able help in improving water retention capacity in the catchment as well as providing the multi-benefits. Such trees could include fruit trees or fodder for animals. These multi-benefit trees could also be planted as filter strip trees and on the edges of contours and terraces in the farms.

Abbreviations

AgMIP: The Agricultural Model Intercomparison and Improvement Project; CFSR: climate forecast and system reanalysis; CMIP5: Climate Model Intercomparison Project 5; CORDEX: Coordinated Regional Climate Downscaling Experiment; DEM: Digital Elevation Model; FAO: Food and Agriculture Organization; GCM: General Circulation Model; RBWO: Rufiji Basin Water Organization; SWAT: Soil and Water Assessment Tool; SWAT-CUP: Soil and Water Assessment Tool-Calibration and Uncertainty Programs; SUFI: sequential uncertainty fitting.

Authors' contributions

LWM and ANM guided the whole research activities and writing of the manuscript. CH did data collection, processing and analysis as well as writing the manuscript. AW and CH calibrated the SWAT model. All authors read and approved the final manuscript.

Author details

[1] Department of Water and Environmental Science and Engineering (WESE), The Nelson Mandela African Institution of Science and Technology (NM-AIST), P.O. Box 447, Arusha, Tanzania. [2] Department of Environmental Planning, Institute of Rural Development Planning, P. O. Box 138, Dodoma, Tanzania. [3] Texas A&M Agrilife Research, Blackland, USA. [4] Department of Geography and Planning, University of Saskatchewan, 9 Campus Drive, Saskatoon, SK S7N 5A5, Canada.

Acknowledgements

The authors appreciate with thanks the technical advice and data received from a number of experts and organizations that made it possible to accomplish this study. We are thankful to Dr.Sixbert Maurice (SUA) and Ruane Alex (NASA) for providing GCMs and climate data processing algorithms.

Competing interests

The authors declare that they have no competing interests.

Availability of data and materials

Landsat images were sourced from the U.S Geological Survey (USGS) at https://glovis.usgs.gov. Specifically, the images were from the three path and rows (path168/row066, path169/row065 and path169/row066). Soil data was downloaded from the FAO Harmonized global soils database at http://www.waterbase.org/download_data.html. The Climate Forecast System Reanalysis (CFSR) global weather data for SWAT were sourced at https://globalweather.tamu.edu. The streamflow and station weather data were provided by Rufiji Basin Water Organisation (RBWO) on request, and cannot be shared publicly without their consent. The CMIP5-GCMs climate data files were provided by Sokoine University of Agriculture in Tanzania and the National Aeronautics and Space Administration Goddard Institute for Space Studies (NASA-GISS) in the USA; the AgMIP project implementing partners.

Funding

This research was funded by Tanzanian Commission for Science and Technology (COSTECH) through Nelson Mandela African Institution of Science and Technology (NM-AIST).

References

Abbaspour KC (2007) SWAT calibration and uncertainty programs-a user manual. Swiss Federal Institute of Aquatic Science and Technology, Duebendorf, p 103

Amogu O et al (2010) Increasing river flows in the Sahel? Water 2:170–199. https://doi.org/10.3390/w2020170

Anaba LA, Banadda N, Kiggundu N, Wanyama J, Engel B, Moriasi D (2017) Application of SWAT to assess the effects of land use change in the Murchison Bay catchment in Uganda. Comput. Water, Energ Environ Eng 6:24–40. https://doi.org/10.4236/cweee.2017.61003

Arnold J et al (eds) (2009) Soil and water assessment tool (SWAT) global applications. World Association of Soil and Water Conservation, Bangkok, p 415

Arnold J et al (2012) SWAT: model use, calibration and validation. T ASABE 55:1491–1508. https://doi.org/10.13031/2013.42256

Arnold J, Kiniry J, Srinivasan R, Williams J, Haney E, Neitsch S (2013) SWAT 2012 input/output documentation. Texa, USA, Texas Water Resources Institute, p 650

Beck L, Bernauer T (2011) How will combined changes in water demand and climate affect water availability in the Zambezi river basin? Global Environ Change 21:1061–1072. https://doi.org/10.1016/j.gloenvcha.2011.04.001

Beeson P, Doraiswamy P, Sadeghi A, Di Luzio M, Tomer M, Arnold J, Daughtry C (2011) Treatments of precipitation inputs to hydrologic models. Transact ASABE 54:2011–2020. https://doi.org/10.13031/2013.40652

Behera M, Borate S, Panda S, Behera P, Roy P (2012) Modelling and analyzing the watershed dynamics using cellular automata (CA)–Markov model-A geo-information based approach. J Earth Syst Sci 121:1011–1024

Carpenter SR, Fisher SG, Grimm NB, Kitchell JF (1992) Global change and freshwater ecosystems. Annu Rev Ecol Syst 23:119–139

Chaturvedi RK, Joshi J, Jayaraman M, Bala G, Ravindranath N (2012) Multi-model climate change projections for India under representative concentration pathways. Curr Sci 103:791–802

Dagar J, Sharma P, Chaudhari S, Jat H, Ahamad S (eds) (2016) innovative saline agriculture. Springer Nature, India, p 519

DeFries R, Eshleman KN (2004) Land-use change and hydrologic processes: a major focus for the future. Hydrol process 18:2183–2186. https://doi.org/10.1002/hyp.5584

Dell M, Jones BF, Olken BA (2012) Temperature shocks and economic growth: evidence from the last half century. Am Econ J 4:66–95. https://doi.org/10.1257/mac.4.3.66

Descroix L, Genthon P, Amogu O, Rajot J-L, Sighomnou D, Vauclin M (2012) Change in Sahelian rivers hydrograph: the case of recent red floods of the Niger River in the Niamey region. Global Planet Change 98:18–30. https://doi.org/10.1016/j.gloplacha.2012.07.009

Dou L, Huang M, Hong Y (2009) Statistical assessment of the impact of conservation measures on streamflow responses in a watershed of the Loess Plateau, China. Water Resour Manage 3:1935–1949. https://doi.org/10.1007/s11269-008-9361-6

Dourte DR (2011) Cropping systems for groundwater security in India: groundwater responses to agricultural land management. Dissertation, University of Florida, USA, pp 106–130

Dwarakish G, Ganasri B (2015) Impact of land use change on hydrological systems: a review of current modeling approaches. Cogent Geosci 1:1–18. https://doi.org/10.1080/23312041.2015.1115691

Eastman JR (2012) IDRISI Selva Tutorial. Clark Labs-Clark University, Worcester

Edenhofer O et al (2011) IPCC special report on renewable energy sources and climate change mitigation. Working Group III of the Intergovernmental Panel on Climate Change, Cambridge, p 37

Elzein AA (2010) The suitability of swat model for land use change impact assessment on streamflows: the case study of Usangu sub-catchment in Tanzania. Dissertation, University of Dar es Salaam, Tanzania, p 112

Feng X, Sun G, Fu B, Su C, Liu Y, Lamparski H (2012) Regional effects of vegetation restoration on water yield across the Loess Plateau, China. Hydrol Earth Syst Sci 16:2617–2628. https://doi.org/10.5194/hess-16-2617-2012

Fiener P, Auerswald K (2003) Effectiveness of grassed waterways in reducing runoff and sediment delivery from agricultural watersheds. J Environ Qual 32:927–936. https://doi.org/10.2134/jeq2003.9270

Fischer S (2013) Exploring a water balance method on recharge estimations in the Kilombero valley. Stockholm University, Stockholm, p 18

Hamlet AF, Salathé EP, Carrasco P (2010) Statistical downscaling techniques for global climate model simulations of temperature and precipitation with application to water resources planning studies. p 28

Hatibu N, Lazaro E, Mahoo H, Rwehumbiza F, Bakari A (1999) Soil and water conservation in semi-arid areas of Tanzania: national policies and local practices, Tanzania. J Agric Sci 2:151–170

Holvoet K, van Griensven A, Seuntjens P, Vanrolleghem P (2005) Sensitivity analysis for hydrology and pesticide supply towards the river in SWAT. Phys Chem Earth 30:518–526. https://doi.org/10.1016/j.pce.2005.07.006

Houet T, Hubert-Moy L Modeling and projecting land-use and land-cover changes with Cellular Automaton in considering landscape trajectories. In: The Proceedings of the 1st EARSeL workshop on land use and land cover, Dubrovnik, Croatia, 2006, vol 1. EARSeL special interest group on land use and land cover, pp 63–76

Hudson N, Ruane A (2013) Guide for running AgMIP climate scenario generation tools with R. AgMIP. http://www.agmip.org/wp-content/uploads/2013/10/Guide-for-Running-AgMIPClimate-Scenario-Generation-with. Accessed 11 Feb 2016

Hyandye C, Martz LW (2017) A Markovian and cellular automata land-use change predictive model of the Usangu catchment. Int J Remote Sens 38:64–81. https://doi.org/10.1080/01431161.2016.1259675

Jensen ME (ed) (1968) Water consumption by agricultural plants. Water deficit and plant growth. Academic Press INC., New York, pp 1–19

Jones M, Singels A, Ruane AC (2015) Simulated impacts of climate change on water use and yield of irrigated sugarcane in South Africa. Agric Syst 139:260–270. https://doi.org/10.1016/j.agsy.2015.07.007

Kangalawe RY, Lyimo JG (2013) Climate change, adaptive strategies and rural livelihoods in semiarid Tanzania. Nat Resour 4:266–278. https://doi.org/10.4236/nr.2013.43034

Kashaigili JJ (2008) Impacts of land-use and land-cover changes on flow regimes of the Usangu wetland and the Great Ruaha river, Tanzania. Phys Chem Earth 33:640–647. https://doi.org/10.1016/j.pce.2008.06.014

Kashaigili J, Majaliwa A (2013) Implications of land use and land cover changes on hydrological regimes of the Malagarasi river, Tanzania. J Agric Sci Appl 2:45–50. https://doi.org/10.14511/jasa.2013.020107

Kashaigili JJ, Mbilinyi BP, Mccartney M, Mwanuzi FL (2006a) Dynamics of Usangu plains wetlands: use of remote sensing and GIS as management decision tools. Phys Chem Earth 31:967–975. https://doi.org/10.1016/j.pce.2006.08.007

Kashaigili JJ, McCartney M, Mahoo HF, Lankford BA, Mbilinyi BP, Yawson DK, Tumbo SD (2006b) Use of a hydrological model for environmental management of the Usangu Wetlands, Tanzania, vol 104. International Water Management Institute, Colombo, p 39

Kashaigili JJ, Rajabu K, Masolwa P (2009) Freshwater management and climate change adaptation: experiences from the Great Ruaha river catchment in Tanzania. Climate Dev 1:220–228. https://doi.org/10.3763/cdev.2009.0025

Kiersch B (2006) Land use impacts on water resources: a literature review. Paper presented at the FAO E-workshop on Land-Water Linkages in Rural Watersheds, Rome, Italy, p 6

Kikula I, Charnley S, Yanda P (1996) Ecological changes in the Usangu plains and their implications on the downstream flow of the Great Ruaha river in Tanzania. Institute of Resource Assessment, University of Dar es Salaam, Dar es Salaam, p 46

Kirby J et al (2016) The impact of climate change on regional water balances in Bangladesh. Climatic Change 135:481–491. https://doi.org/10.1007/s10584-016-1597-1

Koch M, Cherie N (2013) SWAT Modeling of the impact of future climate change on the hydrology and the water resources in the upper Blue Nile river basin, Ethiopia. In: The proceedings of the 6th international conference on water resources and environment research, Koblenz, Germany, 2013. Water and Environmental Dynamics. ICWRER, pp 114–146

Lacey J (2008) Deep-ripping and decompaction. Department of Environmental Conservation, New York, p 12

Laderach P, Eitzinger A (2012) Future climate scenarios for Tanzania's Arabica coffee growing areas. CIAT, Cali, p 22

Lankford B, Franks T (2000) The sustainable coexistence of wetlands and rice irrigation: a case study from Tanzania. J Environ Dev 9:119–137. https://doi.org/10.1177/107049650000900202

Malley Z, Taeb M, Matsumoto T (2009) Agricultural productivity and environmental insecurity in the Usangu plain, Tanzania: policy implications for sustainability of agriculture. Environ Dev Sust 11:175–195. https://doi.org/10.1007/s10668-007-9103-6

McCartney M, Forkuor G, Sood A, Amisigo B, Hattermann F, Muthuwatta L (2012) The water resource implications of changing climate in the Volta river basin, vol 146. IWMI, Colombo, p 40

Moriasi DN, Arnold JG, Van Liew MW, Bingner RL, Harmel RD, Veith TL (2007) Model evaluation guidelines for systematic quantification of accuracy in watershed simulations. Transact ASABE 50:885–900. https://doi.org/10.13031/2013.23153

Mulungu DM, Munishi SE (2007) Simiyu River catchment parameterization using SWAT model. Phys Chem Earth 32:1032–1039. https://doi.org/10.1016/j.pce.2007.07.053

Mutayoba E, Kashaigili JJ (2017) Evaluation for the performance of the CORDEX regional climate models in simulating rainfall characteristics over Mbarali river catchment in the Rufiji Basin. Tanzania. J Geosci Environ Prot 5:139. https://doi.org/10.4236/gep.2017.54011

Mwanukuzi PK (2011) Impact of non-livelihood-based land management on land resources: the case of upland watersheds in Uporoto mountains, South West Tanzania. Geogr J 177:27–34. https://doi.org/10.1111/j.1475-4959.2010.00362.x

Natkhin M, Dietrich O, Schäfer MP, Lischeid G (2015) The effects of climate and changing land use on the discharge regime of a small catchment in Tanzania. Reg Environ Change 15:1269–1280. https://doi.org/10.1007/s10113-013-0462-2

Palamuleni LG, Ndomba PM, Annegarn HJ (2011) Evaluating land cover change and its impact on hydrological regime in Upper Shire river catchment, Malawi. Reg Environ Change 11:845–855. https://doi.org/10.1007/s10113-011-0220-2

Pervez MS, Henebry GM (2015) Assessing the impacts of climate and land use and land cover change on the freshwater availability in the Brahmaputra river basin. J Hydrol 3:285–311. https://doi.org/10.1016/j.ejrh.2014.09.003

Roth V, Lemann T (2016) Comparing CFSR and conventional weather data for discharge and soil loss modelling with SWAT in small catchments in the Ethiopian Highlands. Hydrol Earth Syst Sci 20:921. https://doi.org/10.5194/hess-20-921-2016

Schmalz B, Fohrer N (2009) Comparing model sensitivities of different landscapes using the ecohydrological SWAT model. Adv Geosci 21:91–98

Schuol J, Abbaspour KC, Srinivasan R, Yang H (2008) Estimation of freshwater availability in the West African sub-continent using the SWAT hydrologic model. J Hydrol 352:30–49. https://doi.org/10.1016/j.hydrol.2007.12.025

Serdeczny O et al (2016) Climate change impacts in sub-Saharan Africa: from physical changes to their social repercussions. Reg Environ Change 17:1–16. https://doi.org/10.1007/s10113-015-0910-2

Shu Y, Villholth KG (2012) Analysis of flow and baseflow trends in the Usangu Catchment, Tanzania. IWMI, International Water Management Institute, Pretoria, South Africa. http://www.geus.dk/dk/int_devel_projects/documents/clivet_20120902_copenhagen_shu.pdf. Accessed 10 Dec 2015

SMUWC (2001) Sustainable management of the usangu wetland and its catchment. Baseline. Water resources. University of Dar es Salaam, Dar es Salaam, p 209

Strauch M, Bernhofer C, Koide S, Volk M, Lorz C, Makeschin F (2012) Using precipitation data ensemble for uncertainty analysis in SWAT streamflow simulation. J Hydrol 414:413–424. https://doi.org/10.1016/j.jhydrol.2011.11.014

Subash N, Harbir S, Ruane A, McDermid S, Baigorria G (2016) Uncertainty of GCM projections under different Representative Concentration Pathways (RCPs) at different temporal and spatial scales—reflections from 2 sites

in Indo-Gangetic Plains of India In: the Proceedings of AgMIP 6 Global Workshop, Le Corum, Montepellier, France, 28–30 June, 2016

Taniguchi M (2012) Subsurface hydrological responses to land cover and land use changes. Springer Science & Business Media, New York, p 226

van Griensven A, Ndomba P, Yalew S, Kilonzo F (2012) Critical review of SWAT applications in the upper Nile basin countries. Hydrol Earth Syst Sci 16:3371–3381. https://doi.org/10.5194/hess-16-3371-2012

Wallace J (2000) Increasing agricultural water use efficiency to meet future food production. Agric Ecosyst Environ 82:105–119

Wambura FJ (2014) Stream flow response to skilled and non-linear bias corrected GCM precipitation change in the Wami river sub-basin, Tanzania. Br J Environ Climate Change 4:389–408. https://doi.org/10.9734/BJECC/2014/13457

Wang R, Kalin L (2011) Modelling effects of land use/cover changes under limited data. Ecohydrol 4:265–276. https://doi.org/10.1002/eco.174

Worqlul AW, Maathuis B, Adem AA, Demissie SS, Langan S, Steenhuis TS (2014) Comparison of rainfall estimations by TRMM 3B42, MPEG and CFSR with ground-observed data for the Lake Tana basin in Ethiopia. Hydrol Earth Syst Sci 18:4871–4881. https://doi.org/10.5194/hess-18-4871-2014

Worqlul AW, Yen H, Collick AS, Tilahun SA, Langan S, Steenhuis TS (2017a) Evaluation of CFSR, TMPA 3B42 and ground-based rainfall data as input for hydrological models, in data-scarce regions: the upper Blue Nile Basin, Ethiopia. CATENA 152:242–251

Worqlul AW, Ayana EK, Maathuis BH, MacAlister C, Philpot WD, Leyton JMO, Steenhuis TS (2017b) Performance of bias corrected MPEG rainfall estimate for rainfall-runoff simulation in the upper Blue Nile Basin. Ethiopia J Hydrol 556:1182–1191

PERMISSIONS

LIST OF CONTRIBUTORS

Getachew Workineh Gella
Department of Geography and Environmental Studies, Debre Tabor University, Debre Tabor, Ethiopia
Guna Tana Integrated Field Research and Development Center, Debre Tabor, Ethiopia

Bereket Ayenew
School of Natural Sciences, Department of Chemistry, Madda Walabu University, Bale-Robe, Ethiopia

Abi M. Tadesse
Department of Chemistry, College of Natural and Computational Sciences, Haramaya University, Dire Dawa, Ethiopia

Kibebew Kibret
School of Natural Resources Management and Environmental Sciences, Haramaya University, Dire Dawa, Ethiopia

Asmare Melese
Department of Plant Science, College of Agriculture and Natural Resource Sciences, Debre Berhan University, Debre Berhan, Ethiopia

Kassahun Gashu
Department of Geography and Environmental Studies, University of Gondar, Gondar, Ethiopia

Tegegne Gebre-Egziabher
Department of Geography and Environmental Studies, Addis Ababa University, Addis Ababa, Ethiopia

Eskinder Gidey
Department of Environmental Science, University of Botswana, Gaborone, Botswana
Land Resource Management and Environmental Protection, Mekelle University, Mekelle, Ethiopia 3 Institute of Climate and Society, Mekelle University, Mekelle, Ethiopia

Oagile Dikinya, Reuben Sebego and Eagilwe Segosebe
Department of Environmental Science, University of Botswana, Gaborone, Botswana

Amanuel Zenebe
Land Resource Management and Environmental Protection, Mekelle University, Mekelle, Ethiopia 3 Institute of Climate and Society, Mekelle University, Mekelle, Ethiopia

Rainer Schenk
Rosenberg 17, 06193 Wettin-Löbejün, Germany
Dresden University of Technology, International University Institute Zittau, Sachsen, Germany

Shankar Adhikari
Department of Forests, Ministry of Forests and Environment, Kathmandu, Nepal

Barbara Ozarska
School of Ecosystem and Forests Science, University of Melbourne, Burnley Campus, Melbourne, Australia

E Babatope Faweya
Radiation and Health Physics Division, Department of Physics, Faculty of Science, Ekiti State University, P.M.B 5363, Ado-Ekiti, Nigeria

O Gabriel Olowomofe and H Taiwo Akande
Department of Physics, Faculty of Science, Ekiti State University, Ado-Ekiti, Nigeria

T Adeniyi Adewumi
Department of Physics, Faculty of Science, Federal University, Lafia, Nigeria

Bereket Ayenew
Department of Chemistry, School of Natural Sciences, Madda Walabu University, Bale-Robe, Ethiopia

Abi M. Taddesse
Department of Chemistry, College of Natural and Computational Sciences, Haramaya University, Dire Dawa, Ethiopia

Kibebew Kibret
School of Natural Resources Management and Environmental Sciences, Haramaya University, Dire Dawa, Ethiopia

Asmare Melese
Department of Plant Science, College of Agriculture and Natural Resource Sciences, Debre Berhan University, Debre Birhan, Ethiopia

Alemnew Berhanu Kassegne
Centre for Environmental Science, Addis Ababa University, Addis Ababa, Ethiopia
Department of Chemistry, Debre Berhan University, Debre Berhan, Ethiopia

Tarekegn Berhanu Esho
Central Research Laboratories, Addis Ababa Science and Technology University, Addis Ababa, Ethiopia

Jonathan O. Okonkwo
Department of Environmental, Water & Earth Sciences, Tshwane University of Technology, 175 Nelson Mandela Drive, Arcadia, Pretoria, South Africa

Seyoum Leta Asfaw
Centre for Environmental Science, Addis Ababa University, Addis Ababa, Ethiopia

Mekonnen Birhanie Aregu, Seyoum Leta Asfaw and Mohammed Mazharuddin Khan
Centre for Environmental Sciences, College of Natural Science, Addis Ababa University, Addis Ababa, Ethiopia

Valentine Asong Tellen
Department of Development Studies, Environment and Agricultural Development Program, Pan African Institute for Development-West Africa (PAIDWA), Buea, South West Region, Cameroon

Bernard P. K. Yerima
Department of Soil Science, Faculty of Agronomy and Agricultural Science, University of Dschang Dschang, West Region, Cameroon

Seema Karki
International Centre for Integrated Mountain Development (ICIMOD) Kathmandu, Nepal
Department of Infrastructure Engineering, Melbourne School of Engineering, The University of Melbourne, Parkville, Australia

Aye Myat Thandar
Christian Albrechts University Kiel, Olshausenstr. 75, 24118 Kiel, Germany

Kabir Uddin, Kamal Aryal, Pratikshya Kandel and Nakul Chettri
International Centre for Integrated Mountain Development (ICIMOD) Kathmandu, Nepal

Sein Tun and Win Maung Aye
Inle Lake Biosphere Reserve, Nyaung Shwe, Myanmar

Noah Kyame Asare-Donkor, Japhet Opoku Ofosu and Anthony Apeke Adimado
Department of Chemistry Kwame, Nkrumah University of Science and Technology, Kumasi, Ghana

Canute B. Hyandye
Department of Water and Environmental Science and Engineering (WESE), The Nelson Mandela African Institution of Science and Technology (NM-AIST), Arusha, Tanzania
Department of Environmental Planning, Institute of Rural Development Planning, Dodoma, Tanzania

Abeyou Worqul
Texas A&M Agrilife Research, Blackland, USA

Lawrence W. Martz
Department of Geography and Planning, University of Saskatchewan, 9 Campus Drive, Saskatoon, SK S7N 5A5, Canada

Alfred N. N. Muzuka
Department of Water and Environmental Science and Engineering (WESE), The Nelson Mandela African Institution of Science and Technology (NM-AIST), Arusha, Tanzania

Index

Printed in the USA
CPSIA information can be obtained
at www.ICGtesting.com
LVHW081729210923
758939LV00006B/208

9 781647 401399